OCEANOGRAPHY
and
MARINE BIOLOGY

AN ANNUAL REVIEW

Volume 51

OCEANOGRAPHY
and
MARINE BIOLOGY

AN ANNUAL REVIEW

Volume 51

Editors

R. N. Hughes

Bangor University
Bangor, Gwynedd, United Kingdom
r.n.hughes@bangor.ac.uk

D. J. Hughes

Scottish Association for Marine Science
Oban, Argyll, United Kingdom
david.hughes@sams.ac.uk

I. P. Smith

University Marine Biological Station Millport
Isle of Cumbrae, United Kingdom
philip.smith@millport.gla.ac.uk

CRC Press
Taylor & Francis Group
Boca Raton   London   New York

CRC Press is an imprint of the
Taylor & Francis Group, an **informa** business

International Standard Serial Number: 0078-3218

CRC Press
Taylor & Francis Group
6000 Broken Sound Parkway NW, Suite 300
Boca Raton, FL 33487-2742

© 2013 by R.N. Hughes, D.J. Hughes, and I.P. Smith
CRC Press is an imprint of Taylor & Francis Group, an Informa business

No claim to original U.S. Government works

Printed on acid-free paper
Version Date: 20130430

International Standard Book Number-13: 978-1-4665-6866-2 (Hardback)

Visit the Taylor & Francis Web site at
http://www.taylorandfrancis.com

and the CRC Press Web site at
http://www.crcpress.com

Contents

Preface

The 51st volume of *Oceanography and Marine Biology: An Annual Review* (*OMBAR*) contains five reviews that together reflect an editorial policy of presenting diverse topics to a broadly based readership. *OMBAR* welcomes suggestions from potential authors for topics that could form the basis of appropriate reviews. Contributions from physical, chemical and biological oceanographers that seek to inform both oceanographers and marine biologists are especially welcome. Because the annual publication schedule constrains the timetable for submission, evaluation and acceptance of manuscripts, potential contributors are advised to contact the editors at an early stage of manuscript preparation. Contact details are listed on the title page of this volume.

The editors gratefully acknowledge the willingness and speed with which authors complied with the editors' suggestions and requests and the efficiency of CRC Press, especially Marsha Hecht, Jill Jurgensen, and John Sulzycki, in ensuring the timely appearance of this volume.

Oceanography and Marine Biology: An Annual Review, 2013, **51**, 1-70
© Roger N. Hughes, David Hughes, and I. Philip Smith, Editors
Taylor & Francis

QUATERNARY SEA-LEVEL AND PALAEOTIDAL CHANGES: A REVIEW OF IMPACTS ON, AND RESPONSES OF, THE MARINE BIOSPHERE

JAMES SCOURSE

*School of Ocean Sciences, College of Natural Sciences, Bangor University,
Menai Bridge, Anglesey, LL59 5AB, United Kingdom*
E-mail: j.scourse@bangor.ac.uk

Driven by the glacio-eustatic mechanism, Earth is currently experiencing a phase of high-amplitude sea-level cycles as large as any during its history. Since shelf seas only develop during transgressive highstands, they are transient, yet they are recognized as disproportionately biologically and biogeochemically significant within Earth's system. This review explores the significance of the transience of the shelf seas by considering impacts on, and responses of, the marine biosphere to Quaternary sea-level cycles and to concomitant palaeotidal changes. Geological evidence for sea-level changes and their simulation by glacio-isostatic adjustment (GIA) models is reviewed, enabling the timing, duration and extent of shelf inundation to be assessed. Biological impacts of sea-level change include adjustments in habitat type and extent, species' distributions, biodiversity and reproductive isolation leading to genetic change. Conflict between physical palaeoenvironmental evidence and molecular marker data indicating cryptic northern glacial refugia in the Northern Hemisphere remains unresolved. Changes in tidal amplitudes and tide-dependent parameters (bed stress, seasonal stratification) in response to sea-level changes can be simulated using palaeotidal and palaeowave models constrained by geological data. Palaeotidal changes have significant impacts on productivity, dispersal and benthic habitat distribution. The balance of production and respiration in shelf seas is altered by changing sea-level and tidal dynamics influencing atmospheric $p\mathrm{CO_2}$. The significant potential of GIA, palaeotidal and palaeo-environmental niche modelling approaches for quantifying the biological and biogeochemical impacts of sea-level and palaeotidal change is emphasized.

Introduction

Flooding of the continental shelves by the sea was certainly the most important geologic event of recent time. It initiated the building of modern deltas, many coral reefs, alluviation of river valleys, and the formation of existing beaches and barrier islands. Doubtless, it also had far-reaching effects on climate and on the migrations of marine and terrestrial organisms, including man. (Newell 1961, p. 87)

At the height of the Last Glacial Maximum (LGM), about 26,000 years (26 ka) ago (Peltier & Fairbanks 2006, Clark et al. 2009), global sea level was between 120 and 130 m lower than at present as a consequence of the volume of water exported from the ocean into the great ice sheets (Lambeck et al. 2002, Milne et al. 2002). This major global regression resulted in the emergence of much of the shallow continental shelves (defined as less than 200-m deep) fringing the continents equating to a 7–10% reduction in the areal extent of the global ocean surface and a concomitant increase in the continental land surface (Figure 1). Over the succeeding 18,000 years or so, the ice sheets melted,

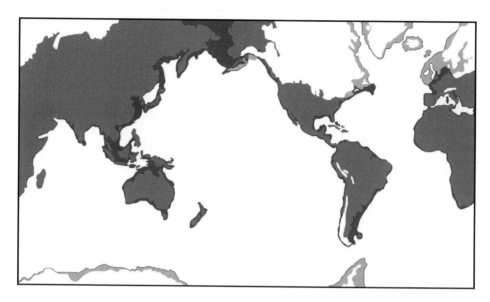

Figure 1 The emergence of the continental shelves at the Last Glacial Maximum. Emergent continental shelves exposed at the –120 to –130 m lowstand are shown in black, ice sheets in white and land surfaces exposed by the lowstand but covered in ice in grey. (Adapted from Ruddiman 2001, with permission from W.H. Freeman & Co. and CLIMAP Project Members 1981, with permission from the Geological Society of America.)

quickly at first, then more slowly, so that the water returning to the oceans caused the sea level to rise and transgress across the continental shelves. This process resulted in the formation of the shelf seas and a great expansion in shallow marine habitat. The areal extent of the shelf seas shallower than 200 m is now 425% greater than during the LGM (Neill et al. 2009). The shelf seas, in their present manifestation, have therefore only been in existence for a few thousand years, in marked contrast to the long-term stability of deep ocean habitats. This pattern of regression and transgression, emergence followed by inundation of the continental shelves, resulting each time in the creation of the shelf seas, has been repeated many times through the history of Earth, sometimes very slowly but at other times remarkably quickly and with a clear cyclic frequency. Earth is currently in a phase of rapid cycling, so that the LGM was only the last in a series of some 49 major regressions within the Quaternary (the last 2.58 million years, Ma; Gibbard et al. 2010). Both peak lowstands and highstands of sea level are temporary and unusual in the context of total Quaternary time. Peak transgressions, with highstand sea levels and maximum extent of the shelf seas, are associated with minima in terrestrial ice volume. Such conditions have persisted within the current cycle only over the last 6000 years or so, and this short duration is typical of each earlier highstand (Tzedakis et al. 2012). The existence of the shelf seas over the last 2.5 Ma has therefore been transient, and the current situation—during which modern civilization has developed—is highly anomalous. For 90% of Quaternary time, the climate has been colder than the current interglacial (Walsh 1988).

Despite their transience, the present-day highstand shelf seas and the shallow marine margins of the continents are recognized as disproportionately significant within the Earth system (Simpson & Sharples 2012). This significance is a function of the nature and magnitude of the physical, biogeochemical and biological processes operating in the shelf seas. Although only constituting 7–10% of their areal extent, these marginal seas serve to dissipate a significant component of the mechanical energy supplied to the global ocean. Swell and tidal wave energies are dissipated by frictional interaction with the shallow bed by breaking waves and tidal currents, respectively. At present, as much as 70–75% (2.6–2.7 TW; 1 terrawatts (TW) = 10^{12} W) of all tidal dissipation in the ocean takes place in the shelf seas, with only 0.9 TW (total oceanic tidal dissipation = 3.5 TW) in the deep ocean through internal tides and waves (Munk & Wunsch 1998, Egbert & Ray 2003, Green et al. 2009). In conjunction with wind stress at the surface, large inputs per unit volume of seawater of solar energy, and significant freshwater inputs via river discharge (Milliman & Farnsworth 2011) and direct rainfall, these dynamic forcings serve to generate vigorous physical mixing and biogeochemical cycling. Despite some estimates as high as 90% (Jennings & Kaiser 1998), the most recent compilations suggest that shelf seas less than 200 m deep host between 16% and 24% of all oceanic primary productivity (Longhurst et al. 1995, Behrenfeld et al. 2005, Jahnke 2010, Simpson & Sharples 2012). The phytoplankton production rate is three to five times that of the open ocean (Simpson & Sharples 2012). This enhanced production derives from high nutrient flux, both from land (via rivers and wind) and via the mixing and upwelling of deep waters, and is well illustrated by global satellite images of surface ocean production (e.g., Behrenfeld & Falkowski 1997). The ecosystems fuelled by this primary production supply 90% of the fish consumed by humanity (Pauly et al. 2002). This enhanced primary productivity has a significant impact on the global carbon cycle; per unit area, the shelf seas have a carbon fixation rate about 2.5 times greater than the open ocean (Simpson & Sharples 2012). This production results in the drawdown of CO_2 from the atmosphere, and a proportion of this carbon is exported into the deep ocean (Walsh 1988), perhaps as much as 47% of the particulate organic carbon export of the total ocean (Jahnke 2010).

In 1844, Edward Forbes famously suggested, based on trawling, that the sea deeper than 300 fathoms (1 fathom = 1.83 m) would be found to be devoid of life, his *azoic* hypothesis. It is unfortunate that Forbes's name is now reputationally anchored to the doomed azoic hypothesis because he was the most influential of nineteenth century scientists to investigate shallow marine systems (Forbes 1859); however, subsequent research has clearly falsified his hypothesis on biodiversity in the deep ocean (Gray 2001). Although comprehensive and reliable data on the relative biodiversity of the shelf seas, as compared with marine deeper habitats, remain elusive (Gray 2001, Kaiser et al. 2005), the present-day highstand shelf sea systems are notably diverse, especially in South-East Asia, the Indo-West Pacific 'hot spot' (Kaiser et al. 2005).

The removal of the shelf seas during the LGM and during each earlier regression resulted in profound reorganization of energy dissipation, tidal dynamics, hydrography, sediment/habitat redistribution, and primary production with implications for the global carbon cycle and biodiversity. Despite the recognized impacts of Quaternary glacial-interglacial cycles on ecosystems, Graham et al. (2010) contended that "the ecological effects of ice ages on benthic marine communities are unknown" (p. 399). The aim of this review is to explore the Earth system implications of the geological transience of shelf seas as a function of glacial-interglacial sea-level change during Quaternary time with a focus on impacts on, and responses of, the marine biosphere. The review explores the most significant data on the magnitude and timing of the sea-level changes and some of the most important empirical evidence on the impacts of these changes on the shelf seas and highlights recent numerical modelling approaches that enable quantitative reconstruction of key shelf sea properties through time. These modelling approaches include glacio-isostatic adjustment (GIA), palaeo-environmental niche modelling (pENM) and palaeotidal modelling (PTM).

Although the review focuses on Quaternary time, epicontinental seas have been at the centre of hypotheses concerning the most important biotic and biogeochemical events since the development of life on Earth. Newell (1967) suggested that six of the most important mass extinctions during Earth history coincide with phases of major marine regression because of eustatic fall of sea level. The mechanism at the centre of Newell's hypothesis is the loss of epicontinental seas, which have been generally assumed to host high benthic biomass and diversity. Thus, the recent events that are the focus of this review have been invoked to interpret the major biotic crises of the geological past. The role of sea level in driving mass extinctions—and of oceanic processes during such events—has been reviewed by Hallam & Wignall (1999) to test Newell's hypothesis. They noted that following the transfer of interest from global regression to bolide impact as the cause of mass extinctions following the discovery of the iridium anomaly at the Cretaceous-Tertiary (K-T) boundary (Alvarez et al. 1980), "a certain amount of scepticism was expressed, for example, at the rarity of extinctions that could be related to the evident glacio-eustatic regressions in the Quaternary" (Hallam & Wignall 1999, p. 218). Hallam & Wignall analysed the sea-level evidence associated with the 'big five' mass extinction events and some minor extinction events and concluded that two, the late Ordovician and terminal Cretaceous (K-T), are associated with "unequivocal, major regressions" and that the biotic crises during the end Guadalupian and end Triassic can also be related to reduction of shelf extent associated with regression (Figure 2). However, the end Permian mass extinction, the most profound in Earth history and the focus of much of Newell's work, is now recognized to have been at a time of transgression, as is the Frasnian–Famennian crisis. Hallam & Wignall concluded that "rapid high amplitude regressive–transgressive couplets are the most frequently observed eustatic changes at times of mass extinction, with the majority of extinctions occurring during the transgressive pulse when anoxic bottom waters often became extensive" (p. 217). This observation has particular relevance for an assessment of the role of Quaternary sea-level cycles on the marine biosphere.

Highstands during 'greenhouse' phases of Earth history have been profoundly important in biogeochemical cycling. During the Cretaceous, the areal extent of the shelf seas may have been three times that of the present (Schopf 1980) as a result of climate warming driven by CO_2 outgassing from volcanic activity (Berner et al. 1983) preventing any significant ice sheets from developing. Not only did the modern oceanic food chain of diatoms, copepods and teleosts develop during this period (Walsh 1988), but also the epicontinental seas of the Cretaceous have been estimated to be the most productive in Earth history, Schopf (1980) estimating a peak of 320 g C m^{-2} yr^{-1} compared to 45 g C m^{-2} yr^{-1} in the Pre-Cambrian and Permian and 188 g C m^{-2} yr^{-1} in the Holocene. During such mega-highstands of the Mesozoic, significant organic carbon accumulation on the continental slopes generated many of the major hydrocarbon fields exploited today. Earth has experienced both extreme highstands and lowstands in the recent geological past, so the significance of shelf seas in mass extinction, speciation, ecosystem functioning and biogeochemical cycling during earlier Earth history provides a rich context for the analysis of the impact of Quaternary change.

Continental margins and sea-level change

The hypsometric curve

The continental shelves represent the shallow submerged extensions of the continents (Figure 3). Their morphology is a function of the interaction of the internal deep structure of Earth with the external processes effecting erosion and deposition. At depth, the shelves are underlain by light, silicic crust, which is similar in composition to the continents and contrasts with the heavy, basaltic/mafic crust of the deep ocean. The variable widths of continental shelves reflect their tectonic setting; passive margins, which represent the trailing edges of moving continents, are characterized by generally extensive shelves ('Atlantic-type' margins), whereas active margins are associated with subducting or transform plate boundaries (Figure 4) and generally very narrow shelves ('Pacific-type' margins). Given the current continental disposition, there are very extensive shelves in both the high latitudes, notably in the Siberian Arctic, and in the low latitudes, notably in South-East Asia; similarly, there are narrow, active, shelves in both high latitudes, such as Alaska, and low latitudes, for instance, equatorial western South America. Continental shelves incline gently from the continents towards the deep ocean, typically at a gradient of between 1.5 and 2 m per km and terminate at the shelf break, an increase in incline that steepens into the continental slope (Figure 3) with a typical gradient of 70 m per km. The shelf break depth is usually around 140 to 160 m, but in some settings can be deeper; the Celtic Sea shelf break is deeper, for instance, at around 200 m (Pantin & Evans

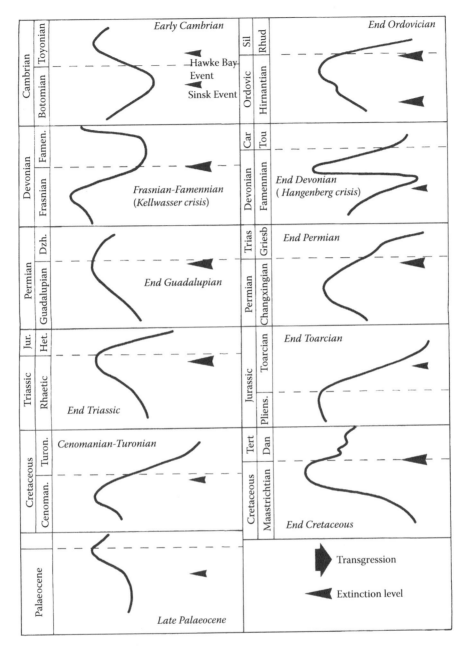

Figure 2 Eustatic sea-level tendencies and changes during intervals of major mass extinction. (Reprinted from Hallam & Wignall 1999, with permission from Elsevier.)

1984), and around actively glaciated continents, such as Antarctica and Greenland, the break depth can be as deep as 300 to 400 m as a result of glacio-isostatic loading. If the areal extent of Earth's surface is plotted between successive elevations, a bimodal distribution is produced that highlights the extent of the continental platforms, including the continental shelf (Figure 3). When plotted as a cumulative frequency distribution, a hypsometric curve is produced that indicates the significance of the shelf break as the natural boundary between the continents and the ocean basins (Figure 5). The wave and tidal energies associated with the highstands, and subaerial erosion by fluvial, glacial

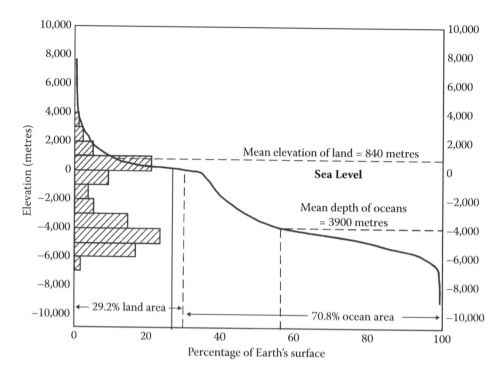

Figure 3 The hypsometric curve showing percentage of Earth's surface by elevation. The change in slope gradient illustrates the significance of the continental shelf break, and the significance of the continental shelf areal extent is demonstrated by the bimodal distribution of Earth surface elevation. (Adapted from Sverdrup et al. 1942.)

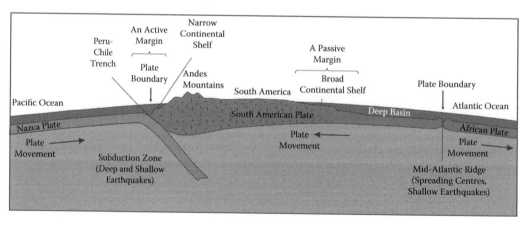

Figure 4 Tectonic control of continental shelf areal extent: narrow continental shelves are found along tectonically active, destructive ('leading edge') plate margins, whereas broad continental shelves characterize tectonically quiescent, passive ('trailing edge') margins. (Adapted from Garrison 1999, Brooks/Cole, a part of Cengage Learning, Inc. Reproduced by permission. http://www.cengage.com/permissions.)

and aeolian agencies during lowstands, result in the planation of the continental margins and contribute to the formation of the shelf. The same agencies deliver sediment from the eroding continents to the shelf break, extending the shelf in the process. The mass of these sediments causes the crust beneath to depress isostatically (see the next section), so the accumulating sediment pile at the shelf edge can be up to 15 km thick and many millions of years old. Thus, the morphology of the

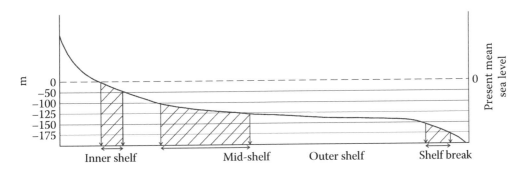

Figure 5 Extent of shallow marine habitat at different stages of sea-level elevation along continental margins as a function of the hypsometric curve. The global extent of shallow marine habitat when sea level is between –100 and –125 m below present mean sea level is much greater than when sea level is either at extreme highstand (between 0 and –50 m) or extreme lowstand.

continental shelves is itself partly a function of the cycles of emergence and inundation modulated by sea-level change that are the focus of this review.

The global sea level of 120 to 130 m lower than present during the LGM resulted in the total or partial emergence of most unglaciated continental shelves. The extent of inundation or emergence as a result of any given vertical change in sea level is, however, a function of the changes of gradient across continental margins represented by the hypsometric curve (Figure 3). Thus, any given vertical change in sea level close to full highstand or full lowstand will result in relatively little change in areal extent of either exposed shelf or sea surface because the gradients of the shelf break and inner shelf are greater than the main extent of the shelf (Figure 5). Simpson & Sharples (2012) noted that an extension from the 200-m to the 500-m isobath only increases the areal extent of the continental shelf by 2%, from 7% to 9%, because this depth range encompasses the steeper gradient of the upper continental slope. The transition from inner shelf to continent is characterized by often-steep, sometimes-cliffed, shorelines representing the reoccupation site or 'notch' of successive interglacial highstands. Any given vertical change in sea level across the midshelf will result in large areal changes in the extent of emergent shelf or sea surface (Figure 5). For ecosystems dependent on specific water depths (such as coral reefs), intertidal habitat or light penetration, this interplay of topographic gradient with sea level exerts a fundamental control. If the transition into the deep ocean from the continents was a linear slope then, at times of sea-level change, the areal extent of habitats and ecosystems would remain constant as they migrate in response to the given vertical change in sea surface. The sharp break of slope at the outer edge of the shelf determines that at times of major regression outer shelf marine habitat extent is compressed or eliminated completely; conversely, during highstands, shallow inner shelf or intertidal habitat is compressed or 'squeezed' against cliff lines and steeper gradients representing former extreme-highstand shorelines.

Relative sea-level change

Although satellites have been able to measure accurately the elevation of the sea surface in relation to Earth's centre of mass over a period of several years, for changes of sea level over geological timescales it is necessary to consider the relative position of the mean sea-surface elevation in relation to a datum elevation defined on any adjacent landmass, known as relative sea level (RSL). RSL changes as a function of two main variables: the elevation of the sea surface itself and the elevation of the adjacent landmass. Transgression (a relative rise of sea level) and regression (a relative fall of sea level) result from the interaction of vertical elevation changes in the mean sea surface, the adjacent land surface, or a combination of both.

The elevation of the sea surface itself—that is the sea-level property measured by satellites—is termed ice-equivalent sea level, commonly known as *eustatic sea level*. Eustatic sea level is effectively a global property and is controlled by a number of variables, the most important of which is the growth and decay of ice sheets, the glacio-eustatic mechanism, which explains why eustatic sea level was 120 to 130 m lower than at present during the LGM; this is equivalent to the transfer of 50 million km^3 of water from the oceans to the ice sheets (Lambeck et al. 2002). The glacio-eustatic mechanism was the most important control on global sea levels during the Neogene and Quaternary. Satellite data indicate that the mean rate of absolute sea-level rise is currently 3.1 ± 0.7 mm yr^{-1} (Intergovernmental Panel on Climate Change (IPCC Assessment Report 4 (AR4); Bindoff et al. 2007), but estimates vary between 2.4 and 3.6 mm yr^{-1} (Cazenave et al. 2009, Leuliette & Miller 2009); most of this, between 2.4 and 3.6 mm yr^{-1}, is accounted for by the melting of glaciers and ice caps, the mass flux effect (Willis et al. 2008, Leuliette & Miller 2009). The other -0.5 to 0.8 mm yr^{-1} is accounted for by the temperature of the ocean (Willis et al. 2008, Leuliette & Miller 2009), which is another eustatic control. As the ocean warms, so its density diminishes, it expands and sea level rises; this thermal expansion of the oceans is known as the steric effect. Over Quaternary timescales, the steric effect is relatively small, and the RSL change is dominated by the glacio-eustatic mechanism.

Over geological timescales, the size of the ocean basins—driven by global tectonics—is a significant control of eustatic sea level. At some times in Earth history, the global ocean basin has been relatively small, with high eustatic sea level, whereas at other times, a volumetric increase in the size of the basin has caused sea level to fall. When rates of seafloor spreading are high, as during the Cretaceous, the enhanced size of the midocean ridges reduces the overall volume of the basin, and eustatic sea level rises; the Cretaceous recorded the highest sea levels of the Cenozoic with some of the most extensive shelf seas of the entire Phanerozoic. When erosion of continents is rapid and this solid mass is transferred into the ocean basins, the volume of the basin is diminished. Other eustatic controls include the addition of so-called juvenile water, water added to the system through volcanic and hydrothermal activity, notably at sites and centres of seafloor spreading, and the amount of water that is held in terrestrial storage, in groundwater, in lakes, and increasingly, in man-made reservoirs. Another important effect is the gravitational attraction between ice sheets and oceans (Milne et al. 2009). In the vicinity of an ice sheet, the mass of the ice sheet generates a gravitational attraction that causes the sea surface to rise up adjacent to the ice sheet. This phenomenon is now being used as a form of sea-level 'fingerprinting' to detect the extent of current melting by the major ice sheets; a major melting of the Greenland Ice Sheet, for instance, is detectable as a rise in sea level in the Southern Hemisphere and a fall adjacent to the ice sheet itself (Mitrovica et al. 2001).

One of the most important sets of factors influencing the vertical elevation of Earth's crust at any point on the surface is linked to loading (Milne et al. 2009). Whenever a load is placed on Earth's surface, it causes the surface to sink in proportion to the mass of the load imposed. Although Earth's crust is relatively rigid, the upper parts of Earth's mantle are viscous and ductile and flow in response to imposed stress. Any load causes flow away from the site of loading in the upper mantle (resulting in subsidence) and transfer to sites beyond the point of that imposed stress (resulting in peripheral uplift). Unloading results in a return to the original equilibrium state via upper mantle transfer. The generic name given to this equilibrium state is *isostasy*; when the loading is effected by the removal or addition of solid rock, the process is known as *denudational isostasy*, by water as *hydro-isostasy* and by ice as *glacio-isostasy*.

In denudational isostasy, erosion of uplands involving transport of material into an adjacent basin of deposition causes release of load in the eroded province, causing uplift, and loading within the basin, causing crustal depression. Climate-driven erosion of the Fennoscandian massif during the Cenozoic, for example, has generated uplift, and the concomitant deposition within the North Sea Basin has caused isostatic subsidence (Nielsen et al. 2009). Because this process was ongoing during the latter Cenozoic, this pattern of highland uplift and basin subsidence is unidirectional;

that is, there has been no reversal in the process. This is not the case with either hydro-isostasy or glacio-isostasy.

The periodic and cyclic flooding and emergence of the continental shelves controlled by glacio-eustasy during the Quaternary have resulted in the loading of the crust forming the shelf by the mass of overlying water proportional to the increase in eustatic sea level. This hydro-isostatic effect causes the depression of the shelf by a few decimetres per 100 m of sea-level change and therefore accentuates the extent of the sea-level rise. This is therefore a positive-feedback mechanism. Conversely, when the sea evacuates the shelf, the load is removed, and the original equilibrium is restored by the shelf returning to its original level. As this effect is controlled by glacio-eustatic cycles, this hydro-isostatic process is a cyclic phenomenon, with the shelf oscillating up and down to the pace of the global ice age cycles.

For the continental shelves, the extent of loading by ice, glacio-isostasy, is two orders of magnitude greater than hydro-isostasy. As an ice sheet grows, it depresses the crust; given the average densities of glacial ice and a mean figure for the viscosity of the upper mantle, the extent of depression approximates to one-quarter of the thickness of the ice sheet. Lambeck (1996a) predicted from numerical models the extent of isostatic depression for three ice sheets of differing size in existence during the LGM: the huge Laurentide (over North America), the Fennoscandian and the British ice sheets. At its thickest, the Laurentide Ice Sheet was almost 4 km thick, causing depression of some 900 m (Figure 6), whereas the much smaller British Ice Sheet was only 1.5 km at its thickest,

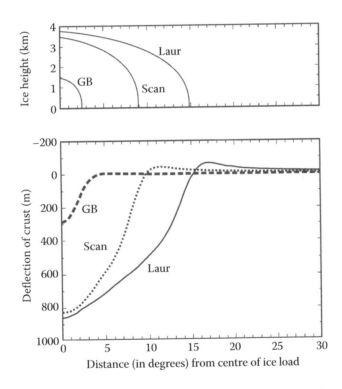

Figure 6 Upper panel: Simulated ice sheet thickness and extent at the Last Glacial Maximum: GB, British-Irish Ice Sheet; SCAN, Fennoscandian Ice Sheet; LAUR, Laurentide Ice Sheet. Lower panel: Simulated regional isostatic adjustment to the respective loadings. Note that isostatic deflection is proportional to ice thickness, and that the zones of maximum depression are compensated by marginal annular forebulges. Hydro-isostatic loading is ignored in these simulations. For effective parameters used in these simulations, see the work of Lambeck (1996a). (Adapted from Lambeck 1996a, with permission from the Geological Society of London.)

possibly less, in the western highlands of Scotland, causing depression of around 300 m. To maintain isostatic equilibrium, the adjacent unloaded areas uplift as a result of lateral flow of material within the upper mantle. Around an ice cap, therefore, there is an annular 'forebulge'. When the ice melts, the loaded province rebounds, and the forebulge collapses back to its original position. This process, as with hydro-isostasy, is ultimately linked to the growth and melting of ice sheets and is therefore cyclic in nature. If the coast or shelf lies in the vicinity of a large ice sheet, this isostatic effect can be profound. The isostatic depression of the crust beneath large ice sheets often brings their margins within the reach of the sea, and in these cases, the ice sheet develops a marine calving margin. This is currently the case around much of the margins of Antarctica and was certainly the case for many of the large ice sheets of the Pleistocene. Before the advent of numerical models of the behaviour of Earth's crust in response to loading, glacio-isostasy was considered to be only of local or regional significance. However, it is now known that large ice sheets influence the vertical position (elevation) of every point on Earth's surface. This effect becomes more significant with proximity to the ice sheet in question; in 'far-field' locations, the effect is small, but nevertheless measurable. Glacio-isostasy is therefore a profoundly important mechanism controlling sea levels at a global scale.

For this review of the impact of Quaternary sea-level cycles on the marine biosphere, by far the most important controls of RSL are the interplays between glacio-eustasy and glacio-isostasy. The integration of these major and other controls of RSL can now be simulated to generate predictions (or hindcasts) depicting the evolution of coastlines and shelf seas through time using GIA models (see the section on GIA modelling).

The Quaternary sea-level record

Sea-level index points

Observational data on RSL change since the LGM are based on the identification and definition of sea-level index points (SLIPs). These are points of known position and elevation that have been precisely and accurately dated, usually using the radiocarbon (^{14}C) method, to provide information on the past position of sea level. Some regions have a much denser coverage of SLIPs than others; north-western Europe and the British Isles and Ireland in particular have a rich SLIP database (Shennan & Horton 2002, Shennan et al. 2006a). SLIPs come from three main settings: drowned former land surfaces, uplifted shorelines and beaches, and 'isolation basins'. Many SLIPs are based on the occurrence of radiocarbon-dateable peat, or other organic sediment of terrestrial origin, either over- or underlying marine sediments.

A common approach to the definition of SLIPs has been formalized by the International Geological Correlation Programme (IGCP) Projects 61 and 200 (e.g., van de Plassche 1986). The general attributes of reliable sea-level observations, with defined uncertainty, which define past changes in sea level relative to the present, are derived from sedimentary sequences containing both minerogenic and organic components and morphological features relating to former coastlines. The sediment sequences must be *in situ*. The four attributes of SLIPs are location, age, altitude and tendency (Shennan et al. 2006a).

Of the four fundamental attributes of SLIPs, location is the geographical coordinates of the site where the samples were collected. Age is usually based on calibrated radiocarbon ages of peat or organic sediment at the point of contact or, exceptionally, on samples of diagnostic microfossils. Most SLIPs consist of sediment sequences in which non-marine sediments are overlain by marine sediments or vice versa. The former are termed *regressive* contacts and often indicate a fall in RSL; the latter are *transgressive* contacts often registering a rise in RSL. Where such contacts can be demonstrated over a wide area, in other words, where a landscape has been inundated (exposed) by the sea and that rise (fall) in RSL can be detected at multiple sites within a vicinity, the terms *regressive overlap* and *transgressive overlap* can be used. One of the most important principles is to establish that the contact is conformable, that there is no hiatus at the point of contact caused by a break in sedimentation or loss of sediment through erosion. Such problems can be overcome by the definition of SLIPs from settings in which erosion or non-deposition has been insignificant; these tend to be associated with salt marshes. In these settings, rising sea level will result in the saltmarsh system migrating landwards and the gradual, conformable, deposition of saltmarsh mud over former terrestrial soil or sediment; in regressive situations, the salt marsh develops over former unvegetated intertidal mudflats. The detection of microfossils (notably diatoms and benthic Foraminifera) or other indicators of saltmarsh sedimentation immediately above or below the point of contact is therefore often a key attribute in the definition of a SLIP (Gehrels et al. 2001).

One of the more intractable attributes of any particular SLIP is its 'altitude attribute'. Whilst it may be straightforward to define the altitude of the key critical contact by surveying, it may be difficult to determine the relationship between that elevation and the palaeo-sea level. Few SLIPs form at the actual level of the palaeo-sea surface; most form higher within the tidal prism, in the upper part of the tidal range, usually at or around mean high water of spring tides (MHWST). This relationship to the former mean sea level is defined as the 'indicative meaning'. Indicative meaning "defines the relationship of the sample to tidal range and allows us to measure relative sea-level change, defined relative to present" (Shennan et al. 2006a, p. 587; van de Plassche 1986). If tidal range is assumed not to have changed through time and if supporting data such as microfossil assemblages are able to indicate the range over which the sediment accumulated in the past, then the palaeo-sea level can be reconstructed. Based on PTM simulations, it is now clear, however, that to assume that tidal range has not varied in the past can introduce significant errors (see section on palaeotidal modelling below).

Another problem in defining altitude is sediment compaction, compression or consolidation. Many SLIPs derive from coastal embayments and estuaries filled with thick sequences of unconsolidated sediments. Over time, these sediments compact; so, following their formation, the SLIPs will subside to a location lower in elevation than their original position. Whilst some quantitative approaches for postdepositional compaction correction have been proposed (e.g., Paul & Barrass 1998, Massey et al. 2006), no universally applicable approach has yet been adopted, and most workers incorporate compaction as an uncertainty term that contributes to the overall error of the sea-level reconstructions. An alternative approach is to focus only on SLIPs in which the dateable horizons rest on substrates with no or minimal compaction, such as bedrock or subglacial till (e.g., Gehrels et al. 2002). Such 'basal peat' contacts provide the highest-quality SLIPs but are only found, or are only readily accessible, in a limited number of shallow estuarine settings; most RSL curves must, of necessity, include SLIPs from compactable contacts. Where compaction has taken place, the palaeo-sea level will have been higher than indicated by the SLIP. It is probable that much of the scatter in the SLIP database is a product of this effect (Shennan et al. 2006a).

Useful information can often be provided by samples whose credentials fail to fully qualify them as SLIPs but which nevertheless indicate that, in a certain place, at a certain time and at a particular elevation, the palaeo-sea surface was either above or below that elevation. Such 'limiting index points' can be either fresh water, indicating that the palaeo-sea level was at a lower elevation, or fully marine, indicating that the palaeo-sea level was at a higher elevation, at that place at that time.

The compilation of SLIP data from different localities demonstrates the course of sea-level change through time in the form of RSL curves (Figure 7; Shennan et al. 2006a). These cover the period since the last deglaciation, although the data are heavily skewed towards the more recent period. It is important to understand exactly what these diagrams show. The x axis of the plots shows time, in this case in calibrated age based on ^{14}C dates; 15,000 cal BP[*] is during the main phase of deglaciation after the LGM, and 11,700 cal BP is the base of the present warm stage, the

---

[*] The abbreviation *cal BP* denotes calibrated radiocarbon ages before present (Reimer et al. 2009).

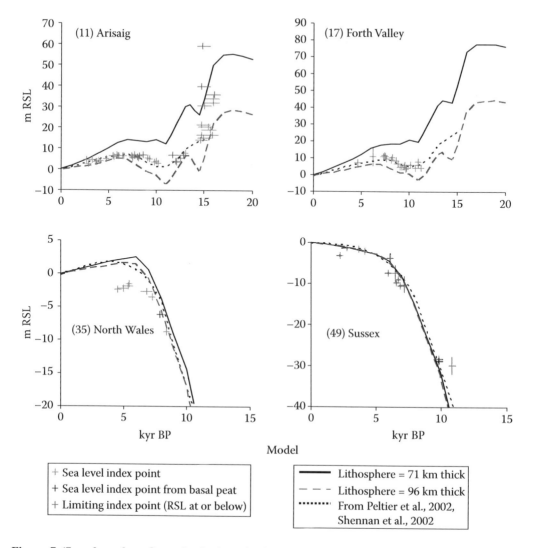

Figure 7 (See also colour figure in the insert) Glacial isostatic adjustment (GIA) model predictions compared with observations of relative sea-level change for four sites in the British Isles (for locations, see Shennan et al. 2006a, Figure 1). Error bars for ages indicate 95% probability range, and vertical bars indicate median calibrated age. GIA results are shown for three simulations, one based on Peltier et al. 2002 and Shennan et al. 2002 and two using different lithosphere thickness; for effective parameters see Shennan et al. (2006a). (Reprinted from Shennan et al. 2006a, with permission from Wiley-Blackwell.)

Holocene (Walker et al. 2009). The *y* axis shows the *present-day* elevation of the dated SLIP in relation to mean sea level, in this case UK Ordnance Datum (OD). The SLIP data therefore show the course of RSL change within the specified regions and highlight any differences in the pattern of sea-level change between regions through time (Shennan et al. 2006a). Note the profound differences in RSL curves between regions that were strongly glacio-isostatically loaded during the last glaciation, such as Scotland, and those that lay beyond the significant influence of ice loading, such as the English Channel coast, which has an RSL history dominated by glacio-eustasy.

SLIPs from eustatically dominated settings are typically found below present sea level, but in glacio-isostatically uplifted regions, they are found above sea level, in the form of uplifted estuarine sequences (such as the 'carselands' of Scotland; Smith et al. 2000), raised beaches or isolation

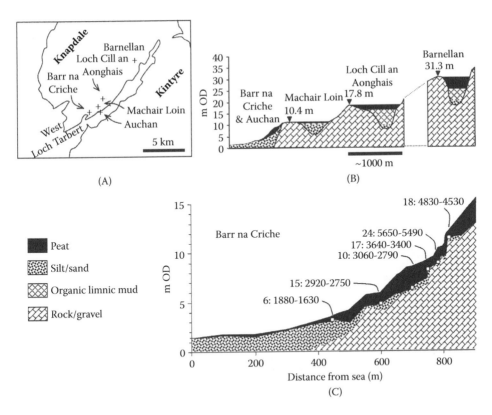

Figure 8 Staircase of isostatically uplifted isolation basins used to generate sea-level index points from north-western Scotland. (A) Site locations in Knapdale. (B) Schematic section showing elevational and stratigraphical relations between sites, including rock sill elevations. (C) Coring transect of the marsh at Barr na Criche: core numbers followed by radiocarbon ages from the base of each peat section (95% probability range, years BP). (Reprinted from Shennan et al. 2006b, with permission from Wiley-Blackwell.)

basins. Raised beaches can be dated using cosmogenic rock-exposure dating (Stone et al. 1996) or, if they contain preserved marine shells, using the ^{14}C method (Smith et al. 2000). Uplifted marine rock basins or sea lochs can become isolated from the sea by isostatic uplift to form isolation basins. The sediments deposited in such basins (Figure 8) show an upwards sequence of first marine, then brackish, and finally fully freshwater sediments (e.g., Shennan et al. 2006b). The diatom assemblages from such sequences, as in many other kinds of SLIPs, are often particularly diagnostic in defining precisely the point in the sequence at which the isolation occurred (Figure 9). If this point of isolation can then be dated using ^{14}C and the elevation of the lowest point of the rock sill established by precise and accurate surveying, then the isolation basin forms a valuable SLIP defining RSL change in the past. Isolation basins also operate in the opposite sense. If, following isolation, the RSL rises again, the sill is overtopped, and marine water once again floods into the basin, marine sediments will be deposited on the surface of the underlying freshwater sediments. This second contact can also be defined using diatoms and other microfossil assemblages and dated to form another SLIP. In this way, multiple SLIPs can be derived from the same isolation basin. There are countless examples of isolation basins spread across the former glaciated terrain of Canada, Fennoscandia, Scotland and southern South America. By judicious selection of isolation basins with sill elevations forming a staircase of different elevations across the landscape, exceptionally detailed records of sea-level change can be established (Figure 8).

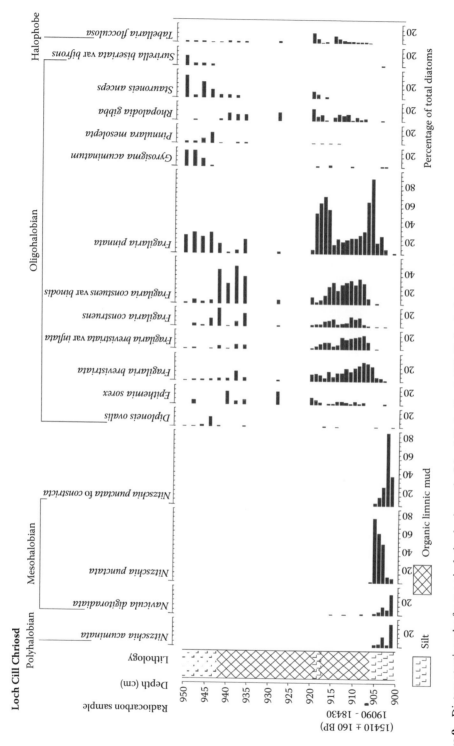

Figure 9 Diatom stratigraphy from an isolation basin on the Isle of Skyle (Loch Cill Chriosd); for location, see Shennan et al. (2006b). Radiocarbon age (±1σ, 95% probability range, years BP), depth (below ground surface) and lithology shown on left. Percentage frequency of diatom species in ecological groupings to right: polyhalobian and mesohalobian classes represent marine taxa, oligohalobian and halophobe classes freshwater taxa. The transition from a marine assemblage to a freshwater assemblage marks the isolation of the basin as a result of isostatic uplift; this transition is also marked by a change in sediment type. (Reprinted from Shennan et al. 2006b, with permission from Wiley-Blackwell.)

Pre-LGM sea-level reconstruction

Prior to the LGM, the reconstruction of RSL change based on direct geological observations such as SLIPs becomes much more difficult. This is primarily because the evidence of earlier RSL cycles has been removed by subsequent erosion as a function of emergence by glaciation, fluvial and aeolian processes. Long-term records of RSL change that extend beyond the LGM are therefore confined to tectonically uplifted, non-glaciated, settings. These are notably characterized by uplifted coral reef terraces in tectonically active margins. The classic example is the uplifted coral terrace sequence from the Huon Peninsula, New Guinea (Chappell 1974, Lambeck & Chappell 2001), but key records also come from the Caribbean, notably Barbados (Bard et al. 1990). If the long-term uplift rate can be accurately estimated, and if fossil coral species with a known relationship to the sea surface (e.g., *Acropora palmata*) are preserved within the coral terrace, then U-series dating of the coral enables a reconstruction of the former RSL. Non-uplifted coral reef sequences from less tectonically active settings are important for providing long-term fixes on global eustatic change (Fairbanks 1989, Stirling et al. 1998).

Such sequences are, however, rare and only provide spatial and temporal snapshots of RSL for the regions concerned. The only means of generating a comprehensive overview of the broad pattern of global sea-level change through the entire Quaternary is indirectly via the deep-sea benthic oxygen isotope record (Lisiecki & Raymo 2005). This record is based on the stable oxygen isotopic composition of fossil benthic Foraminifera, conventionally notated $\delta^{18}O$, which expresses the relative proportion of ^{18}O to ^{16}O in relation to a standard. The $\delta^{18}O$ value of foram tests is partly a function of seawater temperature, but this is overprinted by isotopic composition of the background seawater, denoted δ_w (Shackleton 1967). The most important control on δ_w is the amount of ice stored on land as ice sheets because the water used to build ice sheets is ultimately derived by evaporation from the sea surface. When seawater evaporates, the lighter ^{16}O isotope is preferentially selected over the heavier ^{18}O. Similarly, when water condenses in clouds, the heavier ^{18}O is preferentially selected. Both processes involved in this partitioning (known as Rayleigh distillation) are influenced by temperature; precipitation in the tropics is relatively unselective, but with progressively cooler conditions, ^{16}O is increasingly selected over ^{18}O. The combination of the fractionation during evaporation and during condensation means that precipitation and river water are isotopically lighter than the ocean, progressively more so at higher latitudes. During warm phases in Earth's climate, most of this water flows back into the ocean via rivers, but during cold, or glacial, phases, because ice sheets are isotopically light, particularly those at high latitudes, ^{16}O is preferentially locked in the ice sheets. This in turn causes the ocean δ_w to enrich in heavy ^{18}O. Calcifying organisms such as Foraminifera therefore register isotopically light $\delta^{18}O$ during warm phases and isotopically heavy $\delta^{18}O$ during cold, or glacial, phases. Since benthic $\delta^{18}O_{foram}$ is dominated by δ_w, it is a proxy for global ice volume; this therefore also means that it is a proxy for glacio-eustatic sea level, which is the dominant control of Quaternary sea-level cycles.

The most comprehensive compilation of benthic $\delta^{18}O$ data is the LR04 stack constructed by correlation of 57 individual globally distributed $\delta^{18}O$ records (Lisiecki & Raymo 2005; Figure 10A). This can be treated as a first-order indication of global glacio-eustatic sea-level trends through the last 5.3 million years, although used in this way the record should be treated with caution. The $\delta^{18}O$ values have not been corrected for temperature (Waelbroeck et al. 2002), salinity, diagenetic or vital effects, and there are non-linearities in the relationship between $\delta^{18}O$ and ice volume over time (Mix & Ruddiman 1984), issues highlighted by Chappell & Shackleton (1986) and Shackleton (1987) in their analysis of the relationships between the $\delta^{18}O$ record and the Huon Peninsula RSL record. These effects are accentuated near inflection points in the $\delta^{18}O$ curve (Tzedakis et al. 2012). Furthermore, the relationship between sea levels and ice volumes is not straightforward because of Earth's response (isostasy, gravity, rotation) to changes in loading by ice, water and sediment (Lambeck et al. 2012). Despite these caveats, for the purposes of this review it is possible to treat

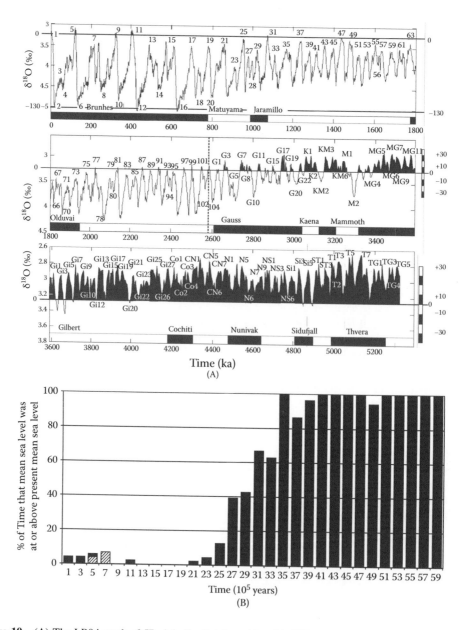

Figure 10 (A) The LR04 stack of 57 globally distributed benthic $\delta^{18}O$ records extending from the present to 6 million years ago. The base of the Quaternary at 2.58 Ma ago (Gibbard et al. 2010) is shown by a vertical dotted line. Present-day highstand sea level is shown by a horizontal line (0) representing $\delta^{18}O$ about 3.25 and extreme lowstand equivalent to the Last Glacial Maximum by $\delta^{18}O$ about 5 shown by horizontal line (-130). Global sea levels at, or exceeding, present-day sea level are shown in black; sea levels lower than present are shown in white. Oxygen isotope stages numbered and palaeomagnetic timescale shown by black and white named chrons. The anomalous nature of the present highstand, the increasing amplitude and wavelength of cycles following the Mid-Pleistocene Revolution, and the gradually falling sea levels of the Neogene are all clear. (Adapted from Lisiecki & Raymo 2005, with permission from the American Geophysical Union.) (B) In black, percentage of time that global sea level was at or above present mean sea level in 200-ka increments from present to 6 Ma ago based on (A). Percentage of time per 200-ka increment that global sea level was at or below the LGM lowstand shown hatched. This indicates that both extreme highstand and extreme lowstand conditions increased in the Late Pleistocene as the amplitude of sea-level cycles increased.

16

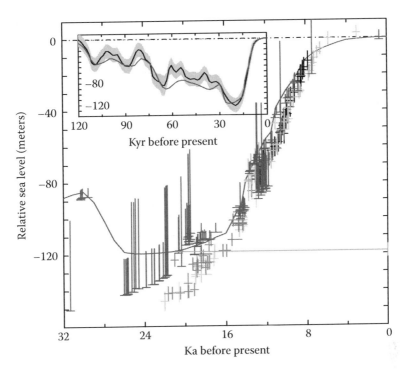

Figure 11 (See also colour figure in the insert) Global sea level since the Last Glacial Maximum (LGM). Main panel: Observations of sea-level history from coral-based record from Barbados shown in blue, with further observations (Lambeck & Chappell 2001) from Barbados (cyan), Huon Peninsula (black), Tahiti (grey) and Sunda Shelf (purple). Red line shows predicted relative sea-level curve for Barbados using ICE-5G (VM2) model. Inset panel: Oxygen isotope-based sea-level history for period since last interglacial (black line with error estimate in grey) based on Waelbroeck et al. (2002). Red line shows predicted relative sea-level curve for Barbados using ICE-5G (VM2) model. (Reprinted from Peltier & Fairbanks 2006, with permission from Elsevier.)

the LR04 data empirically since marine isotope stage (MIS) 1 corresponds to the present, Holocene, interglacial, for which sea level is known, and MIS 2 corresponds to the LGM, for which we have good multiple observational records (Figure 11; Peltier & Fairbanks 2006); this same approach has been used to reconstruct an RSL record for the last four climatic cycles (Waelbroeck et al. 2002) and for Pliocene highstands (Raymo et al. 2009). Therefore, a $\delta^{18}O$ value of about +3.25 equates to present mean sea level, and a $\delta^{18}O$ value of about +5 to an LGM sea level of –130 to –120 below present mean sea level. Extrapolated through the record, this indicates that of recent interglacial stages, MIS 5e, 9 and 11 may have had peak sea levels in excess of the present. For MIS 5e, this is well supported by geological data from Barbados (Stirling et al. 1998) and elsewhere indicating global sea levels about 6–10 m higher than at present (Kopp et al. 2009), probably as a result of significantly reduced Greenland Ice Sheet (Cuffey & Marshall 2000, Otto-Bliesner et al. 2006). Even though global sea levels are now higher than at any time since the inception of the last glaciation, these data indicate that for very short peak highstands during recent interglacials—for MIS 5e between 128 ± 1 and 116 ± 1 ka (Stirling et al. 1998)—the shelf seas were more extensive than they are now. Each highstand is, however, of short duration in relation to total time, reinforcing the anomalous context of the present highstand. MIS 25, 31, 47, 77 and 81 had peak highstands close to the present, and prior to MIS 87, over 2.2 Ma, every highstand exceeded the present. There is a tendency for gradually falling eustatic sea levels through the entire period; prior to about 3.3 Ma, sea level was higher than present for most of the time, and before about 3.7 Ma, only the extreme lowstands were at similar elevations to the present highstand (Figure 10B). The Pliocene shelf seas were therefore much more extensive, and much less spatially variable in extent, than those of the Quaternary, especially those of the late Quaternary (Raymo et al. 2011). The data demonstrate that extreme lowstands are even more anomalous than highstands. There have only been minima as extreme as the LGM (MIS 2) on three occasions during the last 5.3 million years, during MIS 6, 10, 12 and 16. The major change in shelf-sea extent took place with the rapidly falling sea levels of the late Pliocene; prior to 3.4 Ma, the shelf seas were as, or more, extensive than at present for almost all of the time. After 2.1 Ma, shelf seas were only rarely as extensive, or more extensive, than at present. An increase in both extreme high- and lowstands occurred after the Mid-Pleistocene Revolution (~800 ka; Clark et al. 2006) as a function of the transition from low-amplitude, low-wavelength, 40,000-year cycles to high-amplitude, high-wavelength, 100,000-year cycles. This clearly demonstrates that whilst shelf seas of present extent are anomalous, the recent lowstands typified by the LGM are even more anomalous over a period of some millions of years. The "rapid high amplitude regressive–transgressive couplets" identified as most commonly associated with mass extinction events by Hallam & Wignall (1999) are precisely the conditions that have prevailed for the last 800,000 years.

Glacial isostatic adjustment modelling and the reconstruction of continental shelf palaeotopography

Glacial isostatic adjustment modelling

The multiplicity of components that control RSL and the differences in the response times of these various components determine the fundamentals of how coastlines migrate through space and time. The pattern of RSL change since the LGM is therefore the result of the time-dependent interaction of the different controls. These interactions can be explored and analysed using numerical GIA modelling. GIA models integrate physical descriptions of the response of the Earth's crust and mantle (rheology) to loading, and information from glacial geology and ice sheet modelling on how ice sheets evolve through time, with data on ice-equivalent sea level to predict for any defined period of time the extent of isostatic loading and uplift in space (Shennan et al. 2006a). GIA model outputs include predictions of coastline position and, off shore, bathymetry, for defined periods of time. Goodness of fit between data and model can be tested statistically in a number of different ways. The key variables that are used to set up the simulations and are integrated within GIA models—the input parameters—relate to the physical properties of the solid Earth (Earth model parameters), the global changes in ice-equivalent sea level (the sea-level equation; Farrell & Clark 1976), and the glacio-isostatic (un)loading as a function of the location, size, extent and melting histories of ice sheets (the ice sheet model) both close to the region being modelled (near field) and far distant (far field). One important feedback is the hydro-isostatic (un)loading generated by seawater as eustatic sea level (falls) rises across continental shelves. The amplitude of the changes in surface elevation in regions close to the ice sheets during Quaternary time have been large; the ice-equivalent term exceeds 100 m between glacial and interglacial stages and close to centres of ice loading land depression, and rebound via glacio-isostatic (un)loading can be many hundreds of metres (Figure 9).

In integrating these parameters and predicting their interactions through time, GIA model outputs include RSL curves for any point within the model domain and therefore patterns of coastline and bathymetric change through time; in effect, GIA outputs represent palaeotopographies for the successive time steps within the modelled region. The most important ways of testing how successful any particular simulation has been is to compare the predicted RSL curves with observed RSL curves based on the compilation of well-constrained SLIPs (Shennan et al. 2006a; Figure 7). SLIPs are the gold standard observations used to constrain GIA model simulations. The richness of the northern European SLIP database for the period since the LGM combined with the complex patterns of ice sheet growth and decay through time make northern Europe one of the most intensively

THE FUNDAMENTALS OF GIA MODELLING

Shennan et al. (2006a) summarized the fundamentals of GIA modelling using the British Isles as an example. For any location within the model domain, φ, the change in RSL $\Delta\xi_{rsl}$ at time τ can be expressed as

$$\Delta\xi_{rsl}(\tau, \varphi) = \Delta\xi_{eus}(\tau) + \Delta\xi_{iso}(\tau, \varphi) + \Delta\xi_{tect}(\tau, \varphi) + \Delta\xi_{local}(\tau, \varphi)$$

where

$\Delta\xi_{eus}(\tau)$ is the equivalent sea level, the time-dependent eustatic term that varies as a function of global terrestrial ice volume.

$\Delta\xi_{iso}(\tau, \varphi)$ is the total isostatic effect of the (un)loading of near- [$\Delta\xi^B_{iso}(\tau, \varphi)$: British-Irish Ice Sheet, $\Delta\xi^F_{iso}(\tau, \varphi)$: Fennoscandian Ice Sheet] and far-field ice sheets [$\Delta\xi^{f-f}_{iso}(\tau, \varphi)$: Laurentide, Barents, Kara and Antarctic ice sheets] and the hydro-isostatic (un)loading of the continental shelves by $\Delta\xi_{eus}(\tau)$.

$\Delta\xi_{tect}(\tau, \varphi)$ represents any long-term underlying tectonic component of land movement, but for the period since the LGM, this is regarded as negligible in northern Europe (Kiden 1995).

$\Delta\xi_{local}(\tau, \varphi)$ expresses the total effect of local changes within the coastal system (from which SLIP data are generated):

$$\Delta\xi_{local}(\tau, \varphi) = \Delta\xi_{tide}(\tau, \varphi) + \Delta\xi_{sed}(\tau, \varphi)$$

where

$\Delta\xi_{tide}(\tau, \varphi)$ expresses changes in the tidal regime over time and, in particular, the elevation of the sedimentary surface in relation to the tidal amplitude at the time of deposition

$\Delta\xi_{sed}(\tau, \varphi)$ represents sediment compaction and consolidation following deposition.

GIA-modelled regions on Earth (Lambeck 1995, 1996a, Peltier 1998, Peltier et al. 2002, Milne et al. 2006, Shennan et al. 2006a, Bradley et al. 2011). Because GIA model outputs are so important for so many different components of Earth system science—from understanding ice sheet history to global glacio-eustatic (equivalent) sea level to the physical properties of the interior Earth—and because these outputs are now used as input variables for other generations of Earth system models, such as PTMs, it is crucial that the constraining SLIP data are as robust and reliable as possible.

Because the solid Earth surface is everywhere subject to deformation in response to climate-forced ice and water loadings, it is not possible, even in very far-field locations well away from the great ice sheets, to generate sea-level curves that are a pure function of only eustatic, ice-equivalent, sea level. This makes the derivation of $\Delta\xi_{eus}(\tau)$ a problem. However, long records of sea-level change since the LGM from far-field locations are strongly dominated by the eustatic function with only a minor contribution from isostasy (Figure 11); these include key records from Barbados (Fairbanks 1989, Bard et al. 1990), Tahiti (Bard et al. 1996), New Guinea (Chappell & Polach 1991, Cutler et al. 2003), northern Australia (Yokoyama et al. 2000, 2001a) and Indonesia (Hanebuth et al. 2000). Once corrected for local tectonic uplift, these long records provide a basis, along with assessments of the total amount of ice contained in Earth's combined ice sheets during the LGM and their melting history, for solving $\Delta\xi_{eus}(\tau)$, the "generalised sea-level equation" (Mitrovica & Milne 2003).

The observed SLIP data for the shield areas of Fennoscandia and North America (Laurentide Ice Sheet) are sensitive to the description of deep Earth structure within the Earth model. For these large ice sheets, the pattern of loading and rebound is therefore largely controlled by the rheology and deep structure of Earth (Lambeck 1993a,b, 1995, 1996a, Peltier 1998, 2004). However, for the

British–Irish Ice Sheet, the glacio-isostatic component of RSL change is sensitive to the way the shallow Earth is prescribed, in particular the thickness of the relatively rigid lithosphere and the depth profile of viscosity through the upper mantle (Lambeck et al. 1996, Shennan et al. 2000b, Peltier et al. 2002).

Whilst all GIA models use an Earth model that is "a spherically symmetric, self-gravitating, compressible Maxwell viscoelastic body" (Shennan et al. 2006a) that conforms with constraints based on the seismic character of Earth (Dziewonski & Anderson 1981), different approaches have been adopted in describing the components of the Earth model. The lithosphere is usually expressed as a high-viscosity surface layer (e.g., 10^{43} Pa s; Shennan et al. 2006a) that over Quaternary times-cales behaves as an elastic plate, but the thickness of this layer varies between simulations. Recent experiments have used both 71 and 96 km for lithosphere thickness beneath Britain and Ireland (Shennan et al. 2006a). Similarly, the number of layers used to describe the viscosity of the mantle varies, but many runs adopt a simple two-layer upper- and lower-mantle structure (e.g., Lambeck et al. 1996) with an average viscosity of 4×10^{20} Pa s for the upper mantle and 10^{22} Pa s for the lower mantle. Peltier's VM-2 model uses an upper-mantle viscosity of 5×10^{20} Pa s and a lower mantle viscosity of around 2×10^{21} Pa s (Peltier 1998, 2004, Peltier et al. 2002). Recent GIA modelling of the British Isles (Shennan et al. 2006a) adopted a figure of 5×10^{20} Pa s for the upper mantle and 4×10^{22} Pa s for the lower mantle (Bassett et al. 2005). However, Brooks et al. (2008), using the BIM-1 GIA model, based on comparisons with the SLIP and glacial geological evidence for Ireland, suggested that an upper-mantle viscosity of 4×10^{20} Pa s provides the best fit. A recent analysis indicated upper-mantle viscosity in the range 0.3 to 2×10^{21} Pa s for a range of modelled lithospheric thicknesses (Bradley et al. 2009).

A key variable in the ice model is the contribution of the major ice sheets to eustatic sea level and their melting histories. Shennan et al. (2006a) provided a detailed discussion of the problems in deriving a realistic ice model, but many GIA simulations adopt the global ice model proposed by Bassett et al. (2005).

GIA models of coastline and shelf sea evolution

Numerical simulations generate predictions about space beyond the domain of the field data by which they are constrained. Thus, the SLIP database from a few known sites enables extrapolations via GIA models to areas for which no SLIP data exist. GIA models are therefore predictive and generate hypotheses for future field testing. They are key for assessing rates and directions of vertical crustal motion that are required to predict future changes of RSL along coastlines (Gehrels 2010). GIA model output in the form of predicted RSL curves can be compared against observed RSL curves for key localities (Figure 10; Brooks et al. 2008, Bradley et al. 2009, 2011), but the data can also be presented as maps of changing topography, ice sheet extent and RSL—and hence coastline—through time. Peltier (1994, 2004) has produced global GIA reconstructions and Lambeck and coworkers a series of higher-resolution GIA reconstructions for specific shelf seas, notably north-western Europe (Lambeck 1995, 1996a); the Persian Gulf (Lambeck 1996b); Aegean Greece (Lambeck 1996c); the French Atlantic and Channel coasts (Lambeck 1997); French Mediterranean coast (Lambeck & Bard 2000); Australia (Yokoyama et al. 2001b); Italian coast (Lambeck et al. 2004, 2011a); the Mediterranean basin (Lambeck & Purcell 2005); and Red Sea (Lambeck et al. 2011b). The Lambeck GIA predictions for time steps since the LGM over the north-western European shelf seas have been influential and may be taken as a good example of these reconstructions (Figure 12). The simulations are data-constrained maps of the changing geography of Europe—of the migrating position of the shoreline, and of shelf seawater depth—as the eustatic and isostatic controls interact through time to determine the relative elevation of the sea and land surfaces. It is important to remember that GIA simulation maps vary according to the precise inputs that are used to parameterize the model, so there are variations between different simulations and different

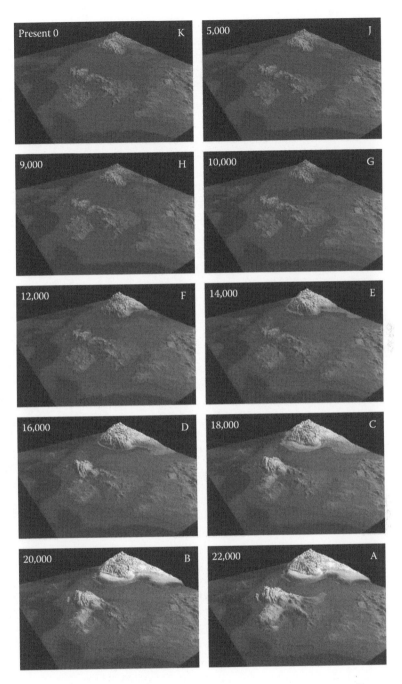

Figure 12 (See also colour figure in the insert)   The flooding of the north-western European continental shelf since the Last Glacial Maximum based on the simulations generated by a modified version of Lambeck (1995) detailed in Uehara et al. (2006). Emergent land shown in green, shelf sea in light blue, deep ocean in dark blue and ice sheets in white-grey: (A) 22, (B) 20, (C) 18, (D) 16, (E) 14, (F) 12, (G) 10, (H) 9, (J) 5 ka, (K) present day (0 ka). Note the redevelopment of an ice cap in Scotland during the Younger Dryas Event (F). Interactive visualizations of the model simulations are available at http://www.vmg.cs.bangor.ac.uk/palaeo.php.

generations of models. It is now widely accepted, for example, that the last British–Irish Ice Sheet was much more extensive than depicted by Lambeck; at the LGM (22 ka in Lambeck's series; Figure 12A), it covered the whole of Ireland, extending to the west as far as the continental shelf break (Sejrup et al. 2005, Scourse et al. 2009a), extended halfway across the Celtic shelf to a marine terminus (Scourse et al. 1990, 2009a, Scourse & Furze 2001), and confluent with the Fennoscandian Ice Sheet across the North Sea Basin (Sejrup et al. 2005, 2009, Bradwell et al. 2008). This evidence is consistent with ice sheet models parameterized to accommodate a deforming, 'slippery', bed (Boulton & Hagdorn 2006, Hubbard et al. 2009). Nevertheless, the major, first-order, changes are consistent between the different simulations, and GIA model outputs provide the best predictions we have on the way the north-western European shelf seas, and the coast, have evolved over the last 24,000 years. The differences of detail largely have an impact on the timing of the main changes depicted rather than their spatial distribution or direction.

It is certain that most of the western, northern and eastern margins of the last British–Irish Ice Sheet were marine at the LGM (Scourse et al. 2009a). The ice sheet was calving into the ocean and generating icebergs. Much of the continental shelf not covered by ice was dry land, connected by a land bridge between Britain and continental Europe. This emergent area was most extensive in the southern North Sea and English Channel and would have constituted a major portion of the catchment of the Fleuve Manche connecting the major rivers of northern Europe into a single drainage basin. Lambeck's model predicts that as the ice sheet started to decay (Figure 12C), a large embayment extended from the north between Shetland–Orkney and the Fennoscandian Ice Sheet into the North Sea Basin, a prediction since supported by field evidence (Bradwell et al. 2008). In the Irish Sea, the simulations predict the emergence of a wide, flat, low-elevation 'Celtic' land bridge linking Britain and continental Europe with Ireland, constrained to the north by the development of an ice lake impounded by the retreating ice sheet northwards. Both the Celtic land bridge and ice lake are predicted to have persisted for several thousand years until around 16 ka ago (Figure 12D), when simultaneous ice retreat across the North Channel and flooding of the Celtic land bridge resulted in marine inundation into the Irish Sea for the first time. There is no really strong geological evidence for the existence of either land bridge or ice lake, but the terrestrial biological evidence that a land bridge existed somewhat later than predicted is convincing (Rowe et al. 2006). That the Irish Sea Ice Stream deglaciated into a water body north of Anglesey is unequivocal (Van Landeghem et al. 2009b; Figure 13), but it is unclear whether this water body was a lake or the sea.

By the time the Irish Sea channel flooded, most of the North Sea remained a flat, unglaciated, plain. Orkney and Shetland formed a continuous northern Scottish peninsula, extending the British mainland to 61° N (Figure 12F). The redevelopment of ice during the Younger Dryas event (Figure 12G) would have had minimal isostatic implications since the ice cap was small and relatively thin. The central and southern North Sea remained as dry land well after the main flooding of the Irish and Celtic seas and the western English Channel was effectively complete. The sill between Orkney and Shetland was inundated at this time. Ireland therefore became an island earlier than Orkney and much earlier than Great Britain. The final severing of the North Sea land bridge was not the Straits of Dover but rather a low-lying connection from East Anglia to The Netherlands (Figure 12H) around 9 ka ago. By this time, the English Channel had flooded through to an enclosed embayment in the Flemish Bight. Once the land bridge had breached, close to 8 ka ago, and the Flemish embayment connected to the North Sea proper, Britain became an island. A residual low-lying island, however, remained in the middle of the North Sea where Dogger Bank is now situated. Doggerland was quite big for a while, initially the size of East Anglia, but was eventually consumed by the rising sea level and was gone by 7 ka ago.

Interactive visualizations of a more recent version of the Lambeck (1995) GIA simulations for north-western Europe (detailed by Uehara et al. 2006) can be downloaded from http://www.vmg. cs.bangor.ac.uk/palaeo.php.

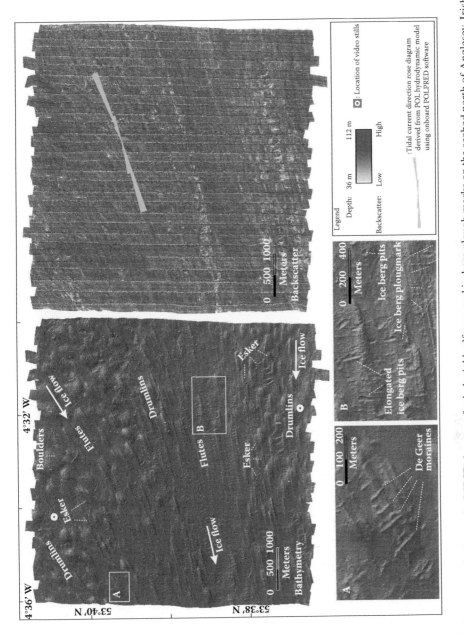

Figure 13 Multibeam swath bathymetric (MBSB) images of subglacial bedforms and iceberg plough marks on the seabed north of Anglesey, Irish Sea (for precise location, see Van Landeghem et al. 2009b). Top left panel shows MBSB image of streamlined subglacial topography, including drumlins and flutes indicating ice flow direction with superimposed eskers. Top right panel shows backscatter intensity of the same area with modelled present-day tidal current directions superimposed. Panel A shows de Geer moraines, and panel B shows iceberg pits and scour marks indicating lift-off of the ice sheet in an aqueous setting. (Reprinted from Van Landeghem et al. 2009b, with permission from Wiley-Blackwell.)

23

Terrestrial environmental responses to Quaternary sea-level change: a brief overview

In the high and middle latitudes during the LGM, and each preceding major regression of the last 800,000 years (Figure 10), the great ice sheets extended as far as the sharp break of slope between the edge of the continental shelf and the continental slope, pumping huge quantities of sediment—much of it originally deposited on the shelves by the preceding transgressive cycle—directly over the edge of the slope in the form of glacial debris flows, forming extensive trough mouth fans (TMFs; Taylor et al. 2002). The Bear Island (Bjørnøya) TMF extends 300 km from the edge of the Barents Sea across the abyssal plain towards the Mid-Atlantic Ridge. As sea level then increased, these packages of sediment were left in precariously unstable situations on the continental slope, sometimes failing as catastrophic submarine landslides, such as the Storegga Slide, in which a major portion of the North Sea TMF failed 8200 years ago (Haflidason et al. 2005). Such failures, including Storegga, are clearly associated in the geological and geomorphological record with evidence for contemporary tsunami (Bondevik et al. 2005). The most significant glaciated shelves are in the Barents Sea, Canadian Arctic and Atlantic margins, Greenland and the northern part of the north-eastern European continental shelf.

River catchments coalesced to form much larger basins; for instance, the Rhine merged with the other great north-western European rivers, including the Elbe, Weser, Meuse and Somme, to form one single, huge catchment, the Channel River or Fleuve Manche (Lericolais et al. 2003). The main trunk of this single basin, fed by glacial headwaters, flowed through the emergent English Channel and delivered sediment to the Celtic and Armorican fans, accumulating on the continental slope of the Biscay margin (Toucanne et al. 2009a,b, 2010). Exposed shelves were subject to erosion by the intense winds of glacial times (Harrison et al. 2001). Emergence and exposure of the continental shelves thus resulted in erosion and removal of the vast majority of any sediments that might have accumulated during the previous highstand(s); the sedimentary record of the continental shelves is therefore fragmentary and at best confined to any lowstand sediments (fluvial, glacial) overlain by sediments related to the subsequent transgression and shallow marine conditions—the "transgressive systems tract" as described by sequence stratigraphers (Miall 1996). In many places swept by intense tidal currents and affected by resuspension by waves, even such transgressive sequences may well be absent, and the eroding seabed surface represents an unconformity in the making (Curry 1989). The sedimentary record of the continental shelves is therefore spatially patchy and at best limited to sequences covering the period from the LGM to present.

The reduction in the areal extent of the ocean surface and in sea-surface temperatures (SSTs; Kucera et al. 2005) slowed the terrestrial hydrological cycle through inhibited evaporation and precipitation, resulting in aridity, particularly in the lower latitudes. In combination with the great expansion of glacial and periglacial (mostly tundra) environments characterized by seasonal melt, this generated a major reorganization of the current pattern of freshwater flux to the ocean margins with generally reduced lower-latitude, and highly seasonal (peaked) high-middle-latitude, discharges.

The most extensive non-glaciated shelves emergent during major regression, typified by the LGM, include most significantly the formation and expansion of Sundaland/Wallacea in South-East Asia, a huge area of exposed seabed connecting mainland South-East Asia with the many-island archipelago of present-day Malaysia and Indonesia (Hanebuth et al. 2000, Cox 2001, Bird et al. 2005). Sundaland/Wallacea served to increase the terrestrial connectivity between Asia and Australasia. The emergent Patagonian shelf linked mainland South America with the Falkland Islands, and much of the eastern Siberian, southern north-western European, eastern Asian (China Sea) and the Bering shelves was exposed. Emergent continental shelves not covered by the major ice sheets or rivers hosted significant terrestrial ecosystems, notably steppic grasslands across the lower midlatitudes and savannah grasslands in the tropics (Wurster et al. 2010). The emergent shelves formed 'land bridges' that constituted barriers inhibiting the migration and dispersal of marine organisms and the transfer of mass and energy via the oceans (Scourse & Austin 1995) but, conversely, facilitating the migration of terrestrial biota, including humans.

The spread of anatomically modern humans, and of earlier hominin species, out of Africa was facilitated by lowered sea levels by the conjoining of East Africa with the Arabian Peninsula (Lambeck et al. 2011b), the Bering land bridge linking Asia with North America and the emergence of Sundaland, although deep-water passages between some of the South-East Asian islands and Australia/New Guinea prevented a continuous connection between Asia and the Australian continent even at the LGM. Although the precise timing of dispersal is hugely controversial (Stringer 2011, Appenzeller 2012) and may have involved sea transport, the current available evidence suggests that *Homo sapiens* was able to reach Asia along a now-submerged 'coastal expressway' 60,000 to 65,000 years ago (Mellars 2006)—though some prefer an inland route over 100,000 years ago (Petraglia et al. 2007)—Australasia via Sundaland 50,000 years ago (Bowler et al. 2003), Europe 45,000 years ago (Higham et al. 2011), and finally the Americas via the Bering land bridge 14,000 to 15,000 years ago (Gilbert et al. 2008, Waters et al. 2011).

Quaternary sea-level change and the marine biosphere: impacts and responses

Geological evidence

Marine habitats vary over a range of frequencies from 10^{-2} s to more than 10^6 year (Figure 14). We are not concerned here with short-term or high-frequency changes to marine habitats, but rather the low-frequency 10^5- to 10^4-year glacial cycles driving Quaternary glacio-eustatic change. The periodic and increasingly extreme emergence of continental shelves during each regression, and the concomitant increase in the size of the great ice sheets with successive glaciations, resulted in the complete removal of shallow marine communities and habitats from the shelves, followed by the development of extensive areas of new marine habitat for niche-filling, colonization and succession during each transgression. Kaiser et al. (2005) noted that the outer shelf communities are therefore likely to be 'old', that is, to have survived extreme lowstands, and the midshelf to inner-shelf communities are 'young', the result of postglacial recolonization: "The oldest and richest fauna might be expected at shelf edges presently at a depth of about 200 to 300 m. The richness of large (mega) benthos and fish increases towards and away from shelf edge depths" (Kaiser et al. 2005, p. 28).

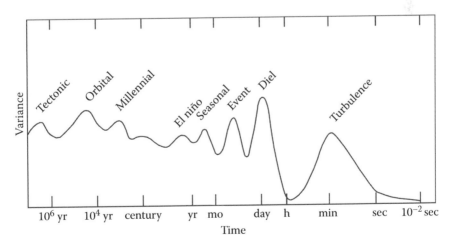

Figure 14 A conceptual view of timescales and variance of habitat change in the marine environment. (Adapted from Monin 1970.)

Norris & Hull (2012) explored the genetic implications of such repeated cycles in habitat extent and fragmentation and noted that "it seems likely that coastal benthic communities have been exposed to large changes in habitat area, distributions, and connectivity on several 1000-year time scales (Rocha 2003) that easily fall within the timescale of speciation" (p. 405).

During sea-level lowstands, habitats and communities shift to lower elevations in the hypsometric curve in parallel with the RSL lowering and then migrate shorewards during transgression. This simple marine equivalent of the 'expansion–contraction' model of Pleistocene biogeography (Provan & Bennett 2008) is complicated by the nonlinear nature of the hypsometric curve (see the previous section on this topic), resulting in both lowstand and highstand 'squeeze' during sea-level extrema (Figure 5), and by the latitudinal migration of communities in response to the parallel changes in seawater properties (notably temperature) and hydrographic reorganizations during Quaternary glacial cycles. Maggs et al. (2008) emphasized the migration of species in response to thermal optima in addition to the elevational restriction of shallow marine benthic communities to the narrow outer shelf/upper continental slope strip during eustatic lowstands. There is therefore at least both an elevational and a latitudinal element to such shifts. The bivalve mollusc *Arctica islandica* (L.), for instance, lives today in waters up to 100 m deep in the cool temperate to boreal shelf seas across the North Atlantic (Dahlgren et al. 2000). During the LGM, the niche for this species was presumably between –120 and –220 below present mean sea level, and its range is known to have extended southwards as far as the western Mediterranean (Froget et al. 1972, Dahlgren et al. 2000); the species became extinct in the early Holocene in its southern domain and is not now found south of Brittany (Froget et al. 1972, Dahlgren et al. 2000). Despite such evidence relating to individual species, and more advanced simulations via pENM (Svenning et al. 2011), there are major difficulties in testing such predictions because direct palaeoecological evidence for shallow marine species is poor, certainly when compared with the richness of the postglacial terrestrial record based on micro- and macrofossil records of plants and animals from limnic and mire settings (e.g., Roberts 1998). This paucity is a function of the limited preservation potential of both shallow marine sequences and organisms. Depositional sequences in which the fossilized remains of benthic marine species might be preserved are extremely rare because shelves are unconformities in the making (see section on terrestrial environmental responses above); furthermore, many of the key foundation species have in any case very low preservation potential (Graham et al. 2010).

Despite these difficulties, postglacial palaeoecological records from rare shelfal basins provide valuable glimpses into the successional changes of benthic marine communities associated with sea-level change. In regions dominated by eustatic rise, these contain transgressive sequences characterized by increasing water depth, and in glacio-isostatically uplifted regions, isolation basins register assemblage shifts associated with the shallowing and salinity changes that follow isolation. The Celtic Deep Basin on the Celtic shelf of north-western Europe, for example, contains a Late-Glacial to Holocene sequence registering benthic foraminiferal succession (Figure 15; Austin & Scourse 1997, Scourse et al. 2002) from around 13,000 to 3000 years ago. This foraminiferal sequence, and other similar examples from the shelf seas (Peacock 1989, Evans et al. 2002, Murray 2004, Erbs-Hansen et al. 2012, Peacock et al. 2012), is rather akin to a postglacial terrestrial pollen diagram. The base of the Celtic Deep sequence, at around –115 m below present mean sea level, is characterized by basal coarse sands and shelly gravels containing molluscan and foraminiferal species indicative of very shallow, even intertidal, conditions representing the transgressing shoreface. The facies then fines upwards into muds with associated foraminiferal assemblages indicating an increase in water depth by calibration with present-day distributions (Scott et al. 2003). Although such sequences are clearly ultimately influenced by changes in water depth as a function of sea level, the proximal causes of habitat and ecological change may be only indirectly linked to, or independent of, sea-level change. For instance, the benthic foraminiferal assemblages in the Celtic Deep sequence are controlled principally by sediment type (i.e., habitat) and nutrient flux (cf. Scott et al. 2003); these factors are linked to the hydrographic evolution of the shelf sea, notably driven by

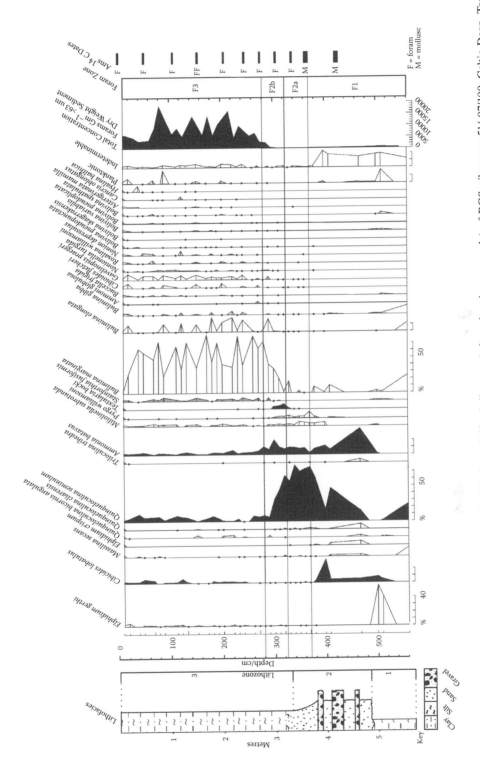

Figure 15 Lithofacies log and percentage frequency benthic foraminiferal diagram (selected major taxa only) of BGS vibrocore 51-07/199, Celtic Deep. Taxa indicative of mixed water (based on Scott et al. 2003) in black, the frontal indicator *Stainforthia fusiformis* in grey, and taxa indicative of stratified water unshaded. Frequencies of planktonic foraminifera and foraminiferal concentrations (gm⁻¹ dry weight sediment) to right. (Adapted from Scourse et al. 2002, with permission from Elsevier.)

tidal dynamics (see section on palaeotidal modelling below), which varies as a function of sea-level change. In addition, the assemblages may be as influenced by changes in seawater temperature in the transition from glacial to interglacial state as changes in water depth/sea level. It is therefore common to find Arctic or cold-water elements in the basal, shallow, facies of such sequences (e.g., Peacock et al. 2012). Many of the species preserved in such sequences are not particularly sensitive to water depth and can be found across a wide range of subtidal elevations. On the other hand, diatom, benthic foraminiferal and testate amoebae assemblages characterize the floral and faunal changes associated with the marine—brackish—freshwater transitions associated with isolation basins (Figure 9; Gehrels et al. 2001) and are therefore precise indicators of RSL. Data from both eustatically and particularly isostatically dominated basins can provide key SLIP or limiting points for sea-level reconstruction (see previous discussion).

Coral reefs are another exception to the rule that shallow marine systems are rarely preserved. Since Darwin's classic work on coral reefs (1842), there has been widespread acceptance of the primary control on reef evolution by changing sea level. The requirement for illumination to drive symbiotic zooxanthellae photosynthesis imposes tight controls on the depth to which individual coral species can live, and it is this relationship that underlies the use of hermatypic coral reef sequences for determining past sea levels (e.g., Chappell 1974, Stirling et al. 1998). In Darwin's model, the evolution of fringing, barrier reefs and atolls is driven by RSL change as a function of volcanic island uplift and subsequent subsidence. By holding the eustatic level of the sea constant, it is vertical crustal movement that controls RSL. However, Quaternary glacio-eustatic cycles interact with vertical crustal motion (see section on relative sea level above) to control RSL, so reef evolution is as profoundly influenced by sea-level change linked to climate as it is to tectonics.

In terms of response to sea-level change, coral reefs are methodologically more tractable than many other shallow marine ecosystems. Not only is the organism–sea-surface relationship tightly constrained, but also reefs have excellent preservation potential, so a record of their response to past sea-level changes is embedded in reef facies architecture. The response of reefs to RSL rise was analysed in an influential contribution by Neumann & Macintyre (1985), who identified three responses: keep up, catch up and give up. Keep-up reefs develop close to sea level and keep pace with any rise in RSL by vertical accretion. McLean & Woodroffe (1994) noted that keep-up reefs are characteristic of uplifting coasts, such as Barbados (Bard et al. 1990) and the Huon Peninsula, New Guinea (Chappell & Polach 1991), but as there is no requirement for reefs to accrete vertically when RSL is falling, such a keep-up response can only occur when eustatic rise outpaces vertical uplift during very rapid eustatic rise, such as Meltwater Pulse 1a (Fairbanks 1989). Cabioch et al. (2003) detailed a switch between keep-up and catch-up responses during Meltwater Pulse 1a in the tectonically uplifting Vanuatu reef system (south-western Pacific) and drowning of reefs during rapid phases of RSL rise in French Polynesia (Cabioch et al. 2008). Keep-up reefs are also found in stable tectonic settings influenced by such rapid eustatic rise (e.g., Collins et al. 1993). McLean & Woodroffe (1994) identified the Great Barrier Reef as an example of a catch-up reef; here, vertical accretion is unable to keep up with RSL rise and only catches up when the rate of rise decreases or stabilizes. If accretionary growth is unable to keep up with RSL rise, then the reef ultimately drowns; McLean & Woodroffe (1994) identified the Chagos Bank (Stoddart 1971) and Saya de Malha (Guilcher 1988) as likely give-up reefs.

Reefs are autochthonous structures. Although reefs may be able to respond to falls in RSL by accreting laterally and basinwards, major Quaternary regressions result in the abandonment of the highstand reef structures, which are subsequently eroded and karstified. McLean & Woodroffe (1994) proposed a model based on the Cocos Islands of atoll development over a typical glacial cycle, using the last 125,000 years as an example. The highstand at MIS 5e 125,000 years ago resulted in atoll development similar to the present. As RSL fell, the atoll evolved into a limestone island subject to intense karstification, resulting in speleothem and solutional pipe development (Lundberg et al. 1990). As the sea level rose once again, reef catch-up and keep-up responses resulted in a series of phases prior to full atoll redevelopment. Solutional pipes are drowned, forming blue holes. McLean & Woodroffe (1994) noted that there is considerable regional variation to this pattern and cited examples of variable Holocene reef growth rates. Kleypas (1997) used a model (ReefHab) to estimate the extent of reef habitat following the LGM based on temperature, salinity, nutrients and depth-attenuated radiation available for photosynthesis. This pENM approach simulated reef area at the LGM 20% that of today and carbonate production of 27% as a function of reduced space (hypsometric control) and reduced temperatures.

The depocentres of the continental slopes, where sedimentation has continued throughout glacial–interglacial cycles unimpeded by emersion, represent potential archives registering changes in the proximal shelf habitats and ecosystems. Such sequences are also particularly valuable in that they coregister changes in adjacent terrestrial and deep-ocean systems. Thus, the post-LGM records from deep-sea fans, such as the Congo in equatorial western Africa, integrate records of marine environmental change via dinoflagellate cyst, foraminiferal and diatom assemblages with records of terrestrial systems in the form of pollen and plant cuticles (Marret et al. 2001). Organic compounds, or biomarkers, specific to particular organisms supplement such records (Hopmans et al. 2004; Versteegh et al. 2004), and palaeotemperature proxies, such as foraminiferal trace element concentrations (Mg/Ca; Weldeab et al. 2007) and oxygen isotopes, provide independent reconstructions of seawater temperatures. Scourse et al. (2005) compared the areal extent of the flooding of the Congo shelf using GIA model output (Peltier 1994) with records of pollen and biomarkers derived from the adjacent mangrove ecosystem from a core recovered from 824 m water depth from the Congo Fan (Figure 16). In this context, *Rhizophora* (mangrove) pollen and taraxerol, a mangrove-derived biomarker, are allochthonous components derived predominantly via river transport to the continental slope; the attribution of these records to the flooding shelf system is, however, accentuated because of the precise intertidal requirements of the mangrove ecosystem. In this study, the temporal changes in taraxerol content and pollen flux are strongly positively correlated with the lateral rate of transgression across the shelf based on GIA model output and the bathymetry of the shelf (Figure 16). The increase in *Rhizophora* pollen and taraxerol are attributed to erosion of progressively flooded mangrove swamp during rapid sea-level rise and the increase in intertidal habitat as a function of shelf inundation following lowstand. These data highlight the large changes in areal extent of significant shallow marine ecosystems as a direct function of RSL change. In this case, the mangrove ecosystem was interpreted to have expanded across the shelf as transgression progressed, reaching a peak in the very early Holocene, at around 12,500 cal BP, before contracting as a function of coastal squeeze (Figure 5) as sea level continued to rise.

Biological evidence

Given the paucity of direct palaeoecological records from shallow marine environments, there are alternative indirect biological approaches for analysing the long-term impact of changing sea level on shallow marine communities and species. One simple approach is to determine the relationship between present-day species richness and time since inundation. The conversion of freshwater lakes into fjords in New Zealand between 9 and 17 ka BP is known from regional RSL curves; based on these data, Smith (2001) identified a strong positive correlation between fjord age and the richness of the present epifaunal invertebrate community. Similar conclusions were reached by the Coordinating Research on the North Atlantic (CORONA) project in assessing the impact of glaciations—and the associated changes in RSL—on the shallow marine communities of eastern North America (Cunningham 2008). There exists considerable potential for similar assessments of species richness in relation to substrate age in other regions for which RSL is well constrained via SLIPs and GIA models.

A more advanced and developing approach is to use pENM. The Southern Californian giant kelp (*Macrocystis pyrifera*) is a foundation species for a significantly productive and diverse ecosystem

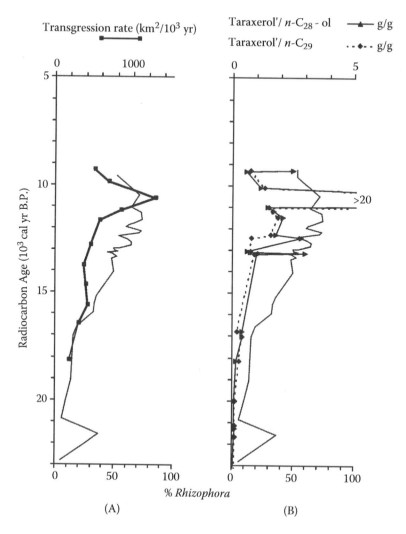

Figure 16 *Rhizophora* (mangrove) pollen percentage frequencies and ratios of taraxerol to C$_{28}$ alcohol and C$_{29}$ alkane (mangrove biomarkers) from Congo Fan core T89-16 and the rate of lateral transgression curve $(km^2(10^3 yr)^{-1})$ for the Congo shelf based on a glacial isostatic adjustment model simulation of relative sea level (Peltier 1994). Ages in calibrated years BP. (Reprinted from Scourse et al. 2005, with permission from Elsevier.)

(Dayton 1985) yet has no direct geological record. Graham et al. (2010) analysed changes in the distribution of kelp forest as a function of geomorphology and oceanography to develop a pENM. The palaeoextent of the kelp forest was reconstructed based on palaeoceanographic records (Kennett & Ingram 1995) and changes in RSL in relation to the topography of the region (Figure 17). In this case, RSL was based on a synthesis of sea-level records (Masters 2006) rather than GIA model output. This pENM simulation predicts high millennial-scale variability in kelp distribution and productivity but a threefold increase in kelp productivity from the LGM to the mid-Holocene followed by a rapid decline of 40–70% to present levels as a function of sand accretion. The changes in substrate concomitant with rising sea level resulted in a shift from rocky shore to infaunal sand-dominated assemblages around 3000 years ago, a shift reflected in invertebrate assemblages from coastal archaeological middens (Graham et al. 2003). These data—and those from the previous

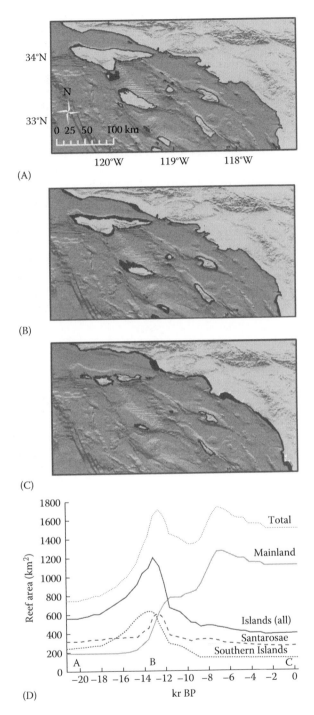

Figure 17 Variation in the Late Quaternary distribution and area capable of supporting giant kelp (*Macrocystis pyrifera*) forests (0–25 m; shown in black) in Southern California at (A) Last Glacial Maximum, 21,500 years BP); (B) 13,500 years BP; and (C) present (0 years BP). (D) Areal extent of giant kelp niche (0–25 m) through time for total (islands + mainland), mainland, and various island groups using 500-year increments with linear interpolations. Letters A, B and C indicate periods corresponding to (A), (B) and (C), respectively. (Reprinted from Graham et al. 2010, with permission from The Royal Society.)

Congo mangrove example—suggest that the extent of both ecosystems appears to have reached a maximum during mid-transgression, with contraction during peak highstands. Nevertheless, the "dependence of kelp forest responses to post-glacial change on nonlinear bathymetries implies that nearshore benthic marine ecosystems generally will not follow the same post-glacial trajectories of abundance, productivity, diversity and connectivity as adjacent terrestrial or pelagic marine habitats" (Graham et al. 2010, p. 405). Niche-modelling approaches, especially when linked to GIA simulation, have significant potential in reconstructing past organism and ecosystem distributions (cf. Svenning et al. 2011). Such data will be critical for testing the evolution of genetic signals in extant populations (Graham et al. 2010) and in archaeology for assessing shifts in intertidal resources and resource allocation in relation to demonstrated changes in human diet, for instance, between the European Mesolithic and Neolithic (Richards et al. 2003, Richards & Schulting 2006).

The third indirect approach is via the genetic analysis of living species. The cyclic inundation and emergence of continental shelves, the waxing and waning of the shelf seas and the formation and coalescence of islands by Quaternary sea-level cycles resulted in isolation and bottlenecks for terrestrial populations, driving genetic divergence and speciation (Hewitt 2000) and profound morphological changes. The formation of new species derives from the reproductive isolation of populations (Norris & Hull 2012), and insularity is an essential precursor to isolation. Such observations constitute central and celebrated planks in evolutionary theory, notably Darwin's observations from Galapagos, and in the development of island biogeography theory (MacArthur & Wilson 1967). In addition to the classic examples of island endemism, the 'island rule' (Foster 1964, Van Valen 1973, Meiri et al. 2008) attempts to explain startling change in body size, including dwarfing of large vertebrates and gigantism of small mammals, as responses of vertebrate species to insularity; such cases include the dwarfing of the terminal woolly mammoth (*Mammuthus primigenius*) population on Wrangel Island in the Siberian Arctic during the Holocene (Vartanyan et al. 1993); of red deer (*Cervus elephas*) on the English Channel Islands during the Last Interglacial (MIS 5e; Lister 1995); and of hippopotamus and elephant (*Stegodon sondaari*) on the island of Flores in Indonesia (Weston & Lister 2009). The Flores dwarf vertebrates were contemporary with the diminutive hominin species *Homo floresiensis* (Brown et al. 2004), which may well have evolved as a result of insularity as a function of sea-level change. The Flores dwarfs were accompanied by a giant tortoise (van den Bergh et al. 2001), and today Flores hosts a large murid rodent (*Papagomys armandvillei*) and the world's largest lizard (*Varanus komodoensis*) (Meiri et al. 2008).

Kaiser et al. (2005) noted that there are "all sorts of islands in the marine environment varying from enclosed coastal lagoons, sea lochs, and fjords" (p. 33) to other deep-water islands, and that the same principles governing terrestrial island biogeography also apply to these 'marine islands'. Sea-level change serves to create and coalesce former marine islands and refugia in the same way as in terrestrial isolation, but usually in the inverse direction. Thus, in their study of the response of Californian kelp forest to RSL change, Graham et al. (2010) noted that after "7500 years the kelp forest decreased rapidly, and they fragmented and increased in isolation distance [as RSL increased], opposite to the commonly observed pattern of expansion and coalescence of terrestrial glacial refugia (Hewitt, 2000)" (p. 404); however, they also noted that "similar to many terrestrial systems (Hewitt, 2000), climate-forced habitat fragmentation/coalescence cycles may provide a useful framework for interpreting genetic signals in the structure of temperate marine species" (p. 404).

The marine equivalence of direct insularity is the isolation of bodies of marine water within topographic basins. Over long, tectonic timescales, the closure and fragmentation of ocean basins through continental convergence, collision and orogenesis (e.g., the progressive and continuing closure of the Tethys-Mediterranean system) clearly results in the isolation of populations and drives speciation (e.g., in littorinid gastropods; Williams & Reid 2004). Over Quaternary timescales, rapid glacio-eustatic sea-level cycles driven by climate result in isolation of marine basins, leading to genetic change. Isolation basins are common in the isostatically uplifted glaciated terrain of the middle and high latitudes (see section on SLIPs above). Bodies of marine water that become isolated in this way quickly transform through freshwater run-off into brackish and finally fully fresh basins with the local extinction of the original marine communities (Figure 12). Isolated marine basins that remain fully saline are extremely rare and depend on some kind of hydraulic connectivity to maintain flow with the ocean (Dawson & Hamner 2005). The Palau Islands (Western Caroline Islands) in the western Pacific host karstified marine lakes in limestone terrain isolated since the LGM from the ocean by changing RSL, but these are maintained at sea level by hydraulic connectivity through tunnels and fissures in fenestrated karst. Despite this subterranean connectivity, phylogenetic analysis indicated the rapid reproductive divergence of three morphs of the *Brachidontes-Hormomya* complex of the Mytilidae (mussel) family (Goto et al. 2011) and of jellyfish (*Mastigias* sp.) as a result of physical isolation during the Holocene (Dawson & Hamner 2005).

The isolation and periodic reconnection of isolation basins with the open sea as a result of the postglacial interplay of isostasy and eustasy across Europe allowed cyclic access and then temporary or more permanent isolation for species of marine and anadromous fish. Complete isolation of basins from the sea has in some cases entrapped and isolated these populations, preventing them from pursuing anadromous behaviour and resulting often in genetic divergence from the parent populations. Classic examples of such species, many of which are genetically plastic and readily show haplotypic variation once isolated, include the sympatric ecomorph species of Arctic charr (*Salvelinus alpinus*; Bernatchez & Wilson 1998, Jonsson & Jonsson 2001), a formerly anadromous salmonid now landlocked by isostatic uplift in lakes across the formerly heavily glaciated parts of northern Europe, and a number of species of coregonids, or lake whitefish, including the Irish pollan, *Coregonus autumnalis*, found only in five Irish lakes, including notably Lough Neagh (Harrod et al. 2001). Genetic analysis of these populations often demonstrates complex histories of isolation and reinvasion—and renewed genetic exchange with the parent population—which map on to the evolution of the landscape driven by RSL change (e.g., Gagnon & Angers 2006). Despite the widespread evidence for genetic change driven by marine isolation, there appears to be a paucity of examples of the island rule (i.e., dwarfing/gigantism) applying to shallow marine organisms.

The connectivity of marine basins as a result of RSL rise is the opposite of isolation; here again, genetics provides key independent data on the impact of such changes on marine species. Over tectonic timescales, the classic example of such connectivity is the trans-Arctic invasion of Pacific species into the Atlantic as a result of the opening of the Bering Strait 5 Ma (Cunningham 2008). This invasion was asymmetric, with between 50% and 80% of the marine biota in the present-day North Atlantic derived from ancestors from the Pacific. Genetic data identify repeated trans-Arctic invasions in littoral bivalves such as in the *Macoma balthica* complex since the connection was established (Luttikhuizen et al. 2003, Väinölä 2003).

Over Quaternary timescales, RSL change informs some key phylogenetic disjunctions. The separation between Indian and Pacific tropical lineages and taxa is a function of the repeated emergence of Sundaland/Wallacea; Hewitt (2000) illustrated this via divergence in butterflyfish, starfish, damselfish, coconut crab, coastal mangroves and fishes (Palumbi 1997, Chenoweth et al. 1998, Duke et al. 1998, Williams & Benzie 1998). A Pleistocene divergence for these taxa is suggested by mitochondrial DNA (mtDNA) and allozyme data, which independently support the geological data for the timing of land barrier emergence. Hewitt (2000) noted: "As with terrestrial and freshwater organisms, the Indo/west Pacific seas are rich in species, which may be due to the many changes in distribution caused by the sea level oscillations in the topographically complex archipelagos. It may function as a species source and refuge in a similar way to mountainous regions in lower latitudes" (p. 912). Thus, the Indo-west Pacific hot spot (Kaiser et al. 2005) may be a function of the legacy of Quaternary sea-level cycles interacting with a shallow and topographically complex and extensive continental shelf. This contrasts, however, with evidence for the long-term genetic stability of coral species in the same region:

In Indo-Pacific shallow-water corals, the frequent sea-level changes during the Quaternary repeatedly laid the continental shelves dry, so that any particular area was available for coral growth for on average 3.2 kyr at a time. These frequent large-scale distributional changes probably prevented β-cladogenesis, and most coral species have little geographical genetic subdivision (Potts 1983). (Jansson & Dynesius 2002, p. 759)

The marine equivalent of the expansion–contraction model (Provan & Bennett 2008), in which lowstand contraction and highstand expansion are supplemented by southern refugia (in the Northern Hemisphere), can be tested via the burgeoning number of genetic investigations of marine taxa and lineages (Provan et al. 2005, Wares & Cunningham 2005, Marko et al. 2010). A diverse picture is emerging. In some settings, notably oceanic islands, the expansion–contraction model is inverted; here, lowstands may result in habitat expansion and increased connectivity between populations so that in this context highstands are conversely associated with reproductive isolation. A good example is the genetic structure of populations of the marine mussel *Brachidontes puniceus* in the Cape Verde Islands (Cunha et al. 2011), which indicates an increase, rather than decrease, in habitat available for larval settlement during the LGM lowstand as a function of topography. Norris & Hull (2012) contrasted speciation on continental margins, in which cryptic marine taxa with overlapping distributions suggest parapatric or sympatric behaviour (Hellberg 1998), with islands that often show primary allopatric or founder speciation in benthic invertebrates (Paulay & Meyer 2002, Meyer et al. 2005, Frey 2010, Malay & Paulay 2010), although islands may be characterized by secondary sympatric distributions (Paulay & Meyer 2002).

On continental margins, there are multiple studies that support the contraction–expansion model. Maggs et al. (2008) summarized the genetic characteristics of this model as having "low genetic diversity in formerly glaciated areas, with a small number of alleles/haplotypes dominating disproportionately large areas, and high diversity including 'private' alleles in glacial refugia. In the Northern Hemisphere, low diversity in the north and high diversity in the south are expected" (p. 108). The CORONA project (Cunningham 2008) noted the impact of glaciations was assumed to be more profound on the eastern North American margin than in north-western Europe, and that there is little hard substrate south of Long Island Sound close to the maximum extent of the Laurentide Ice Sheet at the LGM. Based on this assumption, CORONA tested, amongst other things, the hypothesis that the intertidal rocky community of the north-western Atlantic was colonized by immigrants from Europe (Wares & Cunningham 2001). Much of the genetic data emerging from the project, however, indicated recolonization from southern refugia on the western Atlantic margin (Cunningham 2008), a conclusion supported by other studies (e.g., on *Littorina saxatilis*; Panova et al. 2011). In the eastern North Atlantic, the genetic evidence for northwards spread of thermophilous species following the LGM from southern refugia, notably in Iberia, along the Atlantic littoral, is strong (Figure 18). The Iberian refuge includes the teleosts *Pomataschistus microps* (Gysels et al. 2004), *Pomatoschistus minutus* (Larmuseau et al. 2009), *Raja clavata* (Chevolet et al. 2006) and *Salmo salar* (Consuegra et al. 2002, Langefors 2005); the invertebrates *Celleporella hyalina* (Gómez et al. 2007), *Nucella lapillus* (Colson & Hughes 2007) and *Pollicipes pollicipes* (Campo et al. 2010); the colonial urochordate *Botryllus schlosseri* (Ben-Shlomo et al. 2006); and the plants *Fucus serratus* (Coyer et al. 2003, Hoarau et al. 2007) and *Zostera noltii* (Coyer et al. 2004). A full marine connection was maintained through the Strait of Gibraltar during the LGM, and genetic data indicate a Mediterranean refugium for some species, notably the seagrass *Zostera marina* (Olsen et al. 2004). Refugia in north-western Africa in Macaronesia (Madeiras, Azores and Canaries) have been suggested for *Pollicipes pollicipes* (Campo et al. 2010) and *Raja clavata* (Chevolet et al. 2006) and a 'single southern refugium' for the teleost *Menidia menidia* (Mach et al. 2011).

Molecular marker studies for some species, however, suggest patterns of recolonization from 'glacial' refugia, that is, from localities close to the ice sheets in north-western Europe (Maggs et al. 2008). This parallels the discussion on cryptic 'northern' refugia for terrestrial species (Stewart &

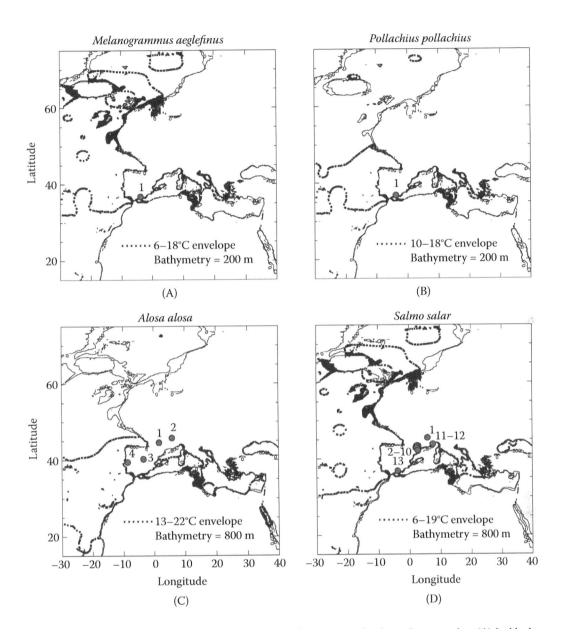

Melanogrammus aeglefinus

(A)

Pollachius pollachius

(B)

Alosa alosa

(C)

Salmo salar

(D)

Figure 18 (See also colour figure in the insert) Predicted ranges for four teleost species: (A) haddock, (B) pollock, (C) allis shad and (D) Atlantic salmon at the Last Glacial Maximum (LGM) based on palaeoenvironmental niche modelling. The ranges are based on calibration via present-day habitat niche and reconstruction of LGM conditions from Multiproxy Approach for the Reconstruction of the Glacial Ocean Surface (MARGO) Project Members (2009). Observations for past distributions are constrained by archaeozoological data from numbered sites. The species' ranges at the LGM are displaced southwards from present distributions and are between 31% and 53% of present-day ranges. LGM palaeotopography based on Peltier (1994) glacial isostatic adjustment simulation. (Reprinted from Kettle et al. 2011, with permission from European Geophysical Union [open access journal].)

Lister 2001, Rowe et al. 2004, Provan & Bennett 2008). Maggs et al. (2008) reviewed the evidence for putative North Atlantic glacial refugia; they noted that if "these periglacial* populations experienced extreme bottlenecks, they could have the low genetic diversity expected in recolonized areas with no refugia, but should have more endemic diversity (private alleles) than recently recolonized areas" (p. 108). Brittany and the Hurd Deep, a deep, tectonically controlled basin on the floor of the English Channel, have been identified as refugial locations for a wide range of taxa (see Figure 8 in Kettle et al. 2011), including the teleost *Pomatoschistus minutus* (Larmuseau et al. 2009); the invertebrates *Carcinus maenas* (Roman & Palumbi 2004), *Celleporella hyalina* (Gómez et al. 2007), *Macoma balthica* (Luttikhuizen et al. 2003, Nikula et al. 2007), *Pectinaria koreni* (Jolly et al. 2005, 2006), *Littorina saxatilis* (Panova et al. 2011) and *Pollicipes pollicipes* (Campo et al. 2010); and the plants *Ascophyllum nodosum* (Olsen et al. 2010), *Fucus serratus* (Coyer et al. 2003, Hoarau et al. 2007) and *Palmaria palmata* (Provan et al. 2005). South-western Ireland has been identified as a glacial refuge for the invertebrates *Macoma balthica* (Luttikhuizen et al. 2003, Nikula et al. 2007), *Pectinaria koreni* (Jolly et al. 2005, 2006) and *Littorina saxatilis* (Panova et al. 2011); and the plants *Fucus serratus* (Coyer et al. 2003, Hoarau et al. 2007) and *Palmaria palmata* (Provan et al. 2005), and possibly the 'Southern North Sea Lake' is a refuge for *Pomataschistus microps* (Gysels et al. 2004) and *Salmo salar* (Consuegra et al. 2002, Langefors 2005). Maggs et al. (2008) concluded that there is significant genetic evidence to support the notion of glacial refugia in these northern locations proximal to the ice sheets. Further, they suggested that "because interpretations of the precise extents of ice sheets and their timing are controversial … and knowledge of past biotas can contribute to reconstructions of glaciations, we highlight examples where our analysis clarifies glacial histories" (p. 109). They considered their conclusions gave "credence to recent climatic reconstructions with less extensive glaciations" (p. 108).

The LGM record of Quaternary glaciation in north-western Europe and of contemporary changes in extraglacial terrestrial and marine environments is incontestably incomplete, but the geological data, including SLIP-constrained GIA modelling of RSL change and direct physical, biogeochemical and biological evidence for terrestrial and marine palaeoenvironments, including ice limits, are more substantial than Maggs et al. (2008) suggested. There is a significant body of independent data that indicates extensive glaciation at the LGM in north-western Europe and that demonstrates that Bowen et al. (2002)—often cited in support of minimal LGM glaciation extent—held a minority view. The genetic evidence for glacial refugia during the LGM is therefore in conflict with the bulk of the geological evidence. At the time of the LGM, the English Channel was a periglacial valley hosting the Fleuve Manche (Bourillet et al. 2003, Lericolais et al. 2003) draining glacial meltwaters from north-western Europe. GIA modelling indicates that the channel was emergent during the LGM (Lambeck 1995). In this context, the Hurd Deep could not have remained a 'marine lake' (Provan et al. 2005) with seasonal ice cover since at lowstand it would have been disconnected from the open ocean, and if hosting an isolated basin, it would have rapidly freshened via meltwater. The adjacent Biscay margin was heavily influenced not only by the delivery of sediment to the Armorican and Celtic fans from the Fleuve Manche (Eynaud et al. 2007, Toucanne et al. 2008) but also by glacimarine conditions associated with the southern margin of the British–Irish Ice Sheet (Scourse & Furze 2001, Scourse et al. 2009b). This margin was also megatidal (see section on palaeotidal modelling; Uehara et al. 2006, Scourse et al. 2009b). Direct micropalaeontological evidence indicates LGM conditions in the Celtic Sea equivalent to the fjords of East Greenland north of 76°N at present (Scourse et al. 1990) with *in situ* ontogenetic series of Arctic ostracod instars. LGM grounded ice (Irish Sea Ice Stream) reached the Isles of Scilly (Scourse 1991, Hiemstra et al. 2006, McCarroll et al. 2010). Evidence from ice-rafted detritus (IRD)

---

* The use of the term *periglacial* for glacial refugial populations is potentially confusing. The term *periglacial* refers to terrestrial environments where frost-action processes dominate with two diagnostic criteria: freezing and thawing of the ground and the presence of permafrost (French, 1976). Hence, such environments cannot constitute marine refugia.

fluxes and sources from the continental slope adjacent to Ireland indicates shelf edge glaciation west and south-west of Ireland (Scourse et al. 2000, 2009a, Haapaniemi et al. 2010) coincident with Heinrich Event 2 at around 24 ka BP, that is, at the height of the LGM. The offshore evidence that the entirety of Ireland was glaciated at the LGM is supported by evidence on ice thickness and vertical offshore extent (Ballantyne et al. 2007, 2008, Ballantyne 2010) and from offshore moraines (Sejrup et al. 2005, O'Cofaigh et al. 2012a), glacial stratigraphy and sedimentology (O'Cofaigh & Evans 2001, 2007, Clark et al. 2012, O'Cofaigh et al. 2012b) and isostatic response (McCabe et al. 2007, Brooks et al. 2008), which are consistent in indicating total ice cover. Similarly, the IRD data are consistent with offshore data indicating shelf edge glaciations north and west of Scotland (Bradwell et al. 2008). Areas of the southern North Sea not directly covered by the Fennoscandian Ice Sheet were periglacial plain or glacial lake.

It is not possible to accommodate the survival of thermophilous marine species in LGM glacial refugia in locations for which there is good independent evidence of ice-infested glacimarine conditions, terrestrial periglacial plains, glacial (freshwater) lakes or ice sheets. The nature of the conditions adjacent to the ice margins would have been inimical to the survival of the species considered. Intertidal and shallow subtidal environments would have been subject to significant scour by sea ice and icebergs. The significance of sea-ice scour in removing shallow marine communities prior to post-LGM immigration has been highlighted in a genetic study of the kelp *Durvillea antarctica* on subantarctic islands (Fraser et al. 2009). Maggs et al. (2008) placed too much emphasis on the potential for genetic data to refine ice limits without considering the existing palaeoenvironmental data for the ice marginal conditions beyond the ice sheets themselves. Though they admitted that "the precise locations of the actual refugia are not necessarily associated with these genetic signatures" (p. S119) and indicated that only direct evidence via sediment from the Hurd Deep "could provide unequivocal evidence" (Maggs et al. 2008, p. S119), on the basis of the weight of geological evidence for ice sheet extent, sea-level and palaeoenvironmental conditions in this region, it is not credible that an isolated refugium existed in the English Channel, in south-western Ireland, or in the southern North Sea during the LGM. These genetic data cannot be used to inform ice limits unless verified by independent palaeoenvironmental evidence. The conflict between the genetic and geological data cannot currently be resolved without accepting that one or another of these datasets has been misinterpreted. Similar controversy exists in the mismatch between the genetic data for cryptic terrestrial glacial refugia and the (lack of) substantive palaeoenvironmental evidence: "There is a ... danger here that circumstantial evidence will be marshalled into an hypothesis which in its totality has some apparent credibility but whose component parts are either capable of alternative explanation or are equivocal" (Scourse 2010, p. 826).

Kettle et al. (2011) employed an integrated archaeozoological and pENM approach to investigate refugia of marine fish in the north-eastern Atlantic during the LGM. In this case, the pENM is based on palaeoceanographic data on sea-surface conditions (MARGO Project Members 2009) and—in the first example of its kind—bathymetry based on GIA output (Peltier 1994). The pENM outputs indicated that the LGM refugial ranges for haddock (*Melanogrammus aeglefinus*), pollock (*Pollachius pollachius*), allis shad (*Alosa alosa*) and Atlantic salmon (*Salmo salar*) are displaced southwards, notably into the Mediterranean (Figure 18). These predictions are supported by archaeozoological data, notably the occurrence of dated fish remains in midden contexts. The conclusions of this study are therefore consistent with the simple contraction–expansion model. As noted by Kettle et al. (2011), the results support the genetic evidence for southern LGM refugia, are not consistent with northern glacial refugia, and emphasize the significance of Mediterranean outer shelf and slope habitats as LGM refugia from the shallow marine biota of north-western Europe. This study demonstrates the effectiveness of GIA-based pENM approaches, independently tested by dated macrofossil assemblages, in highlighting LGM refugial localities and serves as a model for future attempts to define the geological constraints for the interpretation of genetic data.

As "speciation is a process that occurs over time and, as such, can only be fully understood in an explicitly temporal context" (Norris & Hull 2012, p. 393), it is imperative that the results of geological and palaeoenvironmental investigations—including model simulations constrained by such data—are fully integrated into the interpretation of genetic studies. The current gap between the two communities in the marine sphere is not helpful, and unless it is bridged, the synthesis "which united genetics, development, ecology, biogeography, and paleontology" in the study of evolution will falter; "a ... synthetic approach must be taken to further our understanding of the origin of species" (Norris & Hull 2012, p. 393).

There is little evidence to support any suggestion that the high-amplitude sea-level cycles of the Quaternary have increased extinction rates amongst shallow marine taxa (contra Newell 1967). Quaternary sea-level cycles select for generalism and vagility (Dynesius & Jansson 2000), resulting in few oceanic extinctions; "many marine taxa have high fecundity and highly vagile larvae, and these traits are associated with low speciation rates in the fossil record" (Dynesius & Jansson 2000, p. 9117, Jablonski 1986).

The palaeotidal evolution of Quaternary oceans and the marine biosphere: impacts and responses

Time and tide

In addition to the direct impacts of changing RSL on the biology of the shelf seas and shallow marine habitats, the LGM eustatic lowstand—and associated vertical isostatic compensations—served to fundamentally alter the shape of the ocean basins. This dynamic reorganization resulted in palaeotopographic evolution through time characterized by changing coastline shape (Figure 12) and palaeobathymetry. Lacking marginal shelf seas, the ocean at the LGM was of smaller surface area but, on average, deeper. Interglacial oceans, typified by the present, have extensive marginal fringing shelf seas, so the surface area is greater, and the basins are on average shallower. These changes influence the character of the tides. Charles Darwin's son, G.H. Darwin (Darwin 1899), assumed tidal energy dissipation rate as fixed over geological time to determine the history of the separation distance between Earth and Moon, but as noted by Simpson & Sharples (2012), there is no intrinsic reason why dissipation should be fixed over time. Given that around 70–75% of all tidal dissipation in the modern ocean takes place in the marginal shelf seas (Green et al. 2009; see Figures 20 and 21), the logical consequence of the removal of the shelf seas would be to reduce total dissipation. This hypothesis has been tested by global tidal models that integrate palaeotopographies from GIA model output (e.g., Peltier 1994, 2004). Though these simulations are derived from different tidal models, they are consistent in indicating an *increase* in total dissipation at the LGM of between 20% and 33% (Thomas & Sündermann 1999, Egbert et al. 2004, Uehara et al. 2006, Arbic et al. 2008, Griffiths & Peltier 2008, Arbic & Garrett 2010); total dissipation during the LGM was about 4.7 TW (Egbert et al. 2004, Green et al. 2009; Figure 19), compared with 3.5 TW at present. The data indicate about a 2.5 times increase in the rate of energy dissipation in the deep ocean compared to the present (Green et al. 2009). Although this resulted in greater mixing in the glacial ocean with the potential to increase the vigour of the meridional overturning circulation (MOC), it is clear that this effect was masked by a reduction in deep convection through freshwater fluxes to the ocean from melting ice sheets (Green et al. 2009). One implication of this finding is that recovery from freshwater-forced MOC collapse, such as is inferred to have occurred during the Younger Dryas event (11.9–12.6 ka BP; Lowe et al. 2008, Rahmstorf 2002), would have been more rapid under glacial than under the more sluggish interglacial boundary conditions (Green et al. 2009). The counterintuitive increase in LGM tidal dissipation results from the change in the resonance characteristics of the ocean basins. In its LGM configuration, the North Atlantic was, in particular, closer to resonance, resulting in significantly enhanced tides (Figure 19). Tidal amplitudes along

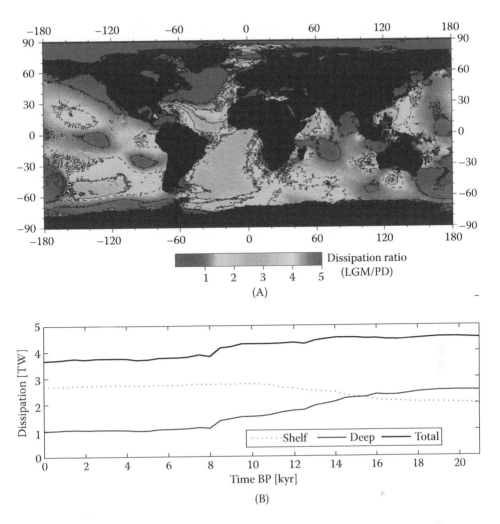

Figure 19 (See also colour figure in the insert) Rates of dissipation of tidal energy between the present day and Last Glacial Maximum (LGM) based on global palaeotidal model simulations. (A) Ratio between LGM and present rates highlighting high LGM dissipation rates in the Arctic Ocean and North Atlantic. (B) Globally integrated dissipation rates from the LGM to present in the total ocean and ocean both shallower and deeper than 1000 m. This highlights greater dissipation rates in the glacial ocean and the increased significance of deep-ocean dissipation during glacial boundary conditions. (Reprinted from Green et al. 2009, with permission from the American Geophysical Union.)

significant shelf edge margins of the LGM ice sheets in the North Atlantic, notably adjacent to the Laurentide Ice Sheet in the Labrador Sea/Hudson Strait, adjacent to the Fennoscandian and British–Irish ice sheets in north-western Europe and in the Arctic Ocean (Egbert et al. 2004, Uehara et al. 2006, Arbic et al. 2008, Griffiths & Peltier 2008), were megatidal with mean spring tidal range in excess of 10 m (Scourse et al. 2011). Tides of this magnitude would have had a significant impact on ice shelf stability (Rosier et al. submitted), tidewater glacier calving rates and ice stream stability; such feedbacks likely contributed to rapid ice sheet collapse (Chiverrell et al. 2013) and may have contributed to the propagation of Heinrich events (Arbic et al. 2008). Tidal energy dissipated by friction stresses acts as torque restraining the rotation of Earth (Thomas & Sündermann 1999) and therefore influences day length. Although other factors influence rotation rate, such as glacial rebound (Dicke 1966) and core-mantle coupling (Johnston & Lambeck 1999), and these are not easily separated from the tidal accelerations (Lambeck 1980), the increase in glacial dissipation will have slowed rotation, increased day length and placed the Moon in a higher orbit (larger Earth–Moon separation) and vice versa during interglacials (Munk 1997). The repeated eustatic cycles of the Quaternary have thus acted as a pulsed torque on Earth's rotation, rather like a driver pumping the brakes of a car.

The changing tides influence the marine biosphere in many different ways. The fundamental attributes of the tide (phase, amplitude) influence intertidal ranges, and the S_2 constituent determines the timing during the day of peak spring low and high spring tides at any particular location. Surface tidal currents influence the dispersal of planktonic organisms, including larvae and propagules, and determine bed shear stress. Bed stress determines sediment entrainment and deposition and therefore controls seabed sediment type with profound implications for habitats. The extent of tidal mixing controls seasonal stratification of the water column, which strongly influences primary productivity, shelf sea ecosystems and carbon cycling (Sharples et al. 2007, Simpson & Sharples 2012). The evolution of these tide and tide-dependent parameters are all amenable to analysis using palaeotidal models. Significant PTM studies of the shelf seas have emerged over the last 20 years that enable quantitative hindcasts and predictions of the evolution of shelf sea systems in response to RSL changes. These have not yet been exploited by the marine biological community to inform and explain present-day patterns of species richness or genetic differentiation or to drive palaeo-ENM simulations.

A history of palaeotidal modelling

Shelf sea palaeotidal simulations operate within a model domain defined by an ocean boundary. The tide within the shelf domain is forced by inputs prescribed across this boundary and by changing sea level through time. Early palaeotidal model simulations used very simplified inputs; generally, the ocean tide was held constant for all time steps, reflecting the present-day situation, and topography was not considered to be dynamic; that is, the changing sea level was prescribed simply in terms of eustatic change since the LGM, with no accommodation for GIA. Belderson et al. (1986) used such a fixed-crust approach to investigate bed stress evolution across the Celtic shelf during the last transgression to attempt to explain the evolution of the Celtic Sea linear tidal sand ridges. Hinton (1995) used this means to provide palaeotidal corrections for SLIPs in eastern England. Hall & Davies (2004) explored changing sediment transport paths in north-western Europe during the Holocene with this approach, and Uehara et al. (2002) investigated the Holocene palaeotidal regime of the Yangtze Estuary, East China Sea and Yellow Sea. Austin (1991) used a 2-dimensional finite-difference model (Flather 1976) of the lunar semidiurnal M_2 constituent applied to the north-western European shelf seas during the Holocene with a fixed, present, ocean tide boundary and simple reductions of sea level to accommodate eustatic change. These simulations provided, for the first time, first-order shelfwide hindcasts of tidal amplitude change, peak bed stress vectors and seasonal stratification.

Following the pioneering study of Scott & Greenberg (1983) to simulate the tidal evolution of the Bay of Fundy, there was a recognition of the need to accommodate isostatic corrections and to use either regional RSL curves or SLIP-constrained GIA simulations of palaeotopographic change rather than simple eustatic change since the GIA for formerly glaciated regions is considerable (Figure 6). Scourse & Austin (1995) corrected for tectonic uplift in the central English Channel in a study investigating enhanced seasonal stratification in this region during Middle Pleistocene highstands, and Gerritsen & Berentsen (1998), Shennan et al. (2000a) and Van der Molen & De Swart (2001) all used GIA output as input terms for palaeotidal models. These simulations all used present-day ocean boundary forcings.

To test the impact of changes in the ocean tide and GIA adjustment on shelf-tide evolution, Uehara et al. (2006) undertook an analysis of the north-western European shelf seas since the

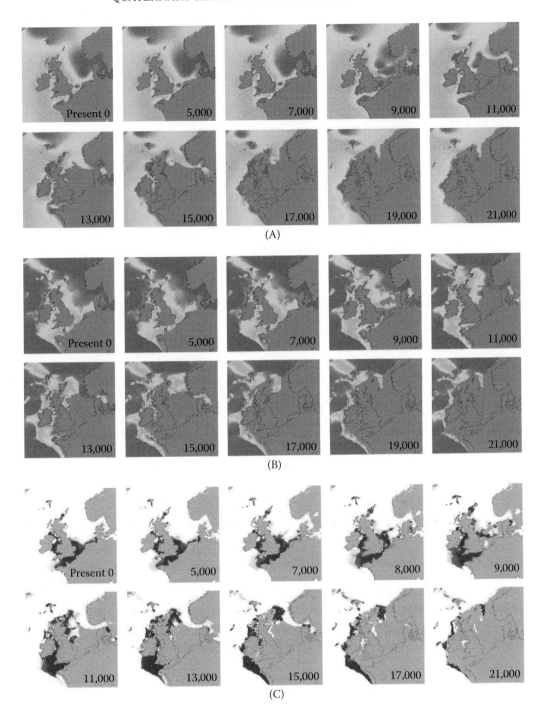

Figure 20 (See also colour figure in the insert) Palaeotides and tide-dependent parameters from the Last Glacial Maximum (LGM) to present for the north-western European shelf seas based on the M_2 simulations generated by Uehara et al. (2006). (A) Tidal amplitudes (half full tidal range) range from zero (amphidromic) in blue to 4 m in red. Time steps shown in black on each panel. Note the increase in megatidal amplitudes along the European margin at the LGM. (B) Surface tidal currents range from 0 (blue) to 1.6 ms^{-1} (red). (C) Stratification; tidally mixed water in black, seasonally stratified water in white, frontal zones (tidal mixing fronts) in red.

41

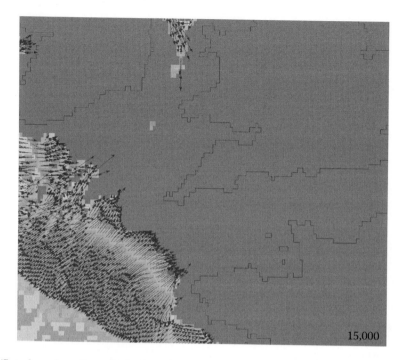

Figure 21 (See also colour figure in the insert)   Palaeotides and tide-dependent parameters from the Last Glacial Maximum (LGM) to present for the north-western European shelf seas based on the M_2 simulations generated by Uehara et al. (2006). Peak bed stress vectors from the Celtic shelf at 15 ka ago. Colours represent vector orientations (e.g., orange towards north-east); arrows represent vector direction and strength; only vectors greater than 2 N/m² shown. Interactive visualizations of the model simulations are available at http://www.vmg.cs.bangor.ac.uk/palaeo.php.

LGM using a 2-dimensional version of the Princeton Ocean Model (Blumberg & Mellor 1987). M_2 palaeotidal simulations were forced by two different GIA models—ICE-4G (VM2) (Peltier 1994) and a revised version of Lambeck (1995)—and the influence of the ocean tide assessed by setting the ocean boundary either fixed to the present state or by incorporating the outputs from a global ocean tidal model. Palaeotopography for the global model was derived from the work of Peltier (1994). The simulations demonstrated that the precise GIA model chosen to define shelf palaeotopography had little impact on tidal dynamics, but that the timing of changes was sensitive to local isostatic effects. The most significant control on shelf sea tidal dynamics was the ocean tide, profoundly altering tidal amplitudes before 10 ka BP and influencing the distribution of seasonal stratification, bed stress and dissipation rates (Figures 20 and 21). Interactive visualizations of the Uehara et al. (2006) simulations can be downloaded (http://www.vmg.cs.bangor.ac.uk/palaeo.php).

Constraining palaeotidal simulations

Given the significance of these palaeotidal simulations for defining the role of the shelf seas within the Earth system, it is clearly important to be able to compare different simulations against geological observations. Many palaeotidal outputs have no proxies in the geological record. It is not possible, for instance, to derive estimates of past tidal range from shallow marine sequences; SLIPs register MHWST rather than MSL and hence require correction for, rather than indicating, past tidal range (Neill et al. 2010). Hitherto the only tide-dependent variable simulated by PTM susceptible to testing using the geological record has been seasonal stratification.

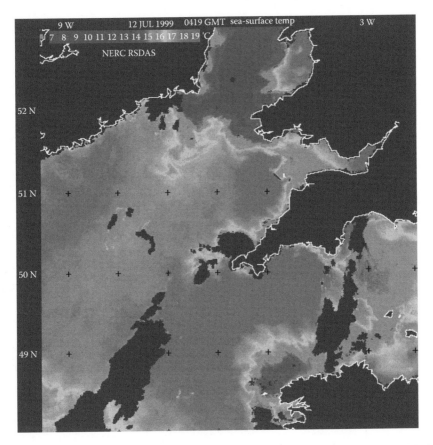

Figure 22 Satellite image (Advanced Very High Resolution Radiometer, AVRHH) of the Celtic Sea showing sea-surface temperatures taken at 0419 GMT on 12 July 1999. Temperature scale (°C) shown top left. The temperature gradient at the Celtic Sea tidal mixing front is shown clearly. Black areas represent land or cloud. (Reproduced with kind permission of the NERC Earth Observation Data Acquisition and Analysis Service [NEODAAS], Plymouth. Reprinted from Scourse & Austin 2002, with permission from Elsevier.)

Heating of the sea surface in summer induces buoyancy, which in some locations is sufficient to overcome the mixing induced by frictional dissipation of tidal energy. The boundaries between the areas that stratify seasonally and those that remain mixed throughout the year are tidal mixing fronts (Figures 20C and 22). The position of tidal mixing fronts can be predicted using the energetic criterion of Simpson & Hunter (1974):

$$\frac{\varepsilon \alpha Q H}{\left(c_D u_T^3\right)}$$

where α is the thermal expansion coefficient, Q is the rate of heat input, ε is the mixing efficiency, H is the water depth, c_D is the bottom drag coefficient and U_T is the tidal current amplitude. Under present boundary conditions, Q is regarded as seasonally but not interannually variable. Since water depth H is controlled by RSL, sea level is a primary determinant of stratification history. Over geological timescales Q does vary (i.e., controlled by insolation receipt as a function of orbital [Milankovitch] cycles; Berger & Loutre 1991), but sensitivity tests (Rippeth et al. 2008) indicated that the position of fronts is insensitive to variability in Q and in α (derived from Schumacher et al. 1979). Tidal mixing fronts are associated with sharp gradients in surface and bottom water

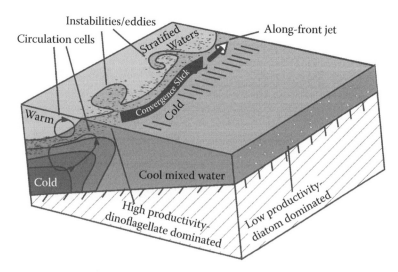

Figure 23 Cartoon of the three-dimensional structure of a tidal mixing front. (Reprinted from Scourse & Austin 2002, with permission from Elsevier.)

temperatures (Figure 23), and the annual thermal profiles of seasonally stratified and mixed regions are markedly different. Fronts are also biogeochemically significant since they are the locus of enhanced primary production as a result of the injection of nutrient-rich waters towards the surface, which fuels higher trophic-level productivity (Simpson & Sharples 2012). The subthermocline layer within the stratified sector is cool, poorly oxygenated and nutrient rich compared with the bottom waters in the mixed layer.

Uehara et al. (2006) predicted the migration of tidal mixing fronts across the north-western European continental shelf following the LGM (Figure 24); they used both fixed (modern) and

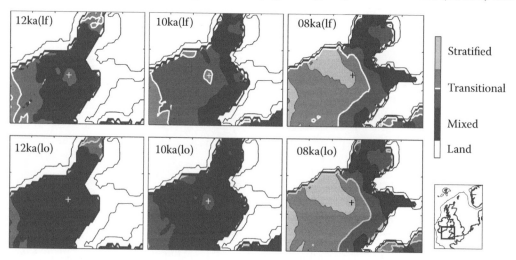

Figure 24 Palaeotidal model simulations of Celtic Sea stratification at 12-, 10- and 8-ka time steps using palaeotopography derived from Lambeck glacial isostatic adjustment model simulations run with fixed ocean boundary (upper panels) and dynamic ocean tide boundary (lower panels). Location of BGS vibrocore 51/-07/199 shown by cross. These simulations show a clear difference in the timing of stratification onset at the core site as a function of the selected ocean boundary input; stratification occurs later in the dynamic boundary simulation, which is consistent with palaeostratification observations from the core (Scourse et al. 2002). (Reprinted from Uehara et al. 2006, with permission from the American Geophysical Union.)

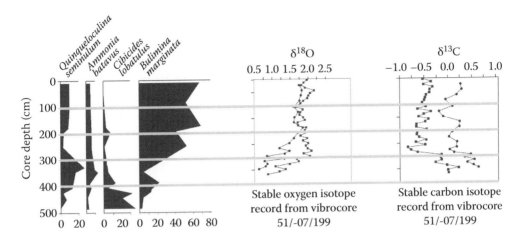

Figure 25 Palaeostratification proxies from the Celtic Deep. Stable isotopic data (squares, *Ammonia batavus*; filled circles, *Quinqueloculina seminulum*) and benthic foraminiferal stratigraphy from BGS vibrocore 51/-07/199 plotted by depth (adapted from Austin & Scourse 1997). The transition from mixed to stratified water is demonstrated by the faunal change and the coeval shifts to positive in the δ^{18}O values and to the negative in the δ^{13}C values. (Reprinted from Scourse et al. 2002, with permission from Elsevier.)

dynamic ocean tide simulations. These indicated a clear difference in the timing of stratification onset across the shelf; in the Celtic Sea, stratification occurred earlier in the fixed than in the dynamic case. These predictions can be tested by geological observations. Austin & Scourse (1997) interpreted stable isotopic and benthic foraminiferal data from a Holocene core (British Geological Survey core 51/-09/199; core 199) recovered from the Celtic Deep in terms of a transition from early Holocene mixed to later Holocene seasonally stratified conditions (Figure 25). As there are no proxies in this record indicative of changes in salinity, and given that the ice volume control on oxygen isotopes is negligible during the Holocene (Fairbanks 1989), the positive trend in oxygen isotopes upcore was attributed to a 4–5°C cooling in bottom water temperature. Such a cooling is consistent with the transition from warmer mixed bottom waters to cooler subthermocline bottom waters during summer conditions (Figure 25). The negative upcore transition in carbon isotopes was similarly attributed to stratification onset, with subthermocline waters enriched in remineralized isotopically light organic matter and hence isotopically lighter than bottom waters in the mixed sector.

To calibrate these data against modern conditions, Scott et al. (2003) reported the control of benthic foraminiferal assemblages in the Celtic Sea by seasonal stratification under present conditions, and Scourse et al. (2004) investigated the stable isotopic content of bottom water and benthic Foraminifera in the same region. Scott et al. (2003) identified three distinct foraminiferal assemblages linked to stratification: (1) a frontal assemblage dominated by *Stainforthia fusiformis*; (2) a mixed water assemblage dominated by *Cibicides lobatulus, Textularia bockii, Spiroplectammina wrightii, Ammonia batavus* and *Quinqueloculina seminulum*; and (3) a stratified assemblage dominated by *Bulimina marginata, Hyalinea balthica, Adercotryma wrighti* and *Nonionella turgida*. Scott et al. (2003) speculated that these assemblages are controlled by food supply and the oxygen concentration of bottom and near-surface porewaters, variables known to be linked to seasonal stratification (Simpson & Sharples 2012) arising from enhanced frontal production, cross-frontal transfer of nutrients and subthermocline depleted oxygen concentrations.

Scourse et al. (2004) provided oxygen isotopic calibration data that supported the interpretation of Austin & Scourse (1997) and, furthermore, importantly demonstrated foraminiferal calcification during the summer months, hence the utility of the foraminiferal proxies (both faunal and isotopic data) as seasonal stratification indicators. Scourse et al. (2004) also identified differences in the timing of calcification between different species, termed a *seasonal effect*, useful for calibrating shelf sea isotopic records based on benthic Foraminifera. Scourse et al. (2002) provided a full benthic foraminiferal record for core 199 (Figure 15) showing clear replacement of the mixed by the stratified assemblages identified by Scott et al. (2003) during the early Holocene; the frontal indicator *Stainforthia fusiformis* was an important component of the record but was transported postmortem from the frontal into the stratified sector by cross-frontal flows. Multiple ^{14}C determinations enabled the timing of stratification onset to be constrained to between 8990 and 8440 yr cal BP (8720 ± 2σ). This timing is consistent with the Uehara et al. (2006) prediction for this location using the dynamic ocean tide boundary but is later than using the fixed boundary. To provide analogous reconstructions of the planktonic environment linked to stratification onset, Marret & Scourse (2002) identified characteristic dinoflagellate cyst assemblages from stratified and mixed sectors in the Celtic Sea, and Marret et al. (2004) identified these same assemblages from core 199 consistent with the interpretations based on benthic foraminiferal isotopes and faunal assemblages.

These geological data, incorporating fully calibrated faunal, floral and stable isotopic proxies, thus indicate that the palaeotidal simulations based on dynamic ocean tide forcing are more realistic or reliable than those based on fixed ocean tide inputs. A number of important points should be noted. First, it is the *timing* of the changes that provides the key test between different simulations. Reliable and robust chronologies are therefore critical. Second, core 199 is an unusual site. Continuous Holocene records from shelf settings are rare, and only one other site has been identified on the north-western European continental shelf that may contain a similar record of changes in seasonal stratification (Evans et al. 2002). Third, precise geological constraint of a single palaeotidal simulation output variable—in this case seasonal stratification—enables confidence to be placed in the other output variables produced by the same simulation. Although it is the pattern and timing of seasonal stratification in the Celtic Sea that is the basis of the constraint, outputs relating to tidal amplitudes, bed stress and tidal currents from the same simulation can be adopted with more confidence than those from simulations for which the constraints are poor. Therefore, it is the simulation based on the dynamic ocean boundary and Lambeck GIA from the Uehara et al. (2006) simulation suite that is regarded as the most consistent with the observational constraints. Fourth, stratification proxies are based on changes in the biology and biogeochemistry of the shelf seas. Finally, and most important, the Celtic Deep sequence provides the first observational support for predictions of dynamical changes in the open ocean tide and tidal dissipation through time. There are no proxies available from the deep ocean to test palaeotidal simulations, and it is unlikely that any will emerge. The history of tides in the global ocean thus relies on model simulations constrained by geological and palaeoceanographic observations from the marginal shelf seas.

It is clearly unsatisfactory that palaeotidal simulations are currently constrained by data from only a single core site, however outstanding and valuable. The stratification proxies described are also time consuming and expensive to generate, so attention has turned to possible alternative proxies. Bed shear stress is an alternative palaeotidal model output susceptible to constraint via the geological record and is under active consideration (Ward et al. in press). Seabed sediment type is a function of bed stress, with mudbanks resulting from low shear stress, and sands and gravels, or even eroded hardground, associated with high shear stress. Sediment grain size evolution (lithostratigraphy) should therefore be a function of changing bed stress through time. As with stratification proxies, interpretations of bed stress change from the sedimentary record will require calibration between bed stress and seabed sediment grain size profiles, and it will be the timing of changes that will provide the key criterion for comparison with model output. Nevertheless, if successful, this approach should be more widely applicable, and less time consuming and expensive, than finding and investigating the rare sedimentary basins preserving stratification proxies.

Palaeotidal simulations and the marine biosphere: the potential

If appropriately constrained by such proxies, palaeotidal models provide an analytical tool for hindcasting and predicting the physical, biological and biogeochemical impact of the periodic inundation and emergence of continental shelves by changing mean sea level during the Quaternary. They also provide a means of predicting the impacts of future rise in sea level. Uehara et al. (2006) provided data on changing tidal amplitudes through time for specific locations (Figure 26), and Pickering et al. (2012) highlighted sectors likely to be significantly impacted by enhanced tidal amplitudes as a function of future rising RSL in Europe. Neill et al. (2010), in a full tidal constituent 3-dimensional analysis using the Proudman Oceanographic Laboratory Coastal Ocean Modelling System (POLCOMS) (Holt & James 2001), demonstrated that mismatches between GIA predictions and SLIP data for north-western Scotland (Arisaig) are resolvable if the higher tidal amplitudes of the deglacial phase are taken into account (Figure 27). Without consideration of palaeotidal correction, the solution to this mismatch demands significant changes to the ice model input to the GIA simulations. There is now an urgent need to provide palaeotidal corrections to the SLIP database, for Great Britain and Ireland, for instance, since uncorrected data are currently used to constrain GIA output; potential circularity can be avoided by using a subset of the SLIP database to calibrate the GIA output. Neill et al. (2009, 2010), incorporating bed stress changes driven by wave activity in addition to the tides, demonstrated a significant reduction since 12,000 years ago in the

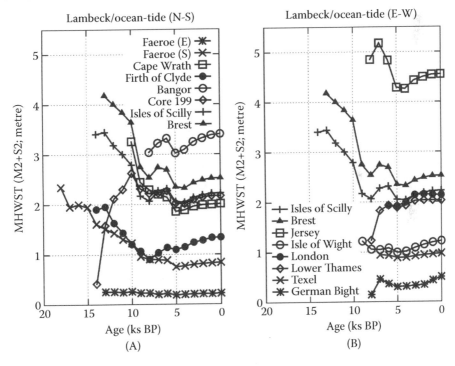

Figure 26 Simulations of tidal amplitude change (mean high water spring tide, MHWST) from the Last Glacial Maximum to present for locations on the north-western European continental shelf based on Lambeck glacial isostatic adjustment simulated palaeotopographies and dynamic ocean boundary. Note that for some locations there is a predicted reduction in tidal amplitude (e.g., Firth of Clyde, South Faeroes) synchronous with predicted increases elsewhere (e.g., Bangor, Lower Thames). (Reprinted from Uehara et al. 2006, with permission from the American Geophysical Union.)

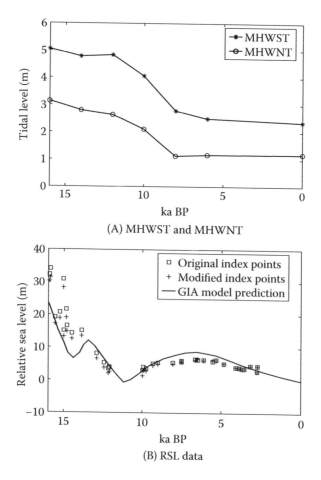

(A) MHWST and MHWNT

(B) RSL data

Figure 27 Palaeotidal correction of sea-level index points (SLIPs). (A) Palaeotidal simulations of mean high water spring tide (MHWST) and mean high water neap tide (MHWNT) amplitudes for Arisaig, north-western Scotland, for the period 15 ka BP to present. (B) Original SLIP data for Arisaig published by Shennan et al. (2000b) and the same data corrected for M_2 and S_2 amplitude change. Solid line shows glacial isostatic adjustment (GIA) model simulation for Arisaig (Shennan et al. 2006b) assuming a lithosphere 76-km thick. Note that the corrected SLIP data are closer to the GIA simulation than the uncorrected data; this highlights that mismatches between SLIP data and GIA simulations may result as much from assuming no changes in tidal amplitude through time as from inadequacies with the ice model inputs to GIA. (Reprinted from Neill et al. 2010, with permission from Springer-Verlag.)

transport of coarse sediment across the north-western European shelf (Figure 28). Because of the strong attenuation by depth of wave orbital velocity, wave-induced bed shear stress is more sensitive to water depth than tide-induced bed shear stress (Neill et al. 2009). Palaeotidal bed stress outputs using the Uehara et al. (2006) simulations help to explain moribund shelf bedforms (sediment waves, dunes, tidal sand ridges) in the Irish (Van Landeghem et al. 2009a) and Celtic (Scourse et al. 2009b) Seas, as originally explored by Belderson et al. (1986), and using POLCOMS, the evolution of sandbanks adjacent to headlands (Neill & Scourse 2009). The location of subglacial palimpsest land systems swept clear of sediments by very high bed stress can be predicted by palaeotidal simulations, such as the remarkable features mapped by multibeam swath bathymetry in the Irish Sea (Van Landeghem et al. 2009b; Figure 13). These reconstructions of seabed sediment and bedform

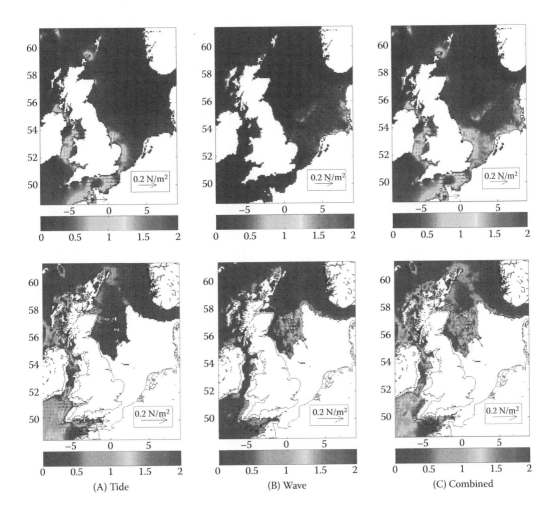

Figure 28 (See also colour figure in the insert) The influence of tide-induced (A), wave-induced (B) and combined (C) changes in mean annual bed shear stress (colour scale in N/m²) for the present day (upper panels) and for 12 ka BP (lower panels). Annual residual bed shear stress vectors shown by arrows. (Reprinted from Neill et al. 2010, with permission from Springer-Verlag.)

evolution have significant implications for benthic habitat evolution (McBreen et al. 2008); the subglacial hardground in the Irish Sea is, for instance, a determinant of *Modiolus modiolus* reefs.

For the period since the LGM, it is now possible, for any location or sector within the north-western European shelf model domain, to predict time-dependent changes in RSL, tidal amplitude, tidal energy dissipation, mean and peak bed shear stress magnitudes (hence bottom sediment type) and vectors, surface tidal current vectors and whether the overlying water column is/was seasonally stratified or mixed. The simulations provide a basis for quantitative predictive analysis of such changes in the physical environment in terms of barriers (tidal mixing fronts, sediment type), isolation and connectivity that may inform studies of genetic diversity. One of the most promising approaches will be to use palaeotidal simulation data as inputs for pENM; this would build on the development adopted by Kettle et al. (2011) but would embed tide and tide-dependent variables into the pENM in addition to simply changing RSL.

The impact of Quaternary sea-level change on atmospheric pCO_2 concentrations

The disproportionately high rates of primary productivity in shelf seas translate into a significant role within the global carbon cycle (Borges 2005), and the periodic flooding and evacuation of the shelf seas have long been regarded as a contender amongst the mechanisms invoked to explain glacial-interglacial changes in atmospheric pCO_2 (cf. Broecker 1982). Seasonally integrated ΔpCO_2 measurements demonstrate that the seasonally stratified temperate shelf seas are significant sinks of carbon, which is hypothesized to be exported into the deep ocean via a 'continental shelf pump' (Tsunogai et al. 1999, Thomas et al. 2004). The stratified northern North Sea, for instance, absorbs 8.5×10^{12} g Ca a^{-1}, whilst the mixed southern North Sea is a weak source of CO_2 (Thomas et al. 2004). This difference arises from the thermocline creating a respiration compartment sealed from exchange from the atmosphere, whereas the mixed water column remains in equilibrium with the atmosphere. Integrating similar measurements from other polar and temperate shelf seas (Tsunogai et al. 1999, Borges et al. 2005, Bates 2006, Cai et al. 2006) indicates a carbon sink equivalent to a net drawdown in atmospheric CO_2 equivalent to 0.17–0.23 ppmv a^{-1} (Borges et al. 2005, Cai et al. 2006).

Rippeth et al. (2008) combined seasonally averaged shelf sea atmosphere CO_2 exchange measurements with palaeotidal model output to assess the impact of the increase in the extent of the temperate shelf seas since the LGM on atmospheric pCO_2 concentrations. The palaeotidal approach of Uehara et al. (2006) was adopted, modified to incorporate palaeotopographies from the ICE-5G GIA model (Peltier 2004). The palaeotidal model outputs predict the areal extent of mixed and seasonally stratified sectors for each time step; an average CO_2 uptake flux of 30 g C m^{-2} a^{-1} was applied for stratified sectors and an outgassing rate of 6 g C m^{-2} a^{-1} for mixed sectors. This enabled the net flux of CO_2 between shelf seas and atmosphere to be compared with the observed changes in atmospheric pCO_2 since the LGM measured from the EPICA (European Project for Ice Coring in Antarctica) Dome C ice core in Antarctica (Monnin et al. 2001; Figure 29).

The results demonstrate that the continental shelf pump of the temperate shelf seas operated as a net sink for carbon during the phase of atmospheric pCO_2 increase from the LGM into the Holocene, implying that without this component of the carbon cycle the rate of increase of pCO_2 would have been higher, that is, the growth of the temperate shelf seas acts as a brake on rising CO_2 similar to the increase in terrestrial biomass. During the initial phase of shelf flooding, the shallow water depths and tidal mixing maintain a net CO_2 source but progressively, and lagging behind the overall increase in the areal extent of the shelf seas, the seasonally stratified sectors develop and initiate a strong net sink. This indicates the requirement to integrate seasonal stratification dynamics into such estimates.

Rippeth et al. (2008) thus offered an analytical template based on PTM to quantify the impact of shelf sea biogeochemistry on the carbon cycle. However, the results are only as good as the observational database on present shelf sea CO_2 fluxes, so as new and refined observations emerge (e.g., Laruelle et al. 2010, Xue et al. 2011), they can be integrated with palaeotidal reconstructions to generate improved estimates. There is also a need to assess more fully the role of sea ice in modulating air–sea CO_2 fluxes (Rippeth et al. 2008) and to integrate data from tropical shelf seas, which the available observations suggest act as strong CO_2 sources (Borges et al. 2005, Bates 2006, Cai et al. 2006). The growth of the tropical shelf seas following the LGM would therefore have acted in the opposite direction to the temperate and polar shelf seas in contributing to, rather than counteracting, the deglacial increase in atmospheric pCO_2.

As RSL controls the extent of coral reef growth, carbon sequestered or released by reef systems has fuelled discussion of the 'coral reef hypothesis' to explain glacial–interglacial variations in atmospheric pCO_2. Though precipitation of calcium carbonate within reefs results in the sequestering of carbon, an accompanying shift in pH results in release of CO_2 (Ware et al. 1992, Gattuso et al. 1999). Coral reefs are therefore sources and not sinks of CO_2, and the growth in reef extent

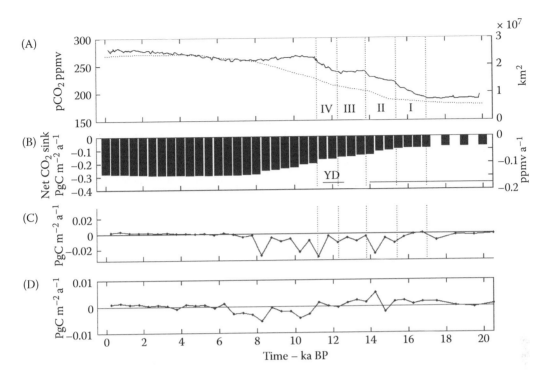

Figure 29 The influence of the sea-level change on atmospheric pCO_2 since the Last Glacial Maximum (LGM). (A) Change in atmospheric pCO_2 since the LGM from EPICA Dome C ice core (solid line; Monnin et al. 2001) and change in the areal extent of the shelf seas (dotted line). (B) Magnitude of the temperate shelf sea CO_2 sink (-ive represents uptake of CO_2) based on palaeotidal simulations of seasonal stratification (Uehara et al. 2006) using ICE-5G global palaeotopography (Peltier 2004) and integrated air-sea CO_2 fluxes (Thomas et al. 2004). Periods of significant multiyear ice cover shown by horizontal black line. YD, Younger Dryas. (C) Changes in the annual CO_2 source (+) and sink (-) for the global temperate shelf seas. (D) Impact of seasonal stratification in the temperate shelf seas on the continental shelf CO_2 pump. (Reprinted from Rippeth et al. 2008, with permission from the American Geophysical Union.)

consequent on the deglacial flooding of the continental shelves has been invoked to explain the increase in atmospheric pCO_2 since the LGM (Opdyke & Walker 1992). Kleypas's (1997) pENM of reef extent since the LGM suggests a reefal CO_2 source equivalent to the total increase in atmospheric pCO_2 increase from the LGM to preindustrial Holocene, though an alternative modelling approach (Lerman et al. 2011) indicates that the reef contribution is no more than 24 ppmv of the total 100 ppmv increase. Nevertheless, Kleypas's (1997) approach is complementary to the work of Rippeth et al. (2008) in providing an analytical tool to assess the role of the shelf sea carbonate system in the global carbon cycle. Any integrated assessment of the impact of shelf inundation and emergence on atmospheric pCO_2 requires the quantification of carbon storage and outgassing from both marine and terrestrial (Montenegro et al. 2006) domains.

Conclusions

The present sea-level highstand is highly anomalous within the context of recent geological time, yet it is from this anomalous context that our understanding of patterns and processes in the marine biosphere has developed. Global sea-level cycles, driven by the glacio-eustatic mechanism, have resulted in about 49 transgressive–regressive cycles during the last 2.58 million years, the Quaternary. The most extreme lowstands and highstands of this period have been experienced during the last

700,000 years as a result of the high-amplitude and high-wavelength glacial–interglacial cycles characteristic of the late Quaternary. Shelf seas developed during each transgression, but these have been present for only a small proportion of total recent geological time. During the preceding Pliocene period, shelf seas were quasi-permanent. As a disproportionately large amount of total oceanic primary production and carbon fixation takes place in shelf seas, their recent and increasing transience has implications for the marine biosphere and the carbon cycle. Although it has been suggested that significant regressions had a part to play in the major mass extinctions of Earth history, no strong tendency towards the extinction of biota within shallow marine environments can be detected as a result of the extreme Quaternary sea-level cycles, implying that sea-level change may only have played an ancillary role in driving mass extinction events, and that other factors, such as changes in ocean oxygen levels, CO_2, pH, and temperature, may have been more important (Payne & Clapham 2012).

Relative sea level is strongly controlled by glaciation through the interplay of glacio-eustasy and glacio-hydroisostasy, an interaction that can be simulated through space and time using GIA models. GIA simulations are constrained by SLIPs. Biological data are critical in the definition of SLIPs. SLIPs are confined to the period since the LGM as a result of preservational constraints and the practical limit of radiocarbon dating. The deep-sea oxygen isotope record thus serves as a proxy for global eustatic sea level for the entirety of Quaternary and Pliocene time and can be used to determine the timing and duration of eustatic highstands when shelf seas were in existence and of lowstands when they were absent.

During each transgressive–regressive cycle, shallow marine habitats migrate in response to both sea-level changes and other major coacting environmental controls, notably seawater temperature. Such habitat shifts are strongly modulated by the nature of the hypsometric curve; for some habitats, the largest areal expansions may occur during mid-transgression (e.g., mangrove, kelp forest) rather than at lowstand or highstand, where steeper gradients result in coastal squeeze. For others (e.g., coral reefs), it is the rate of RSL change that is important. The pattern of habitat expansion and contraction with sea-level cycles on continental margins is inverted in oceanic islands, where major regressions are often associated with the physical conjoining of island archipelago and habitat expansion. Apart from coral reefs, direct geological evidence for such elevational shifts is rare as a result of the poor preservation potential of transgressive sequences and many key species. Palaeo-environmental niche modelling, especially when constrained by GIA model inputs, is an important and exciting development in reconstructing past habitats and species' ranges given knowledge of the major environmental variables (palaeoceanography). Although the suggestion that the 'oldest' shallow marine ecosystems may be those surviving at the shelf edge remains an untested hypothesis, the 'age' of marine ecosystems as determined from sea-level records, in flooded fjords, for instance, appears to correlate positively with biodiversity.

Quaternary sea-level changes have resulted in the isolation of shallow marine refugial populations and the creation of bottlenecks that have driven genetic change. Although physically isolated marine basins quickly revert to freshwater systems, causing local extinction of marine species, some entrapped marine and anadromous teleost species, notably Arctic charr and coregonids, adapt to the changed conditions and show evidence of haplotypic variation. Unusual isolated marine lakes in karstified contexts maintaining hydraulic connectivity with the ocean in the Pacific Palau Islands host genetically modified mussel and jellyfish populations as a result of reproductive isolation. Physical separation of ocean basins during regression, notably the emergence of Sundaland/Wallacea in South-East Asia separating the Indian and Pacific Oceans, has resulted in the genetic divergence of many species, except, apparently, hermatypic corals.

There are significant marine genetic data supporting the single 'expansion–contraction' model of southern refugia in the Northern Hemisphere. In the North Atlantic, this model is supported by genetic data from teleosts, invertebrates and plants and by recent GIA-informed pENM of four teleost species (haddock, pollock, allis shad, Atlantic salmon) independently supported via archaeo-zoological finds from well-dated coastal archaeological middens. There is strong evidence that the LGM refugia of many of these species may have extended into the Mediterranean Basin as a result of the maintenance of a full marine connection between the Atlantic and Mediterranean even at extreme lowstand. The interpretation of molecular marker data from many other species, however, suggests the existence of cryptic northern glacial refugia, similar to recent interpretations of genetic data from some terrestrial species. It has been argued that such data can be used to refine ice sheet limits, in north-western Europe, for instance. The northern marine glacial refugia hypothesis is contradicted by multiple lines of independent palaeoenvironmental evidence that indicate that the putative refugial locations are completely incompatible with even the most basic requirements for a thermophilous marine refuge; that is, the areas were dry land, under ice sheets, fresh water or in glacimarine settings scoured by sea ice and icebergs. This incompatibility implies that either the genetic or the palaeoenvironmental data have been misinterpreted and highlights a requirement for the marine genetic and palaeoenvironmental communities to work more closely together.

Changes in RSL alter the shape, morphology and bathymetry of ocean basins and coasts, resulting in significant changes in tides. Sea-level modulated palaeotidal changes exert a strong control on the marine biosphere in addition to the direct impact of changes in mean sea level. These changes can be simulated by palaeotidal models using GIA simulations as palaeotopographic inputs. Multiple independent palaeotidal simulations indicated that tidal dissipation was significantly higher during the LGM, and during earlier regressions, than at present. The glacial ocean was more vigorously mixed by mechanical tidal energy, and the resonant glacial North Atlantic and Arctic oceans were megatidal with significant implications for ice–ocean interaction (ice shelf stability, iceberg calving rates, rate of ice transfer to ocean). As RSL has changed, so have tidal amplitude and phase and tide-dependent parameters, including surface tidal currents, bed stress, and seasonal stratification, reflected in the locale of tidal mixing fronts. Palaeotidal models have been constrained by floral, faunal and stable isotopic proxies for seasonal stratification dynamics; bed stress is another candidate output potentially valuable for constraining palaeotidal models via sediment core grain size evolution. When constrained by geological observations, palaeotidal simulations are important for providing elevational corrections for SLIPs, and there is significant potential for the use of palaeotidal simulations for exploring temporal changes in benthic habitat extent, type and connectivity.

Observations on air–sea CO_2 flux in present shelf seas indicated that polar and temperate seasonally stratified shelf seas are net carbon sinks, whereas tropical and fully mixed temperate shelf seas, and coral reef systems, are net carbon sources. Integrations of such observations with GIA and palaeotidal model simulations enable estimates of the contribution of the shelf seas to atmospheric pCO_2 over the period since the LGM. These can be compared with the record of atmospheric pCO_2 from Antarctic ice cores. Without the growing carbon sink of the expanding temperate and polar shelf seas during the last deglaciation, the rise in atmospheric CO_2—the crucial greenhouse feedback—would have been less attenuated. Sea-level change therefore modulates the rate and patterns of production and respiration in shallow marine systems; this is a significant Earth system feedback affecting the carbon cycle.

The synchronous development of GIA, palaeotidal and pENM provides a rich opportunity to develop an integrated Earth systems approach to the simulation of the impacts of RSL change on the marine biosphere (Figure 30). When appropriately constrained by palaeoenvironmental and palaeoceanographic data, such simulations provide a powerful means of predicting the impacts of future sea-level rise on the marine biosphere. GIA simulations are required to provide the palaeo-topographic inputs for both palaeotidal and pENM simulations. All three classes of model require observational constraints. Palaeotidal models are currently constrained via palaeostratification proxies and potentially by bed stress proxies, and pENM by palaeoenvironmental data on habitat type and extent, and (micro)palaeontological data on past species' distributions. As palaeotidal

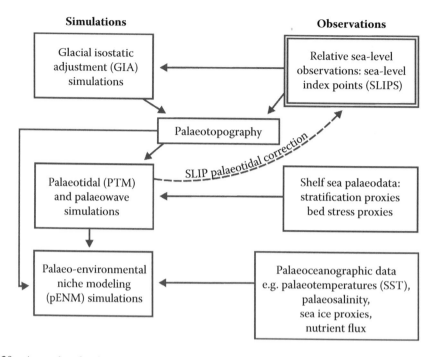

Figure 30 A template for the integrated model-data comparison of the impacts of relative sea-level change on the marine system. The primacy of sea-level index point (SLIP) observations is emphasized (double margin). SLIP data constrain glacial isostatic adjustment (GIA) model simulations, which in turn provide palaeotopographies for both palaeotidal modelling (PTM) and wave simulations and palaeo-environmental niche model (pENM) simulations. PTM simulations require observational constraints from shelf sea proxy data, and pENM simulations need them from palaeoceanographic proxy data. PTM simulations have not yet been used as pENM inputs. The correction of SLIP data for tidal range change using PTM output is an urgent research priority.

and pENM models depend on GIA model output, SLIPs emerge as *the* fundamental observational constraint. The generation of new SLIP data and their systematic palaeotidal correction is thus an urgent research priority.

Relative sea level determines responses in the marine biosphere, including biodiversity, habitat extent, resource distribution, isolation, bottlenecks, genetic change, and rates of production and respiration with impacts on the global carbon cycle. Relative sea-level change in turn results from the interaction of glaciation (climate change) and tectonics. Glaciation controls sea-level cycles through glacio-eustasy and glacio-hydroisostasy, and these cycles are played out against a topographic backdrop—the hypsometric curve—that is primarily of tectonic origin.

Acknowledgements

As the research field reviewed is necessarily multidisciplinary and collaborative, I must therefore acknowledge the help, advice and insight of colleagues and students—many associated with the School of Ocean Sciences (Menai Bridge) at Bangor University—without whom some of the research described and the ideas generated would never have reached fruition. The aim of reconstructing the seasonal dynamics of the shelf seas grew out of the fundamental work on shelf sea dynamics by John Simpson and colleagues and led to early palaeotidal simulations undertaken by Roger Austin and supervised by Ed Hill (now National Oceanography Centre Southampton) and myself. In generating palaeo proxies to constrain seasonal stratification dynamics, I must acknowledge my former research students Bill Austin (now University of St. Andrews) and Gillian Scott (now National University of Ireland, Maynooth), postdoctoral assistant Fabienne Marret (now University of Liverpool), colleague Hilary Kennedy and my long-standing assistant, Brian Long. My interest in sea-level change and the relationships between sea-level and tidal dynamics was stimulated by Roland Gehrels (Plymouth University) and my former student and current colleague Mike Roberts. Research on the glacial history of the mid-latitude shelves has been stimulated by research students Mark Furze (now Grant McEwen University, Edmonton, Canada) and Anna Haapaniemi; former student and current colleague Anna Pieńkowski; and current initiatives coordinated by the BRITICE (British–Irish Ice Sheet) Consortium led by Chris Clark (University of Sheffield). For taking PTM to a new level, the contribution of Katsuto Uehara (Kyushu University, Japan) has been seminal, and I thank Katsuto, Grant Bigg (University of Sheffield) and my former and current colleagues in Menai Bridge, Kevin Horsburgh (now National Oceanography Centre Liverpool), Mattias Green, Simon Neill, Tom Rippeth and Katrien Van Landeghem for their enthusiastic and productive response to the opportunities offered by linking shelf sea processes with the recent geological past. Adam Wainwright deserves special thanks for bringing GIA and palaeotidal simulations to life with a series of stunning visualizations. Palaeotidal and tidal modelling continues with a remarkable group of current Menai Bridge research students—Sophie Ward, Sebastian Rosier, Sophie Wilmes and Holly Pelling—who ask challenging questions on a daily basis. My interest in the impacts on the marine biosphere of sea-level and palaeotidal change was stimulated by proposal writing with Steve Hawkins (now University of Southampton). For practical help with this review, I must acknowledge Mattias Green, Tom Rippeth, Katrien Van Landeghem, Simon Neill, Katsuto Uehara and Paul Butler for answering questions and supplying figures; Brian Long and David Roberts for help with drafting figures and graphics; and editor Roger Hughes for his encouragement and advice.

This review is a contribution to the Climate Change Consortium of Wales (C3W), and the research has been supported by a series of grants, notably the EU (European Union) SHELF and BALTEEM projects; EU Marie Curie Fellowships (F. Marret and A. Pieńkowski); the Leverhulme Trust; the Cemlyn Jones Trust; the Royal Society/Japan Association for the Promotion of Science (K. Uehara); Finnish Graduate School (A. Haapaniemi); Fujitsu-High Performance Computing Wales (S. Wilmes); UK Natural Environment Research Council (NERC) (Small Grant GR9/02631, Standard Grant GR3/11467, Standard Grant NER/A/S/2001/01189); and a series of awards from the NERC Radiocarbon Facility.

References

Alvarez, L.W., Alvarez, W., Asaro, F. & Michel, H.V. 1980. Extraterrestrial cause for the Cretaceous-Tertiary extinction: experimental results and theoretical interpretation. *Science* **208**, 1095–1108.

Appenzeller, T. 2012. Eastern odyssey. *Nature* **485**, 24–26.

Arbic, B.K. & Garrett, C. 2010. A coupled oscillator model of shelf and ocean tides. *Continental Shelf Research* **30**, 564–574.

Arbic, B.K., Mitrovica, J.X., MacAyeal, D.R. & Milne, G.A. 2008. On the factors behind large Labrador Sea tides during the last glacial cycle and the potential implications for Heinrich events. *Paleoceanography* **23**, PA3211.

Austin, R.M. 1991. Modelling Holocene tides on the NW European continental shelf. *Terra Nova* **3**, 276–288.

Austin, W.E.N. & Scourse, J.D. 1997. Evolution of seasonal stratification in the Celtic Sea during the Holocene. *Journal of the Geological Society* **154**, 249–256.

Ballantyne, C.K. 2010. Extent and deglacial chronology of the last British-Irish Ice Sheet: implications of exposure dating using cosmogenic isotopes. *Journal of Quaternary Science* **25**, 515–534.

Ballantyne, C.K., McCarroll, D. & Stone, J.O. 2007. The Donegal ice dome, northwest Ireland: dimensions and chronology. *Journal of Quaternary Science* **22**, 773–783.

Ballantyne, C.K., Stone, J.O. & McCarroll, D. 2008. Dimensions and chronology of the last ice sheet in Western Ireland. *Quaternary Science Reviews* **27**, 185–200.

Bard, E., Hamelin, B., Arnold, M., Montaggioni, L., Cabioch, G., Faure, G. & Rougerie, F. 1996. Deglacial sea-level record from Tahiti corals and the timing of global meltwater discharge. *Nature* **382**, 241–244.

Bard, E., Hamelin, B. & Fairbanks, R.G. 1990. U-Th ages obtained by mass spectrometry in corals from Barbados—sea-level during the past 130,000 years. *Nature* **346**, 456–458.

Bassett, S.E., Milne, G.A., Mitrovica, J.X. & Clark, P.U. 2005. Ice sheet and solid earth influences on far-field sea-level histories. *Science* **309**, 925–928.

Bates, N.R. 2006. Air-sea CO_2 fluxes and the continental shelf pump of carbon in the Chukchi Sea adjacent to the Arctic Ocean. *Journal of Geophysical Research* **111**, C10013.

Behrenfeld, M.J., Boss, E., Siegel, D.A. & Shea, D.M. 2005. Carbon-based ocean productivity and phytoplankton physiology from space. *Global Biogeochemical Cycles* **19**, GB1006.

Behrenfeld, M.J. & Falkowski, P.G. 1997. Photosynthetic rates derived from satellite-based chlorophyll concentration. *Limnology and Oceanography* **42**, 1–20.

Belderson, R.H., Pingree, R.D. & Griffiths, D.K. 1986. Low sea-level tidal origin of Celtic Sea sand banks— evidence from numerical modelling of M2 tidal streams. *Marine Geology* **73**, 99–108.

Ben-Shlomo, R., Paz, G. & Rinkevich, B. 2006. Postglacial-period and recent invasions shape the population genetics of botryllid ascidians along European Atlantic coasts. *Ecosystems* **9**, 1118–1127.

Berger, A. & Loutre, M.F. 1991. Insolation values for the climate of the last 10 million years. *Quaternary Science Reviews* **10**, 297–317.

Bernatchez, L. & Wilson, C.C. 1998. Comparative phylogeography of nearctic and palearctic fishes. *Molecular Ecology* **7**, 431–452.

Berner, R.A., Lasaga, A.C. & Garrels, R.M. 1983. The carbonate-silicate geochemical cycle and its effect on atmospheric carbon dioxide over the last 100 million years. *American Journal of Science* **283**, 641–683.

Bindoff, N., Willebrand, J., Artale, V., Cazenave, A., Gregory, J., Gulev, S., Hanawa, K., Qu, C.L., Levitus, S., Nojiri, Y., Shum, C., Talley, L. & Unnikrishnan, A. 2007. Observations: oceanic climate change and sea level. In *The Physical Science Basis. Contribution of Working Group I to the Fourth Assessment Report of the Intergovernmental Panel on Climate Change*, S. Solomon, D. Qin, M. Manning, Z. Chen, M. Marquis, K.B. Averyt, M. Tignor & H.L. Miller (eds). Cambridge, UK: Cambridge University Press.

Bird, M.I., Taylor, D. & Hunt, C. 2005. Environments of insular Southeast Asia during the Last Glacial Period: a savannah corridor in Sundaland? *Quaternary Science Reviews* **24**, 2228–2242.

Blumberg, A.F. & Mellor, G.L. 1987. A description of a three dimensional coastal ocean circulation model. In *Three-Dimensional Coastal Ocean Models, Coastal Estuarine Studies*, N.S. Heaps (ed.). Washington, DC: AGU, **4**, 1–16.

Bondevik, S., Løvholt, F., Harbitz, C., Mangerud, J., Dawson, A. & Svendsen, J.I. 2005. The Storegga Slide tsunami—comparing field observations with numerical simulations. *Marine and Petroleum Geology* **22**, 195–208.

Borges, A.V. 2005. Do we have enough pieces of the jigsaw to integrate CO_2 fluxes in the coastal ocean. *Estuaries* **28**, 3–27.

Borges, A.V., Delille, B. & Frankignoulle, M. 2005. Budgeting sinks and sources of CO_2 in the coastal ocean: Diversity of ecosystems counts. *Geophysical Research Letters* **32**, L14601, doi:10.1029/2005GL023053.

Boulton, G.S. & Hagdorn, M., 2006. Glaciology of the British Isles Ice Sheet during the last glacial cycle: form, flow, streams and lobes. *Quaternary Science Reviews* **25**, 3359–3390.

Bourillet, J.F., Reynaud, J.Y., Baltzer, A. & Zaragosi, S. 2003. The 'Fleuve Manche': the submarine sedimentary features from the outer shelf to the deep-sea fans. *Journal of Quaternary Science* **18**, 261–282.

Bowen, D.Q., Phillips, F.M., McCabe, A.M., Knutz, P.C. & Sykes, G.A. 2002. New data for the Last Glacial Maximum in Great Britain and Ireland. *Quaternary Science Reviews* **21**, 89–101.

Bowler, J., Johnston, H., Olley, J.M., Prescott, J.R., Roberts, R.G., Shawcross, W. & Spooner, N.A. 2003. New ages for human occupation and climatic change at Lake Mungo, Australia. *Nature* **421**, 837–840.

Bradley, S.L., Milne, G.A., Shennan, I. & Edwards, R. 2011. An improved glacial isostatic adjustment model for the British Isles. *Journal of Quaternary Science* **26**, 541–552.

Bradley, S.L., Milne, G.A., Teferle, F.N., Bingley, R.M. & Orliac, E.J. 2009. Glacial isostatic adjustment of the British Isles: new constraints from GPS measurements of crustal motion. *Geophysical Journal International* **178**, 14–22.

Bradwell, T., Stoker, M.S., Golledge, N.R., Wilson, C.K., Merritt, J.W., Long, D., Everest, J.D., Hestvik, O.B., Stevenson, A.G., Hubbard, A.L., Finlayson, A.G. & Mathers, H.E. 2008. The northern sector of the last British Ice Sheet: maximum extent and demise. *Earth Science Reviews* **88**, 207–226.

Broecker, W.S. 1982. Glacial to interglacial changes in ocean chemistry. *Progress in Oceanography* **11**, 151–197.

Brooks, A.J., Bradley, S.L., Edwards, R.J., Milne, G.A., Horton, B. & Shennan, I. 2008. Postglacial relative sea-level observations from Ireland and their role in glacial rebound modelling. *Journal of Quaternary Science* **23**, 175–192.

Brown, P., Sutikna, T., Morwood, M.J., Soejono, R.P., Jatmiko, Saptomo, E.W. & Due, R.A. 2004. A new small-bodied hominin from the Late Pleistocene of Flores, Indonesia. *Nature* **431**, 1055–1061.

Cabioch, G., Banks-Cutler, K.A., Beck, W.J., Burr, G.S., Corrège, T., Edwards, R.L. & Taylor, F.W. 2003. Continuous reef growth during the last 23 cal kyr BP in a tectonically active zone (Vanuatu, southwest Pacific). *Quaternary Science Reviews* **22**, 1771–1786.

Cabioch, G., Montaggioni, L., Frank, N., Seard, C., Salle, E., Payri, C., Pelletier, B. & Paterne, M. 2008. Successive reef depositional events along the Marquesas foreslopes (French Polynesia) since 26 ka. *Marine Geology* **254**, 18–34.

Cai, W.-J., Dai, M. & Wang, Y. 2006. Air-sea exchange of carbon dioxide in ocean margins: a province-based synthesis. *Geophysical Research Letters* **33**, L12603, doi:10.1029/2006GL026219.

Campo, D., Molares, J., Garcia, L., Fernandez-Rueda, P., Garcia-Gonzales, C. & Garcia-Vazquez, E. 2010. Phylogeography of the European stalked barnacle (*Pollicipes pollicipes*): identification of glacial refugia. *Molecular Biology* **157**, 147–156.

Cazenave, A., Dominh, K., Guinehut, S., Berthier, E., Llovel, W., Ramillien, G., Ablain, M. & Larnicol, G. 2009. Sea level budget over 2003–2008: a re-evaluation from GRACE space gravimetry, satellite altimetry and Argo. *Global and Planetary Change* **65**, 83–88.

Chappell, J. 1974. Geology of coral terraces, Huon Peninsula, New Guinea: study of Quaternary tectonic movements and sea-level changes. *Bulletin of the Geological Society of America* **85**, 553–570.

Chappell, J. & Polach, H. 1991. Post-glacial sea-level rise from a coral record at Huon Peninsula, Papua New Guinea. *Nature* **349**, 147–149.

Chappell, J. & Shackleton, N.J. 1986. Oxygen isotopes and sea-level. *Nature* **324**, 137–140.

Chenoweth, S.F., Hughes, J.M., Keenan, C.P. & Lavery, S. 1998. When oceans meet: a teleost shows secondary intergradation at an Indian-Pacific interface. *Proceedings of the Royal Society* **B265**, 415–420.

Chevolet, M., Hoarau, G., Rijnsdorp, A.D. & Stam, W.T. 2006. Phylogeography and population structure of thornback rays (*Raja clavata* L., Rajidae). *Molecular Ecology* **15**, 3693–3705.

Chiverrell, R.C., Thrasher, I., Thomas, G.S.P., Lang, A., Scourse, J.D., Van Landeghem, K.J.J., McCarroll, D., Clark, C.D., Ó'Cofaigh, C., Evans, D.J.A. & Ballantyne, C. 2013. Bayesian modelling the retreat of the Irish Sea Ice Stream. *Journal of Quaternary Science* **28**, 200–209.

Clark, C.D., Hughes, A.L.C., Greenwood, S.L., Jordan, C. & Sejrup, H.P. 2012. Pattern and timing of retreat of the last British-Irish Ice Sheet. *Quaternary Science Reviews* **44**, 112–146.

Clark, P.U., Archer, D., Pollard, D., Blum, J.D., Rial, J.A., Brovkin, V., Mix, A.C., Pisias, N.G. & Roy, M. 2006. The Middle Pleistocene transition: characteristics, mechanisms and implications for long-term changes in atmospheric pCO_2. *Quaternary Science Reviews* **25**, 3150–3184.

Clark, P.U., Dyke, A.S., Shakun, J.D., Carlson, A.E., Clark, J., Wohlfarth, B., Mitrovica, J.X., Hostetler, S.W. & McCabe, A.M. 2009. The Last Glacial Maximum. *Science* **325**, 710–714.

CLIMAP Project Members. 1981. Seasonal reconstruction of the Earth's surface at the Last Glacial Maximum. *Geological Society of America Map and Chart Series* **MC-36**.

Collins, L., Zhu, Z.R., Wyrwoll, K.H., Hatcher, B.G., Playford, P., Chen, J.H., Eisenhauser, A. & Wasserburg, G.J. 1993. Late Quaternary facies characteristics and growth history of a high latitude reef complex: the Abrolhos carbonate platforms, eastern Indian Ocean. *Marine Geology* **111**, 203–212.

Colson, I. & Hughes, R.N. 2007. Contrasted patterns of genetic variation in the dogwhelk *Nucella lapillus* along two putative post-glacial expansion routes. *Marine Ecology Progress Series* **343**, 183–191.

Consuegra, S., Garcia de Leáiz, C., Serdio, A., González Morales, M., Straus, L.G., Knox, D. & Verspoor, E. 2002. Mitochondrial DNA variation in Pleistocene and modern Atlantic salmon from the Iberian glacial refugium. *Molecular Ecology* **11**, 2037–2048.

Cox, C.B. 2001. The biogeographic regions reconsidered. *Journal of Biogeography* **28**, 511–523.

Coyer, J.A., Diekmann, O.E., Serrão, E.A., Procaccini, G., Milchakova, N., Pearson, G.A., Stam, W.T. & Olsen, J.L. 2004. Population genetics of dwarf eelgrass *Zostera noltii* throughout its biogeographic range. *Marine Ecology Progress Series* **281**, 51–62.

Coyer, J.A., Peters, A.F., Stam, W.T. & Olsen, J.L. 2003. Post-ice age recolonization and differentiation of *Fucus serratus* L. (Phaeophyceae; Fucaceae) populations in Northern Europe. *Molecular Ecology* **12**, 1817–1829.

Cuffey, K.M. & Marshall, S.J. 2000. Substantial contribution to sea-level rise during the last interglacial from the Greenland ice sheet. *Nature* **404**, 591–594.

Cunha, R.L., Lopes, E.P., Reis, D.M. & Castilho, R. 2011. Genetic structure of *Brachidontes puniceus* populations in Cape Verde archipelago shows signature of expansion during the Last Glacial Maximum. *Journal of Molluscan Studies* **77**, 175–181.

Cunningham, C.W. 2008. Lessons learned from coordinating research on the North Atlantic (CORONA). *Ecology* **89**, S1-S2.

Curry, D. 1989. The rock floor of the English Channel and its significance for the interpretation of marine unconformities. *Proceedings of the Geologists' Association* **100**, 339–352.

Cutler, K.B., Edwards, R.L., Taylor, F.W., Cheng, H., Adkins, J., Gallup, C.D., Cutler, P.M., Burr, G.S. & Bloom, A.L. 2003. Rapid sea-level fall and deep-ocean temperature change since the last interglacial period. *Earth and Planetary Science Letters* **206**, 253–271.

Dahlgren, T.G., Weinberg, J.R. & Halanych, K.M. 2000. Phylogeography of the ocean quahog (*Arctica islandica*): influences of paleoclimate on genetic diversity and species range. *Marine Biology* **137**, 487–495.

Darwin, C. 1842. *The Structure and Distribution of Coral Reefs. Being the First Part of the Geology of the Voyage of the* Beagle, *Under the Command of Capt. Fitzroy, R.N. During the Years 1832 to 1836.* London: Smith Elder.

Darwin, G.H. 1899. *The Tides and Kindred Phenomena in the Solar System.* Boston: Houghton.

Dawson, M.N. & Hamner, W.M. 2005. Rapid evolutionary radiation of marine zooplankton in peripheral environments. *Proceedings of the National Academy of Sciences USA* **102**, 9235–9240.

Dayton, P.K. 1985. Ecology of kelp communities. *Annual Review of Ecology and Systematics* **16**, 215–245.

Dicke, R.H. 1966. The secular acceleration of the Earth's rotation and cosmology. In *The Earth Moon System*, B.G. Marsden & A.G.W. Cameron (eds). New York: Plenum Press, 98–164.

Duke, N.C., Benzie, J.A.H., Goodall, J.A. & Ballment, E.R. 1998. Genetic structure and evolution of species of the mangrove genus *Avicennia* (Avicenniaceae) in the Indo-West Pacific. *Evolution* **52**, 1612–1626.

Dynesius, M. & Jansson, R. 2000. Evolutionary consequences of changes in species' geographical distributions driven by Milankovitch climate oscillations. *Proceedings of the National Academy of Sciences USA* **97**, 9115–9120.

Dziewonski, A.M. & Anderson, D.L. 1981. Preliminary reference Earth model. *Physics of the Earth and Planetary Interiors* **25**, 297–356.

Egbert, G.D. & Ray, R.D. 2003. Semidiurnal and diurnal tidal dissipation from TOPEX/POSEIDON altimetry. *Geophysical Research Letters* **30**, 1907, doi:10.1029/2003GL017676.

Egbert, G.D., Ray, R.D. & Bills, B.G. 2004. Numerical modeling of the global semidiurnal tide in the present day and in the Last Glacial Maximum. *Journal of Geophysical Research* **109**, C03003, doi:10.1029/2003JC001973.

Erbs-Hansen, D.R., Knudsen, K.L., Gary, A.C., Gyllencreutz, R. & Jansen, E. 2012. Holocene climatic development in Skagerrak, eastern North Atlantic: foraminiferal and stable isotopic evidence. *The Holocene* **22**, 301–312.

Evans, J.R., Austin, W.E.N., Brew, D.S., Wilkinson, I.P. & Kennedy, H.A. 2002. Holocene shelf sea evolution offshore northeast England. *Marine Geology* **191**, 147–164.

Eynaud, F., Zaragosi, S., Scourse, J.D., Mojtahid, M., Hall, I.R., Penaud, A., Locascio, M. & Reijonen, A. 2007. Deglacial laminated facies on the NW European continental margin: the hydrographic significance of British Irish Ice Sheet disintegration and Fleuve Manche paleoriver discharges. *Geophysics, Geochemistry, Geosystems* **8**, Q06019, doi:10.1029/2006GC001496.

Fairbanks, R.G. 1989. A 17,000-year glacio-eustatic sea-level record—influence of glacial melting rates on the Younger Dryas event and deep-ocean circulation. *Nature* **342**, 637–642.

Farrell, W.E. & Clark, J.A. 1976. On postglacial sea level. *Geophysical Journal of the Royal Astronomical Society* **46**, 647–667.

Flather, R.A. 1976. A tidal model of the north-west European continental shelf. *Mémoires de la Société Royale des Sciences Liège, Ser.* **6**, **10**, 141–164.

Forbes, E. 1844. Report on the Mollusca and Radiata of the Aegean Sea, and on their distribution, considered as bearing on geology. *Report of the British Association for the Advancement of Science for 1843*, 129–193.

Forbes, E. 1859. *The Natural History of the European Seas*. Edited and continued by R.A.C. Godwin-Austen. London: Van Voorst.

Foster, J.B. 1964. Evolution of mammals on islands. *Nature* **202**, 234–235.

Fraser, C.I., Nikula, R., Spencer, H.G. & Waters, J.M. 2009. Kelp genes reveal effects of subantarctic sea ice during the Last Glacial Maximum. *Proceedings of the National Academy of Sciences USA* **106**, 3249–3253.

French, H.N. 1976. *The Periglacial Environment*. London: Longman.

Frey, M.A. 2010. The relative importance of geography and ecology in species diversification: evidence from a tropical marine intertidal snail (*Nerita*). *Journal of Biogeography* **37**, 1515–1528.

Froget, C., Thommeret, J. & Thommeret, Y. 1972. Mollusques septentrionaux en Méditerranée occidentale: datation par le ^{14}C. *Palaeogeography, Palaeoclimatology, Palaeoecology* **12**, 285–293.

Gagnon, M.C. & Angers, B. 2006. The determinant role of temporary proglacial drainages on the genetic structure of fishes. *Molecular Ecology* **15**, 1051–1065.

Garrison, T. 1999. *Oceanography*. Pacific Grove, California: Wadsworth, 3rd edition.

Gattuso, J.P., Frankignoulle, M. & Smith, S.V. 1999. Measurement of community metabolism in the coral reef CO_2 source-sink debate. *Proceedings of the National Academy of Sciences USA* **96**, 13017–13022.

Gehrels, W.R. 2010. Late Holocene land- and sea-level changes in the British Isles: implications for future sea-level predictions. *Quaternary Science Reviews* **29**, 1648–1660.

Gehrels, W.R., Belknap, D.F., Black, S. & Newnham, R.M. 2002. Rapid sea-level rise in the Gulf of Maine, USA, since AD 1800. *The Holocene* **12**, 383–389.

Gehrels, W.R., Roe, H.M. & Charman, D.J. 2001. Foraminifera, testate amoebae and diatoms as sea-level indicators in U.K. saltmarshes: a quantitative multiproxy approach. *Journal of Quaternary Science* **16**, 201–220.

Gerritsen, H. & Berentsen, C.W.J. 1998. A modelling study of tidally induced equilibrium sand balances in the North Sea during the Holocene. *Continental Shelf Research* **18**, 151–200.

Gibbard, P.L., Head, M.J., Walker, M.J.C. & the Subcommission on Quaternary Stratigraphy. 2010. Formal ratification of the Quaternary System/Period and the Pleistocene Series/Epoch with a base at 2.58 Ma. *Journal of Quaternary Science* **25**, 96–102.

Gilbert, M.T.P., Jenkins, D.L., Gotherstrom, A., Naveran, N., Sanchez, J.J., Hofreiter, M., Thomsen, P.F., Binladen, J., Higham, T.F.G., Yohe, R.M., Parr, R., Cummings, L.S. & Willerslev, E. 2008. DNA from pre-Clovis human coprolites in Oregon, North America. *Science* **320**, 786–789.

Gómez, A., Hughes, R.N., Wright, P.J., Carvalho, G.R. & Lunt, D.H. 2007. Mitochondrial DNA phylogeography and mating compatibility reveal marked genetic structuring and speciation in the NE Atlantic bryozoans *Celleporella hyalina*. *Molecular Ecology* **16**, 2173–2188.

Goto, T.V., Tamate, H.B. & Hanzawa, N. 2011. Phylogenetic characterization of three morphs of mussels (Bivalvia, Mytilidae) inhabiting isolated marine environments in Palau Islands. *Zoological Science* **28**, 588–579.

Graham, M.H., Dayton, P.K. & Erlandson, J.M. 2003. Ice ages and ecological transitions on temperate coasts. *Trends in Ecology and Evolution* **18**, 33–40.

Graham, M.H., Kinlan, B.P. & Grosberg, R.K. 2010. Post-glacial redistribution and shifts in productivity of giant kelp forests. *Proceedings of the Royal Society* **B277**, 399–406.

Gray, J.S. 2001. Marine diversity: the paradigms in patterns of species richness examined. *Scientia Marina* **65**, 41–56.

Green, J.A.M., Green, C.L., Bigg, G.R., Rippeth, T.P., Scourse, J.D. & Uehara, K. 2009. Tidal mixing and the strength of the Meridional Overturning Circulation from the Last Glacial Maximum. *Geophysical Research Letters* **36**, L15603, doi:10.1029/2009GL039309.

Griffiths, S.D. & Peltier, W.R. 2008. Megatides in the Arctic Ocean under glacial conditions. *Geophysical Research Letters* **35**, L08605, doi:10.1029/2008GL033263.

Guilcher, A. 1988. *Coral Reef Geomorphology*. Chichester, UK: Wiley.

Gysels, E.S., Hellemans, B., Pampoulie, C. & Volckaert, F.A.M. 2004. Phylogeography of the common goby, *Pomatoschistus microps*, with particular emphasis on the colonization of the Mediterranean and the North Sea. *Molecular Ecology* **13**, 403–417.

Haapanimei, A.I., Scourse, J.D., Peck, V.L., Kennedy, D.P., Kennedy, H., Hemming, S.R., Furze, M.F.A., Pieńkowski-Furze, A.J., Walden, J., Wadsworth, E. & Hall, I.R. 2010. Source, timing, frequency and flux of ice-rafted detritus to the Northeast Atlantic margin, 30–12 ka: testing the Heinrich precursor hypothesis. *Boreas* **39**, 576–591.

Haflidason, H., Lien, R., Sejrup, H.P., Forsberg, C.F. & Bryn, P. 2005. The dating and morphometry of the Storegga Slide. *Marine and Petroleum Geology* **22**, 123–136.

Hall, P. & Davies, A.M. 2004. Modelling tidally induced sediment transport paths over the northwest European shelf: the influence of sea-level reduction. *Ocean Dynamics* **54**, 126–141.

Hallam, A. & Wignall, P.B. 1999. Mass extinctions and sea-level changes. *Earth-Science Reviews* **48**, 217–250.

Hanebuth, T., Stattegger, K. & Grootes, P.M. 2000. Rapid flooding of the Sunda Shelf: a Late-Glacial sea-level record. *Science* **288**, 1033–1035.

Harrison, S.P., Kohfeld, K.E., Roelandt, C. & Claquin, T. 2001. The role of dust in climate changes today, at the Last Glacial Maximum and in the future. *Earth-Science Reviews* **54**, 43–80.

Harrod, C., Griffiths, D., McCarthy, T.K. & Rosell, R. 2001. The Irish pollan, *Coregonus autumnalis*: options for its conservation. *Journal of Fish Biology* **59**, 339–355.

Hellberg, M.E. 1998. Sympatric sea shells along the sea's shore: the geography of speciation in the marine gastropod *Tegula*. *Evolution* **52**, 1311–1324.

Hewitt, G. 2000. The genetic legacy of the Quaternary ice ages. *Nature* **405**, 907–913.

Hiemstra, J., Evans, D.J.A., Scourse, J.D., Furze, M.F.A., McCarroll, D. & Rhodes, E. 2006. New evidence for a grounded Irish Sea glaciation of the Isles of Scilly, U.K. *Quaternary Science Reviews* **25**, 299–309.

Higham, T., Compton, T., Stringer, C., Jacobi, R., Shapiro, B., Trinkhaus, E., Chandler, B., Groning, F., Collins, C., Hillson, S., O'Higgins, P., FitzGerald, C. & Fagan, M. 2011. The earliest evidence for anatomically modern humans in northwestern Europe. *Nature* **479**, 521–524.

Hinton, A.C. 1995. Holocene tides of The Wash, U.K.: the influence of water-depth and coastline-shape changes on the record of sea-level change. *Marine Geology* **124**, 87–111.

Hoarau, G., Coyer, A.A., Veldsink, J.H., Stam, W.T. & Olsen, J.L. 2007. Glacial refugia and recolonization pathways in the brown seaweed *Fucus serratus*. *Molecular Ecology* **16**, 3606–3616.

Holt, J.T. & James, I.D. 2001. An s coordinate density evolving model of the Northwest European continental shelf: 1 model description and density structure. *Journal of Geophysical Research* **106**, 14015–14034.

Hopmans, E.C., Weijers, J.W.H., Schefuss, E., Herfort, L., Damste, J.S.S. & Schouten, S. 2004. A novel proxy for terrestrial organic matter in sediments based on branched and isoprenoid tetraether lipids. *Earth and Planetary Science Letters* **224**, 107–116.

Hubbard, A., Bradwell, T., Golledge, N., Hall, A., Patton, H., Sugden, D., Cooper, R. & Stoker, M. 2009. Dynamic cycles, ice streams and their impact on the extent, chronology and deglaciation of the British-Irish Ice Sheet. *Quaternary Science Reviews* **28**, 758–776.

Jablonski, D. 1986. Larval ecology and macroevolution in marine invertebrates. *Bulletin of Marine Science* **39**, 565–587.

Jahnke, R.A. 2010. Global synthesis. In *Carbon and Nutrient Fluxes in Continental Margins: A Global Synthesis*, K.K. Liu et al. (eds). Berlin: Springer-Verlag, 597–615.

Jansson, R. & Dynesius, M. 2002. The fate of clades in a world of recurrent climatic change: Milankovitch oscillations and evolution. *Annual Review of Ecology and Systematics* **33**, 741–777.

Jennings, S. & Kaiser, M.J. 1998. The effects of fishing on marine ecosystems. *Advances in Marine Biology* **34**, 201–352.

Johnston, P. & Lambeck, K. 1999. Postglacial rebound and sea level contributions to changes in the geoid and the Earth's rotational axis. *Geophysical Journal International* **136**, 537–558.

Jolly, M.T., Jollivet, D., Gentil, F., Thiébaut, E. & Viard, F. 2005. Sharp genetic break between Atlantic and English Channel populations of the polychaete *Pectinaria koreni*, along the north coast of France. *Heredity* **94**, 23–32.

Jolly, M.T., Viard, F., Gentil, F., Thiébaut, E. & Jollivet, D. 2006. Comparative phylogeography of two coastal polychaete tubeworms in the North East Atlantic supports shared history and vicariant events. *Molecular Ecology* **15**, 1841–1855.

Jonsson, B. & Jonsson, N. 2001. Polymorphism and speciation in Arctic charr. *Journal of Fish Biology* **58**, 605–638.

Kaiser, M.J., Attrill, M.J., Jennings, S., Thomas, D.N., Barnes, D.K.A., Brierley, A.S., Polunin, N.V.C., Raffaelli, D.G. & Williams, P.J. le B. 2005. *Marine Ecology: Processes, Systems and Impacts.* Oxford, UK: Oxford University Press.

Kennett, J.P. & Ingram, B.L. 1995. A 20,000-year record of ocean circulation and climate change from the Santa Barbara Basin. *Nature* **377**, 510–512.

Kettle, A.J., Morales-Muñiz, A., Roselló-Izquierdo, E., Heinrich, D. & Vøllestad, L.A. 2011. Refugia of marine fish in the northeast Atlantic during the Last Glacial Maximum: concordant assessment from archaeozoology and palaeotemperature reconstructions. *Climate of the Past* **7**, 181–201.

Kiden, P. 1995. Holocene relative sea-level change and crustal movement in the southwestern Netherlands. *Marine Geology* **124**, 21–41.

Kleypas, J.A. 1997. Modeled estimates of global reef habitat and carbonate production since the Last Glacial Maximum. *Paleoceanography* **12**, 533–545.

Kopp, R.E., Simons, F.J., Mitrovica, J.X., Maloof, A.C. & Oppenheimer, M. 2009. Probabilistic assessment of sea level during the last interglacial stage. *Nature* **462**, 863–867.

Kucera, M., Rosell-Melé, A., Schneider, R., Waelbroeck, C. & Weinelt, M. 2005. Multiproxy approach for the reconstruction of the glacial ocean surface (MARGO). *Quaternary Science Reviews* **24**, 813–819.

Lambeck, K. 1980. *The Earth's Variable Rotation: Geophysical Causes and Consequences.* Cambridge, UK: Cambridge University Press.

Lambeck, K. 1993a. Glacial rebound of the British Isles—I. Preliminary model results. *Geophysical Journal International* **115**, 941–959.

Lambeck, K. 1993b. Glacial rebound of the British Isles—II. A high-resolution, high-precision model. *Geophysical Journal International* **115**, 960–990.

Lambeck, K. 1995. Late Devensian and Holocene shorelines of the British Isles and North Sea from models of glacio-hydro-isostatic rebound. *Journal of the Geological Society* **152**, 437–448.

Lambeck, K. 1996a. Glaciation and sea-level change for Ireland and the Irish Sea since Late Devensian/Midlandian time. *Journal of the Geological Society* **153**, 853–872.

Lambeck, K. 1996b. Shoreline reconstructions for the Persian Gulf since the Last Glacial Maximum. *Earth and Planetary Science Letters* **142**, 43–57.

Lambeck, K. 1996c. Sea-level change and shore-line evolution in Aegean Greece since upper palaeolithic time. *Antiquity* **70**, 588–611.

Lambeck, K. 1997. Sea-level change along the French Atlantic and channel coasts since the time of the Last Glacial Maximum. *Palaeogeography, Palaeoclimatology, Palaeoecology* **129**, 1–22.

Lambeck, K., Antonioli, F., Anzidei, M., Ferranti, L., Leoni, G., Scicchitano, G. & Silenzi, S. 2011a. Sea level change along the Italian coast during the Holocene and projections for the future. *Quaternary International* **232**, 250–257.

Lambeck, K., Antonioli, F., Purcell, A. & Silenzi, S. 2004. Sea-level change along the Italian coast for the past 10,000 yr. *Quaternary Science Reviews* **23**, 1567–1598.

Lambeck, K. & Bard, E. 2000. Sea-level change along the French Mediterranean coast for the past 30,000 years. *Earth and Planetary Science Letters* **175**, 203–222.

Lambeck, K. & Chappell, J. 2001. Sea level change through the last glacial cycle. *Science* **292**, 679–686.

Lambeck, K., Johnston, P., Smither, C. & Nakada, M. 1996. Glacial rebound of the British Isles—III. Constraints on mantle viscosity. *Geophysical Journal International* **125**, 340–354.

Lambeck, K. & Purcell, A. 2005. Sea-level change in the Mediterranean Sea since the LGM: model predictions for tectonically stable areas. *Quaternary Science Reviews* **24**, 1969–1988.

Lambeck, K., Purcell, A. & Dutton, A. 2012. The anatomy of interglacial sea levels: the relationship between sea levels and ice volumes during the Last Interglacial. *Earth and Planetary Science Letters* **315**, 4–11.

Lambeck, K., Purcell, A., Flemming, N.C., Vita-Finzi, C., Alsharekh, A.M. & Bailey, G.N. 2011b. Sea level and shoreline reconstructions for the Red Sea: isostatic and tectonic considerations and implications for hominin migration out of Africa. *Quaternary Science Reviews* **30**, 3542–3574.

Lambeck, K., Yokoyama, Y. & Purcell, T. 2002. Into and out of the Last Glacial Maximum: sea-level change during oxygen isotope stages 3 and 2. *Quaternary Science Reviews* **21**, 343–360.

Langefors, A.H. 2005. Adaptive and neutral genetic variation and colonization history of Atlantic salmon, *Salmo salar. Environmental Biology of Fishes* **74**, 297–308.

Larmuseau, M.H.D., Van Houdt, J.K.J., Guelinckx, J., Hellemans, B. & Volckaert, F.A.M. 2009. Distributional and demographic consequences of Pleistocene climate fluctuations for a marine demersal fish in the north-eastern Atlantic. *Journal of Biogeography* **36**, 1138–1151.

Laruelle, G.G., Durr, H.H., Slomp, C.P. & Borges, A.V. 2010. Evaluation of sinks and sources of CO_2 in the global coastal ocean using a spatially-explicit typology of estuaries and continental shelves. *Geophysical Research Letters* **37**, L15607, doi:10.1029/2010GL043691.

Lericolais, G., Auffret, J.-P. & Bourillet, J.-F. 2003. The Quaternary Channel River: seismic stratigraphy of its palaeo-valleys and deeps. *Journal of Quaternary Science* **18**, 245–260.

Lerman, A., Guidry, M., Andersson, A.J. & Mackenzie, F.T. 2011. Coastal ocean Last Glacial Maximum to 2100 CO_2-carbonic acid-carbonate system: a modelling approach. *Aquatic Geochemistry* **17**, 749–773.

Leuliette, E.W. & Miller, L. 2009. Closing the sea level rise budget with altimetry, Argo, and GRACE. *Geophysical Research Letters* **36**, L04608, doi:10.1029/2008GL036010.

Lisiecki, L.E. & Raymo, M.E. 2005. A Pliocene-Pleistocene stack of 57 globally distributed benthic $\delta^{18}O$ records. *Paleoceanography* **20**, PA1003, doi:10.1029/2004PA001071.

Lister, A.M. 1995. Sea-levels and the evolution of island endemics: the dwarf red deer of Jersey. *Special Publication of the Geological Society* **96**, 151–172.

Longhurst, A., Sathyendranath, S., Platt, T. & Caverhill, C. 1995. An estimate of global primary production in the ocean from satellite radiometer data. *Journal of Plankton Research* **17**, 1245–1271.

Lowe, J.J., Rasmussen, S.O., Bjorck, S., Hoek, W.Z., Steffensen, J.P., Walker, M.J.C. & Yu, Z.C. 2008. Synchronisation of palaeoenvironmental events in the North Atlantic region during the Last Termination: a revised protocol recommended by the INTIMATE group. *Quaternary Science Reviews* **27**, 6–17.

Lundberg, J., Ford, D.C., Schwarcz, H.P., Dickin, A.P. & Li, W.-X. 1990. Dating sea level in caves. *Nature* **343**, 217–218.

Luttikhuizen, P.C., Drent, J. & Baker, A.J. 2003. Disjunct distribution of highly diverged mitochondrial lineage clade and population subdivision in a marine bivalve with pelagic larval dispersal. *Molecular Ecology* **12**, 2215–2229.

MacArthur, R.H. & Wilson, E.O. 1967. *The Theory of Island Biogeography*. Princeton, New Jersey: Princeton University Press.

Mach, M.E., Sbrocco, E.J., Hice, L.A., Duffy, T.A., Conover, D.O. & Barber, P.H. 2011. Regional differentiation and post-glacial expansion of the Atlantic silverside, *Menidia menidia*, an annual fish with high dispersal potential. *Marine Biology* **158**, 515–530.

Maggs, C.A., Castilho, R., Foltz, D., Henzler, C., Jolly, M.T., Kelly, J., Olsen, J., Perez, K.E., Stam, W., Väinölä, R., Viard, F. & Wares, J. 2008. Evaluating signatures of glacial refugia for North Atlantic benthic marine taxa. *Ecology* **89** Supplement, S108–S122.

Malay, M.C.D. & Paulay, G. 2010. Peripatric speciation drives diversification and distributional pattern of reef hermit crabs (Decapoda: Diogenidae: Calcinus). *Evolution* **64**, 634–662.

MARGO (Multiproxy Approach for the Reconstruction of the Glacial Ocean Surface) Project Members. 2009. Constraints on the magnitude and patterns of ocean cooling at the Last Glacial Maximum. *Nature Geoscience* **2**, 127–132.

Marko, P.B., Hoffman, J.M., Emme, S.A., McGovern, T.M., Keever, C.C. & Cox, L.N. 2010. The 'expansion-contraction' model of Pleistocene biogeography: rocky shores suffer a sea change? *Molecular Ecology* **19**, 146–169.

Marret, F. & Scourse, J.D. 2002. Control of modern dinoflagellate cyst distribution in the Irish and Celtic seas by seasonal stratification dynamics. *Marine Micropalaeontology* **47**, 101–116.

Marret, F., Scourse, J.D. & Austin, W.E.N. 2004. Holocene shelf sea seasonal stratification dynamics: a dinoflagellate cyst record from the Celtic Sea. *The Holocene* **14**, 689–696.

Marret, F., Scourse, J.D., Versteegh, G., Jansen, J.H.F. & Schneider, R. 2001. Integrated marine and terrestrial evidence for abrupt Congo River palaeodischarge fluctuations during the last deglaciation. *Journal of Quaternary Science* **16**, 761–766.

Massey, A.C., Paul, M.A., Gehrels, W.R. & Charman, D.J. 2006. Autocompaction in Holocene coastal back-barrier sediments from south Devon, southwest England, UK. *Marine Geology* **226**, 225–241.

Masters, P.M. 2006. Holocene sand beaches of southern California: ENSO forcing and coastal processes on millennial scales. *Palaeogeography, Palaeoclimatology, Palaeoecology* **232**, 73–95.

McBreen, F., Wilson, J.G., Mackie, A.S.Y. & Aonghusa, C.N. 2008. Seabed mapping in the southern Irish Sea: predicting benthic biological communities based on sediment characteristics. *Hydrobiologia* **606**, 93–103.

McCabe, A.M., Clark, P.U. & Clark, J. 2007. Radiocarbon constraints on the history of the western Irish ice sheet prior to the Last Glacial Maximum. *Geology* **35**, 147–150.

McCarroll, D., Stone, J., Ballantyne, C.K., Scourse, J.D., Fifield, L.K., Evans, D.J.A. & Hiemstra, J.F. 2010. Exposure-age constraints on the extent, timing and rate of retreat of the last Irish Sea ice stream. *Quaternary Science Reviews* **29**, 1844–1852.

McLean, R.F. & Woodroffe, C.D. 1994. Coral atolls. In *Coastal Evolution*, R.W.G. Carter & C.D. Woodroffe (eds). Cambridge, UK: Cambridge University Press, 267–302.

Meiri, S., Cooper, N. & Purvis, A. 2008. The island rule: made to be broken? *Proceedings of the Royal Society* **B275**, 141–148.

Mellars, P. 2006. Going east: new genetic and archaeological perspectives on the modern human colonization of Eurasia. *Science* **313**, 796–800.

Meyer, C.P., Geller, J.B. & Paulay, G. 2005. Fine scale endemism on coral reefs: archipelagic differentiation in turbinid gastropods. *Evolution* **59**, 113–125.

Miall, A.D. 1996. *The Geology of Stratigraphic Sequences*. Berlin: Springer-Verlag.

Milliman, J.D. & Farnsworth, K.L. 2011. *River Discharge to the Global Ocean*. Cambridge, UK: Cambridge University Press.

Milne, G.A., Gehrels, W.R., Hughes, C.W. & Tamisea, M.E. 2009. Identifying the causes of sea-level change. *Nature Geoscience* **2**, 471–478.

Milne, G.A., Mitrovica, J.X. & Schrag, D.P. 2002. Estimating past continental ice volume from sea-level data. *Quaternary Science Reviews* **21**, 361–176.

Milne, G.A., Shennan, I., Youngs, B.A.R., Waugh, A.I., Teferle, F.N., Bingley, R.M., Bassett, S.E., Cuthbert-Brown, C. & Bradley, S.L. 2006. Modelling the glacial isostatic adjustment of the U.K. region. *Philosophical Transactions of the Royal Society* **A364**, 931–948.

Mitrovica, J.X. & Milne, G.A. 2003. On post-glacial sea level: I. General theory. *Geophysical Journal International* **154**, 253–267.

Mitrovica, J.X., Tamisea, M.E., Davis, J.L. & Milne, G.A. 2001. Recent mass balance of polar ice sheets inferred from patterns of global sea-level change. *Nature* **409**, 1026–1029.

Mix, A.C. & Ruddiman, W.F. 1984. Oxygen isotope analysis and Pleistocene ice volumes. *Quaternary Research* **21**, 1–20.

Monin, A.S. 1970. Weather and climate oscillation. In *Scientific Exploration of the South Pacific*, W.S. Wooster (ed.). Washington, DC: National Academy of Sciences, 5–15.

Monnin, E., Indermühle, A., Dallenbach, A., Flückiger, J., Stauffer, B., Stocker, T.F., Raynaud, D. & Barnola, J.M. 2001. Atmospheric CO_2 concentrations over the last glacial termination. *Science* **291**, 112–114.

Montenegro, A., Eby, M., Kaplan, J.O., Meissner, K.J. & Weaver, A.J. 2006. Carbon storage on exposed continental shelves during the glacial-interglacial transition. *Geophysical Research Letters* **33**, L08703, doi:10.1029/2005GL025480.

Munk, W. 1997. Once again: once again—tidal friction. *Progress in Oceanography* **40**, 7–35.

Munk, W. & Wunsch, C. 1998. Abyssal recipes II: energetics of tidal and wind mixing. *Deep-Sea Research I* **45**, 1977–2010.

Murray, J.W. 2004. The Holocene palaeoceanographic history of Muck Deep, Hebridean shelf, Scotland: has there been a change of wave climate in the past 12,000 years? *Journal of Micropalaeontology* **23**, 153–161.

Neill, S.P. & Scourse, J.D. 2009. The formation of headland/island sandbanks. *Continental Shelf Research* **29**, 2167–2177.

Neill, S.P., Scourse, J.D., Bigg, G.R. & Uehara, K. 2009. Changes in wave climate over the northwest European shelf seas during the last 12,000 years. *Journal of Geophysical Research* **114**, C06015, doi:10.1029/2009JC005288.

Neill, S.P., Scourse, J.D. & Uehara, K. 2010. Evolution of bed shear stress distribution over the northwest European shelf seas during the last 12,000 years. *Ocean Dynamics* **60**, 1139–1156.

Neumann, A.C. & MacIntyre, I.G. 1985. Reef response to sea level rise: keep-up, catch-up or give-up. *Proceedings of the Fifth International Coral Reef Congress, Tahiti* **3**, 105–110.

Newell, N.D. 1961. Recent terraces of tropical limestone shores. *Zeitschrift für Geomorphologie Supplement Band* **3**, 87–106.

Newell, N.D. 1967. Revolutions in the history of life. *Geological Society of America, Special Paper* **89**, 63–91.

Nielsen, S.B., Gallagher, K., Leighton, C., Balling, N., Svenningsen, L., Jaconsen, B.H., Thomsen, E., Nielsen, O.B., Heilmann-Clausen, C., Egholm, D.L., Summerfield, M.A., Clausen, O.R., Piotrowski, J.A., Thorsen, M.R., Huuse, M., Abrahamsen, N., King, C. & Lykke-Andersen, H. 2009. The evolution of western Scandinavian topography: a review of Neogene uplift versus the ICE (isostasy-climate-erosion) hypothesis. *Journal of Geodynamics* **47**, 72–95.

Nikula, R., Strelkov, P. & Väinölä, R. 2007. Diversity and trans-Arctic invasion history of mitochondrial lineages in the North Atlantic *Macoma balthica* complex (Bivalvia; Tellinidae). *Evolution* **61**, 928–941.

Norris, R.D. & Hull, P.M. 2012. The temporal dimension of marine speciation. *Evolutionary Ecology* **26**, 393–415.

O'Cofaigh, C., Dunlop, P. & Benetti, S. 2012a. Marine geophysical evidence for Late Pleistocene ice sheet extent and recession off northwest Ireland. *Quaternary Science Reviews* **44**, 147–159.

O'Cofaigh, C. & Evans, D.J.A. 2001. Sedimentary evidence for deforming bed conditions associated with a grounded Irish Sea glacier, southern Ireland. *Journal of Quaternary Science* **16**, 435–454.

O'Cofaigh, C. & Evans, D.J.A. 2007. Radiocarbon constraints on the age of the maximum advance of the British-Irish Ice Stream in the Celtic Sea. *Quaternary Science Reviews* **26**, 1197–1203.

O'Cofaigh, C., Telfer, M.W., Bailey, R.M. & Evans, D.J.A. 2012b. Late Pleistocene chronostratigraphy and ice sheet limits, southern Ireland. *Quaternary Science Reviews* **44**, 160–179.

Olsen, J.L., Stam, W.T., Coyer, J.A., Reusch, T.B.H., Billingham, M., Boström, C., Calvert, E., Christie, H., Granger, S., La Lumière, R., Milchakova, N., Oudot-Le Secq, M.-P., Procaccini, G., Sanjabi, B., Serrão, E., Veldsink, J., Widdicombe, S. & Wyllie-Echverria, S. 2004. North Atlantic phylogeography and large-scale population differentiation of the seagrass *Zostera marina* L. *Molecular Ecology* **13**, 1923–1941.

Olsen, J.L., Zechman, F.W., Hoarau, G., Coyer, J.A., Stam, W.T., Valero, M. & Åberg, P. 2010. The phylogeographic architecture of the fucoid seaweed *Ascophyllum nodosum*: an intertidal 'marine tree' and survivor of more than one glacial-interglacial cycle. *Journal of Biogeography* **37**, 842–856.

Opdyke, B.N. & Walker, J.C.G. 1992. Return of the coral-reef hypothesis—basin to shelf partitioning of $CaCO_3$ and its effect on atmospheric CO_2. *Geology* **20**, 733–736.

Otto-Bliesner, B.L., Marshall, S.J., Overpeck, J.T., Miller, G.H. & Hu, A.X. 2006. Simulating Arctic climate warmth and icefield retreat in the last interglaciation. *Science* **311**, 1751–1753.

Palumbi, S.R. 1997. Molecular biogeography of the Pacific. *Coral Reefs* **16**, S47–S52.

Panova, M., Blakeslee, A.M.H., Miller, A.W., Mäkinen, T., Ruiz, G.M., Johannesson, K. & André, C. 2011. Glacial history of the North Atlantic marine snail, *Littorina saxatilis*, inferred from distribution of mitochondrial DNA lineages. *PLoS one* **6**, e17511, doi:10.1371/journal.pone.0017511.

Pantin, H.M. & Evans, C.D.R. 1984. The Quaternary history of the Central and Southwestern Celtic Sea. *Marine Geology* **57**, 259–293.

Paul, M.A. & Barrass, B.F. 1998. A geotechnical correction for post-depositional sediment compression: examples from the Forth Valley, Scotland. *Journal of Quaternary Science* **13**, 171–176.

Paulay, G. & Meyer, C. 2002. Diversification in the tropical Pacific: comparisons between marine and terrestrial systems and the importance of founder speciation. *Integrative and Comparative Biology* **42**, 922–934.

Pauly, D., Christensen, V., Guenette, S., Pitcher, T.J., Sumaila, U.R., Walters, C.J., Watson, R. & Zeller, D. 2002. Towards sustainability in world fisheries. *Nature* **418**, 689–695.

Payne, J.L. & Clapham, M.E. 2012. End-Permian mass extinction in the oceans: an ancient analog for the twenty-first century. *Annual Review of Earth and Planetary Sciences* **40**, 89–111.

Peacock, J.D. 1989. Marine molluscs and late Quaternary environmental studies with particular reference to the late-glacial period in northwest Europe: a review. *Quaternary Science Reviews* **8**, 179–192.

Peacock, J.D., Horne, D.J. & Whittaker, J.E. 2012. Late Devensian evolution of the marine offshore environment of western Scotland. *Proceedings of the Geologists' Association* **123**, 419–437.

Peltier, W.R. 1994. Ice-age palaeotopography. *Science* **265**, 195–201.

Peltier, W.R. 1998. Postglacial variations in the level of the sea: implications for climate dynamics and solid-earth geophysics. *Reviews of Geophysics* **36**, 603–689.

Peltier, W.R. 2004. Global glacial isostasy and the surface of the ice-age earth: the ICE-5G (VM2) model and GRACE. *Annual Review of Earth and Planetary Sciences* **32**, 111–149.

Peltier, W.R. & Fairbanks, R.G. 2006. Global glacial ice volume and Last Glacial Maximum duration from an extended Barbados sea level record. *Quaternary Science Reviews* **25**, 3322–3337.

Peltier, W.R., Shennan, I., Drummond, R. & Horton, B. 2002. On the postglacial isostatic adjustment of the British Isles and the shallow viscoelastic structure of the Earth. *Geophysical Journal International* **148**, 443–475.

Petraglia, M., Korisettar, R., Boivin, N., Clarkson, C., Ditchfield, P., Jones, S., Koshy, J., Lahr, M.M., Oppenheimer, C., Pyle, D., Roberts, R., Schwenninger, J.-L., Arnold, L. & White, K. 2007. Middle Palaeolithic assemblages from the Indian subcontinent before and after the Toba super-eruption. *Science* **317**, 114–116.

Pickering, M.D., Wells, N.C., Horsburgh, K.J. & Green, J.A.M. 2012. The impact of future sea-level rise on the European Shelf tides. *Continental Shelf Research* **35**, 1–15.

Potts, D.C. 1983. Evolutionary disequilibrium among Indo-Pacific corals. *Bulletin of Marine Science* **33**, 619–632.

Provan, J. & Bennett, K.D. 2008. Phylogeographic insights into cryptic glacial refugia. *Trends in Ecology and Evolution* **23**, 564–571.

Provan, J., Wattier, R.A. & Maggs, C.A. 2005. Phylogeographic analysis of the red seaweed *Palmaria palmata* reveals a Pleistocene marine glacial refugium in the English Channel. *Molecular Ecology* **14**, 793–803.

Rahmstorf, S. 2002. Ocean circulation and climate during the past 120,000 years. *Nature* **419**, 207–214.

Raymo, M.E., Hearty, P., de Conto, R., O'Leary, M., Dowsett, H.J., Robinson, M.M. & Mitrovica, J.X. 2009. PLIOMAX: Pliocene maximum sea level project. *PAGES News* **17**(2), 58–59.

Raymo, M.E., Mitrovica, J.X., O'Leary, M.J., DeConto, R.M. & Hearty, P.L. 2011. Departures from eustasy in Pliocene sea-level records. *Nature Geoscience* **4**, 328–332.

Reimer, P.J., Baillie, M.G.L., Bard, E., Bayliss, A., Beck, J.W., Blackwell, P.G., Ramsey, C.B., Buck, C.E., Burr, G.S., Edwards, R.L., Friedrich, M., Grootes, P.M., Guilderson, T.P., Hajdas, I., Heaton, T.J., Hogg, A.G., Hughen, K.A., Kaiser, K.F., Kromer, B., McCormac, F.G., Manning, S.W., Reimer, R.W., Richards, D.A., Southon, J.R., Talamo, S., Turney, C.S.M., vander Plicht, J. & Weyhenmeye, C.E. 2009. INTCAL09 and Marine09 radiocarbon age calibration curves, 0–50,000 years cal BP. *Radiocarbon* **51**, 1111–1150.

Richards, M.P. & Schulting, R.J. 2006. Touch not the fish: the Mesolithic-Neolithic change of diet and its significance. *Antiquity* **80**, 444–456.

Richards, M.P., Schulting, R.J. & Hedges, R.E.M. 2003. Sharp shift in diet at onset of Neolithic. *Nature* **425**, 366 only.

Rippeth, T., Scourse, J.D., Uehara, K. & McKeown, S. 2008. The impact of sea-level rise over the last deglacial transition on the strength of the continental shelf CO_2 pump. *Geophysical Research Letters* **35**, L24604, doi:10.1029/2008GL035880.

Roberts, N. 1998. *The Holocene: an Environmental History*. Oxford, UK: Blackwell, 2nd edition.

Rocha, L.A. 2003. Patterns of distribution and processes of speciation in Brazilian reef fishes. *Journal of Biogeography* **457**, 837–842.

Roman, J. & Palumbi, S.R. 2004. A global invader at home: population structure of the green crab, *Carcinus maenas*, in Europe. *Molecular Ecology* **13**, 2891–2898.

Rosier, S.H.R., Green, J.A.M. & Scourse, J.D. Submitted. On the role of ocean tides in West Antarctic Ice Sheet collapse. *Earth and Planetary Science Letters*.

Rowe, G., Harris, D.J. & Beebee, T.J.C. 2006. *Lusitania* revisited: a phylogeographic analysis of the natterjack toad *Bufo calamita* across its entire biogeographical range. *Molecular Phylogenetics and Evolution* **39**, 335–346.

Rowe, K.C., Hesk, E.J., Brown, W. & Paige, K.N. 2004. Surviving the ice: northern refugia and postglacial colonization. *Proceedings of the National Academy of Sciences USA* **101**, 10355–10359.

Ruddiman, W.F. 2001. *Earth's Climate: Past and Future*. New York: Freeman.

Schopf, T.J. 1980. *Paleoceanography*. Cambridge, MA: Harvard University Press.

Schumacher, J.D., Kinder, T.H., Pashinski, D.J. & Charnell, R.L. 1979. A structural front over the continental shelf of the eastern Bering Sea. *Journal of Physical Oceanography* **9**, 79–87.

Scott, D.B. & Greenberg, D.A. 1983. Relative sea level rise and tidal development in the Fundy tidal system. *Canadian Journal of Earth Sciences* **20**, 1554–1564.

Scott, G.A., Scourse, J.D. & Austin, W.E.N. 2003. The distribution of benthic Foraminifera in the Celtic Sea: the significance of seasonal stratification. *Journal of Foraminiferal Research* **33**, 32–61.

Scourse, J.D. 1991. Late Pleistocene stratigraphy and palaeobotany of the Isles of Scilly. *Philosophical Transactions of the Royal Society* **B334**, 405–448.

Scourse, J.D. 2010. Comment on: "A Last Glacial Maximum pollen record from Bodmin Moor showing a possible cryptic northern refugium in southwest England" by Kelly et al., 2010, *JQS* **25**, 296–308. *Journal of Quaternary Science* **25**, 826–827.

Scourse, J.D. & Austin, R.A. 1995. Palaeotidal modelling of continental shelves: marine implications of a land-bridge in the Straits of Dover during the Holocene and Middle Pleistocene. *Special Publication of the Geological Society* **96**, 75–88.

Scourse, J.D. & Austin, W.E.N. 2002. Quaternary shelf sea palaeoceanography: recent developments in Europe. *Marine Geology* **191**, 87–94.

Scourse, J.D., Austin, W.E.N., Bateman, R.M., Catt, J.A., Evans, C.D.R., Robinson, J.E. & Young, J.R. 1990. Sedimentology and micropalaeontology of glacimarine sediments from the central and southwestern Celtic Sea. *Special Publications of the Geological Society of London* **53**, 329–347.

Scourse, J.D., Austin, W.E.N., Long, B.T., Assinder, D.J. & Huws, D. 2002. Holocene evolution of seasonal stratification in the Celtic Sea: refined age model, mixing depths and foraminiferal stratigraphy. *Marine Geology* **191**, 119–145.

Scourse, J.D. & Furze, M.F.A. 2001. A critical review of the glaciomarine model for Irish Sea deglaciation: evidence from southern Britain, the Celtic shelf and adjacent continental slope. *Journal of Quaternary Science* **16**, 419–434.

Scourse, J.D., Haapaniemi, A.I., Colmenero-Hidalgo, E., Peck, V.L., Hall, I.R., Austin, W.E.N., Knutz, P.C. & Zahn, R. 2009a. Growth, dynamics and deglaciation of the last British-Irish Ice Sheet: the deep-sea ice-rafted detritus record. *Quaternary Science Reviews* **28**, 3066–3084.

Scourse, J.D., Hall, I.R., McCave, I.N., Young, J.R. & Sugdon, C. 2000. The origin of Heinrich layers: evidence from H2 for European precursor events. *Earth and Planetary Science Letters* **182**, 187–195.

Scourse, J.D., Kennedy, H., Scott, G.A. & Austin, W.E.N. 2004. Stable isotopic analyses of modern benthic foraminifera from seasonally stratified shelf seas; disequilibria and the 'seasonal effect'. *The Holocene* **14**, 758–769.

Scourse, J.D., Marret, F., Versteegh, G.J.M., Jansen, J.H.F., Schefuss, E. & van der Plicht, J. 2005. High resolution last deglaciation palynological record from the Congo Fan reveals significance of mangrove pollen and biomarkers as indicators of shelf transgression. *Quaternary Research* **64**, 57–69.

Scourse, J.D., Neill, S., Green, M. & Uehara, K. 2011. Megatides in the glacial North Atlantic and their role in rapid deglaciation. *XVIII INQUA Congress Bern, Switzerland Abstract 2071*.

Scourse, J.D., Uehara, K. & Wainwright, A. 2009b. Celtic Sea linear tidal sand ridges, the Irish Sea Ice Stream and the Fleuve Manche: palaeotidal modelling of a transitional passive margin depositional system. *Marine Geology* **259**, 102–111.

Sejrup, H.-P., Hjelstuen, B.O., Dahlgren, K.I.T., Haflidason, H., Kuijpers, A., Nygard, A., Praeg, D., Stoker, M.S. & Vorren, T.O. 2005. Pleistocene glacial history of the NW European continental margin. *Marine and Petroleum Geology* **22**, 1111–1129.

Sejrup, H.P., Nygård, A., Hall, A.M. & Haflidason, H. 2009. Middle and Late Weichselian (Devensian) glaciation history of south-western Norway, North Sea and eastern U.K. *Quaternary Science Reviews* **28**, 370–380.

Shackleton, N.J. 1967. Oxygen isotope analysis and Pleistocene temperatures re-assessed. *Nature* **215**, 15–17.

Shackleton, N.J. 1987. Oxygen isotopes, ice volume and sea-level. *Quaternary Science Reviews* **6**, 183–190.

Sharples, J., Tweddle, J.F., Green, J.A.M., Palmer, M.R., Kim, Y.N., Hickman, A.E., Holligan, P.M., Moore, C.M., Rippeth, T.P., Simpson, J.H. & Krivtsov, V. 2007. Spring-neap modulation of internal tide mixing and vertical nitrate fluxes at a shelf edge in summer. *Limnology and Oceanography* **52**, 1735–1747.

Shennan, I., Bradley, S., Milne, G., Brooks, A., Bassett, S. & Hamilton, S. 2006a. Relative sea-level changes, glacial isostatic modelling and ice-sheet reconstructions from the British Isles since the Last Glacial Maximum. *Journal of Quaternary Science* **21**, 585–599.

Shennan, I., Hamilton, S., Hillier, C., Hunter, A., Woodall, R., Bradley, S., Milne, G., Brooks, A. & Bassett, S. 2006b. Relative sea-level observations in western Scotland since the Last Glacial Maximum for testing models of glacial isostatic land movements and ice-sheet reconstructions. *Journal of Quaternary Science* **21**, 601–613.

Shennan, I. & Horton, B. 2002. Holocene land- and sea-level changes in Great Britain. *Journal of Quaternary Science* **17**, 511–526.

Shennan, I., Lambeck, K., Flather, R., Horton, B., McArthur, J., Innes, J., Lloyd, J., Rutherford, M. & Kingfield, R. 2000a. Modelling western North Sea palaeogeographies and tidal changes during Holocene. *Special Publication of the Geological Society* **166**, 299–319.

Shennan, I., Lambeck, K., Horton, B., Innes, J., Lloyd, J., McArthur, J., Purcell, T. & Rutherford, M. 2000b. Late Devensian and Holocene records of relative sea-level changes in northwest Scotland and their implications for glacio-hydro-isostatic modelling. *Quaternary Science Reviews* **19**, 1103–1135.

Shennan, I., Peltier, W.R., Drummond, R. & Horton, B. 2002. Global to local scale parameters determining relative sea-level changes and the postglacial isostatic adjustment of Great Britain. *Quaternary Science Reviews* **21**, 397–408.

Simpson, J.H. & Hunter, J.R. 1974. Fronts in the Irish Sea. *Nature* **250**, 404–406.

Simpson, J.H. & Sharples, J. 2012. *Introduction to the Physical and Biological Oceanography of Shelf Seas.* Cambridge, UK: Cambridge University Press.

Smith, D.E., Cullingford, R.A. & Firth, C.R. 2000. Patterns of isostatic land uplift during the Holocene: evidence from mainland Scotland. *The Holocene* **10**, 489–501.

Smith, F. 2001. Historical regulation of local species richness across a geographical region. *Ecology* **82**, 792–801.

Stewart, J.R. & Lister, A. 2001. Cryptic northern refugia and the origins of the modern biota. *Trends in Ecology and Evolution* **16**, 608–613.

Stirling, C.H., Esat, T.M., Lambeck, K. & McCulloch, M.T. 1998. Timing and duration of the Last Interglacial: evidence for a restricted interval of widespread coral reef growth. *Earth and Planetary Science Letters* **160**, 745–762.

Stoddart, D.R. 1971. Environment and history in Indian Ocean reef morphology. *Symposium of the Zoological Society of London* **28**, 3–38.

Stone, J., Lambeck, K., Fifield, L.K., Evans, J.M. & Cresswell, R.G. 1996. A lateglacial age for the main rock platform, western Scotland. *Geology* **24**, 707–710.

Stringer, C. 2011. *The Origin of Our Species.* London: Lane.

Svenning, J.C., Flojgaard, C., Marske, K.A., Nogues-Bravo, D. & Normand, S. 2011. Applications of species distribution modelling to paleobiology. *Quaternary Science Reviews* **30**, 2930–2947.

Sverdrup, H.U., Johnson, M.W. & Fleming, R.W. 1942. *The Oceans: Their Physics, Chemistry and General Biology.* Englewood Cliffs, New Jersey: Prentice-Hall.

Taylor, J., Dowdeswell, J.A., Kenyon, N.H., O'Cofaigh, C., 2002. Late Quaternary architecture of trough-mouth fans: debris flows and suspended sediments on the Norwegian margin. *Special Publications of the Geological Society of London* **203**, 55–71.

Thomas, H., Bozec, Y., Elkalay, K. & de Baar, H.J.W. 2004. Enhanced open ocean storage of CO_2 from shelf sea pumping. *Science* **304**, 1005–1008.

Thomas, M. & Sündermann, J. 1999. Tides and tidal torques of the world ocean since the Last Glacial Maximum. *Journal of Geophysical Research* **104**(C2), 3159–3183.

Toucanne, S., Zaragosi, S., Bourillet, J.F., Cremer, M., Eynaud, F., Van Vliet-Lanoë, B., Penaud, A., Fontanier, C., Turon, J.L., Cortijo, E. & Gibbard, P.L., 2009a. Timing of massive 'Fleuve Manche' discharges over the last 350 kyr: insights into the European ice-sheet oscillations and the European drainage network from MIS 10 to 2. *Quaternary Science Reviews* **28**, 1238–1256.

Toucanne, S., Zaragosi, S., Bourillet, J.F., Gibbard, P.L., Eynaud, F., Giraudeau, J, Turon, J.L., Cremer, M., Cortijo, E., Martinez, P. & Rossignol, L. 2009b. A 1.2 Ma record of glaciation and fluvial discharge from the West European Atlantic margin. *Quaternary Science Reviews* **28**, 2974–2981.

Toucanne, S., Zaragosi, S., Bourillet, J.F., Marieu, V., Cremer, M., Kageyama, M., Van Vliet-Lanoë, B., Eynaud, F., Turon, J.L. & Gibbard, P.L. 2010. The first estimation of Fleuve Manche palaeoriver discharge during the last deglaciation: evidence for Fennoscandian ice sheet meltwater flow in the English Channel ca. 20–18 ka ago. *Earth and Planetary Science Letters* **290**, 459–473.

Toucanne, S., Zaragosi, S., Bourillet, J.F., Naughton, F., Cremer, M., Eynaud, F. & Dennielou, B. 2008. Activity of the turbidite levees of the Celtic–Armorican margin (Bay of Biscay) during the last 30,000 years: imprints of the last European deglaciation and Heinrich events. *Marine Geology* **247**, 84–103.

Tsunogai, S., Watanabe, S. & Sago, T. 1999. Is there a 'continental shelf pump' for the absorption of atmospheric CO_2? *Tellus* Series B, **51**, 701–712.

Tzedakis, P.C., Wolff, E.W., Skinner, L.C., Brovkin, V., Hodell, D.A., McManus, J.F. & Raynaud, D. 2012. Can we predict the duration of an interglacial? *Climate of the Past Discussions* **8**, 1057–1088.

Uehara, K., Saito, Y. & Hori, K. 2002. Paleotidal regime in the Changjiang (Yangtze) Estuary, the East China Sea, and the Yellow Sea at 6 ka and 10 ka estimated from a numerical model. *Marine Geology* **183**, 179–192.

Uehara, K., Scourse, J.D., Horsburgh, K.J., Lambeck, K. & Purcell, A. 2006. Tidal evolution of the northwest European shelf seas from the Last Glacial Maximum to the present. *Journal of Geophysical Research* **111**, C09025, doi:10.1029/2006JC003531.

Väinölä, R. 2003. Repeated trans-Arctic invasions in littoral bivalves: molecular zoogeography of the *Macoma balthica* complex. *Marine Biology* **143**, 935–946.

van den Bergh, G.D., de Vos, J. & Sondaar, P.Y. 2001. The Late Quaternary palaeogeography of mammal evolution in the Indonesian archipelago. *Palaeogeography, Palaeoclimatology, Palaeoecology* **171**, 385–408.

van de Plassche, O. 1986. *Sea-Level Research: A Manual for the Collection and Evaluation of Data*. Norwich, UK: GeoBooks.

Van der Molen, J. & De Swart, H.E. 2001. Holocene tidal conditions and tide-induced sand transport in the southern North Sea. *Journal of Geophysical Research* **106**(C5), 9339–9362.

Van Landeghem, K., Uehara, K., Wheeler, A.J., Mitchell, N. & Scourse, J.D. 2009a. Post-glacial sediment dynamics in the Irish Sea and sediment wave morphology: data-model comparisons. *Continental Shelf Research* **29**, 1723–1736.

Van Landeghem, K.J.J., Wheeler, A.J. & Mitchell, N.C. 2009b. Seafloor evidence for palaeo-ice streaming and calving of the grounded Irish Sea Ice Stream: implications for the interpretation of its final deglaciation phase. *Boreas* **38**, 119–131.

Van Valen, L.M. 1973. Pattern and balance of nature. *Evolutionary Theory* **1**, 31–49.

Vartanyan, S.L., Garutt, V.E. & Sher, A.V. 1993. Holocene dwarf mammoths from Wrangel Island in the Siberian Arctic. *Nature* **362**, 337–340.

Versteegh, G., Schefuss, E., Dupont, L., Marret, F., Sinninghe Damsté, J.S. & Jansen, J.H.F. 2004. Taraxerol and *Rhizophora* pollen as proxies for tracking past mangrove ecosystems. *Geochimica and Cosmochimica Acta* **68**, 411–422.

Waelbroeck, C., Labeyrie, L., Michel, E., Duplessy, J.C., McManus, J.F., Lambeck, K. & Balbon, E. 2002. Sea-level and deep water temperature changes derived from benthic Foraminifera isotopic records. *Quaternary Science Reviews* **21**, 295–305.

Walker, M., Johnsen, S., Rasmussen, S.O., Popp, T., Steffensen, J.-P., Gibbard, P., Hoek, W., Lowe, J., Andrews, J., Björck, S., Cwynar, L., Hughen, K., Kershaw, P., Kromer, B., Litt, T., Lowe, D.J., Nakagawa, T., Newnham, R. & Schwander, J. 2009. Formal definition and dating of the GSSP (Global Stratotype Section and Point) for the base of the Holocene using the Greenland NGRIP ice core, and selected auxiliary records. *Journal of Quaternary Science* **24**, 3–17.

Walsh, J.J. 1988. *On the Nature of Continental Shelves*. London: Academic Press.

Ward, S., Neill, S. & Scourse, J.D. In press. Impacts of past and future sea-level rise on shelf sea sediment dynamics. Young Coastal Scientists and Engineers Conference 2013. *Proceedings of the Institution of Civil Engineers*.

Ware, J.R., Smith, S.V. & Reakakudla, M.L. 1992. Corals reefs—sources or sinks of atmospheric CO_2? *Coral Reefs* **11**, 127–130.

Wares, J.P. & Cunningham, C.W. 2001. Phylogeography and historical ecology of the North Atlantic intertidal. *Evolution* **55**, 2455–2469.

Wares, J.P. & Cunningham, C.W. 2005. Diversification before the most recent glaciation in *Balanus glandula*. *Biological Bulletin* **208**, 60–68.

Waters, M.R., Stafford, T.W., McDonald, H.G., Gustafson, C., Rasmussen, M., Cappellini, E., Olsen, J.V., Szklarczyk, D., Jensen, L.J., Gilbert, M.T.P. & Willerslev, E. 2011. Pre-Clovis mastodon hunting 13,800 years ago at the Manis Site, Washington. *Science* **334**, 351–353.

Weldeab, S., Lea, D.W., Schneider, R.R. & Andersen, N. 2007. 155,000 years of West African monsoon and ocean thermal evolution. *Science* **316**, 1303–1307.

Weston, E.M. & Lister, A.M. 2009. Insular dwarfism in hippos and a model for brain size reduction in *Homo floresiensis*. *Nature* **459**, 85–88.

Williams, S.T. & Benzie, J.A.H. 1998. Evidence of a biogeographic break between populations of a high dispersal starfish: congruent regions within the Indo-West Pacific defined by color morphs, mt DNA, and allozyme data. *Evolution* **52**, 87–99.

Williams, S.T. & Reid, D.G. 2004. Speciation and diversity on tropical rocky shores: a global phylogeny of snails of the genus *Echinolittorina*. *Evolution* **58**, 2227–2251.

Willis, J.K., Chambers, D.P. & Nerem, R.S. 2008. Assessing the globally averaged sea-level budget on seasonal to interannual timescales. *Journal of Geophysical Research* **113**, C06015, doi:0.1029/2007JC004517.

Wurster, C.M., Bird, M.I., Bull, I.D., Creed, F., Bryant, C., Dungait, J.A.J. & Paz, V. 2010. Forest contraction in north equatorial Southeast Asian during the Last Glacial Period. *Proceedings of the National Academy of Sciences USA* **107**, 15508–15511.

Xue, L., Zhang, L.J., Cai, W.J. & Jiang, L.Q. 2011. Air-sea CO_2 fluxes in the southern Yellow Sea: an examination of the continental shelf pump hypothesis. *Continental Shelf Research* **31**, 1904–1914.

Yokoyama, Y., De Deckker, P., Lambeck, K., Johnston, P. & Fifield, L.K. 2001a. Sea-level at the Last Glacial Maximum: evidence from northwestern Australia to constrain ice volumes for oxygen isotope stage 2. *Palaeogeography, Palaeoeclimatology, Palaeoecology* **165**, 281–297.

Yokoyama, Y., Lambeck, K., De Deckker, P., Johnston, P. & Fifield, L.K. 2000. Timing of the Last Glacial Maximum from observed sea-level minima. *Nature* **406**, 713–716.

Yokoyama, Y., Purcell, A., Lambeck, K. & Johnston, P. 2001b. Shore-line reconstruction around Australia during the Last Glacial Maximum and Late Glacial Stage. *Quaternary International* **83–5**, 9–18.

Oceanography and Marine Biology: An Annual Review, 2013, **51**, 71-192
© Roger N. Hughes, David Hughes, and I. Philip Smith, Editors
Taylor & Francis

OCEANS AND MARINE RESOURCES
IN A CHANGING CLIMATE*

JENNIFER HOWARD[1,*,†], ELEANORA BABIJ[2,†], ROGER GRIFFIS[1,†], BRIAN HELMUTH[3,†],
AMBER HIMES-CORNELL[4,†], PAUL NIEMIER[1,†], MICHAEL ORBACH[5,†],
LAURA PETES[1,†], STEWART ALLEN[6], GUILLERMO AUAD[7], CAROL AUER[1],
RUSSELL BEARD[8], MARY BOATMAN[7], NICHOLAS BOND[9], TIMOTHY BOYER[1],
DAVID BROWN[10], PATRICIA CLAY[1], KATHERINE CRANE[11], SCOTT CROSS[12],
MICHAEL DALTON[4], JORDAN DIAMOND[13], ROBERT DIAZ[14], QUAY DORTCH[15],
EMMETT DUFFY[14], DEBORAH FAUQUIER[1], WILLIAM FISHER[16], MICHAEL
GRAHAM[17], BENJAMIN HALPERN[18], LARA HANSEN[19], BRYAN HAYUM[2], SAMUEL
HERRICK[20], ANNE HOLLOWED[4], DAVID HUTCHINS[21], ELIZABETH JEWETT[1], DI
JIN[22], NANCY KNOWLTON[23], DAWN KOTOWICZ[6], TROND KRISTIANSEN[1,24], PETER
LITTLE[4], CARY LOPEZ[1], PHILIP LORING[25], RICK LUMPKIN[26], AMBER MACE[27],
KATHRYN MENGERINK[13], J. RU MORRISON[28], JASON MURRAY[29], KARMA NORMAN[30],
JAMES O'DONNELL[31], JAMES OVERLAND[4], ROST PARSONS[8], NEAL PETTIGREW[6],
LISA PFEIFFER[4], EMILY PIDGEON[32], MARK PLUMMER[30], JEFFREY POLOVINA[33],
JOSIE QUINTRELL[34], TERESSA ROWLES[1], JEFFREY RUNGE[34], MICHAEL RUST[30],
ERIC SANFORD[35], UWE SEND[36], MERRILL SINGER[37], CAMERON SPEIR[38],
DIANE STANITSKI[11], CAROL THORNBER[39], CARA WILSON[40] & YAN XUE[41]

*[1]National Oceanic and Atmospheric Administration,
1315 East West Hwy., Silver Spring, MD 20910, USA*
E-mail: Jennifer.Howard@noaa.gov (corresponding author)
[2]United States Fish and Wildlife Service, 4401 N. Fairfax Dr., Arlington, VA 22203, USA
[3]Northeastern University, Marine Science Center, Nahant, MA 01908, USA
*[4]National Oceanic and Atmospheric Administration,
7600 Sand Point Way NE, Seattle, WA 98115, USA*
[5]Duke University, 135 Duke Marine Lab Road, Beaufort, NC 28516, USA
*[6]National Oceanic and Atmospheric Administration,
1601 Kapiolani Blvd., Honolulu, HI 96814, USA*
[7]Bureau of Ocean Energy Management, 1849 C Street NW, Washington, DC 20240, USA
*[8]National Oceanic and Atmospheric Administration, National Coastal Data Development Center,
Stennis Space Center, MS 39529, USA*
[9]University of Washington, 7600 Sand Point Way NE, Seattle, WA 98115, USA
[10]National Oceanic and Atmospheric Administration, 819 Taylor St., Fort Worth, TX 76102, USA
*[11]National Oceanic and Atmospheric Administration,
1100 Wayne Ave., Silver Spring, MD 20910, USA*

---

* This document is one of the foundational Technical Input Reports for the 2013 National Climate Assessment conducted by the US Global Change Research Program.
† Primary authors.

[12]National Oceanic and Atmospheric Administration,
219 Ft. Johnson Rd., Charleston, SC 29412, USA
[13]Environmental Law Institute, 2000 L Street NW, Washington, DC 20036, USA
[14]Virginia Institute of Marine Science, 1375 Greate Rd., Gloucester Point, VA 23062, USA
[15]National Oceanic and Atmospheric Administration,
1305 East West Hwy., Silver Spring, MD 20910, USA
[16]Environmental Protection Agency, National Health and Environment
Effects Research Laboratory, Gulf Breeze, FL 32561, USA
[17]Moss Landing Marine Laboratories, 8272 Moss Landing Rd., Moss Landing, CA 95039, USA
[18]National Center for Ecological Analysis and Synthesis,
735 State Street, Santa Barbara, CA 93101, USA
[19]EcoAdapt, P.O. Box 11195, Bainbridge Island, WA 98110, USA
[20]National Oceanic and Atmospheric Administration,
8604 La Jolla Shores Dr., La Jolla, CA 92037, USA
[21]University of Southern California, Department of Biological Sciences,
Los Angeles, CA 90089, USA
[22]Woods Hole Oceanographic Institution, Mailstop 41, Woods Hole, MA 02543, USA
[23]Smithsonian Institute, P.O. Box 37012, MRC 163, Washington, DC 20013, USA
[24]Institute of Marine Research, P.O. Box 1870 Nordnes, 5817 Bergen, Norway
[25]University of Alaska, Fairbanks, P.O. Box 755910, Fairbanks, AK 99775, USA
[26]National Oceanic and Atmospheric Administration,
4301 Rickenbacker Cswy., Miami, FL 33149, USA
[27]University of California, Davis, One Shields Avenue, Davis, CA 95616, USA
[28]North East Regional Association of Coastal and Ocean Observing Systems,
570 Ocean Blvd., Rye, NH 03870, USA
[29]University of South Carolina, 1705 College Street, Columbia, SC 29208, USA
[30]National Oceanic and Atmospheric Administration,
2725 Montlake Blvd. East, Seattle, WA 98112, USA
[31]University of Connecticut, 1080 Shennecossett Rd., Groton, CT 06340, USA
[32]Conservation International, 2011 Crystal Dr., Arlington, VA 22202, USA
[33]National Oceanic and Atmospheric Administration, 2570 Dole St., Honolulu, HI 96822, USA
[34]National Federation of Regional Associations for Coastal Observing,
205 Oakledge Rd., Harpswell, ME 04079, USA
[34]University of Maine, 350 Commercial St., Portland, ME 04101, USA
[35]University of California, Davis, Bodega Marine Laboratory, Bodega Bay, CA 94923, USA
[36]Scripps Institution of Oceanography, 9500 Gilman Dr., La Jolla, CA 92093, USA
[37]University of Connecticut, 354 Mansfield Rd., Storrs, CT 06269, USA
[38]National Oceanic and Atmospheric Administration,
110 Shaffer Rd., Santa Cruz, CA 95060, USA
[39]University of Rhode Island, 120 Flagg Rd., Kingston, RI 02881, USA
[40]National Oceanic and Atmospheric Administration,
1352 Lighthouse Ave., Pacific Grove, CA 93950, USA
[41]National Oceanic and Atmospheric Administration, 5200 Auth Rd., Suitland, MD 20746, USA

The United States is an ocean nation—our past, present, and future are inextricably connected to and dependent on oceans and marine resources. Marine ecosystems provide many important services, including jobs, food, transportation routes, recreational opportunities, health benefits, climate regulation, and cultural heritage that affect people, communities, and economies across the United States and internationally every day. There is a wealth of information documenting the strong linkages between the planet's climate and ocean systems, as well as how changes in the climate system can produce changes in the physical, chemical, and biological characteristics of ocean ecosystems on a variety of spatial and temporal scales. There is relatively little information on how these climate-driven changes in ocean ecosystems may have an impact on ocean services and uses, although it is predicted that ocean-dependent users, communities, and economies will likely become increasingly vulnerable in a changing climate. Based on our current understanding and future projections of the planet's ocean systems, it is likely that marine ecosystems will continue to be affected by anthropogenic-driven climate change into the future. This review describes how these impacts are set in motion through a suite of changes in ocean physical, chemical, and biological components and processes in US waters and the significant implications of these changes for ocean users and the communities and economies that depend on healthy oceans. US international partnerships, management challenges, opportunities, and knowledge gaps are also discussed. Effectively preparing for and responding to climate-driven changes in the ocean will require both limiting future change through reductions of greenhouse gases and adapting to the changes that we can no longer avoid.

Introduction

Marine ecosystems under US sovereignty, including areas under state and federal jurisdiction, generally extend from the shore to 203 nautical miles seawards. The area under federal jurisdiction spans 3.4 million square nautical miles of ocean—an area referred to as the US Exclusive Economic Zone (EEZ). The United States has the largest EEZ in the world, an area 1.7 times the land area of the continental United States and encompassing 11 different large marine ecosystems (LMEs). In 2004, the ocean-dependent economy generated $138 billion, or 1.2% of US gross domestic product (GDP) (Kildow et al. 2009). US ocean areas are also inherently connected with the nation's vital coastal counties, which make up only 18% of the US land area but are home to 36% of the US population and account for over 40% of the national economic output (Kildow et al. 2009).

These valuable marine ecosystems and the services they provide are increasingly at risk from a variety of pressures, including climate change and the related issue of ocean acidification (Osgood 2008, Doney et al. 2012). These pressures are affecting ocean physical, chemical, and biological systems, as well as human uses of these systems. Increasing levels of atmospheric carbon dioxide (CO_2) are one of the most serious problems because the effects are globally pervasive and irreversible on ecological timescales (National Research Council [NRC] 2011). The present CO_2 concentration is the highest on record in at least the last 800,000 years (based on ice core data) (Lüthi et al. 2008).

The two primary direct consequences of increased atmospheric CO_2 in marine ecosystems are increased ocean temperatures (Bindoff et al. 2007) and higher acidity (Doney et al. 2009). Increased acidity of the ocean is directly related to oceanic absorption of CO_2 from the atmosphere. The CO_2 reacts with seawater to change the chemical environment of the oceans fundamentally (Feely et al. 2010). However, oceanic absorption of CO_2 is dependent on water temperature and pH, and these mechanisms are likely to become less efficient as waters warm and pH decreases under future climate scenarios. Increasing temperatures produce a variety of ocean changes, including rising sea level, increased ocean stratification, decreased extent of sea ice, and altered patterns of ocean circulation, storms, precipitation, and freshwater input (Doney et al. 2012). In addition, non-climatic stressors resulting from a variety of human activities, including pollution, fishing impacts, and overuse, can interact with and exacerbate impacts of climate change. These impacts are expected to increase in the future with continued changes in the global climate system and increases in human population levels. These and other changes in ocean physical and chemical conditions (such as changes in oxygen concentrations and nutrient availability) are having an impact on a variety of ocean biological features (e.g., primary production, phenology, species distribution, species interactions and community composition), which in turn can have an impact on vital ocean services across the nation (Figure 1).

Projections of future change indicate that it is likely that marine ecosystems under US jurisdiction, and US activities and partnerships internationally, will continue to be affected by anthropogenic-driven climate change, and that those impacts will likely vary by magnitude and by region. Uncertainty regarding the rate and magnitude of climate-related changes in biophysical aspects of marine resources limits our current ability to assess socioeconomic impacts. However, the evidence that many climate-related changes are already occurring lends urgency to the need to prepare for additional change in the future.

Climate-driven physical and chemical changes in marine ecosystems

Covering more than two-thirds of Earth's surface, the oceans are a central component of the global climate system. The oceans help control the timing and regional distribution of Earth's response to climate change, primarily through their absorption of CO_2 and heat, resulting in observed changes to the physical and chemical properties of the oceans (Figure 2). The International Panel on Climate Change (IPCC) assessment released in 2007 projected that due to the persistence of greenhouse gases in the atmosphere, it is highly likely that the oceans will continue to warm, and these impacts will continue to be felt for centuries (IPCC 2007). Expected physical and chemical consequences for the oceans include increased sea-surface temperature, accelerated melting of Arctic ice, increased ocean acidity, sea-level rise, increased stratification of the water column, and changes in ocean circulation, climate patterns, and salinity.

Research shows that between 1961 and 2003, the average temperature of the upper 700 m of water increased by 0.2°C (Bindoff et al. 2007), arctic sea ice volume has shrunk by 75% over the last decade (Laxton et al. 2013), incidences of hypoxia (a condition by which an aquatic environment has oxygen concentrations that are insufficient to sustain most animal life) within US estuaries increased 30-fold since 1960 (Diaz & Rosenberg 2008), and ocean acidity increased by 30% over the past century (Feely et al. 2004). Warming of ocean waters increases the available energy used to create short-lived storms, and while the frequency of hurricanes and typhoons may not change, it is likely that a warming ocean will result in increased storm intensity (Knutson et al. 2010). As the ocean surface warms, stratification increases, resulting in warmer water remaining at the surface instead of mixing with cooler water below. Warm water is not as efficient at absorbing CO_2, and while this might have a slowing effect on ocean acidification, consequences include potential reductions in uptake of atmospheric CO_2 by the oceans. While there is some variability in salinity levels globally, recent analyses of water density and atmospheric data collected from 1970 to 2005 suggest that there are overriding changes, including acceleration, in the global hydrological cycle (Helm et al. 2010).

The physical and chemical changes taking place in the global oceans set the stage for subsequent effects on marine organisms, US communities and economies dependent on marine services, US governance and interactions with neighbouring countries, and potential adaptation strategies. A great deal of uncertainty remains with respect to how rapidly the physical and chemical attributes of the oceans will change in the future, as well as the magnitude of specific impacts and what, if any, potential feedbacks will occur. Key to advancing knowledge and projections of change will be sustained, long-term monitoring of the physical and chemical components of the world's oceans. Research and modelling are also needed to improve understanding and projection of oceanic properties at various temporal and spatial scales. While historical data can provide important insight into patterns, trends, and trajectories of change, additional information collected over the coming years will be vital for predicting future responses to climate change.

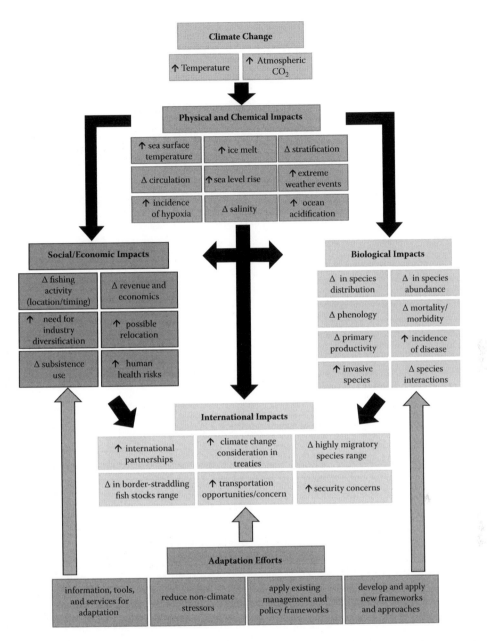

Figure 1 (See also colour figure in the insert) Impacts of climate change on marine ecosystems. Conceptualized diagram depicting how climate change impacts, such as increased temperature and ocean acidification, set in motion a suite of changes in the oceans' physical, chemical, and biological components and processes, resulting in significant implications for ocean users and the communities and economies that depend on healthy oceans. Changes in the physical and chemical makeup of the ocean is having an impact on marine organisms and ecosystems, local economies such as fishing and tourism, as well as becoming an ever-increasing topic of consideration in international conservation communities. Adaptation efforts are beginning to be put in place to try to counter climate change impacts as well as help build ecosystem and socioeconomic resiliency. Black arrows represent impacts driven by climate change either directly or indirectly. Gray arrows represent countering effects of various adaptation efforts. ↑ indicates where climate change is predicted to increase the incidence or magnitude of that attribute, and Δ indicates attributes where the impact of climate change on that attribute is variable.

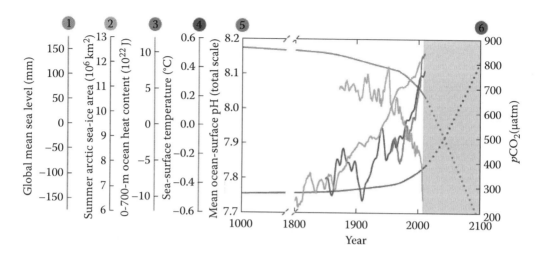

Figure 2 (See also colour figure in the insert) Observed changes to the physical and chemical properties of the oceans. Changes in (1) global mean sea level (teal line; Jevrejeva et al. 2008); (2) summer Arctic sea-ice areas (yellow line; Walsh & Chapman 2001); (3) 0- to 700-m ocean heat content (orange line; Levitus et al. 2009); (4) sea-surface temperature (brown line; Rayner et al. 2006); (5) mean ocean surface pH (blue line, National Research Council 2010); and (6) pCO_2 (red line; Petit et al. 1999). Light purple-shaded region denotes projected changes in pH and pCO_2 consistent with the IPCC's twenty-first-century A2 emissions scenario with rapid population growth. (Modified from Doney et al. 2012.)

Ocean temperature and heat trapping

According to the IPCC (2007), "most of the observed increase in globally averaged temperatures since the mid-twentieth century is very likely due to the observed increase in anthropogenic greenhouse gas concentrations," and current modelling systems predict that it is highly likely that the rate of warming will accelerate over the next few decades. Air temperature and sea-surface temperature are strongly correlated, and just as increasing concentrations of atmospheric CO_2 and other greenhouse gases lead to increased air temperatures, ocean temperatures are also expected to increase. Indeed, estimates show that from 1955 to 2008, approximately 84% of the added atmospheric heat was absorbed by the oceans (Levitus et al. 2009). The oceans are now experiencing some of the highest temperatures on record (Bindoff et al. 2007).

Increases in sea-surface temperature are likely to accelerate over the next few decades, along with a predicted increase in global mean surface air temperature between 1.1°C (under low-CO_2 emission scenario B1; IPCC 2007) and 6.4°C (under high-CO_2 emission scenario A1FI; IPCC 2007) by the end of the twenty-first century (Meehl et al. 2007). Even now, historical data show that the global ocean surface temperatures for January 2010 were the second warmest on record since 1880 (when records began), and in 2009, sea-surface temperatures from June to August reached 0.58°C above the average global temperature for the twentieth century (Hoegh-Guldberg & Bruno 2010).

It is important to note that changes in ocean temperature are not uniform, and while the average temperature has increased globally, localized temperature decreases have also been observed. The differences in temperature observed locally are mostly due to the atmospheric and oceanic processes that govern both the gains and the losses in sea-surface temperatures (Deser et al. 2010). Dominant atmospheric factors driving ocean temperature include wind speed, air temperature, cloudiness, and humidity; dominant oceanic factors include heat transport by currents and vertical mixing (Figure 3) (Deser et al. 2010). Oceanic and atmospheric processes play off one another, and heat exchange between the oceanic and atmospheric environments is a driving force of atmospheric circulation.

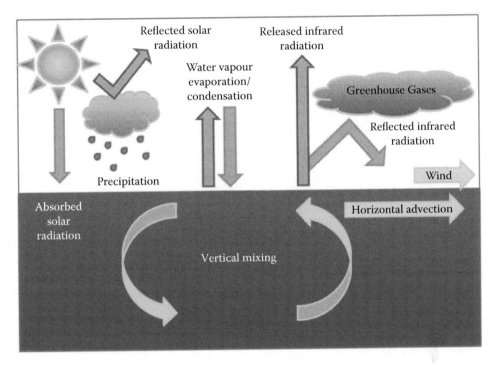

Figure 3 Exchange of heat between the ocean and the atmosphere. Idealized depiction of how solar energy is absorbed by the earth's surface, causing the earth to warm and to emit infrared radiation. The greenhouse gases then trap the infrared radiation, thus warming the atmosphere. Wind-induced turbulence, horizontal advection, and vertical mixing of the ocean waters aid in heat exchange by constantly mixing the surface layer of the ocean.

Evaporation rates are expected to increase as a result of climate change, leading to an increase in atmospheric water vapour. Water vapour is itself a potent greenhouse gas (much more so than carbon dioxide); atmospheric water vapour reflects heat emitted from Earth in the form of infrared radiation back down to the surface, which increases the amount of heat that is retained in the atmosphere (Randall et al. 2007). The potential exists for a cycle of increased water vapour stimulating increased warming and thus continued production of water vapour, leading to a 'runaway greenhouse effect' (Trenberth et al. 2009). In addition, water vapour stimulates cloud formations as it cools and condenses in the upper atmosphere, and clouds can cause both warming and cooling conditions. They warm by reflecting Earth's heat back down as mentioned, and they cool by reflecting solar radiation back into space (Stephens et al. 2008). It is uncertain which cloud effect, warming or cooling, will have the greatest impact, but the preponderance of evidence indicates that it will be an increase in warming (e.g., Dessler 2011), which is likely to cause upper ocean temperatures to increase at a faster rate than observed during the last few decades (Friedlingstein et al. 2001).

The interplay between oceanic and atmospheric heat exchanges has important implications for global weather and climate patterns. For example, sea-surface temperatures in the Pacific Ocean influence winter precipitation in the south-western United States (Wagner et al. 2010). Observations indicate that changes in sea-surface temperature in the tropical North Atlantic affect the precipitation in North and Central America, leading to increased incidence of drought throughout the United States and Mexico (Kushnir et al. 2010). Tropical storms form over warm ocean waters, which supply the energy for hurricanes and typhoons to grow and move (Trenberth et al. 2007). Conversely, the oceans have tremendous thermal inertia, which can slow and dampen the rate of climate change (Schewe et al. 2010). The significantly larger heat capacity of the deep ocean is particularly important when looking at timescales of decades to millennia, which are relevant for long-term climate change. Ocean currents and mixing by gyres, winds, and waves can transport and redistribute heat to deeper ocean layers. Heat energy can reside in this deep reservoir for centuries, further stabilizing Earth's climate and slowing the effects of climate change (Hansen et al. 2005).

Perhaps one of the most destructive effects of ocean warming comes in the form of sea-level rise. Most of the sea-level rise observed over the past several centuries can be accounted for by two major variables: the amount of water that is being released by landlocked glaciers and ice sheets and the thermal expansion or contraction of the oceans. Complete loss of arctic glaciers and small ice caps has the potential to raise future sea levels by about 0.2–0.7 m (IPCC 2007). Loss of Greenland and Antarctic ice sheets could increase that estimate by several metres; however, there is greater uncertainty surrounding those estimates. In the case of thermal expansion, given an equal mass, the total volume of ocean waters decreases when ocean temperatures drop and expands when temperatures increase. Thermal expansion was responsible for approximately 30% of the rise in sea level for the period 1961–2003 (Cazenave & Llovel 2010). Thermal expansion associated with increased temperature due to CO_2 emissions predicts an irreversible global average sea-level rise of at least 0.4–1.0 m if twenty-first century CO_2 concentrations exceed 600 parts per million by volume (ppmv) (current level is 385 ppmv) and as much as 1.9 m for a peak CO_2 concentration exceeding about 1000 ppmv (IPCC 2007), and these estimates are in addition to sea-level rise associated with ice melt described previously.

As global temperatures continue to increase, sea-level rise will become more extensive, with impacts on US coastal communities and ecosystems.

Loss of Arctic ice

As a result of warming temperatures, Arctic sea ice has been decreasing in extent throughout the second half of the twentieth century and the early twenty-first century (Figure 4A) (Maslanik et al. 2007, Nghiem et al. 2007, Comiso & Nishio 2008, Deser & Teng 2008, Alekseev et al. 2009, Arctic Marine Shipping Assessment [AMSA] 2009). Arctic sea ice naturally extends surface coverage each winter and recedes each summer, but the rate of overall loss since 1978, when satellite records began, has accelerated (IPCC 2007). The summer of 2007 saw a record low, when sea ice extent shrank to approximately 3 million km^2, approximately 1 million km^2 less than the previous minima of 2005 and 2006. Every year since then, September ice extent has been lower than in years prior to 2007, with 2011 extent being second lowest compared with 2007 (Perovich et al. 2011, Stroeve et al. 2011). Overall, the observed sea ice extent for the years 1979 to 2006 indicates an annual loss of approximately 3.7% per decade (Comiso et al. 2008) and an overall 42% decrease in thick, multi-year ice over a similar time period (Figure 4B) (Giles et al. 2008, Kwok & Untersteiner 2011). Over the last decade, arctic sea ice volume has shrunk by 75% (Laxton et al. 2013). The relative impacts of the loss of thicker, older ice versus the younger, thinner ice are currently not well understood.

In recent decades, the magnitude of warming in Arctic near-surface air temperatures has been approximately twice as large as the global average, a phenomenon known as 'Arctic amplification' (Serreze & Francis 2006, G. Miller et al. 2010), an effect largely due to diminishing sea ice (Screen & Simmonds 2010). Sea ice is efficient at reflecting solar energy; the much darker ocean surface is more effective at absorbing solar energy. Therefore, as reflective sea ice is lost, further warming of the ocean occurs, leading to additional ice melt (Serreze & Barry 2011). Decreases in snow and ice cover may ultimately decrease global albedo (the reflection of solar radiation), therefore amplifying warming, particularly at high latitudes (Bony et al. 2006, Soden & Held 2006, Randall et al. 2007). The roles of reductions in snow and changes in atmospheric and oceanic circulation, cloud cover and water vapour in Arctic amplification are still uncertain. Nevertheless, evidence suggests that strong positive ice-temperature feedbacks have emerged in the Arctic, increasing the chances of further rapid warming and sea-ice loss.

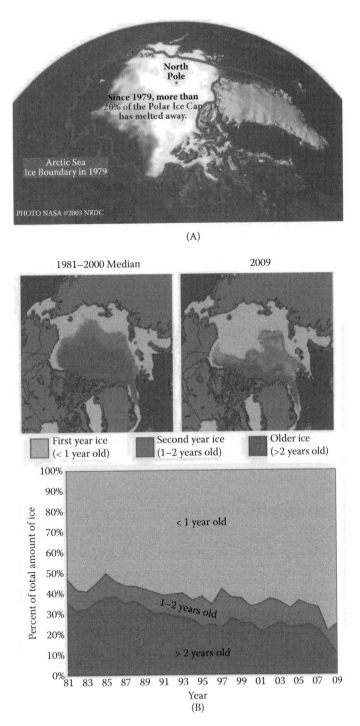

Figure 4 (See also colour figure in the insert) Changes in Arctic sea-ice extent and thickness. (A) Change in Arctic sea-ice boundaries from 1979 to 2007. (Photo NASA © NRDC.) (B) At top left, median image of sea-ice thickness at the end of each February cycle for the 1981–2000 time frame. On top right, the sea ice thickness for 2009. Bottom, percentage of ice that is new (<1 year old), 1–2 years old, and more than 2 years old per year from 1981 to 2009. (From the National Snow and Ice Data Center, courtesy J. Maslanik and C. Fowler, University of Colorado.)

These changes have consequences for polar ecosystems as well as human communities and activities. Decreasing Arctic sea ice also affects continental shelves downstream of the Arctic. The north-eastern US continental shelf has experienced a freshening over the past 30 years as a result of the advective supply of fresh water from the Arctic or Labrador Seas (Greene et al. 2008). Perhaps the most devastating consequence of melting polar ice will also be on sea-level rise, as noted in the preceding section.

The Arctic is likely moving toward a new state as more and more old ice is removed through increased melt rates and new ice becomes the dominant feature. The IPCC (2007) projections suggest that the Arctic may be virtually ice free in the summer by the late twenty-first century. However, others predict that reductions may happen even more rapidly, as previous projections utilized coupled air-sea-ice climate models that tend to overestimate ice thickness, and hence an ice-free Arctic in summer could occur as early as 2030 (Stroeve et al. 2008).

Salinity

Salinity refers to the salt content of the oceans. Contributors to salinity include precipitation and freshwater input (resulting in decreased salinity) and evaporation and terrestrial mineral deposits (resulting in increased salinity). Ocean salinity changes are an indirect but potentially sensitive indicator for detecting changes in precipitation, evaporation, river run-off, and ice melt. Thus, salinity may function as a proxy for identifying climate-driven changes in Earth's hydrological cycle (Helm et al. 2010).

The globally averaged surface salinity change to date is small, whereas basin averages are more noteworthy: increased salinity for the Atlantic, freshening in the Pacific, and a near-neutral result for the Indian Ocean (Durack & Wijffels 2010). Cravatte et al. (2009) uncovered large changes to ocean properties in the tropical western Pacific, indicating that a freshening and density decrease had occurred in the 1955–2003 period. Despite overall increased salinity in the Atlantic, studies conducted in the Scotian Shelf and Gulf of Maine waters showed a decrease in salinity (Greene et al. 2008). One of the main reasons for this is the melting of Arctic sea ice and resulting freshwater input into the Labrador Current. Another influence on decreased salinity in the Gulf of Maine is increased precipitation. Many climate models predict that there will be an increase in all forms of precipitation for the Gulf of Maine area (Wake et al. 2006), increasing river flow and terrestrial run-off. This input of fresh water will be especially large in the spring, when there is typically high river flow as pack ice melts and travels through the watershed (Wake et al. 2006). These inputs of fresh surface water inhibit the vertical mixing of the upper portion of the water column within the Gulf of Maine, and the resulting increased stratification prevents nutrients from being brought into the surface layer. Reductions in available nutrients affect phytoplankton productivity, the base of the food web, and these effects can cascade up to larger organisms like fish and marine mammals.

Changes to the global hydrological cycle are anticipated as a consequence of anthropogenic climate change (Held & Soden 2006, IPCC 2007). Even a small variation in the hydrological cycle, and resulting impacts on ocean salinity, can have a dramatic effect on the marine environment (Holland et al. 2001). Changes in salinity affect density parameters, as salt water is denser than fresh water, which in turn affects stratification. Ocean circulation, driven by temperature and salinity, may also be impacted. As cold, high-salinity water sinks at the poles, warmer, lower-salinity water is pulled from the equator to replace it (Broecker 1991), and if salinity declines due to climate-related increases in precipitation and glacial melt, it is possible that ocean circulation will begin to slow (Bindoff et al. 2007). Slowing of the oceans' circulation may result in dramatic changes to Earth's climate.

Stratification

Stratification, or layering, of ocean water is a naturally occurring phenomenon that is important to water column structure, circulation, and marine productivity. Water density is strongly influenced by temperature and salinity, with less-dense, warmer surface waters floating on top of denser, colder waters. Warming of the upper layers of the oceans enhances the density difference between the surface mixed layer and the deeper waters beneath. All else being equal, as waters warm, this increased density difference strengthens the vertical stratification of the oceans and suppresses vertical mixing across the density gradient. The result is contrasting effects on nutrient and light availability for phytoplankton growth. On the one hand, stratification reduces nutrient influx from deep, nutrient-rich waters into the surface mixed layer, thus limiting the availability of nutrients for phytoplankton growth (Behrenfeld et al. 2006, Huisman et al. 2006). On the other hand, stratification keeps phytoplankton in the surface mixed layer, resulting in improved light conditions (Huisman et al. 1999, Berger et al. 2007). Increased stratification and its effect on phytoplankton productivity can have negative consequences for coastal ecosystems. For example, warming of the surface waters in Southern California between 1951 and 1993 resulted in an 80% decrease in phytoplankton and zooplankton biomass, as well as decreased coastal upwelling and nutrient availability (Palacios et al. 2004).

Stratification varies regionally due to mixing, upwelling, and the uneven distribution of sea-surface temperatures. Many waters in the tropics and subtropics, such as those surrounding the Pacific Islands, are permanently stratified. Nutrient concentrations in the surface mixed layer of these waters are strongly depleted and are characterized by extremely low primary production. Climate-ocean models predict that, by the year 2050, due to ocean warming, the ocean area covered by permanent stratification will have expanded by 4.0% and 9.4% in the Northern and Southern Hemispheres, respectively (Sarmiento et al. 2004), thereby reducing overall ocean productivity (Behrenfeld et al. 2006). However, these predictions are surpassed by recent observations, which indicate a much faster expansion of the ocean's least-productive waters between 1998 and 2006 (Polovina et al. 2008).

In the temperate zone and at high latitudes, waters are not permanently stratified, and deep mixing during winter or spring provides nutrients into the surface layer. In these regions, phytoplankton growth is often light-limited in winter due to short day lengths as well as deep vertical mixing. Warming temperatures lead to earlier onset of stratification in spring, which retains phytoplankton in the well-lit surface layer, where they can take advantage of nutrients that have not yet been depleted, thereby favouring their growth. This leads to an earlier spring bloom and the potential for a substantially longer growing season in the temperate zone (Winder & Schindler 2004, Peeters et al. 2007).

Stratification can also be affected through changes in precipitation. Climate-induced precipitation increases will introduce fresh water into nearshore environments, either directly through rainfall or indirectly through increased run-off from the terrestrial environment. This freshening will tend to increase stratification, resulting in decreased diffusion of oxygen and nutrients. Decreasing oxygen concentrations will negatively affect biological communities and secondary production and disrupt biogeochemical cycles (Rabalais et al. 2009). Thus, climate-related changes in stratification are likely to cause subsequent impacts on inshore and nearshore US ecosystems.

Changes in precipitation and extreme weather events

Climate change influences winds and precipitation may be moderate but vary regionally (Trenberth 2011). Ironically, because of increased evaporation (as mentioned), some areas will experience intense surface drying (throughout the subtropics) such that there is an increased risk of flooding when intense storms occur (Trenberth 2011). In other areas, increased humidity may lead to more intense storms and hence more extreme precipitation and wind events. It is likely that warming temperatures will lead to some shifts in precipitation from snow to rain, resulting in decreased snow-pack and earlier snow melts (Trenberth 2011). Taken together, climate change is likely to influence Earth's hydrological cycle and the character of weather events.

Winds

Winds can have a major influence on marine ecosystems. In fact, wind changes may be more important than temperature changes for inducing the stratification effects mentioned; however, wind predictions have higher uncertainty. Global wind stress has increased since 1999 and shows variable impacts regionally (Ecosystem Assessment Program [EAP] 2009). These changes in wind stress may be linked to climate regimes (such as the North Atlantic Oscillation [NAO], Pacific Decadal Oscillation [PDO], and the El Niño Southern Oscillation [ENSO]), as well as a northwards shift in the location of the jet stream (Archer & Caldeira 2008). If wind patterns or intensities change, currents and their effects on coastal waters might change, with potential consequences for oxygen concentrations. For example, off the Oregon and Washington coasts, wind-driven shifts in the California Current region occurred in 2002 and subsequent years, altering upwelling dynamics and resulting in extensive hypoxia along the inner continental shelf (Grantham et al. 2004, Chan et al. 2008). Climate-related changes to wind and ocean currents are also likely to interact with non-climatic stressors (e.g., the addition of excessive nutrients from terrestrial run-off), further increasing the incidence and severity of hypoxic (reduced oxygen) and anoxic (lack of oxygen) events.

Precipitation

Precipitation is an important part of Earth's climate system, linking the global water and energy cycles through condensational heating of the atmosphere and providing a link between the hydrological cycle and radiative processes, such as cloud feedback. Precipitation forms as water vapour is condensed, usually in rising air that expands and cools. Climate change impacts on the frequency of precipitation events will likely be moderate; however, precipitation events are predicted to become more intense (Trenberth et al. 2003). Changes in precipitation can adversely affect the supply of water for humans, agriculture, and ecosystems. Modelling predictions of future precipitation changes currently have high uncertainty over certain regions and particularly at small regional scales. However, advances in modelling show that a robust characteristic of anthropogenic climate change is the shifting of precipitation patterns and storm tracks (Held & Soden 2006, Seager et al. 2007). Also, as the climate warms, the percentage of precipitation that falls as snow is likely to decrease. Implications of decreased snowfall include earlier snowmelt, diminished snowpack, and thus a temporal change in stratification due to the timing of freshwater input into coastal and marine environments (Trenberth et al. 2003).

Storms

There is growing concern about the potential for climate-related impacts on storm events. The record-breaking hurricane season in the North Atlantic in 2005 had the largest number of named storms (28), the largest number of hurricanes (15), the most intense hurricane in the Gulf of Mexico (Rita), and the most damaging hurricane on record (Katrina), with Katrina the deadliest in the United States since 1928 (IPCC 2007). Uncertainty remains regarding the effects of climate change on storm intensity. Multidecadal variability and the lack of high-quality tropical cyclone records prior to routine satellite observations (which began in approximately 1970) complicate the detection of long-term trends in extreme weather events (Blunden et al. 2011). However, using downscaling based on the ensemble mean of 18 global climate change projections, Bender et al. (2010) predicted a decrease in the total number of tropical cyclones, but of the storms that do manifest they predicted a doubling of category 4 and 5 storms by the end of the twenty-first century. The largest increase was projected to occur in the Western Atlantic, north of 20° N. However, not all of the individual models showed such an increase (Bender et al. 2010).

In addition to changes in severe storm activity, climate models indicate that the environment will be less favourable for development of weaker storms (Vecchi & Soden 2007); thus, there may be fewer storms overall (Knutson et al. 2010). This has the potential to affect water supplies because the weaker storms, which now are much more frequent than what is predicted under future climate scenarios, are an important source of rainfall for coastal regions. However, rainfall may increase by as much as 20% in an individual storm (Knutson et al. 2010), leading to larger, short-term pulses of fresh water.

During a tropical storm, strong surface winds not only take heat out of the ocean but also mix the ocean at depths from tens to hundreds of metres, cooling the surface and creating a cold wake (Walker et al. 2005, Trenberth & Fasullo 2007, Trenberth et al. 2007). Hence, tropical storm activity depends not only on sea-surface temperatures but also on subsurface temperatures; this is especially important for setting the stage for the next storm and thus an entire active season. Better understanding of these feedbacks should improve ocean model predictions of hurricanes. However, surface fluxes are highly uncertain for strong winds, and the roles of ocean spray and ocean mixing are also uncertain. An example of this theory that is already occurring naturally is in the case of the warm and deep 'Loop Current' in the Gulf of Mexico, which appeared to play a key role in the intensification of Hurricanes Ivan (e.g., Walker et al. 2005), Katrina (Figure 5), and Rita (Trenberth et al. 2007).

Storm events in the Aleutian Islands are already having an impact on marine fisheries, commerce, and coastal ecosystems. In November 2011, the worst storm in 40 years hit the Alaskan coast, with hurricane-strength winds and large storm surges (Ulbrich et al. 2008). The increasing frequency of extreme events may be amplified by the loss of sea ice, which typically buffers the effect of winter storm surge on Alaskan coastlines. In addition, added heat to the lower atmosphere can generate storms in newly ice-free areas (Inoue & Hori 2011), as well as have impacts over larger regions (Overland et al. 2011).

Though modelling projections for climate change and storms predict fewer storms overall, a small increase of the most destructive storms will have a large impact because (1) damage increases exponentially with wind speed; (2) coastal population and infrastructure will increase over the coming century, resulting in greater vulnerability to strong storms; (3) storm surge has historically been responsible for the greatest loss of life; and (4) expected rise in sea level will likely produce greater storm surges.

Ocean circulation

Wind, heat, and freshwater fluxes at the ocean surface, together with tidal and other energy sources, are responsible for global ocean circulation, mixing, and the formation of a broad range of water masses. On the global scale, individual shallow and deep-ocean currents form an interconnected pattern known as the thermohaline circulation (THC), sometimes referred to as the 'global conveyor belt' (Broecker 1991). The path of the THC is generally described as originating in the Northern Atlantic Ocean, where cold, dense water sinks and travels across ocean basins to the tropics, where it warms and upwells to the surface. The warmer, less-dense, tropical waters are then drawn to polar latitudes to replace the cold, sinking water. However, in practice, the THC interacts with other currents, and mixing occurs among the waters travelling along these intersecting pathways, creating a more complex situation (Broecker 1991). The THC plays an important role in transferring heat from the oceans to the atmosphere, causing the water to become colder and denser, thus renewing the cycle (Lumpkin & Speer 2007, Kanzow et al. 2010). The THC is responsible for much of the

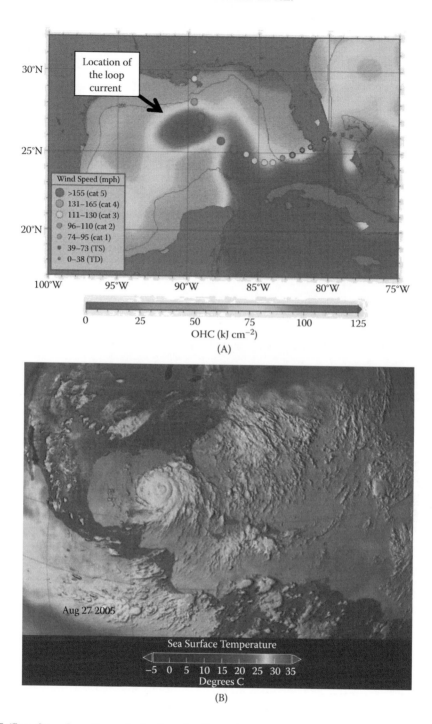

Figure 5 (See also colour figure in the insert) Ocean heat content and sea-surface temperature during Hurricane Katrina. (A) The ocean heat content in the prestorm environment for Hurricane Katrina. The storm intensity and position are indicated by the circles. (Modified from Mainelli et al. 2008.) (B) This image was created from AMSR-E data on NASA's Aqua satellite and shows a 3-day average of actual sea-surface temperatures (SSTs) for the Caribbean Sea and Atlantic Ocean, from 25–27 August 2005. Yellow, orange or red areas are 82°F or above (80°F is needed to maintain hurricanes). The position of Katrina is from 27 August. (From NASA Goddard's Scientific Visualization Studio.)

distribution of heat energy from the equatorial oceans to the polar regions and has a large influence on Earth's climate.

It is possible that the THC may weaken by the end of the twenty-first century as a result of climate change (Bindoff et al. 2007). Melting of polar ice will reduce the salinity and thus density of polar waters, which could weaken the rate at which this water sinks, possibly impairing circulation (Hu et al. 2011). However, there is large uncertainty related to the quantity of freshwater input necessary for such slowing of the THC (Kuhlbrodt et al. 2009). The pattern of temperature change as a consequence of THC slowing is complex, with predicted cooling over the North Atlantic (Wood et al. 2003, 2006, Vellinga & Wu 2004) and warming occurring further east (Stouffer et al. 2006). Another possible impact of weakening circulation may be increases in sea-level rise as a result of circulation-induced pressure gradients (Sturges 1974), especially on the East Coast of the United States, where models predicted a possible sea-level rise of 20 cm above and beyond what is predicted as a result of thermal expansion and glacial melt (Yin et al. 2010).

The following is a brief overview of possible impacts of climate change on two ocean current systems, the California Current and the Gulf Stream. Additional currents important to the United States (such as the Alaska Current) may also experience changes in behaviour as a result of climate change but are not discussed here.

California Current

The California Current spans strong physical gradients in circulation and water column structure (King et al. 2011). There is considerable evidence for physical changes, including significant warming and concomitant changes in water column stratification across the California Current over the past century (Palacios et al. 2004). For example, declines in oxygen concentrations due to changes in upwelling have already been measured along the coasts of California, Oregon and Washington (Grantham et al. 2004, Bograd et al. 2008, Chan et al. 2008). Despite this finding, climate-related effects on the California Current are highly uncertain. Some observations indicate that upwelling events are becoming less frequent, stronger, and longer in duration in accordance with climate change, with consequences for primary productivity along the US West Coast (King et al. 2011). Others suggest only moderate oceanographic changes, such as mild surface warming accompanied by relatively minor increases in upwelling-favourable winds in northern portions of the California Current, proposing that natural variability may overshadow climate change signals (Wang et al. 2010). Determining if the observed changes in upwelling reflect anthropogenic changes in variability of the California Current climate system and determining the possible impacts of climate-driven delayed onset of upwelling on future regional oceanic production in the California Current (Di Lorenzo et al. 2005) have emerged as two critical questions for further research.

Gulf Stream

The waters of the Gulf Stream travel from the Gulf of Mexico, pass through the Straits of Florida, travel off the US coast, and move through the Mid-Atlantic Bight until reconnecting to the coast at the Grand Banks off Newfoundland. Long-term observations of the Gulf Stream exist at only a few locations along this path. As stated, numerical climate models have projected slowing of the THC in the Atlantic over the next several decades. The Gulf Stream is one component of the THC, so this slowing could result in a potential reduction in Gulf Stream strength. However, the Gulf Stream is also influenced by the subtropical wind-driven gyre, and the possibility cannot yet be excluded that a slowdown of the THC might be accompanied by an increase in the wind-driven gyre, leading to no net impact on the Gulf Stream. If the Gulf Stream were to decrease in strength, there might be a rise in sea level along the US eastern seaboard as a result of relaxation of the pressure gradient associated with the Gulf Stream (Kelly et al. 1999). A decrease in the Gulf Stream volume and heat transport might also have significant impacts on precipitation patterns, hurricane tracks, and surface air temperatures throughout the Northern Hemisphere. The long-term variations of the Gulf Stream are not yet completely understood, warranting future observation and study.

Climate regimes

Regime shifts refer to a broad set of often-basinwide changes in the characteristic behaviour of the physical environment, such as persistent increases in ocean and atmospheric temperatures or shorter-term perturbations related to climatic events. These shifts have impacts on climate conditions in oceans and on land. Climate change and the subsequent warming of the atmosphere and oceans will likely have an impact on regional climate regimes, such as those encapsulated by the NAO, PDO, and ENSO. However, the nature of these oscillations over varying timescales creates challenges for climate prediction models, resulting in a moderate level of uncertainty with regard to climate change impacts. Climate model outputs are routinely used in a variety of climate change impact studies and assessment products, including the reports of the IPCC. There are, however, a number of global climate model limitations that must be carefully considered when interpreting and assessing uncertainty in regional climate-marine ecosystem impact studies (Stock et al. 2011). Establishing a clear set of considerations for assessing uncertainty in regional climate change impact assessments is vital for enabling an informed response to potential climate risks to Earth's oceans on multidecadal-to-century timescales.

The following is a brief overview of possible impacts of climate change on three climate regimes, the NAO, the PDO, and the ENSO. Additional regimes important to the United States (such as the North Pacific Gyre Oscillation) may also experience changes in behaviour as a result of climate change but are not discussed here.

North Atlantic Oscillation

The climate of the Atlantic region exhibits considerable variability on a wide range of timescales. A substantial portion of that variability is associated with the NAO (Stenseth et al. 2002), a measure of the fluctuation in the sea-level pressure difference between the Icelandic low (Stykkishólmur, Iceland) and the Azores high (Lisbon, Portugal). The NAO fluctuates between one of two states, positive or negative (Czaja et al. 2003, Arzel et al. 2011). When the NAO index is high (positive NAO state), there is an increase in precipitation over the eastern seaboard of the United States, and ocean temperatures are relatively warm. Conversely, when the NAO index is low (negative NAO state), decreased storminess, drier conditions, and relatively cool ocean temperatures occur in the eastern United States (Hurrell & Deser 2010). While the NAO index varies from year to year, it also exhibits a tendency to remain in one phase for intervals lasting more than a decade. An unusually long period of positive phase from 1970 to 2000 led to the theory that climate change was affecting the behaviour of the NAO (Goodkin et al. 2008). Uncertainty remains regarding whether climate change is influencing the timing of the NAO phases, but evidence indicates that the strength of its variability is increasing, as phases are becoming more strongly positive and negative (Rind et al. 2005).

Pacific Decadal Oscillation

The two principal modes of sea-surface temperature anomalies acting in the North Pacific Ocean are the longer-term PDO events, which occur on interdecadal timescales of roughly 25–30 years, and the shorter-term ENSO (discussed in El Niño/Southern Oscillation) events, which occur at a timescale of approximately 3–7 years (Mantua et al. 1997). The PDO, like the NAO, fluctuates between positive and negative phases. During the positive phase of the PDO, stronger-than-normal downwelling winds along the northern California and Alaska regions generate a local convergence of water masses at the coast that result in higher sea-surface height, warmer sea-surface temperatures, and anomalous polewards circulation near the coast (King et al. 2011). The opposite occurs during the negative phase. Overland & Wang (2007) suggested that under the A1B (middle range) CO_2 emission scenario, the change in winter sea-surface temperatures due to anthropogenic influences will surpass the natural variability in most of the North Pacific in less than 50 years.

El Niño/Southern Oscillation

The ENSO events exhibit a seesaw pattern of reversing surface air pressure and surface ocean temperature between the eastern and western tropical Pacific. The ENSO cycle is comprised of two interacting climate regimes, El Niño, which produces warmer sea-surface temperatures in the eastern Pacific, and La Niña, which produces cooler temperatures in that region. A balance of amplifying and damping feedbacks in ocean-atmosphere exchange controls year-to-year ENSO variability, and one or more of the physical processes responsible for determining the characteristics and global impacts of ENSO will likely be modified by climate change. For example, climate-driven changes in the upwelling cycle, with delayed and weak seasonal upwelling, have been documented in the central California Current region during El Niño years (Bograd et al. 2009). Higher-than-normal uplifting of the thermocline and strengthened polewards circulation in the proximity of the Southern California coast occurred during the most recent (2010–2011) La Niña event (Nam et al. 2011). These events influence ocean productivity, as well as atmospheric circulation, and consequently regional rainfall rates and extreme weather events at interannual timescales. However, despite considerable progress in our understanding of the potential impacts of climate change on many of the processes that contribute to ENSO variability, it is not yet possible to say whether ENSO activity will be enhanced or damped or if the frequency of events will be altered under climate change (Collins et al. 2010).

Carbon dioxide absorption by the oceans

The annual accumulation of atmospheric CO_2 has been increasing, and in 2010, global atmospheric CO_2 concentration was 39% above the concentration at the start of the Industrial Revolution in 1750 (Global Carbon Project 2011). The IPCC indicated that an 85% reduction in current CO_2 deposited into the atmosphere by the year 2050 would prevent exceeding a global mean temperature increase of 2.0°C, a temperature increase that could result in a tipping point toward extreme global changes (IPCC 2007). CO_2 reductions can be achieved both through reducing anthropogenic sources of CO_2 and supporting CO_2 uptake and storage through the conservation of natural ecosystems with high carbon sequestration rates and capacity (Canadell & Raupach 2008).

Ocean water holds approximately 50 times more CO_2 than the atmosphere, with the majority being held in the deeper, colder waters, but the ability of oceans to absorb CO_2 is not infinite. Currently, the oceans absorb more CO_2 than they release; however, CO_2 is less soluble in warmer waters, so as the sea-surface temperature increases, a decrease in oceanic uptake of CO_2 from the air is likely (IPCC 2007). CO_2 uptake also depends on the pH of ocean water, which decreases as more CO_2 is absorbed. Thus, as the pH decreases, so does the buffering capacity, or the ability of the ocean to continue to take up CO_2 (Revelle & Seuss 1957, Broecker & Takahashi 1966, Stumm & Morgan 1970, Skirrow & Whitfield 1975, Andersen & Malahoff 1977). However, future projections of oceanic sink strength and regional distribution are highly uncertain (Doney et al. 2009).

'Blue carbon' is a term used to describe the biological carbon sequestered and stored by marine and coastal organisms, with a significant fraction stored in sediments, coastal seagrasses, tidal marshes, and mangroves. Earth's marine and coastal ecosystems (mangrove forests, seagrass beds, salt marshes) are proportionately (on a per acre basis) more effective in sequestering carbon dioxide than are terrestrial ecosystems (Donato et al. 2011, McLeod et al. 2011, Fourqurean et al. 2012). When degraded or disturbed, these systems release carbon dioxide into the atmosphere or ocean. Currently, blue carbon sinks lose between about 0.7% to 7% of their area annually (McLeod et al. 2011). Global percentage cover of mangrove areas declined by 20% between 1980 and 2005

(Giri et al. 2010, Spalding et al. 2010). Carbon continues to be lost from the most organic soils in coastal areas. For instance, analysis of the agricultural soils of Sacramento's San Joaquin Delta, a diked and drained former tidal wetland, documents emissions of CO_2 at rates of 5 to 7.5 million tCO_2 each year, or 1% of California's total greenhouse gas emissions. Each year, an inch of organic soil evaporates from these drained wetlands, leading to a total release of approximately 1 billion tCO_2 over the past 150 years (Crooks et al. 2009, Deverel & Leighton 2010, Hatala et al. 2012). Similar emissions are likely occurring from other converted wetlands along the East and Gulf Coasts of the United States. Conservation and improved management of these systems bring climate change mitigation benefits in addition to increasing their significant adaptation value (Crooks et al. 2011, McLeod et al. 2011). Developing a better understanding of blue carbon science and ecosystem management issues has implications for future climate adaptation and mitigation strategies, as well as coastal habitat conservation.

Ocean acidification

Ocean chemistry is changing in response to the absorption of CO_2 from the atmosphere at a rate unprecedented over perhaps more than 50 million years (Hönisch et al. 2012). Ocean acidification refers to the decrease in the pH of Earth's oceans associated with uptake of atmospheric CO_2 and the subsequent chemical reactions. Ocean acidification is related to, but distinct from, climate change. However, it is important to include in this review because both climate change and ocean acidification share a common cause: increasing carbon dioxide concentrations in the atmosphere. Moreover, ocean acidification and climate change frequently have interactive effects on organisms and ecosystems. Particularly, ocean acidification has recently caught considerable attention as it affects the health of calcifying organisms and the rates of biogeochemical processes in the oceans (see reviews by Riebesell et al. 2007 and Doney et al. 2009). While its impacts on marine ecosystems have the potential to be severe, discussions regarding mitigation of CO_2 emissions and adaptation to climate change often ignore ocean acidification.

Surface waters of the ocean are estimated to have absorbed approximately 25% of all anthropogenically generated carbon since 1800 (Sabine et al. 2004). In the past, it was believed that the oceans would offset the effects of greenhouse gas emissions, but it is now understood that while absorption of CO_2 by oceans slows the atmospheric greenhouse effect, CO_2 reacts with seawater to fundamentally change the chemical environment of the oceans (Feely et al. 2010). Carbon occurs naturally and in abundance in seawater, simultaneously as a suite of multiple compounds or ions, including dissolved carbon dioxide [$CO_{2(aq)}$], carbonic acid (H_2CO_3), bicarbonate ions (HCO_3^-), and carbonate ions (CO_3^{2-}). The relative proportions of these compounds and ions adjust to maintain the ionic charge balance in the ocean (see the following steps).

Step 1: Air-sea exchange: $CO_{2(atmos)} \leftrightarrow CO_{2(aq)}$
Step 2: Hydrogen ion production: $CO_{2(aq)} + H_2O \leftrightarrow H_2CO_3 \leftrightarrow H^+ + HCO_3^- \leftrightarrow 2H^+ + CO_3^{2-}$
Step 3: Production of calcium carbonate: $Ca^{2+} + CO_3^{2-} \leftrightarrow CaCO_3$

Once uptake from the atmosphere by the oceans has occurred, aqueous CO_2 reacts with water to form carbonic acid (H_2CO_3); however, most of the H_2CO_3 disassociates to form hydrogen (H^+) and bicarbonate (HCO_3^-) ions. The hydrogen ions produced through this pathway lower the overall pH of the ocean and react with carbonate ions in the water to produce additional bicarbonate. The more hydrogen there is to react with carbonate ions, the fewer carbonate ions there are to produce calcium carbonate, a major building block for many shell-forming organisms. The average pH of the upper layers of the world's oceans has already declined by 0.1 units (from an average value of 8.2 to 8.1) over the past century. Given that pH is measured on a logarithmic scale, this represents a 30% increase in ocean acidity (Caldeira & Wickett 2003, 2005, Feely et al. 2004). Under current CO_2 emission rates, a further decline in pH of 0.3 to 0.4 units could occur by the year 2100 (Orr et al. 2005).

US ocean regions will respond differently to the rapidly changing carbon chemistry. Regions with strong upwelling regimes (such as the US West Coast) could be more vulnerable because upwelled waters are naturally higher in CO_2 (and thus lower in pH); these waters are already showing signs of anthropogenically augmented CO_2 levels (Feely et al. 2008). Coastal regions with high freshwater input (such as areas of the Chesapeake Bay) may also be more vulnerable to acidification because fresh water generally has a decreased ability to neutralize acids and may carry other acidifying solutes (Salisbury et al. 2008, Kelly et al. 2011). Ocean acidification will also be exacerbated in polar ecosystems (e.g., Arctic Ocean) because CO_2 is more soluble in colder water and the loss of sea ice in summer results in greater exposure of seawater to the atmosphere, allowing more exchange of CO_2 across the ocean–atmosphere interface. Freshwater dilution from ice melt at high latitudes also exacerbates acidification and results in undersaturation of carbonate minerals (Denman et al. 2011). Finally, predictions indicate that surface waters in the Arctic will be undersaturated with respect to aragonite, the more soluble form of calcium carbonate, as soon as within the next decade—sooner if other climate-related factors are taken into account (shrinking sea ice, increased fresh water) (Orr et al. 2005, Steinacher et al. 2009). Similarly, projections for the Southern Antarctic Ocean surface waters suggest that this region also is likely to become undersaturated with respect to aragonite by the year 2050 (Figure 6) (Orr et al. 2005).

Acidification can also be exacerbated by changes in oceanic circulation. The California Current has been identified as an area of particular concern because coastal upwelling brings to the sea-surface 'deep older waters' that are naturally low in pH. These deep waters carry the cumulative signature of decomposition of organic matter through respiration processes that have taken place over hundreds of years. When this very old water upwells in the California Current, it is naturally rich in CO_2 and has a high concentration of nutrients, low oxygen concentration, and low pH. Few measurements have been made of CO_2 and pH in upwelled waters of the California Current, but

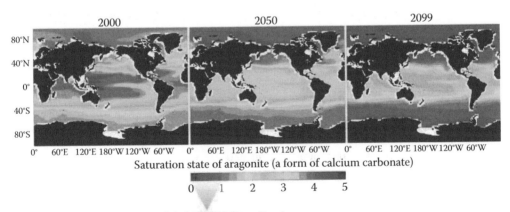

Saturation state of aragonite (a form of calcium carbonate)

0   1   2   3   4   5

Shells and skeletons likely to dissolve

Figure 6 (See also colour figure in the insert) Aragonite levels predicted to drop as ocean acidifies. Calculated saturation states of aragonite, a form of calcium carbonate often used by calcifying organisms. Shades of red indicate areas where levels are so low that organisms may be unable to make new shells or skeletons and where most unprotected aragonite structures will dissolve. By the end of this century, polar and temperate oceans may no longer be conducive for the growth of calcifying organisms such as some molluscs, crustaceans, and corals. (Modified from Feely et al. 2009.)

available data indicate that upwelled waters are undersaturated with respect to aragonite and have a pH of 7.6 to 7.7 (Feely et al. 2008), values already lower than those expected for global oceans by 2100. This also makes the coastal and marine ecosystems of the US West Coast particularly vulnerable to the ocean acidification impacts described previously, as well as the impacts of ocean acidification on biological resources and ocean services shown to negatively affect oyster hatcheries in the US Pacific North-west.

It should be noted that low-latitude systems are not immune to such changes. In fact, the greatest rate of change in carbonate mineral saturation state has unfolded within the Atlantic tropical waters. Tropical water ecosystems will remain supersaturated with respect to aragonite for the foreseeable future (Feely et al. 2009); however, the effects of ocean acidification on marine organisms could also potentially manifest themselves in the form of tipping points, causing shifts in ecosystem dynamics. The location and likelihood of such events remain uncertain.

One of the more remarkable effects of the ocean's rapidly changing pH is the impact on low-frequency sound absorption (Hester et al. 2008). Sound is produced as a by-product of many anthropogenic activities (e.g., shipping, oil and gas exploration) and by natural sources (e.g., marine mammals, wind, earthquakes). A decline in pH of approximately 0.3 causes a 40% decrease in the intrinsic sound absorption properties of surface seawater (Hester et al. 2008). It has been suggested that sounds at frequencies important for marine mammals and for naval and industrial interests will travel some 70% farther with the reduction in ocean pH expected from a doubling of CO_2 (Brewer & Hester 2009). More recent modelling suggests, however, that due to the complexities of sound travel through the ocean, actual increases in background noise are likely to range from negligible (Joseph & Chiu 2010, Reeder & Chiu 2010) to a few decibels within the next 100 years (Udovydchenkov et al. 2010). With the magnitude of potential impacts uncertain and generating debate among researchers, the effect of ocean acidification on background sound levels in the ocean is an area that deserves further study.

Hypoxia

Hypoxia has been recognized as one of the most important water quality problems worldwide. The term *hypoxia* refers to a condition in which an aquatic environment has oxygen concentrations that are insufficient to sustain most animal life (Diaz & Rosenberg 1995, Vaquer-Sunyer & Duarte 2008) and can lead to 'dead zones'. The number of water bodies with recorded and published accounts of hypoxia from around the globe increased from fewer than 50 in the 1960s to approximately 400 by 2008 (Diaz & Rosenberg 2008). The number of water bodies in the United States with documented hypoxia follows the same trend, increasing from 12 prior to 1960 (Bricker et al. 2007) to more than 300 by 2008 (Committee on Environment and Natural Resources [CERN] 2010).

Hypoxia naturally develops when the water column becomes stratified, isolating an oxygen-depleted layer of bottom water and sediments (due to decomposition of organic matter) from a usually well-oxygenated surface layer (due to interactions with the atmosphere). Development and maintenance of hypoxic regions are strongly affected by water column mixing. Seasonal mixing due to temperature and tidal changes can destabilize stratification, leading to relatively abrupt mixing or 'turnover' of the water column that eliminates hypoxia. Hypoxia is especially threatening in estuaries, such as Mobile Bay, Alabama, and Pensacola Bay, Florida, which have low-amplitude tides (Hagy & Murrell 2007), as well as others, such as the Albermarle–Pamlico Estuarine System on the North Carolina coast, that are virtually tideless (Luettich et al. 2002) and therefore do not have as great a benefit from tidal mixing. These estuaries, which are also located in a warmer climate that results in warmer waters, are particularly susceptible to stratification and thus hypoxia (Reynolds-Fleming & Luettich 2004, Hagy & Murrell 2007, Park et al. 2007).

While hypoxia is a natural occurrence, human activities are exacerbating the frequency, duration, and intensity of hypoxic events (Cooper & Brush 1991, Helly & Levin 2004, Diaz & Rosenberg 2008). Eutrophication, an increase in the rate of supply of organic matter to an ecosystem, is most often associated with anthropogenic nutrient enrichment of coastal and ocean waters from urban and agricultural land run-off, wastewater treatment plant discharges, and air deposition of nutrients (Bricker et al. 2007, Galloway et al. 2008). Changes in precipitation patterns due to climate change, as stated previously, may have a significant impact on the amount of nutrients introduced to the marine environment through increased run-off. Eutrophication encourages algal blooms and subsequently the population growth of organisms that feed on algae; together, these organisms may utilize more oxygen than is being mixed into the water. This net loss of oxygen results in the development of hypoxic waters.

The second-largest eutrophication-related hypoxic area in the world (after the Baltic Sea, which is approximately 80,000 km^2) (Karlson et al. 2002, Hansen et al. 2007) occurs in the United States and is associated with the discharge from the Mississippi/Atchafalaya Rivers in the northern Gulf of Mexico (Alexander et al. 2008). The northern Gulf of Mexico hypoxic area has increased substantially in size since the mid-1980s, when it was first measured at about 4000 km^2 (Rabalais et al. 2007). In 2008, the 'dead zone' encompassed 20,719 km^2, the second largest hypoxic area on record for the Gulf of Mexico (CERN 2010). The most commonly reported eutrophication-related problems include hypoxia, losses of submerged grasses, excessive algal blooms, and numerous occurrences of nuisance and toxic harmful algal blooms (HABs; CERN 2010). Changes in temperature, precipitation, and winds will likely lead to long-term ecological changes that favour progressively earlier onset and duration of hypoxia each year.

Upwelling of nutrient-rich deep-ocean water into shallow areas can also support large blooms of phytoplankton (Chan et al. 2008) and may result in hypoxia. Climate-related changes in regional wind patterns might already be enabling the extent and severity of hypoxia off the coast of Oregon, a system dominated by upwelling (Chan et al. 2008). Subtler upwelling events have also been observed along the New Jersey coast and have been implicated in development of nearshore hypoxia (Glenn et al. 1996, 2004).

Climate change will almost certainly exacerbate both naturally occurring and eutrophication-related hypoxia, as well as the incidence of HABs. In general, the expected long-term ecological changes favour progressively earlier onset of hypoxia each year and, possibly, longer overall duration (Boesch et al. 2007). Increasing average water temperature is one mechanism by which climate change may increase susceptibility of systems to hypoxia. Higher water temperatures promote increased water column stratification, decreased solubility of oxygen, and increased metabolic rates for marine organisms, leading to increased oxygen consumption and nutrient recycling.

Climate predictions also suggest large changes in precipitation patterns, but with significant uncertainty regarding what changes will occur in any given watershed (Christensen et al. 2007). Increased precipitation can be expected to promote increased run-off of nutrients to estuarine and coastal ecosystems and increased water column stratification within these systems, contributing to more severe oxygen depletion (Justić et al. 2007, Karl et al. 2009). Climate predictions for the Mississippi River basin suggest a 20% increase in river discharge (Miller & Russell 1992), which is expected to increase the average extent of hypoxia on the northern Gulf of Mexico shelf (Greene et al. 2009).

A great deal of uncertainty remains about how rapidly the physical and chemical attributes of the oceans will change in the future, as well as the location and magnitude of specific impacts and what, if any, potential feedbacks will occur. The physical and chemical changes taking place in the global oceans set the stage for subsequent effects on marine organisms and US communities and economies dependent on marine services. Change to these systems may have larger consequences for US international partnerships and potential adaptation strategies.

Impacts of climate change on marine organisms

Climate change effects on marine organisms and ecosystems are occurring throughout the United States. These effects are occurring across scales, ranging from changes in physiology of individuals, to alterations in interactions between species, to ecosystem regime shifts. Climate-related impacts on ocean ecosystems include altered growth, reproduction, and survival; shifts in species phenology and ranges; increases in species invasions and disease; and changes in the abundance, productivity and diversity of marine plants and animals, among others. Observations and research have demonstrated high variability in the vulnerability and responses of organisms to changes in climate, resulting in 'winners' (i.e., species positively impacted) and 'losers' (i.e., species negatively impacted).

It is likely that these climate change effects on marine organisms and ecosystems will increase based on projected changes in the magnitude and variability of temperature, ocean pH, and other environmental parameters. These impacts are a result of both changes in average conditions and the occurrence of rare but extreme events. In addition, there may be unprecedented effects due to the complexity of marine ecosystems and the likelihood of non-linear interactions among stressors.

Climate change can serve as a threat multiplier, having an impact on marine organisms by interacting with non-climatic stressors, such as nutrient pollution and fishing pressure. While in many cases these 'multiple stressors' are simply additive in their impacts, synergistic (more than the sum of the individual effects) and antagonistic (less than the sum of the individual effects) interactions are also common. Scientific studies that address both climatic and non-climatic stressors can provide critical insight into these interactions and feedbacks. From a policy and management perspective, reducing non-climatic stressors (e.g., nutrient loading, habitat destruction, overharvesting) at local-to-regional scales can provide an opportunity to enhance the resilience of marine ecosystems to climate change and ocean acidification.

There is strong evidence that stressors can have both direct physiological effects on organisms and indirect effects through impacts on the species with which they interact. In addition, these responses are often non-linear, making threshold effects ('tipping points') an area of concern. Key to advancing knowledge and projections of change will be the sustained, long-term monitoring of ecological responses, with concomitant measurements of physical drivers and associated socioeconomic impacts. Research is also needed to improve understanding of the processes and mechanisms by which changing conditions affect organisms and ecosystems. Past and current responses of marine ecosystems to climate change can provide important insight into patterns, trends, and trajectories of change, and progress is being made in forecasting the ecological responses of ocean systems to climate change. However, the development of robust methods for projections into the future (including potentially novel environments) remains a challenge. Extrapolations to future responses must be made with caution given the likelihood of novel environments and the high uncertainty about the degree to which organisms can acclimate and populations can genetically adapt to climate-related changes. Thus, understanding the underlying mechanisms by which climate change and non-climatic stressors affect organisms and ecosystems is critical.

Physiological responses

The ability of marine organisms to grow, reproduce, and survive is affected by their environment, as well as by the other organisms and species with which they interact. Virtually all physiological processes are affected by an organism's body temperature (e.g., Somero 2011). Recent studies have also addressed the important impacts of ocean acidification (Hoegh-Guldberg et al. 2007, Widdicombe & Spicer 2008, Hofmann et al. 2010), alterations in salinity (Gedan & Bertness 2010, Lockwood & Somero 2011a), and changes in food resources (Lesser et al. 2010) on physiology. Physiological responses of marine organisms to environmental change include effects on growth (Menge et al. 2008) and metabolism (Jansen et al. 2007), survival (Jones et al. 2009), changes in the timing (Carson et al. 2010) and magnitude (Petes et al. 2008) of reproduction, as well as other consequences, such as increased susceptibility to disease (Mikulski et al. 2000, Anestis et al. 2010a). Importantly, climatic and non-climatic stressors, including pollution (Sokolova & Lannig 2008), physical disturbance (Bussell et al. 2008), and overharvesting (Hsieh et al. 2008, Sumaila et al. 2011), interact with one another, both in their physiological effects on individuals (Hutchins et al. 2007, Hofmann & Todgham 2010) and in their cumulative impacts on ecological communities (Crain et al. 2008, Pandolfi et al. 2011, Dijkstra et al. 2012). Thus, while considerable progress has been made in understanding physiological and ecological responses to climate change, and in making predictions about the likelihood of future physiological and ecological responses (Helmuth 2009, Mislan & Wethey 2011, Nye et al. 2011), additional work is needed to more clearly understand the impacts of the temporally and spatially complex changing environment on marine organisms and ecosystems. Advances in molecular technology offer one promising approach for improved understanding of the genetic underpinnings and physiological mechanisms associated with organismal stress responses (e.g., Trussell & Etter 2001, Dahlhoff 2004, Hofmann & Place 2007, Pörtner 2010, Tomanek & Zuzow 2010, Somero 2011, Tomanek 2011, Place et al. 2012).

One area of uncertainty is the ability of marine organisms to acclimatize, or populations to adapt locally, to new and rapidly changing environmental conditions (Trussell & Etter 2001, Bell & Collins 2008, Hofmann & Todgham 2010, Sorte et al. 2011). Some evidence exists for local adaptation of marine organisms to high-stress environments. For example, marine snails on the Oregon coast experience higher levels of aerial thermal stress (due to local environmental conditions experienced at low tide) than do southern populations of the same species in California (Kuo & Sanford 2009). Adult snails in Oregon have higher thermal tolerance than do their counterparts at cooler California sites, and they transmit this high tolerance to their offspring, suggesting that local populations have genetically adapted to the more extreme conditions that they experience (Kuo & Sanford 2009). Most research on local adaptation in marine systems to date has been conducted on organisms with fast reproductive cycles, such as microorganisms (Collins & Bell 2004, Bell & Collins 2008) and copepods (Kelly et al. 2012). Additional research is needed to elucidate better the potential roles of acclimatization and local adaptation to current and future change in marine environments (Schmidt et al. 2008, Kelly et al. 2012).

Effects of temperature change

Temperature has diverse effects on physiological processes in marine organisms (Somero 2011), including changes in metabolic rate (Jansen et al. 2007), as well as the functioning of critical enzymes and other cellular functions (Somero 2011). Thermal stress can lead to an increase in metabolic oxygen demand by organisms and ultimately to oxygen deficiency at the cellular level (Pörtner & Farrell 2008, Pörtner 2010). Metabolic oxygen deficiency has already been documented in commercially important species (e.g., Atlantic cod; Sartoris et al. 2003, Pörtner et al. 2008) and may become increasingly problematic in the future as increasing ocean temperatures are expected to exacerbate ambient low-oxygen concentrations through decreased oxygen solubility and increased oxygen demand by algae (Hofmann et al. 2011).

All organisms display tolerance limits that, when exceeded, lead to impacts on metabolism, growth, reproduction, or survival. Endothermic (i.e., 'warm-blooded') organisms (e.g., mammals and birds) must maintain a relatively constant body temperature, and changes in the ambient temperature outside their preferred range therefore require additional expenditures of energy. If temperatures become either too warm or too cold to maintain body temperature within tolerable limits, sublethal and lethal effects can occur. For example, manatees living in Florida experience a cold stress syndrome when water temperatures fall below 20°C for several days; consequences can include emaciation, immunosuppression, and increased mortality (Bossart et al. 2002). A recent unusual mortality event occurred from January to April 2010, when a total of 480 manatees (70% of which were juveniles) were found dead, with greater than 50% of the mortality attributed to cold stress (Barlas et al. 2011) associated with a record-setting negative phase of the Arctic Oscillation (National Climate Data Center [NCDC] 2010). Increasing severity of extreme weather events or alterations in average winter or summer temperatures can have negative impacts on endothermic marine species, and repeated mortality events resulting from thermal stress can lead to population decreases.

The vast majority of marine organisms (other than birds and mammals) are ectothermic (i.e., 'cold blooded'): Their internal temperatures are driven by ambient environmental conditions. Some marine organisms already live close to or at their thermal tolerance limits, whereas others exhibit broad tolerances (Hochachka & Somero 2002, Somero 2011). Extreme or prolonged high- or low-temperature events can lead to sublethal effects, such as reduced growth and changes in the timing and magnitude of reproductive output (Petes et al. 2008, Anestis et al. 2010b, Dijkstra et al. 2012), as well as mortality (Harley 2003, Petes et al. 2007, Harley & Paine 2009). Elevated temperature has also been linked to risk of physical disturbance. For example, the dislodgement of mussels by waves may be increased, potentially because byssal threads, the animals' primary anchoring system, decay more quickly in warmer water (Moeser & Carrington 2006). Loss of mussels would translate into a reduction in local diversity because hundreds of invertebrate species rely on mussel beds for habitat (Smith et al. 2006). Declines in such physiologically vulnerable 'foundation species' may exacerbate the detrimental impacts of climate change on ocean ecosystems (Gedan & Bertness 2010). In contrast, warm-adapted species may increase in abundance and range as they are able to invade new territory due to warming temperatures (Urian et al. 2011).

Reef-building corals around the world, including those in US states and territories, have been negatively impacted by increasing water temperatures. Corals are among the most vulnerable organisms to even slight changes in temperature. When water temperatures exceed 'normal' summer extremes by as little as 1–2°C for 3–4 weeks (Gleeson & Strong 1995), corals eject their symbiotic dinoflagellates (zooxanthellae), on which most coral species depend for their metabolic requirements and the ability to form skeletons (Hoegh-Guldberg et al. 2007). This phenomenon, 'coral bleaching', is not necessarily immediately fatal, but it can lead to severe reductions in reef health and resilience (Figure 7) (Hoegh-Guldberg et al. 2007). If populations of zooxanthellae are unable to reestablish themselves within host coral tissue, corals often suffer mortality in the mid- to long term (Pandolfi et al. 2011). Corals that have recently bleached may also be more susceptible to disease outbreaks (Miller et al. 2009). While debate on the ability of corals to adapt to environmental stress still exists (Hoegh-Guldberg et al. 2011, Pandolfi et al. 2011), a recent report by the World Resources Institute (Burke et al. 2011) indicated that 75% of the world's coral reefs, including almost all of the reefs in Florida and Puerto Rico, are threatened due to the interactive effects of climate change, ocean acidification, and local sources of stress, such as nutrient pollution. The same report projected that roughly 50% of the world's reefs will experience severe bleaching due to thermal stress by the 2030s, and more than 95% by the 2050s, based on current trajectories of greenhouse gas emissions (Burke et al. 2011). Carpenter et al. (2008) suggested that one-third of all reef-building coral species are at risk of extinction due to the combined effects of climate change and local stressors. Loss of coral cover and reef 3-dimensional complexity leads to losses of the many species of associated fishes and invertebrates that depend directly and indirectly on corals for habitat and food (Graham et al. 2006, Idjadi & Edmunds 2006, Alvarez-Filip et al. 2009). Therefore, continued loss of coral reefs is highly likely to have cascading effects on diversity, structure, function, and the valuable ecosystem services on which humans depend (e.g., Mumby & Steneck 2011).

Interactions between thermal stress and food availability can also affect physiology. For some marine animal species, increasing food supply can result in higher levels of thermal tolerance (Schneider et al. 2010). It is anticipated that climate change will have an impact on individual nutrition status as prey species shift geographic and depth ranges, altering food web dynamics (e.g., Bluhm & Gradinger 2008; see Figure 8). These impacts could be either negative, in the cases

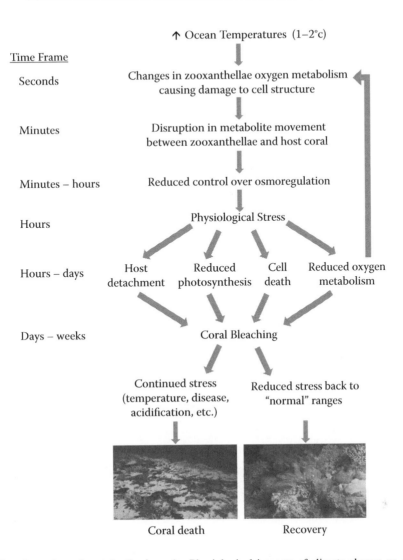

↑ Ocean Temperatures (1–2°c)

Time Frame

Seconds — Changes in zooxanthellae oxygen metabolism causing damage to cell structure

Minutes — Disruption in metabolite movement between zooxanthellae and host coral

Minutes – hours — Reduced control over osmoregulation

Hours — Physiological Stress

Hours – days — Host detachment — Reduced photosynthesis — Cell death — Reduced oxygen metabolism

Days – weeks — Coral Bleaching

Continued stress (temperature, disease, acidification, etc.)

Reduced stress back to "normal" ranges

Coral death

Recovery

Figure 7 (See also colour figure in the insert) Physiological impacts of climate change on coral reefs. Impact of prolonged heat stress on the physiology of corals and their symbiotic zooxanthellae. (Modified from Mayfield & Gates 2007.)

of organisms that depend on specific prey items, or neutral or positive, for species that have generalized diets. Food depletion and resultant nutritional stress are well-known causes of immune suppression across several taxa (Burek et al. 2008). Increased exposure to unfavourable environmental conditions may exacerbate nutritional deficiencies, thus leaving individuals weakened and more susceptible to stress.

Ocean acidification impacts

Ocean acidification causes a reduction in the solubility in ocean waters of calcium carbonate, which many organisms use to create shells. Biological processes known to be affected by ocean acidification include calcification, photosynthesis, nitrogen fixation and nitrification, ion transport, enzyme activity, and protein function (Hutchins et al. 2009, Hofmann et al. 2010, Gattuso & Hansson 2011).

A growing number of laboratory and field studies have documented the negative impacts of ocean acidification on calcifying organisms (Fabry 2008, Doney et al. 2009, Gattuso & Hansson

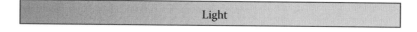

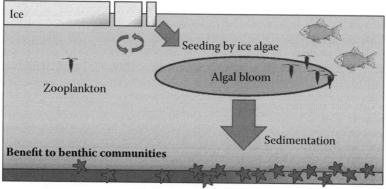

(A) Current/late ice retreat

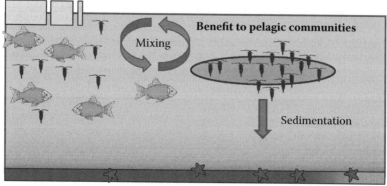

(B) Future/early ice retreat

Figure 8 (See also colour figure in the insert)   Schematic representation of two scenarios that relate seasonal cycles of marine production to ice cover. (A) In years of abundant sea ice (and thereby cold surface waters), herbivorous zooplankton are less abundant early in the season and have little grazing impact on the ice algal and marginal ice zone blooms, resulting in primary production being largely exported to the benthic community. This supports a benthic-dominated food web, including bottom-feeding mammals and birds. (B) In years with less ice (and thereby warm surface waters), a later-occurring phytoplankton bloom dominates over sea-ice-related blooms. These phytoplankton blooms may be efficiently grazed by abundant zooplankton, which in turn are capable of supporting pelagic larval and juvenile fishes. (Modified from Bluhm & Gradinger 2008.)

2011). A recent meta-analysis (Kroeker et al. 2010) found an overall negative effect for the many types of organisms studied; however, when separated into taxonomic groups, the responses were variable. Physiological studies have likewise shown that organisms can vary greatly in their responses to decreasing pH, even among closely related species (Fabry et al. 2008, Doney et al. 2009, Ries et al. 2009, Hofmann et al. 2010, Byrne 2011). One challenge to enhanced understanding is that there are currently few natural modern analogs to a world under increased acidification (Hönisch et al. 2012). Therefore, recent studies have examined natural pH gradients surrounding underwater volcanic vents (Hall-Spencer et al. 2008, Fabricius et al. 2011, Porzio et al. 2011) and acidification in the fossil record (Ries 2010, Crook et al. 2011, Pandolfi et al. 2011, Hönisch et al. 2012) to gain insight into acidified conditions.

Although the biological implications of acidification for open ocean ecosystems remain an active area of research, some general trends are emerging. Calcification by the planktonic algal group coccolithophores, which form massive open-ocean blooms, will almost certainly be greatly reduced (Riebesell et al. 2000, Feng et al. 2008, Beaufort et al. 2011, Hutchins 2011), although responses to increased CO_2 do appear to vary between species (Fabry 2008). Zooplankton groups, such as Foraminifera and pteropods, that produce calcium carbonate shells will also be adversely affected (Orr et al. 2005, Moy et al. 2009, Lombard et al. 2010). Pteropods are an especially critical link in high-latitude food webs as commercially important species (e.g., salmon) depend heavily on them as prey items (Fabry et al. 2009).

In addition to calcification, another critical biogeochemical process that appears to be strongly inhibited by ocean acidification is nitrification. This process is a key link in the ocean's nitrogen cycle by which certain prokaryotes convert ammonia to nitrate, thereby making oxidized nitrogen forms available to marine biota (Beman et al. 2011). Nitrogen fixation rates of some dominant marine cyanobacteria increase substantially at low pH (Hutchins et al. 2007, Fu et al. 2008), as does cellular toxin production by some HAB species (Sun et al. 2011).

Certain species and ecosystems are particularly vulnerable to ocean acidification. Because coral skeletons are formed of calcium carbonate, the future of coral reefs under ocean acidification has received considerable attention. Studies have suggested that reef accretion stops at atmospheric CO_2 concentrations of 480 ppm (Kleypas & Langdon 2006), and climate models predict that at atmospheric CO_2 levels of 550 ppm, the worldwide dissolution of coral reefs is possible (Silverman et al. 2009). A study based on cores from the Great Barrier Reef in Australia showed that coral calcification rates declined 21% between 1988 and 2003 (Cooper et al. 2008). The observed decrease exceeds that predicted by changes in pH alone, suggesting potential effects of multiple and interacting stressors, such as ocean acidification, temperature, and nutrient stress (Cooper et al. 2008, Doney et al. 2009). Studies of reef communities at varying distances from natural CO_2 seeps have documented large declines in coral colony size, coral species richness, and coralline algae (the preferred settlement substrate of coral larvae) at low pH (Fabricius et al. 2011).

Research has also demonstrated the negative effects of ocean acidification on calcification by shellfish (Gazeau et al. 2007), particularly in areas such as the US Pacific North-west, where acidified waters caused wild oysters and oyster growers to suffer persistent production failures (Feely et al. 2008). Larval oysters are sensitive to the carbonate chemistry of the water, particularly as they transition from having no shell at all to having 70% of their mass consist of shell mineral material. During this period, there is a greater dependence on carbonate drawn from seawater, rather than from internal reserves, for shell carbon than during later stages. These results paint a picture of larval development that depends on favourable ambient conditions during critical and energetically expensive early-growth bottlenecks, with results that do not express themselves clearly until later in the organisms' lives. Hatchery managers who embrace quality measurement technology can optimize operations for favourable conditions or even control conditions with active manipulation of inflow water chemistry. Natural populations, however, will be subject to stress from additional acidification as CO_2 levels rise.

The geological past shows that the abundance and diversity of calcifying organisms are reduced when large amounts of CO_2 are released rapidly into the atmosphere (Zachos et al. 2005). However, this evidence is only partially helpful in understanding potential impacts of ocean acidification in the future, as even relatively abrupt geological changes in CO_2 levels in the past likely occurred over thousands of years, allowing the ocean enough time to buffer these increases chemically (Gattuso & Hansson 2011). The rate of acidification is much more rapid today (Gattuso & Hansson 2011), creating a magnitude of ocean change that is potentially unparalleled in at least the past approximately 300 million years of Earth's history (Hönisch et al. 2012).

Notably, to date, most studies of the physiological impacts of ocean acidification have been based on short-term experiments that range in duration from hours to weeks (Doney et al. 2009). Therefore, understanding the biological effects of chronic exposure to decreased pH and how ocean acidification interacts with other stressors in intact ecosystems remain relatively underexplored but critical areas of research.

Exposure to toxicants

Toxicants are poisonous substances that can be produced by organisms (biotoxins), released from geologic stores (e.g., heavy metals, some hydrocarbons), or result from a variety of anthropogenic sources, such as persistent organic pollutants, petroleum hydrocarbons, heavy metals, and radio-active substances (Burek et al. 2008). Climate change can alter toxicant exposure levels for marine organisms through changes in the distribution, frequency, and toxicity of HABs and other toxins. Climate-related changes can also occur through alterations in ocean currents that carry both toxi-cants and organisms into novel environments, increased precipitation that can lead to additional run-off of toxicants into estuaries, and changes in feeding ecology that propagate toxicants through-out the food web (Macdonald et al. 2005). Responses of marine organisms and ecosystems to changes in toxicant exposure are difficult to predict and will largely depend on the chemical features of the specific toxicants to which they are exposed (Segner 2011). Toxicant exposure levels may also change as animals alter their diets based on shifts in their prey items (Burek et al. 2008). Moreover, in an example of interactions between climatic and non-climatic stressors, organisms experiencing thermal stress may be more susceptible to the effects of contaminants (Schiedek et al. 2007).

Effects on life-history trade-offs and larval dispersal

Physiological trade-offs occur when resources are limited as each organism has a certain amount of energy available to maintain physiological processes, such as growth, reproduction, metabolism, and immune function (e.g., Roff 1992, Stearns 1992). The production of gametes and offspring, which is energetically costly, may therefore be compromised under stressful conditions as energy is redirected toward defence and survival mechanisms (Wingfield & Sapolsky 2003). Evidence of these life-history trade-offs has been documented for US West Coast intertidal mussels, which exhibited reduced relative allocation of energy towards growth and reproduction and increased energy towards physiological defences under high-stress conditions (Petes et al. 2008). In the west-ern North Pacific, differences in optimal temperatures for growth of larvae of Japanese anchovy (22°C) and sardine (16°C) lead to contrasting fluctuations in larval growth rates of these two species based on ambient temperature (Takasuka et al. 2007). Such preferences for thermal regimes create out-of-phase stock oscillations for anchovies and sardines off California; that is, when conditions are optimal for sardine growth, they are suboptimal for anchovies and vice versa (Chavez et al. 2003). Changes in water temperature may alter these oscillations and therefore the relative abun-dance of these species in the future.

Phenology, the timing of annual life-history events (e.g., migration, breeding) can provide valu-able insight into the impacts of climate change. Thermal stress has been found to alter the timing of spawning events in marine organisms, leading to mismatches of larval production with the peak in phytoplankton that serves as their food supply (Philippart et al. 2003, Edwards & Richardson 2004, Durant et al. 2007). These mismatches can lead to starvation, lower growth and development rates, reduced survival probabilities, and decreased recruitment (Cushing 1996). Impaired repro-duction can have large, negative consequences for population dynamics and, in the most extreme cases, can lead to species collapse (Beaugrand et al. 2003). Differential changes in timing of repro-duction across biogeographic gradients can have subsequent effects on patterns of larval dispersal (Carson et al. 2010).

Exchange rates of adults, juveniles, larvae, and gametes determine levels of connectivity between populations and can drive both local processes and meta-population dynamics (Erlandsson

& McQuaid 2004, Gouhier et al. 2010). Over large spatial scales, larval supply can determine bio-geographic range boundaries (Herbert et al. 2009) as well as the colonization and spread of invasive species (US Environmental Protection Agency [EPA] 2008, Zardi et al. 2011). Thus, changing patterns of oceanic circulation are likely to have a significant influence on the ecology and population genetics of marine organisms. Temperature-dependent metabolism leads to an inverse relationship between temperature and the duration of planktonic larvae in the water column, suggesting that some species may develop more quickly under elevated temperatures, spending less time as larvae in the water column and therefore potentially dispersing shorter distances (O'Connor et al. 2007). On the other hand, faster growth and developmental rates under increased temperatures may increase survival probabilities through the larval stage (Hare & Cowen 1997, Kristiansen et al. 2011), provided that prey resources are adequate to meet the elevated energy requirements due to increased metabolism. Larval stages of some marine organisms are more vulnerable to stress than are their corresponding juvenile stages (e.g., Zippay & Hofmann 2010, Talmage & Gobler 2011) or adult (e.g., Matson & Edwards 2007). These findings emphasize the importance of considering the relative vulnerabilities of different life-history stages to climate change to understand and predict changes in future population sizes better (Russell et al. 2011).

Population and community responses

There is strong evidence that climate-driven changes in environmental conditions are affecting the survival, growth, and reproduction of diverse marine species, resulting in alterations in population sizes, which in turn affect marine community dynamics. Shifts in the distribution of many marine species that are consistent with changes in climate have been observed in coastal waters of all US regions. In general, warm-adapted species are moving polewards (e.g., Parmesan & Yohe 2003), although there is high variability in these responses due to the impacts of local environmental conditions, including non-climatic drivers (Helmuth et al. 2006a). Population size and distribution are also indirectly affected by climate-related changes in species interactions, such as competition and predation. In addition, strong evidence indicates that many marine species appear to be more vulnerable to disease when exposed to climate-related environmental stress, such as elevated ocean temperature. Collectively, these impacts are leading to observed changes in community composition and ecosystem processes. Exploring the relative sensitivity of marine species and their interactions to changing environmental conditions within an ecological context is key to advancing understanding and projections of future change.

Primary productivity

Marine primary productivity, by both microscopic and macroscopic photosynthetic organisms, forms the base of most of the ocean's food webs. The majority of marine primary producers are phytoplankton, a diverse suite of microscopic photosynthetic organisms. Macroalgae (i.e., seaweeds) and seagrasses are important primary producers that also provide nearshore habitat and food sources to diverse marine organisms. Shifts in primary productivity are frequently linked with patterns of oceanic circulation or changes in ocean temperature. Due to the complex linkages between air and water temperature, oceanic circulation, and atmospheric conditions, the consequences of climate change on primary productivity are often non-intuitive in both coastal and open-ocean systems. In the open ocean, primary productivity can be affected through the impacts of increased water temperatures on metabolic rates (increased productivity; e.g., Doney et al. 2012), as well as through increased stratification (decreased productivity; e.g., Behrenfeld et al. 2006). Warming can also lead to dominance by small-celled phytoplankton (picophytoplankton), reducing energy flow to higher trophic levels (Hare et al. 2007, Morán et al. 2010). A recent model projected changes in phytoplankton primary production, total density, and size structure for the North Pacific over the twenty-first century, with the dominant response being a shift towards smaller-size phytoplankton, which alters food chain length (Polovina et al. 2011).

It is uncertain whether marine primary productivity will increase or decrease under future climate change scenarios. On a global scale, a recent study suggested that the past several decades have shown an overall increase in marine primary productivity (Chavez et al. 2011). Primary productivity in the central and southern California Current system has increased over the past three decades (Chavez et al. 2011), correlated with increases in the intensity and duration of wind-driven coastal upwelling events (Garcia-Reyes & Largier 2010). In contrast, satellite-derived time series of chlorophyll have shown significant changes in phytoplankton, notably that the most chlorophyll-poor areas have been expanding (McClain et al. 2004, Behrenfeld et al. 2006, Polovina et al. 2008, Irwin & Oliver 2009), indicating reductions in primary productivity. Comparing the output from different earth system models, Steinacher et al. (2010) projected reductions in global primary production of 2–20% by 2100, with declines in mid-to-low latitudes due to reduced nutrient input into the euphotic zone, and gains in polar regions due to warmer temperatures and less sea ice. Additional observations, research, and modelling efforts will be necessary for improving understanding of the complex relationships between climate change and marine primary productivity.

Harmful algal blooms, of macroscopic and microscopic (single-cell) species, have been recorded on nearly all of the world's coastlines. Over the past 10 years, HABs have been reported in all major US coastal locales (Anderson 2012). HABs have increased in duration, number, and species diversity since the 1980s (Anderson 1989, 2009, Hallegraeff 1993) and can have large negative ecological and socioeconomic consequences. Some microalgal HABs cause direct impacts through production of toxins harmful or lethal to consumers, including shellfish, seabirds, and humans. Levels of cellular toxin production by some HAB species increase dramatically under high-CO_2 conditions, especially when growth is limited by nutrients (Fu et al. 2010, Sun et al. 2011); these findings have implications for increased HAB impacts in a more acidified, stratified future ocean. HABs involving the red tide organism *Karenia brevis*, a dinoflagellate that produces potent neurotoxins called brevetoxins, occur frequently along Florida's south-western coast, causing episodes of high mortality in fish, sea turtles, birds, bottlenose dolphins, and manatees (Gunter et al. 1948, Bossart et al. 1998, Flewelling et al. 2005, Kreuder et al. 2005, Landsberg et al. 2009). Although brevetoxin exposure increases during *K. brevis* blooms, the persistence of the toxin in the food web and the long-term effects of exposure on marine mammals are unclear (Fire et al. 2007). Blooms of 'nuisance' macroalgae may shade out other benthic primary producers (either seagrasses or perennial macroalgae) and negatively impact coral reefs through competitive interactions and reductions in coral larval settlement (Taylor et al. 1995, Lapointe et al. 2005, Hughes et al. 2007, Diaz-Pulido et al. 2011). In addition, when blooms of micro- or macroalgae senesce, their decomposition may cause large-scale mortalities of benthic and pelagic organisms due to lowered water-column oxygen levels (Deacutis et al. 2006, Lopez et al. 2008). When reef herbivores (e.g., fish and urchins that graze on macroalgae) are abundant and nutrient concentrations are low, corals are the competitive dominants on tropical hard bottoms. Local human activities tend to shift the competitive balance in favour of macroalgae at the expense of corals by removing herbivores that would normally keep macroalgal abundance relatively low. Increased nutrient levels in coastal waters can also favour macroalgal growth and lead to stress in corals. On many reefs, these processes have already resulted in a phase shift, which may be sudden or gradual, from coral-dominated reefs to algal-dominated reefs (Hughes et al. 2010).

There are many potential mechanisms responsible for the expansion of algal blooms into new areas and their extended duration in pre-existing areas. Increased nutrient inputs, such as those resulting from sewage treatment plants and fertilizer run-off, may be partially responsible for increases (Valiela et al. 1997, Teichberg et al. 2008, Thornber et al. 2008), whereas reductions in nutrient inputs may result in decreases in bloom density (Johansson 2002). Although nutrient increases (i.e., eutrophication) are not responsible for all HAB events (Anderson 2009, 2012), it can be difficult to separate the relative and interacting effects of eutrophication and climate change on HABs (Heisler et al. 2008, Rabalais et al. 2009).

Macroalgae and seagrasses can be important primary producers in shallow coastal waters. Kelps are among the largest and most conspicuous macroalgae, supporting some of the most diverse and productive ecosystems along the US coast (Mann 1973, Dayton 1985, Graham et al. 2008), primarily due to the provision of energy and complex habitat by the kelps themselves (Graham et al. 2007). Climate change is affecting the productivity of kelp forests across a variety of temporal and spatial scales, providing insight into macroalgal responses. Climate-induced variability in kelp distribution and abundance can affect the distribution of associated kelp forest fauna (Holbrook et al. 1997, Harley et al. 2006), as well as the productivity and diversity of kelp-associated communities (Dayton 1985, Graham 2004).

Most studies of the effects of annual-to-decadal variability of environmental factors on kelp systems have focused on the impacts of rising water temperature (and the generally concomitant decrease in coastal upwelling) on their growth and survival (Dayton et al. 1999, Broitman & Kinlan 2006, Reed et al. 2008). Shorter (e.g., 1-month) periods of exposure to anomalously warm, nutrient-poor ocean conditions can cause deterioration of kelp biomass, whereas prolonged (e.g., yearly-to-decadal) exposure to such conditions can lead to high mortality and distributional shifts of kelp taxa (Schiel et al. 2004). In some kelp systems, a shift may occur in the identity of the dominant kelp taxa according to species-specific environmental tolerances (Schiel et al. 2004), whereas in other systems, kelps and their associated communities may disappear altogether, resulting in an alternative habitat state (e.g., the formation of sea urchin barrens; Ebeling et al. 1985, Harrold & Reed 1985, Ling et al. 2009). Although the global response of kelp systems to rising temperatures (and decreasing nutrients) may appear ubiquitous, the specific response in any given region will depend on the biogeography and environmental tolerances of the local kelp taxa (e.g., Martínez et al. 2003, Wernberg et al. 2010, 2011a, Merzouk & Johnson 2011). Monthly-to-decadal climate-related changes in wave disturbance can also have dramatic negative impacts on kelp forest distribution, abundance, and structure (Dayton et al. 1999, Reed et al. 2008, Byrnes et al. 2011). Furthermore, it has been predicted that kelp systems will be similarly impacted by changes in ocean conditions over millennial timescales, with kelp forest optima occurring during cool, nutrient-rich, and well-illuminated periods (Graham et al. 2010).

Local-to-regional scale variability in ocean pH and both atmospheric and oceanic CO_2 concentrations are also likely to affect macroalgae and their associated communities, yet direct studies of these environmental impacts are generally lacking. The survivorship of calcified macroalgae, which are present in temperate and tropical habitats throughout US coastal waters, is greatly reduced under ocean acidification scenarios (e.g., Anthony et al. 2008). The combined impacts of ocean acidification and warming can increase skeletal dissolution rates (Diaz-Pulido et al. 2012) or lead to necrosis (Martin & Gattuso 2009) for calcified macroalgae. In contrast, non-calcified macroalgae may have higher tolerance to ocean acidification (Diaz-Pulido et al. 2011). While one study suggests that certain kelp life-history stages may be sensitive to ocean acidification (Roleda et al. 2012), others predict enhanced seaweed performance with rising CO_2 concentrations (Harley et al. 2006, Connell & Russell 2010). Canopy-forming kelps that can directly access atmospheric CO_2 may be particularly sensitive to increases in CO_2 concentrations, but the degree of carbon limitation in kelps is relatively unstudied (Harley et al. 2006). The impacts of ocean acidification and other stressors on calcified and non-calcified macroalgae are an important opportunity for future research.

Shifts in species distribution

Climate-related changes have been shown to influence the local and geographic ranges of many marine species (Hoegh-Guldberg & Bruno 2010, Burrows et al. 2011). Analyses of shifts in species distributions have demonstrated that marine systems appear to be changing substantially faster than terrestrial ecosystems (Helmuth et al. 2006b, Sorte et al. 2010a, Burrows et al. 2011). Studies have shown range shifts in response to both gradual changes in the environment (Findlay et al. 2010, Lockwood & Somero 2011b), as well as the lasting, sometimes multidecadal, impacts of rare-but-extreme events (Denny et al. 2009, Harley & Paine 2009, Firth et al. 2011, Wethey et al. 2011). Climate-related shifts often occur at range boundaries, but due to the importance of local environmental factors (Helmuth et al. 2006a, Burrows et al. 2011), responses such as decreased growth and increased physiological stress and mortality can also occur well within a species' range boundaries (Harley 2008, Place et al. 2008, Beukema et al. 2009). As temperatures warm, current range limits at polewards range boundaries may shift, and warm-adapted species (including certain invasive species) may become able to invade new territory (Urian et al. 2011). Forecasts of future responses to climate change based on observations of present-day changes and on knowledge of physiological responses strongly suggest that changes in species distribution will continue (Runge et al. 2010, Nye et al. 2011). As indicated, the pace and precise location of these changes remain uncertain due to the interactive effects of multiple stressors, the species-specific effects of these changes on interacting organisms, spatial and temporal heterogeneity in environmental drivers, and the ability of organisms to acclimatize or adapt to changing conditions (Sagarin & Gaines 2002, Denny et al. 2009, Nye et al. 2011, Sanford & Kelly 2011).

Evidence of temperature-driven shifts in species ranges is emerging across diverse marine primary producers and invertebrates. Temperature increases are already thought to be altering the distributions of major phytoplankton groups, with numerous observations of polewards range extensions for temperate species (Peperzak 2003, Merico et al. 2004, Hays et al. 2005, Cubillos et al. 2007, Hallegraeff 2010). Jones et al. (2009) reported high mortality of US East Coast intertidal mussels at their southern range boundary in North Carolina as a result of warming temperatures between 1956 and 2007. A study of the marine snail *Kelletia kelletii* in California demonstrated that the northern range boundary had extended northwards by over 400 km between the late 1970s and early 1980s, the first recorded extension north of Point Conception (Zacherl et al. 2003). This distributional shift was consistent with an observed gradient in seawater temperature and the confluence of two major ocean currents. A study using trace-elemental fingerprinting of larval mussel shells demonstrated autumn polewards movement and spring equatorwards movement in the larvae of two species of mussels on the coast of California (Carson et al. 2010). These results suggest that effects of climate change on larval dispersal due to changes in currents and alterations in the timing of reproduction may lead to shifts in species distributions. Temperature change is also leading to shifts in depth distributions of benthic invertebrates. For example, water temperature increases can be particularly acute in shallower, nearshore waters and have led to shifts of surf clams into deeper, cooler waters (Figure 9) (Weinberg 2005).

Climate change can also affect the distribution and abundance of marine fish species through diverse physical and biological processes and mechanisms, the relative importance of which varies across space and time (Ottersen et al. 2010, Overland et al. 2010). For example, fish stocks may shift geographically with changes in ocean temperature, oceanic circulation, or distribution of their prey (Humston et al. 2004, Hsieh et al. 2008, Barange & Perry 2009, Cheung et al. 2009; Figure 9). Climate-related shifts in the abundance and distribution of commercially important species can have consequences for associated fisheries (Cheung et al. 2010). Evidence of climate-induced shifts in the distribution of marine fish has been recorded in several regions of the US EEZ. Fodrie et al. (2010) documented changes in assemblages of fish within seagrass beds in the northern Gulf of Mexico between the 1970s and 2007, reporting the addition of numerous fish species that had previously not been observed. Nye et al. (2009) examined the spatial distribution of 36 species of marine fish found in the bottom trawl surveys of the National Oceanic and Atmospheric Administration (NOAA) along the continental shelf off the north-eastern coast of the United States from 1968 to 2007 and compared shifts to increases in bottom temperature. A significant polewards shift was

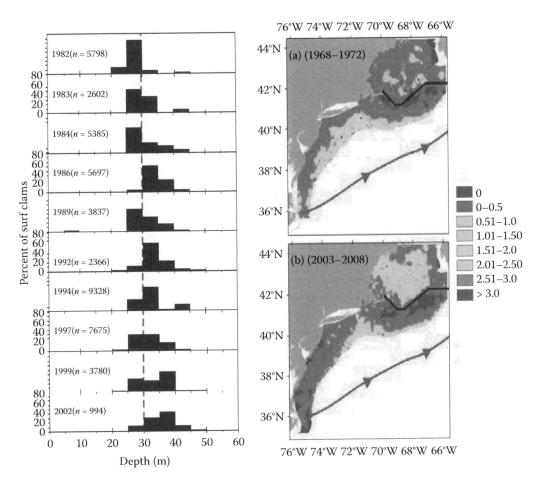

Figure 9 (See also colour figure in the insert) Changing spatial distribution for two marine species. Left: As water temperatures increase, an impact especially acute in shallower nearshore waters, surf clams have shifted their range into deeper, cooler waters. The graphs depict the percentage of surf clams captured at various depths (in 5-m intervals) from 1982 to 2002. n = Total number of surf clams captured (From Weinberg 2005. With permission from Oxford University Press.) Right: Changing spatial distribution of silver hake. (a) Past (1968–1972) and (b) recent (2003–2008) spatial distribution of silver hake biomass in the spring (March–May). Contoured colours represent the log weight (kg) per tow. Note the polewards shift in biomass. Inverse distance weighting was used to smooth the biomass of silver hake. The black line demarcates the boundary between the northern and southern populations of silver hake. The red line indicates the mean position of the Gulf Stream. (From Nye et al. 2011. With permission from Nature Publishing Group.)

found in 17 stocks, a southern shift in 4 stocks, and significant range expansion in 10 stocks (Nye et al. 2009). Shifts in the strength of the polewards undercurrent and ocean temperatures influence the spatial distribution of Pacific hake along the West Coast of North America, with a greater proportion of fish feeding off the coast of Canada in warm conditions (Agostini et al. 2006, 2007). It is unclear whether these shifts are responses to changes in oceanographic conditions or to changes in zonal and seasonal variations in the zooplankton community (Peterson & Keister 2003, Hooff & Peterson 2006). Shifts in thermal habitats and prey fields are expected to influence the distribution of Pacific tuna and other fish populations (Lehodey et al. 2003, 2010, Polovina 2007, Su et al. 2011). In the Arctic, where the rate and relative magnitude of change in ocean conditions is accelerated, differences in topography and currents suggest that there is a higher likelihood of range expansions from the Atlantic side than from the Pacific side (Sigler et al. 2011). These expansions are likely to occur in response to shifts in population density and productivity (Wassmann 2011). The ability of subarctic species to compete with species that are uniquely adapted to survive in the conditions of the Arctic is unknown.

Seabirds are also exhibiting shifts in species ranges, as well as changes in foraging behaviour resulting from shifts in prey distribution and abundance. For example, high ocean temperatures around Alaska's Pribilof Islands have led to prey shortages for least auklets (Springer et al. 2007). In the same region, reduced sea ice and increasing temperatures have led to breeding phenology shifts in kittiwakes over a 32-year period (Byrd et al. 2008). In addition, evidence indicates a possible range expansion of the razorbill in the Canadian Arctic in response to range expansion of favoured prey (Gaston & Woo 2008), as well as climate-related mismatches of prey availability and timing of breeding in thick-billed murres (Gaston et al. 2009).

Marine diseases

Over recent decades, there has been a significant increase in reported disease outbreaks in corals, urchins, molluscs, marine mammals and turtles (Ward & Lafferty 2004), and several disease-causing pathogens that were once thought only to occur on land are now known to have marine counterparts. The impacts of climate change on disease emergence and transmission are likely to act through a combination of several mechanisms, including host and pathogen range shifts, changes in contact frequency, changes in the proportion of individuals carrying disease vectors, introductions from terrestrial systems into marine environments, impacts on pathogen ability to reproduce, and increased environmental stress that leads to increased susceptibility of hosts to infection (Mills et al. 2010).

Evolved balances between disease agents, vectors, and hosts will likely be altered by climate change. In some cases, these changes could limit disease; in other cases, diseases could increase, particularly in stressed populations (Harvell et al. 1999, Altizer et al. 2003). Pathogens (macro- and microparasites) are in a constant state of change, and pathogen selection or alteration may affect host species or the course of an outbreak. Trends in infectious disease correlate with host-pathogen-environmental interactions, as either the host becomes more susceptible to disease or the pathogen's virulence increases. Variations in species' ranges may alter pathogen distribution, and warmer winters due to climate change can increase pathogen overwinter survivorship (Harvell et al. 2009). The protistan parasite *Perkinsus marinus*, which causes Dermo disease in oysters, proliferates at high water temperatures and high salinities. In oyster populations within Delaware Bay, epidemics followed extended periods of warm winter weather; these trends in time are mirrored by the northwards spread of Dermo up the eastern seaboard as water temperatures have warmed (Ford 1996, Cook et al. 1998). There is also evidence that increased water temperature is responsible for the enhanced survival of certain marine *Vibrio* bacteria, which can cause seafood-borne illness in humans (Martinez-Urtaza et al. 2010). Similarly, a survey of shell disease in American lobsters conducted in Massachusetts suggested that higher-than-average water temperatures between 1993 and 2003 led to increased disease prevalence (Glenn & Pugh 2006).

As terrestrial species expand their range or increased run-off from land occurs due to increased precipitation, pathogens novel to marine organisms can enter coastal and ocean systems. For example, faecal waste from the invasive Virginia opossum on the US West Coast has resulted in an increase in the spread of *Sarcocystis neurona*, a protozoan parasite that infects and kills marine mammals, including sea otters (M.A. Miller et al. 2010). In addition, the emergence and pathogenesis of the disease leptospirosis has been associated with environmental variability. Leptospirosis causes mortality of California sea lions (Gulland et al. 1996, Lloyd-Smith et al. 2007), as well as effects in harbour seals and northern elephant seals (Colegrove et al. 2005, Kik et al. 2006). Increased leptospirosis has been associated with increases in precipitation and flooding during El Niño events (Levett 2001, Storck et al. 2008). In Hawaii, increased cases of leptospirosis have been linked to flooding and have also shifted to wetter months of the year (Gaynor et al. 2007, Katz et al. 2011). *Leptospira* bacteria have been found to survive longer in fresh water and many of the sea lions stranded near freshwater estuaries, thereby increasing the possibility of transmission of the bacteria to domestic animals, terrestrial wildlife, or humans (Meites et al. 2004, Monahan et al. 2009, Zuerner et al. 2009). In the future, the combination of human population increase and urbanization, increasing populations and expansions of the ranges of marine mammals, and changes in environmental conditions such as extreme weather events, increased flooding, and increased temperatures may increase the exposure to and incidence of leptospirosis in both humans and marine mammals (Lau et al. 2010).

The impacts of climate change on future rates of marine disease are uncertain. Changes in environmental conditions may lead to range shifts of macro- and microparasites, but those shifts do not necessarily result in increased disease spread. New habitats may contain physical, physiological, or ecological (e.g., due to competition or predation) barriers to the spread of disease (Slenning 2010). As with their hosts, pathogens and vectors are susceptible to climate-related stressors (Lafferty 2009). Parasites that release gametes or larvae into the open marine environment or utilize intermediaries to complete various stages of their life cycle are particularly sensitive to climate change because their success is dependent on environmental conditions or on the availability and responses of their intermediate host species (Burek et al. 2008, Macey et al. 2008). Rising sea levels, warming ocean temperatures, and changes in ocean circulation and estuarine salinity may alter fish parasite composition and biogeography (Palm 2011). Reductions in non-climatic stressors provided by protected areas may potentially reduce disease prevalence. For example, a survey of 94 oyster reefs found significantly higher densities of oysters and significantly lower disease prevalence and severity inside sanctuaries (Powers et al. 2009).

In many cases, the lack of integrated, long-term data on marine diseases limits the ability to predict future climate-related changes in infection prevalence and intensity, emphasizing the need for enhanced and sustained surveillance. Pathogen discovery and identification is in a relatively nascent stage for marine systems. As molecular techniques become more accessible, source tracking is allowing scientists to better understand the connections between marine, terrestrial, and freshwater systems, as well as the evolution of marine pathogens. Improved understanding of these relationships will provide insight into current and future impacts of climate change on marine diseases.

Invasive species

Invasions by non-native species are widely recognized as significant threats to native biodiversity (e.g., Carlton 1996, Ruiz et al. 2000, Stachowicz et al. 2002a, Rahel & Olden 2008). The frequency of introductions has increased dramatically over the past two centuries, and species introductions have been documented in most marine habitats worldwide (e.g., Ruiz et al. 1999, 2000). San Francisco Bay is one of the most heavily impacted coastal systems, with over 230 non-indigenous species (Cohen & Carlton 1998). While only a fraction of invasive species have had significant impacts on established food webs and trophic linkages, their ecological and economic impacts can be profound.

The majority of marine species introductions are caused by shipping (ballast water and hull fouling) or fisheries effects (Ruiz et al. 2000), and invasions consistent with climatic drivers (e.g., changes in temperature) have also been reported (e.g., Reid et al. 2007, Firth et al. 2011). Climate-related impacts on tourism, commerce, and recreation could also create indirect effects on the frequency of invasions (Hellmann et al. 2008). Potential changes in shipping patterns and new routes resulting from loss of Arctic sea ice could lead to new introductions through ballast water or hull fouling (Pyke et al. 2008). In addition, climate change-induced shifts in the distributions of invasive species that are already established could lead to their movement into habitats that were previously uninvaded (de Rivera et al. 2011, Doney et al. 2012). For example, climate change is predicted to result in the movement of many planktonically dispersing, fast-growing Pacific species (e.g., molluscs) into Atlantic waters via the Bering Strait and Arctic Ocean (Vermeij & Roopnarine

2008), although existing physical and physiological barriers to movement (i.e., seasonal ice cover and cold bottom waters) indicate that the number of species capable of invading the polar region may be limited (Sigler et al. 2011). On the Florida coast, evidence suggests that northwards range shifts of the introduced Asian green mussel, *Perna viridis*, may currently be limited by cold temperatures (Urian et al. 2011). Therefore, increasing temperatures could potentially allow for range expansion of this invasive species (Urian et al. 2011).

Once species become established in new areas, climate change may facilitate their subsequent success (Hellmann et al. 2008) as many invasive species have wider temperature tolerance ranges than their native counterparts (Stachowicz et al. 2002a, Braby & Somero 2006, Sorte et al. 2010b, Abreu et al. 2011, Lockwood & Somero 2011b). Climate-mediated invasions and range shifts may also alter species interactions as superior competitors (Stachowicz et al. 2002a) and predators (Smith et al. 2011) move into temperate and polar latitudes. In addition, introduced species may affect the distribution of diversity among trophic levels; in several coastal food webs, introduced species have been at least partially responsible for community-wide shifts towards lower trophic levels (Byrnes et al. 2007). These consequences of invasions present a major threat to the persistence and interactions of native marine species in a changing climate.

Protected species

There is strong evidence that climate change is already affecting a variety of marine protected species, such as marine mammals, sea turtles, and seabirds, and it is likely that these impacts will increase in the future. The effects on these species are expected primarily from shifts in productivity and prey availability; changes in critical habitats, such as sea ice (due to climate warming) and nesting and rearing beaches (due to sea-level rise); and increases in diseases and biotoxins due to warming temperatures and shifts in coastal currents. Predicting the consequences of climate change on marine protected species is difficult due to the relative paucity of data and uncertainties regarding how they will respond if numbers and densities are reduced (Simmonds & Isaac 2007, Hoegh-Guldberg & Bruno 2010, Kaschner et al. 2011, Wassmann et al. 2011).

Many protected marine species are highly mobile or migratory, occupying and utilizing a wide range of habitats and resources throughout their life history. Animal migration is closely connected to climatic factors, and as a result, these species are in many ways more vulnerable due to the differential impacts they may experience at various life stages. For example, marine turtles may cross entire ocean basins throughout their lifetimes and can occupy diverse habitats, such as sandy beaches, mangroves, and seagrass beds (Musick & Limpus 1997, Hawkes et al. 2006, Polovina et al. 2006, Shillinger et al. 2008). In the Atlantic Ocean, warmer years would mean a stronger-than-average Gulf Stream current, helping juvenile turtles get to the North Atlantic Gyre and leading to increased productivity and population size. However, in the Pacific Ocean, loggerheads perform best under anomalously cold conditions. Therefore, available climate data indicate the potential for significant population declines of the Pacific population by 2040 due to warming temperatures (Van Houtan & Halley 2011). Projected increases in sea level and extreme event intensity, coupled with fortification of coastal areas, could erode shorelines and compromise the availability of suitable nesting beaches (Hawkes et al. 2009). In addition, increased beach temperatures have led to altered sex ratios (higher female-to-male ratios) in hatchlings of marine turtles, for which nest temperature determines the sex of offspring. In parts of the southern United States, hatchlings of loggerhead sea turtles are currently female biased, and even moderate further increases in temperature could lead to a severe lack of males (Hawkes et al. 2007), which can reduce population viability (Poloczanska et al. 2009). Anomalously cold temperatures can also affect turtles. Sea turtles along the US Atlantic Coast and Gulf of Mexico experience episodic, cold-stunning events when water temperatures drop below 10°C (Witherington & Ehrhart 1989, Morreale et al. 1992, Foley et al. 2007). During these events, hundreds of cold-stunned turtles float listlessly on the water or are washed on to shore. An open question remains regarding how temperature fluctuations, which are expected to increase under climate change, are likely to affect turtle populations (Neuwald & Valenzuela 2011), including sea turtles.

Changes in thermal regimes are affecting the abundance, distribution, feeding, and phenology of protected seabirds (Bertram & Kaiser 1993, Chastel et al. 1993, Montevecchi & Myers 1997, Grémillet & Boulinier 2009). For example, declines in oceanic productivity around the north-western Hawaiian Islands in the 1980s led to a 50% reduction in the survival of red-footed booby and red-tailed tropicbird eggs and chicks (Polovina & Haight 1999). Reduced productivity and warmer temperature in the South-east Farallon Islands (e.g., Sydeman et al. 2009), as well as low prey abundance (Sydeman & Thompson 2010), have led to delayed breeding and reduced offspring numbers in Cassin's auklet (Wolf et al. 2009). Wolf et al. (2010) have projected additional climate-related population declines of 11–45% by the end of the century. The common murre has also exhibited a declining trend in reproductive success in the South-east Farallon Islands, reflecting reduced availability of their preferred prey item (rockfish). In 2009, reproductive success of common murres was among the lowest observed in the previous 38 years and the lowest ever recorded during a non–El Niño year (Warzybok & Bradley 2010).

Warming water temperatures and loss of sea ice are fundamentally changing the behaviour, condition, survival, and interactions of Arctic marine mammals (Kovacs et al. 2010, Heide-Jørgenson et al. 2011, Thomas & Laidre 2011, Wassmann et al. 2011), and these changes are expected to continue. Cetaceans, including grey whales (Moore et al. 2003, Stafford et al. 2007, Moore 2008); orcas (Higdon & Ferguson 2009); and sei, fin, and minke whales (Norwegian Polar Institute Marine Mammal Sighting Data Base: http://www.npolar.no), have been sighted further north or at higher northern densities than normal. Similar impacts are occurring for pinnipeds; harbour porpoises are appearing in northern areas, and harp seals are being sighted in northern locations during abnormal times of the year (Norwegian Polar Institute Marine Mammal Sighting Data Base). Major declines in pup production and abundance have been documented for hooded seals in the North-east Atlantic, ringed seals in Hudson Bay, and harp seals in the White Sea (Ferguson et al. 2005, Chernook & Boltnev 2008). Polar bears are spending more time on land due to reduced ice cover, resulting in declines in survival, condition, body size, and reproductive rates (Stirling et al. 1999, Stirling & Parkinson 2006). In addition, landwards shifts of polar bear dens (Fischbach et al. 2007) and declines in the condition and survival of polar bear cubs have occurred (Regehr et al. 2006, 2010). Pacific walrus females and pups are also being forced to spend more time resting on land, and abandoned calves at sea suggest nutritional stress (Cooper et al. 2006, Garlich-Miller et al. 2011, Kavry et al. 2008) due to separation from feeding areas and the loss of sea-ice resting platforms (Kovacs et al. 2010). These examples illustrate some of the challenges facing marine protected species and their managers in a changing climate.

Ecosystem structure and function

Shifts in species distributions and interactions are also beginning to create novel, 'no-analog' ecosystems consisting of species with little or no shared evolutionary history (Hobbs et al. 2006, Williams & Jackson 2007), and it is likely that this will continue with unprecedented environmental change in the future. While progress is being made in forecasting future responses, complex, non-linear effects of changing environmental conditions on marine communities present additional uncertainty and challenges for managers (Crain et al. 2008).

Particularly problematic is obtaining an understanding of how complex feedback interactions between changes in the physical environment will affect ecological processes. Warming, ocean acidification, stratification, and other climate-related parameters can be both synergistic and antagonistic in their effects on marine organisms, making whole-ecosystem predictions difficult with the current state of knowledge (Boyd et al. 2008, Pörtner 2008, Hutchins et al. 2009, Hofmann et al. 2010). Ongoing research efforts are targeting these interactive, multistressor effects, but the fact that many environmental factors are simultaneously in flux makes accurate forecasting of ecosystem-level responses a challenging undertaking.

Species interactions and trophic relationships

Marine ecosystems are influenced not only by the direct effects of climate change on individuals and populations but also through indirect effects, as environmental change alters the strength of species interactions, including competition, predation, parasitism, and mutualism (reviewed by Kordas et al. 2011). These indirect effects can arise via different mechanisms.

Environmental change can alter an organism's physiology and behaviour and therefore its per capita effect on the species with which it interacts. Changing ocean temperature and chemistry can affect the per capita feeding rate of an individual consumer (Sanford 1999, Pincebourde et al. 2008, Gooding et al. 2009, O'Connor 2009) or modify an individual's competitive ability through effects on its growth rate (Stachowicz et al. 2002b, Wethey 2002, Sorte et al. 2010b). Differential impacts of thermal stress on predators and their prey can lead to altered species interactions (Yamane & Gilman 2009). For example, due to temperature-related effects on metabolism, exposure to warm water can increase the feeding rates of US West Coast sea stars (*Pisaster ochraceus*) on their mussel prey (Sanford 1999) until temperatures exceed thermal optima and feeding rates are reduced due to stress (Pincebourde et al. 2008). Studies have further shown that the interactive effects of increased water temperatures and increased aerial body temperature during low tide significantly affect rates of predation by *P. ochraceus*, in that these predators are more strongly affected by stressors that occur out of phase with one another, that is, when animals are constantly under stress due to elevated water and aerial temperatures (Pincebourde et al. 2012). Changes in hydrodynamic conditions can affect the ability of prey to detect predators, as has been shown by experiments examining the behavioural responses of whelks to predatory crabs under different flow conditions (Large et al. 2011).

Climate change alters species interactions via changes in the population density of interacting species. Environmental changes affect species differentially, and the resulting increases or decreases in population abundance can trigger chains of indirect effects (Poloczanska et al. 2008, O'Connor et al. 2009). For example, changes in ocean temperature and carbonate chemistry frequently alter the relative abundance of macroalgal species, which can in turn affect the abundance of herbivores that feed on them (Schiel et al. 2004, Kroeker et al. 2011). Often, a few key species interactions contribute disproportionately to maintaining community structure and ecosystem function (Paine 1992). For example, the saltmarsh grass *Spartina patens* reduces salinity stresses acting on species living within the plant canopy, and thus the removal of this structural species can have cascading effects on marsh communities (Gedan & Bertness 2010). If these interactions are sensitive to environmental conditions, they may act as 'leverage points' through which small changes in climate are amplified to produce large changes at the community and ecosystem levels (Sanford 1999, Kordas et al. 2011, Monaco & Helmuth 2011). Similarly, when the direct effects of climate change have a negative impact on the abundance of habitat-forming species such as coral, kelps, and mussels, there are often cascading effects on ecosystem function due to loss of the services that these foundation species provide (Schiel et al. 2004, Pratchett et al. 2008, Wootton et al. 2008, Wernberg et al. 2011b). Such rippling effects are often unpredictable due to the complexity of food webs (Schiel et al. 2004, Doney et al. 2012).

Climate-related shifts in the geographic distribution of marine species are altering biogeographic patterns of co-occurrence and interaction (Sorte et al. 2010a, Kordas et al. 2011). Analogous shifts in the vertical distribution of sessile intertidal prey species have increased their overlap with, and vulnerability to, predatory sea stars (Harley 2011). Similarly, as described previously, ocean warming can alter the timing of life-history events such as spawning, leading to temporal mismatches, or sometimes increased overlap, between consumers and their food sources (Philippart et al. 2003, Edwards & Richardson 2004, Kristiansen et al. 2011). Climate-related shifts in species dominance have also been observed throughout the United States, including the California Current, the Gulf of Alaska, the Bering Sea (Hare & Mantua 2000), and the North Atlantic (Auster & Link 2009). However, these shifts do not always result in changes in ecological processes. For example, Auster and Link (2009) found that while climate-induced shifts in species dominance were observed in the Georges Bank ecosystem, these often involved switching between species that occupied redundant roles within trophic guilds. This redundancy may buffer against ecosystem reorganization under climate change, but the extent of buffering is unknown in most systems.

Currently, the ability to project the impacts of climate change on trophic linkages within marine ecosystems is limited primarily to conceptual models of marine ecosystem organization (Hunt et al. 2011, King et al. 2011, Monaco & Helmuth 2011). As understanding of the range of complex responses of marine species to environmental disturbance improves and coupled biophysical models of marine ecosystems become available, the ability to predict with higher certainty the likely implications of climate change on marine ecosystems will be enhanced. In the near term, observations and monitoring systems provide the best method of detection and attribution of changes in the trophic structure of marine ecosystems and of validating models of population- and ecosystem-level responses (Helmuth et al. 2006b, Wethey & Woodin 2008, Helmuth 2009, Wethey et al. 2011). Integrated, sustained observations of the abundance, diets, distribution, and physiological condition of marine species, coupled with field and laboratory studies that identify species responses to ecosystem change, will be critical. There is a need to understand patterns of genetic variance (Trussell & Etter 2001, Schmidt et al. 2008), particularly that which underlies traits that influence susceptibility to environmental stress (Place et al. 2008). Over time, insight gained through these efforts will provide the data and understanding needed to more reliably model complex ecosystem responses to climate change.

Biodiversity

Several efforts have been implemented to document, quantify, and assess biodiversity of US marine ecosystems through initiatives such as the Census of Marine Life (e.g., Fautin et al. 2010) to gain baseline understanding and monitor changes through time. Climate-related distribution shifts have already altered community composition and biodiversity of many systems and taxa, including phytoplankton (Peperzak 2003, Merico et al. 2004, Hallegraeff 2010), pelagic copepods (Beaugrand et al. 2002), rocky intertidal invertebrates (Barry et al. 1995, Southward et al. 2005, Helmuth et al. 2006b, Wethey et al. 2011), fishes (Perry et al. 2005, Fodrie et al. 2010, Nye et al. 2009, Last et al. 2011), and seabirds (Hyrenbach & Veit 2003), among others. These changes in community composition are a function of both local extinction of species and invasions from elsewhere.

Ocean acidification is also affecting biodiversity and community structure in marine ecosystems. In a north-eastern Pacific rocky shore community, declining pH over 8 years corresponded with gradual shifts from the mussel-dominated communities typical of such temperate shores to communities more dominated by fleshy algae and barnacles (Wootton et al. 2008). Similarly, in shallow benthic communities near natural CO_2 seeps, calcareous corals and algae are replaced by non-calcareous algae, and juvenile molluscs are sharply reduced in number or absent altogether (Hall-Spencer et al. 2008).

Importantly, not only are levels of biodiversity affected by climate change but also levels of biodiversity can influence the resilience of marine ecosystems to climate change. Experiments show that more diverse communities tend to be more stable (i.e., less susceptible to disturbance and variability through time) (Jiang & Pu 2009). Analyses of global fisheries time series also support the hypothesis that marine biodiversity increases ecosystem stability and resilience to perturbations (Worm et al. 2006). Biocomplexity and diversity of fishes have been shown to decrease variability in stock productivity (Hilborn et al. 2003) and increase profitability to resource users (Schindler et al. 2010). Further research is needed to understand the causes and consequences of biodiversity change in marine ecosystems, including impacts on resilience, stability, and the provisioning of ecosystem services on which humans depend. Nevertheless, there is now sufficient evidence from a wide range of ecosystems to conclude with confidence that, on average, loss of biodiversity reduces ecosystem productivity and stability (Stachowicz et al. 2007, Cardinale et al. 2011).

If species cannot migrate or adapt to a changing environment, they face local or even global extinction. Humans have directly caused the global extinction of more than 20 described marine species, including seabirds, marine mammals, fishes, invertebrates, and algae—and many others have likely disappeared unnoticed (Sala & Knowlton 2006). Under projected climate change, marine species extinctions are expected to be most frequent in subpolar regions, the tropics, and semienclosed seas (Cheung et al. 2009). Cold- and ice-adapted species are especially vulnerable as ocean warming degrades their preferred habitats and invasions occur from temperate regions. For example, warming waters have recently allowed large lithodid 'king' crabs to invade the Antarctic shelf for the first time in 14 million years, where they have reduced benthic diversity and appear to have driven certain species locally extinct (Smith et al. 2011).

Quantitative estimates of species losses based on historical comparisons, measured and projected habitat loss, and demographic trajectories of wild populations suggest that Earth is now approaching (if not already in the midst of) the sixth mass extinction in its history, with rates of species loss two to five orders of magnitude above the average over geologic time (Pimm et al. 1995, Dirzo & Raven 2003, Pimm 2008, Butchart et al. 2010, Pereira et al. 2010, Barnosky et al. 2011). The principal drivers of the current extinction wave are habitat loss, overexploitation, pollution, and impacts of invasive species (Purvis et al. 2000), but changing climate has contributed to several mass extinction events in the past (Barnosky et al. 2011), and today's much more rapidly changing climate is expected to exacerbate the impacts of these other drivers in the coming century (Brook et al. 2008).

Regime shifts and tipping points

As a result of environmental and ecological complexity in responses to climatic and non-climate stressors, rapid changes in ecosystem structure and function are a particular area of concern. Evidence of rapid phase shifts (or regime shifts) is emerging across diverse US geographic locations and ocean ecosystems (Hoegh-Guldberg & Bruno 2010). Regime shifts occur when dominant populations of an ecological community respond gradually and continuously to changes in environmental conditions until a particular threshold or 'tipping point' is reached, beyond which the community rapidly shifts to a new dominant species or suite of species (Scheffer et al. 2001, 2009, Scheffer & Carpenter 2003). In many instances, these 'replacement' assemblages are less 'desirable' from a human standpoint, such as when coral reefs are replaced by fast-growing macroalgae (e.g., Dudgeon et al. 2010) due to a combination of stressors, such as nutrient pollution, overharvesting of herbivorous fishes, disease, and thermal stress (Hoegh-Guldberg et al. 2007, Dudgeon et al. 2010, Hughes et al. 2010). Regime shifts thus have significant implications for ecosystem functioning and services, with consequences for associated ecological, economic, and human social systems (Mumby et al. 2011b). Systems that are already degraded and depleted by non-climatic stressors often have lower resilience and are therefore more susceptible to climate-related regime shifts and tipping points (Folke et al. 2004). Tipping points can be difficult to predict because physiological and ecological thresholds, and significant declines in ecosystem services, can theoretically be reached prior to any associated large-scale changes in the environment (Harley & Paine 2009, Monaco & Helmuth 2011, Mumby et al. 2011b).

Certain US marine systems are on a trajectory for rapid change, and others have already crossed a tipping point (Hoegh-Guldberg & Bruno 2010). In the Chesapeake Bay, eelgrass (*Zostera marina*) died out almost completely during the record hot summers of 2005 and 2010, evidently because too many days exceeded the species' tolerance threshold of 30°C (Moore & Jarvis 2008). Gardner et al. (2003) reported an 80% reduction in Caribbean coral cover, from 50% cover to only 10% cover, in less than three decades. In the northern California Current LME, severe low-oxygen (hypoxic) events have recently emerged as a novel phenomenon due to changes in the timing and duration of coastal upwelling (Barth et al. 2007, Chan et al. 2008). These events have led to high mortality of benthic invertebrates such as Dungeness crabs (Grantham et al. 2004), as well as the loss of rockfish from low-oxygen areas (Chan et al. 2008), with consequences for local fisheries. In many instances, it is unknown whether reversing these rapid, climate-related trajectories of ocean disruption and decline will be possible. However, evidence indicates that reducing non-climatic stressors, such as overharvesting and pollution, can potentially prevent tipping points from occurring (Hsieh et al. 2008, Diaz-Pulido et al. 2009, Sumaila et al. 2011).

One of the most critical approaches to addressing tipping points will be improving the ability to detect and anticipate regime shifts before they occur to enhance preparedness and response efforts. Early-warning signals exist for ecosystems, indicating if a critical threshold is approaching (Scheffer et al. 2009). For example, decreases in the growth, recruitment, and reproduction of key species have been linked to climate-related stress (Philippart et al. 2003, Petes et al. 2008, Beukema et al. 2009). Long-term, integrative observations and monitoring data provide a critical foundation for understanding and documenting early-warning signs of regime shifts (Scheffer et al. 2009), as well as for testing the accuracy of predictive models (e.g., Wethey & Woodin 2008). Enhancing the ability to anticipate tipping points will require integration of long-term observations with experimental and modelling approaches (Scheffer & Carpenter 2003). Advances in determining the underlying physiological and ecological mechanisms responsible for ecosystem regime shifts (Monaco & Helmuth 2011) would help to inform sustainable management of ocean resources under environmental change (Polovina 2005).

Ocean resource managers can no longer expect 'smooth' patterns of change, as evidenced by sudden and non-linear regime shifts. Instead, decision makers and managers should expect surprises and work to anticipate and prevent tipping points whenever possible (Lubchenco & Petes 2010). The utility of enhanced understanding and anticipation of tipping points will depend on the ability of management to respond rapidly and effectively (Biggs et al. 2009). Ocean management must include consideration of multiple interacting stressors (Crain et al. 2008), as well as various life-history stages (Runge et al. 2010). Although climate change consistently ranks as a top pressure to marine ecosystems, at global (Halpern et al. 2008a) and regional areas within the United States (e.g., California Current: Halpern et al. 2009b; north-western Hawaiian Islands: Selkoe et al. 2008, 2009), many other stressors play significant roles in impacting overall condition. The potential for the appearance of tipping points in ocean ecosystems lends urgency to minimizing stressors over which there is more direct control at the local scale (e.g., overfishing, nutrient pollution) to enhance resiliency to climate change and ocean acidification (Hsieh et al. 2008, Lubchenco & Petes 2010, Kelly et al. 2011).

Impacts of climate change on human uses of the ocean and ocean services

US marine ecosystems are highly valuable and provide a wide variety of resources and services that support a diverse array of activities, businesses, communities and economies across the nation. Commercial and recreational fisheries in the United States represent an annual multibillion dollar industry (National Marine Fisheries Service [NOAA Fisheries] 2010). Subsistence fishing, defined as fishing for direct consumption or barter without the product entering a market, also contributes significantly to the health and well-being of fishing-dependent communities and local economies across the United States. Most of the effects of climate change on US fisheries will stem from the changes to the fish stocks brought about by direct and indirect climate impacts on productivity and distribution; others stem from impacts that climate has on the fisheries themselves, as well as fishing-dependent communities across the country.

Although not well documented across all marine regions of the United States, evidence to date suggests that substantial socioeconomic impacts to marine resource-dependent communities and economies worldwide are likely. Moving forward, an interdisciplinary perspective will be crucial for analyzing the interwoven impacts of climate change on the socioeconomic uses of marine resources. That said, data are available regarding the extent of human uses of marine resources, as well as the biophysical effects of climate change on marine resources on which those uses depend. There are many potential consequences of climate change on human uses of oceans, such as possible displacement of fishing fleets from their traditional fishing grounds, increased access to polar region environments for navigation and mineral exploration, and changes in the growth and distribution of waterborne pathogens.

The biophysical impacts of climate change on oceans also affect humans and human systems that interact with the ocean. For example, fishing-dependent communities and the national economy are affected by climate-related impacts on populations of marine resources. Understanding climate impacts to fish and shellfish stocks enables improved assessment of the impacts of those changes on fishing behaviours, industries, infrastructure, and communities.

The term *ocean services* refers both to the quantifiable monetary and non-market value that use of the ocean provides and to the currently unquantifiable but identifiable benefits that the ocean provides to humans. The US Commission on Ocean Policy (USCOP) characterized the value of the ocean sector to the United States in the following way:

> The ocean economy, the portion of the economy that relies directly on ocean attributes … in 2000 … contributed more than $117 billion to American prosperity and supported well over two million jobs. Roughly three quarters of the jobs and half the economic value were produced by ocean-related tourism and recreation. For comparison, ocean-related employment was almost 1½ times larger than agricultural employment in 2000, and total economic output was 2½ times larger than that of the farm sector. (USCOP 2004, p. 31; Figure 10)

The report also noted that, to date, the current governmental standards used to measure ocean economy are insufficient because they do not take into account "the intangible values associated with healthy ecosystems, including clean water, safe seafood, healthy habitats, and desirable living and recreational environments" (USCOP 2004, p. 31). Without a greater understanding of these intrinsic values, Americans are dramatically underestimating the value of the oceans and coasts.

Substantial socioeconomic effects in specific areas, some positive and some negative, are likely to result from changes in marine resources due to climate change. Identifying the crucial areas and potential directions of socioeconomic effects is of vital importance. This review discusses three approaches to exploring the impacts of climate change on human uses of the ocean. The first approach is to combine a baseline of current human uses of marine resources with case studies of currently documentable changes occurring in specific marine resources and their associated socioeconomic impacts. The second approach is construction of generally expected impacts given certain changes in specific marine resources and environments. Finally, the implications of all of these changes for marine resource governance systems will be explored.

Climate effects on capture fisheries

Commercial and recreational fisheries in the United States represent an annual multibillion-dollar industry (Tables 1 and 2). Subsistence fishing also contributes significantly to the health and well-being of fishing-dependent communities and local economies across the United States. In the United States, fisheries managed by the federal government are generally defined as fishing activities that take place between 3 and 200 nautical miles from the coastline. Nationwide NOAA

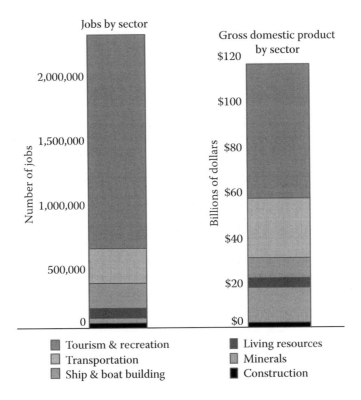

Figure 10 Job numbers and gross domestic product attributable to ocean resources. The ocean economy includes activities that rely directly on ocean attributes or that take place on or under the ocean. On the left, ocean economy is broken down by number of jobs. On the right, ocean economy is broken down by industries' contribution to the gross domestic product in 2000. In 2000, the tourism and recreation sector was the largest in the ocean economy, providing approximately 1.6 million jobs. (Modified from US Commission on Ocean Policy [USCOP] 2004.)

Table 1 2009 Economic impacts of the United States seafood industry

	Jobs	Sales ($1,000s)	Income ($1,000s)	Value Added ($1,000s)
Total Impacts	1,029,542	116,224,548	31,556,643	48,282,319
Commercial harvesters	135,466	10,349,446	3,435,027	5,340,116
Seafood processors and dealers	183,895	25,240,441	7,965,719	11,073,240
Importers	178,387	49,070,476	7,864,480	14,958,830
Seafood wholesalers and distributors	47,405	6,505,383	2,137,714	3,058,777
Retail	484,389	25,058,802	10,153,704	13,851,356

Source: National Marine Fisheries Service. 2010. Fisheries Economics of the United States, 2009. US Dept. Commerce, NOAA Tech. Memo. NMFS-F/SPO-118, 172 p.

Fisheries, the primary federal entity with authority over US fisheries management, oversees the management of 230 major fish stocks or stock complexes that comprise 90% of the nation's commercial harvest. In addition, individual US states retain management authority over fishing activities within 3 (or in some cases up to 9) nautical miles of their coasts or in their inland waters, such as Puget Sound (Washington) or Chesapeake Bay (Maryland).

Table 2 2009 Economic impacts of recreational fishing expenditures

		Jobs	Sales ($1,000s)	Income ($1,000s)	Value Added ($1,000s)
Total Impacts		327,123	49,811,961	14,574,464	23,196,422
	For hire	17,217	1,915,452	606,983	1,039,705
	Private boat	31,176	4,243,541	1,253,804	2,158,414
	Shore	35,293	4,312,850	1,319,865	2,243,036
	Durable equipment	243,438	39,340,118	11,393,812	17,755,268

Source: National Marine Fisheries Service. 2010. Fisheries Economics of the United States, 2009. US Dept. Commerce, NOAA Tech. Memo. NMFS-F/SPO-118, 172p.

Changes in the ocean's physical, chemical, and biological processes due to climate change potentially will have an impact on human community reliance on fisheries resources as well as fisheries governance systems. Most of these effects will stem from changes to the fish stocks brought about by direct and indirect climate impacts on stock productivity and distribution; others will stem from impacts that climate has on the fisheries themselves, as well as on fishing-dependent communities. Extreme weather events may also disrupt fishing operations and land-based infrastructure, and sea-level rise may have an impact on both fisheries infrastructure and fishing-dependent communities. However, because the management of fisheries in the United States is partly based on metrics that depend on productivity, such as maximum sustainable yield (MSY), the effects of climate change on fisheries will substantially depend on how fisheries managers respond to those changes.

Alterations in the biophysical characteristics of fish stocks can sometimes mean changing gear (which is often expensive) or learning new fishing grounds and species' habits. Fishers often rely on social networks for information sharing while fishing (e.g., Palmer 1990, Kitts et al. 2007, St. Martin & Hall-Arber 2008, Holland et al. 2010); thus, changing species may result in the need to cultivate new networks. Species whose range extends further north or south may result in fishers making longer trips or relocating their home base, either of which has effects on families and communities. Conversely, commercially important fish that extend their range into new habitat could be beneficial to fishers in the new location. Fishers often choose day versus trip fishing based on family considerations (Maurstad 2000). Where trips are longer, household dynamics change, affecting time with spouse and children and ability to participate in community and school events. Where households relocate, family as well as fishers' social networks are lost, and part of a community's economic base disappears (though the communities where the fishers move gain). Gentrification is creating pressure on small fishing-dependent communities (Clay & Olson 2008, NOAA Fisheries 2009), making coastal property less affordable. Any climate change-related loss of fishing households could exacerbate this trend. The exact degree or even direction of any of these economic impacts for commercial fishers depends on which specific climate impacts occur, factors affecting market dynamics at that point in time (Markowski et al. 1999), and choices based on social and cultural factors.

In addition to commercial fishers, recreational and subsistence fishers will likely be affected by the impacts of climate change. Recreational fishers will largely change target species, with unclear economic impacts, as many aspects of the recreational fishing experience are unrelated to specific species (Fedler & Ditton 1994). Subsistence fishers generally fish a wider range of species than those fishing for pure recreation and so would likely be able to adapt, provided enough species remain accessible (Steinback et al. 2007). However, to the extent that multiple species become unavailable, these fishers may experience negative nutritional consequences, especially since they are also more likely to collect non-fin-fish marine resources (Steinback et al. 2009), such as squid, seaweed, kelp

(Ling et al. 2009) or shellfish, the last of which are expected to be impacted by increasing ocean acidification (Cooley & Doney 2009).

Effects on the productivity and location of fish stocks

The most direct potential effects of climate change on fisheries will come through changes in the productivity and location of the fish stocks that are the targets of those fisheries. How climate change effects will ultimately manifest themselves in the fish stocks is uncertain, in part because the complexity of trophic relationships makes predictions difficult (Brander 2010). Thus, determining which individual fisheries are likely to suffer significant adverse effects from climate change and which are likely to benefit is challenging. Fisheries that target stocks adversely affected by climate change may be able to target alternate stocks that benefit from climate change, so the ultimate impacts depend strongly on the capacity of particular fisheries to adapt to changes (Brander 2010). Similarly, fisheries that target fish stocks with evolving spatial locations will experience changes in the required amount of fuel and other fishing inputs, time at sea, and exposure to ice (Mahon 2002, Badjeck et al. 2010), but whether these changes will be detrimental or beneficial depends on exactly how those locations change and the exact prices (social and economic) associated with the adaptation strategies available to fishers, their families, their communities and fisheries-dependent industries.

Economic effects on commercial fisheries and fishing-dependent communities

Climate change can affect the quantity and quality of yields through biophysical impacts, and the magnitude of these impacts will depend on responses to these changes by harvesting and processing sectors. These responses will be reflected in fish and seafood markets through changes in prices and yield values and through changes in the costs of fishing (e.g., fuel prices). Taken together, it is the net value of fish (i.e., sales revenues minus costs) that will determine incomes of fishers and economic value of fish stocks to fishing-dependent communities.

ENSO-induced climate variability on fisheries can serve as a partial proxy for what could happen to individual fisheries as a result of climate change (Sumaila et al. 2011). In general, ENSO events are associated with a warming of sea-surface temperature in the tropical Eastern Pacific. Dalton (2001) used an ENSO event (1981 to 1999) to estimate the impact of climate change on fisheries in Monterey Bay for sablefish (*Anoplopoma fimbria*), albacore tuna (*Thunnus alalunga*), Chinook salmon (*Oncorhynchus tshawytscha*), and market squid (*Loligo opalescens*). These four fisheries were chosen because together they account for approximately 50% of the revenues associated with landings at Monterey Bay ports. Results showed a 60% decrease in active sablefish vessels in Monterey Bay accompanied by a 25% decrease in ex-vessel price (Dalton 2001). The albacore fishery showed temporary increases in active vessels and ex-vessel prices of approximately 20% in response to a major ENSO event. The Chinook fishery showed no change in numbers of vessels but did exhibit a substantial decrease in ex-vessel prices. Application of the same ENSO model for the market squid fishery showed an increase in ex-vessel price, together with a drastic decrease in the number of active vessels. In fact, there were no recorded landings of market squid at Monterey Bay ports during the major 1998 ENSO event, which had a sea-surface temperature anomaly of 1.9°C.

Unlike ocean warming, which has a robust literature on fisheries impacts (Sumaila et al. 2011), relatively few studies have assessed the wider impacts of ocean acidification on fisheries. Moore (2011) explored the economic impact of ocean acidification on US mollusc production and estimated the economic loss to be approximately $10 million per year ($0.07 per US household) in 2020 and to increase to almost $300 million per year ($1.78 per US household) in 2100. Based on regression results from a study by Ries et al. (2009) for 18 selected species of marine calcifiers, this impact represents a cumulative cost in net present value terms (with discount rate of 5%) of $734 million. However, ocean acidification impacts are difficult to predict and not all species will be impacted to the same degree. For example, the effect on blue crabs may not be as negative as that estimated for molluscs. Ries et al. (2009) found that blue crab, *Callinectes sapidus*, did not exhibit significant ocean acidification-related effects. However, there is already some evidence that ocean acidification can have an impact on other crab species (Walther et al. 2009, 2010), and that commercially important crab stocks in the North Pacific, for example, are vulnerable. Furthermore, the small set of previous studies on ocean acidification impacts did not differentiate effects on different life-history stages, including early life stages, which may be especially sensitive (Gazeau et al. 2010); it did not apply to animals where demographic factors are a critical feature of population dynamics, which is clearly the case with many, perhaps most, commercially important species.

Recent scientific concern about ocean acidification is turning to discussions of socioeconomic security (NRC 2010b), particularly to the potential negative impacts of ocean acidification on certain commercial fisheries (Cooley & Doney 2009, Cooley et al. 2009). According to a 2010 NRC report, "Ocean acidification may result in substantial losses and redistributions of economic benefits in commercial and recreational fisheries," adding that, "although fisheries make a relatively small contribution to the total economic activity at a national and international level, the impacts at the local and regional level and on particular user groups could be quite important" (NRC 2010b, p. 89).

In addition to the effects of climate change on fish stocks, both fishery operations and fishing-dependent communities are likely to be directly affected. Extreme weather events can disrupt fishing operations and damage the community-based infrastructure, such as landing sites, boats and gear, that supports the fisheries (Jallow et al. 1999, Westlund et al. 2007, Badjeck et al. 2010). Climate change that manifests itself in increased fluctuations in fishery production and income can affect communities through their choice of livelihoods and other social outcomes (Sarch & Allison 2000, Coulthard 2008, Iwasaki et al. 2009, Badjeck et al. 2010).

Fishing-dependent communities in the United States and elsewhere are diverse in economic and social characteristics. The effects of climate change on fisheries may be felt more acutely by those communities that are more dependent on fishing (i.e., those with fewer alternative economic activities or higher reliance on fisheries—especially for subsistence) and those that are more dependent on one or a few fish stocks (i.e., those less diversified in their target fisheries) (Phillips & Morrow 2007).

Overall, it is projected that the impact of climate change and ocean acidification on commercial fisheries will result in decreased job security for workers employed in commercial fishing gear manufacturing and sales, vessel loading, vessel construction and repair, fish and shellfish processing, wholesale and retail, commercial docks, ice suppliers to commercial fishing vessels, and other support industries, possibly resulting in income decline and job loss. Workers in recreational fishing gear and vessel sales, recreational outfitting, marinas, and other recreational support industries could also be affected.

Regional effects of climate change on fisheries

Although the specific effects of climate change on particular marine ecosystems and fish populations are difficult to predict, on a global and regional basis there is sufficient research to indicate that many, but not all, of these impacts will be negative (Grafton 2010). Some fish stocks are experiencing shifting distributions; for others, their overall abundance or population characteristics are fluctuating due to climate-induced shifts in marine ecosystems. The following presents a review of this research regarding the regional effects that are known or expected.

Subsistence fishing and hunting in the North Pacific   Alaskan communities and local economies depend on, and are engaged in, subsistence harvesting of marine resources more than any other region in the United States. Regional climatic and environmental changes are already having a notable (though unpredictable and often non-linear) effect on subsistence activities in the ocean environment, through changes in hydrology, seasonality and phenology, and fish and wildlife abundance and distributions (White et al. 2007, Loring & Gerlach 2009, McNeeley 2009, Rattenbury et al.

2009, Loring et al. 2011). Residents of rural Alaska are already reporting unprecedented changes in the geographic distribution and abundance of fish and marine mammals, increases in the frequency and intensity of storm surges in the Bering Sea, changes in the distribution and thickness of sea ice, and increases in coastal erosion. When combined with ongoing social and economic change, changes in climate, weather, and the biophysical system interact in a complex web of feedbacks, making life in rural Alaska extremely challenging.

Climate change-related effects on alterations in sea ice and weather patterns are also already creating numerous new environmental challenges for those who harvest marine species. The most striking change in the Arctic marine environment in recent years has been the rapid loss of summer sea ice (Perovich et al. 2011). In the Bering, Chukchi, and Beaufort Seas off Alaska's coast, this physical change has led to many ecological impacts (Moore & Gill 2011, Mueter et al. 2011b) and altered physical access to the region (AMSA 2009, Arctic Monitoring and Assessment Program [AMAP] 2008), affecting human use of marine resources.

For example, residents of Alaska Native communities rely on sea ice to ease their travel to the hunting grounds for whales, ice seals, walrus and polar bears. Krupnik et al. (2010) identified numerous effects of climate change that challenge and threaten local adaptive strategies, including times and modes of travel for hunting, fishing and foraging. In addition to the stress on marine mammal and polar bear populations, possibly resulting in lower quality and reduced number of prey, hunters will have to travel farther and longer to reach haul-outs and will have to travel over open water for greater distances, both of which will increase the risks associated with hunting for subsistence-dependent populations (Gearheard et al. 2006). Fuel and vessel maintenance costs associated with subsistence hunting will also increase as hunters need to travel greater distances (Callaway et al. 1999).

The impacts of climate change on Alaskans are also seen in shifts in the abundance and distribution of culturally important species. Salmon, which has been described as the cultural keystone food of Alaska, has become a less-dependable subsistence resource than in the past, with direct implications for food security (Loring & Gerlach 2010). A closure of the king salmon fishery on the Yukon River in 2009, for example, resulted in empty storage facilities, empty smokehouses and barren fish racks from Stevens Village to Fort Yukon and beyond. The 2009 closure produced a 'perfect storm' for a food security crisis, especially in combination with low harvest rates of moose and other terrestrial resources in some areas, the high price of fuel, and climate-driven changes in hydrology and water resources (Loring & Gerlach 2010).

Climate change effects on commercial fishing in the North Pacific The Eastern Bering Sea groundfish fishery, from which 14% of the total value of the fisheries of the United States is taken, is conducted north of the Alaskan Peninsula and Aleutian Islands (Hiatt et al. 2010). Climate change-related shifts in atmospheric conditions, ocean properties, and ecosystem interactions have the potential to greatly affect this multibillion-dollar industry. However, little concrete information is available regarding how these fisheries will be affected. Potential climate change impacts on North Pacific fisheries have been studied in only the past few years. Specific studies have examined how Pacific salmon, walleye pollock and Pacific cod populations are expected to react. Salmon populations are expected to be impacted by increased snowmelt and water flows, causing fall/winter floods that could affect salmon eggs laid in gravel beds (Low 2008). In the summer, higher average summer temperatures could diminish the oxygen content of the water in streams where smolts (juvenile fish) live, thus increasing smolt mortality. Warmer temperatures could also affect migration of smolts and lead to timing mismatches with their zooplankton prey base, timing that is critical as they enter salt water. While climate change impacts on fisheries in the North Pacific have not yet been well studied, it is anticipated that they could be significant. Studies have found that while there has been a northwards shift in the distribution of pollock and cod fishing in recent years (2006–2009), the northwards shifts are associated with colder-than-average years in the Bering Sea

(Haynie & Pfeiffer 2012, Pfeiffer & Haynie 2012). A large ice and cold pool extent concentrates fish populations in the northern region of the fishing grounds, giving fishers in the north an advantage over those in the south. The redistribution has occurred in both the winter and summer seasons of the Pacific cod fishery and in the summer pollock fishery. However, there has been little redistribution of effort in the winter pollock fishery, which is driven by the pursuit of valuable roe-bearing fish that spawn in the southern part of the Eastern Bering Sea. This large difference in value per fish in the roe fishery means that harvesters are unlikely to shift to the north for marginal increases in catchability.

Climate change effects on subsistence and commercial fishing on the West Coast   For tribes in the US Pacific North-west, questions have been raised recently about how climate change will affect the maintenance and reproduction of indigenous rights for salmon and other marine species whose distributions may change with a changing ocean environment (Colombi 2009). The right to harvest marine resources on traditional fishing grounds is guaranteed to these tribes through government-to-government treaties (NOAA Fisheries 2009). However, allocation of catch is based on allowable catch quantities, and treaty rights to harvest have referred to tribal 'usual-and-accustomed' fishing areas. Since some predictions of climate change impacts involve target species range shifts (Mantua et al. 2010), as water temperatures and prey ranges shift, the implications for geographically bounded tribal fishing rights are uncertain and of great to concern to North-west 'treaty tribes'.

Similarly, non-tribal fishers relying on personal use of marine resources are potentially impacted by changes in the abundance and range habitats of targeted nearshore species, as well as by climate-based shoreline shifts that may disrupt shoreside infrastructure. In Los Angeles County, more than three-quarters of a million low-income adults live with hunger or make daily decisions about whether to eat or pay for other essential needs, such as shelter or clothing (Harrison et al. 2007). To the extent that urban extraction of locally caught seafood represents a coping strategy for such food insecurity, potential nearshore climate change impacts present livelihood and nutritional issues for a number of pier-based fishers.

Little has been documented regarding how climate change is affecting fisheries along the West Coast of the United States. However, the largest effects on fisheries will likely be due to changes in the distributions and the abundance of stocks, as is documented in other regions. Perhaps the best example of known effects of climate variability on fish stocks in the Pacific is with the California sardine fishery; during a warm regime, biomass of Pacific sardine increases, and conversely, a cold-water regime results in a decrease in abundance of the sardine stock. In response to fluctuations in productivity due to climate variability, the US Pacific sardine fishery is managed using an environmentally based harvest control rule to determine the annual harvest level. The harvest control rule is intended to prevent overfishing, sustain consistent yield levels (Herrick et al. 2007), and reduce the exploitation rate if stock biomass decreases or if ocean conditions become cooler and less favourable for the stock. However, given the general lack of existing research and modelling capabilities on socioeconomic impacts of climate change on fisheries, the impacts on Pacific sardine fisheries are uncertain.

Climate change effects on subsistence and commercial fishing in the Pacific Islands   Much of the fishing effort in the Pacific Islands region depends on species and habitat associated with coral reefs (Bell et al. 2011). Coral reef habitats, and therefore the fish species that depend on them, are threatened by changes to water temperature, acidification of the ocean, sea-level rise, and possibly more severe cyclones and storms (Bell et al. 2011). The loss of live corals results in local extinctions and a reduced number of reef fish species (Karl et al. 2009). Declining coral reefs will have an impact on coastal communities, tourism, fisheries, and overall marine biodiversity; abundance of commercially important shellfish species may decline, and negative impacts on finfish may occur (Fletcher

2010). Reduced catches of reef-associated fish will widen the expected gap between the availability of fish and the protein needed for food security.

A recent assessment published by the Secretariat of the Pacific Community (SPC) assessed the vulnerability of tropical Pacific fisheries and aquaculture to climate change (Bell et al. 2011). The assessment did not include Hawaii but covered 22 Pacific Island countries and territories. Across the region, fish provide 51–94% of the animal protein in the diet in rural areas and 27–83% in urban areas. The great majority of fish needed for food security in the region is derived from coastal subsistence fishing; in 14 of the countries and territories, 52–91% of the fish eaten in rural areas is caught from coral reefs and other coastal habitats, and high levels of subsistence fishing are common in urban areas on many of the smaller island areas (Pratchett et al. 2011).

Nearly 70% of the world's annual tuna harvest, approximately 3.2 million tons, comes from the Pacific Ocean. Climate change is projected to cause a decline in tuna stocks and an eastwards shift in their location (IPCC 2007). On balance, the Pacific Island countries and territories (including the United States) appear to be in a better position than nations in other regions to cope with the implications of climate change for fisheries and aquaculture. Expected effects for the region as a whole are among the better possible outcomes worldwide. In particular, the Pacific Island commonwealths and territories (PICTs) with the greatest dependence on tuna (e.g., Kiribati, Nauru, Tuvalu and Tokelau) are likely to receive greater benefits as the fish move east, whereas the projected decreases in production occur in those PICTs where industrial fishing and processing make only modest contributions to GDP and government.

However, projections that storms (including cyclones, hurricanes and typhoons) could become progressively more intense would pose increased risk of damage to shore-based facilities and fleets for domestic tuna fishing and processing (Bell et al. 2011), as well as increased risk to safety at sea. The increased costs associated with repairing and relocating shore-based facilities and addressing increased risks to fishers' safety, could affect the profitability of domestic fishing operations. Aquaculture impacts could also occur; for example, changing patterns of precipitation and more intense storms could damage aquaculture ponds or make small pond farming more difficult due to more frequent droughts (Bell et al. 2011). There could also be higher financial risks associated with coastal aquaculture as a result of greater damage to infrastructure from rising sea levels and more severe storms.

For island fisheries sustained by healthy marine ecosystems and coral reefs, climate change impacts could exacerbate stresses such as overfishing, affecting both fisheries and tourism that depend on abundant and diverse reef fish (Karl et al. 2009). This context suggests that how society responds and adapts to the impacts of climate change not only may reduce some effects but also could exacerbate others, especially in the short term. One approach to addressing climate change impacts to subsistence fisheries is to reduce other types of fishing (i.e., commercial and recreational). Another likely response to reducing climate change effects on coral reefs and associated fish species will be the establishment of marine protected area (MPA) networks (Mumby et al. 2011a), which can enhance the resilience of marine resources to climate change and could protect certain areas from additional fishing pressure. When climate change poses risks to protected species such as monk seals (Baker et al. 2006) or loggerhead turtles (Van Houtan & Halley 2011), the resulting measures could potentially include reduced access to fisheries.

Climate change effects on fisheries in the South-east   Given limitations on the current knowledge of the biophysical effects of climate change in the south-eastern region, little is known about how the ocean services provided by the South Atlantic and Gulf of Mexico are or will be impacted. However, one of the most pronounced effects of climate change in the Gulf of Mexico is likely to be the increased intensity of hurricanes, whose impacts already include the loss of wetlands and barrier islands that protect or serve as nursery grounds for marine resources. The Gulf Coast represents the region with the highest potential for annual hurricane seasons that disrupt all types of fishing

Table 3 Known or expected direction of social and economic impacts on some major north-eastern commercial and recreational species

Species	Direction of impact
Atlantic cod (*Gadus morhua*)	Negative
Atlantic croaker (*Micropogonias undulatus*)	Positive
Atlantic lobster (*Homarus americanus*)	Ambiguous, but perhaps more negative
Atlantic sea scallop (*Placopecten magellanicus*)	Negative
Blue crab (*Callinectes sapidus*)	Negative

Source: Based on Hare et al. 2010; Fogarty et al. 2008; Frumhoff et al. 2007.

(NOAA Fisheries 2009). Loss of coastal habitat along the Gulf Coast has been well documented (Ingles & McIlvaine-Newsad 2007), and barrier island and wetland losses are projected to increase in the future. In addition, with the possible increases in hurricane intensity, storm surge and high winds, communities that rely on coastal marine resources for subsistence are likely to be increasingly limited in their ability to undertake harvesting activities.

Climate change effects on subsistence and commercial fishing in the North-east   For North-east fishers and the families, households, firms and communities that depend on them, the most relevant changes are those occurring to the ocean of the North-east US shelf ecosystem (NEUS) and its denizens. Water temperatures are rising, surface seawater pH is decreasing, precipitation is increasing, salinities are decreasing, and stratification is increasing (EAP 2012). All of these changes have an impact on marine life (Table 3).

Some North-east species, such as Atlantic cod (*Gadus morhua*), will likely move into Canadian waters and out of the range of North-east fishers due to warming water temperatures (Fogarty et al. 2008). Others, such as Atlantic croaker (*Micropogonias undulatus*), will likely see an increase in biomass as well as a range shift northwards from the Mid-Atlantic into southern New England, thus providing New England fishers with a larger stock to fish on but leaving Mid-Atlantic fishers with less easy access (Hare & Able 2007, Hare et al. 2010). Yet other species, such as American lobster, will likely also see their ranges move northwards, leaving the waters of New York and Rhode Island and increasing their presence in Maine; however, warmer waters may also lead to increases in 'lobster shell disease' (Frumhoff et al. 2007), making the impact on fishers more difficult to judge. Increased acidity could affect shellfish, including scallops, lobsters, and blue crab—three of the North-east's highest-value species, so economic and social impacts are potentially high (Cooley & Doney 2009, McCay et al. 2011). Certain shellfish (e.g., lobster) could also suffer from sea-level rise if the coastal wetlands necessary to their juvenile stages are flooded (Frumhoff et al. 2007).

However, marine ecological changes are not the only climate change issues affecting fishers and fishing-dependent industries. Sea-level rise will also flood coastal infrastructure, especially docks and other fishing-related structures that are on the very edge of the current coastline. In the North-east, many smaller ports have already lost infrastructure to gentrification (Gale 1991, Colburn & Jepson 2012), among other causes. With vital infrastructure such as boat repair facilities concentrated in fewer ports, loss in any of the remaining hubs could have important negative impacts on the entire region's fishing fleet (NOAA 1997a,b, Robinson & Gloucester Community Panel 2003, Robinson et al. 2005).

Implications of climate change for aquaculture

The United States imports 86% of its seafood, and approximately half of that is from aquaculture production. Two-thirds of marine aquaculture is molluscan shellfish, such as oysters, clams, and mussels, and the remainder is shrimp and salmon, with lesser amounts of barramundi, sea bass, sea bream, and other species. The impacts of climate change on global aquaculture are not yet fully known. Unlike capture fisheries, organisms being reared in captivity are subject to more controlled environments that may allow for acclimation or adaptation. Captive breeding can also lead to reduction in genetic heterogeneity, which can make populations even more vulnerable to stress (given lower diversity/resilience).

As in commercial, recreational and subsistence capture fisheries, climate impacts of aquaculture are likely to be both positive and negative, arising from direct (e.g., through physical and physiological processes) and indirect impacts (e.g., through impacts on the natural resources required for aquaculture), the major issues being water, land, seed, feed and energy (De Silva & Soto 2009).

Direct impacts of climate change on aquaculture

A rise in sea-surface temperatures may trigger the growth of HABs that can extend the spatial or temporal scope of a bloom or release toxins into the water and kill cultured fish and shellfish, particularly for fish in cage-based aquaculture systems and shellfish beds. Higher water temperatures may also result in increased disease incidence and parasites, which may develop more rapidly in warmer waters and higher salinities and threaten the aquaculture sector. Species cultured in temperate regions, predominantly salmon and cod species, have a relatively narrow range of optimal temperatures for growth. For example, temperatures over 17°C would be detrimental to the salmon-farming sector due to heat stress, causing feed intake to drop and feed utilization to be reduced (De Silva & Soto 2009).

Certain aspects of aquaculture may benefit from climate-related changes. Higher water temperatures may also increase the availability of new culture sites, especially in areas previously too cold to support aquaculture. An increase in water temperature may also have a positive effect on metabolism and stimulate growth of cultured species, as long as the change is gradual and within the thermal tolerance range of the species.

Aquaculture does have some advantages for dealing with climate-related impacts. The tool of selective breeding could potentially give aquacultured stocks an adaptive advantage since changes in preferred genetic traits can be as high as 10% per generation for selective breeding programmes (Gjøen & Bentsen 1997). Most aquacultured species still contain a great deal of genetic diversity, which means they should be adaptable to the direct impacts of climate change. Likewise, aquacultured organisms can be treated for parasites and diseases (Moffitt et al. 1998, DIPNET 2007). Vaccination, selective breeding, and better nutrition have all improved the resistance of farmed fish to wild diseases, and this trend is likely to continue (Torrissen et al. 2011).

Indirect impacts of climate change on aquaculture

The dependence of aquaculture on fishmeal and fish oil becomes an important issue under most climate change scenarios (De Silva & Soto 2009). Tacon et al. (2011) estimated on a global basis that in 2008 the aquaculture sector consumed 3.72 million metric tons of fishmeal (60.8% of the total global fishmeal production) and 0.78 million metric tons of fish oil (73.8% of the total reported global fish oil production in 2008). Industrial fishmeal and fish oil production is typically based on a few, fast-growing, short-lived, productive stocks of small pelagic fish in the subtropical and temperate regions. The major stocks that contribute to this global industry, of which the United States is a major exporter, are the Peruvian anchovy, capelin, sandeels, and sardines.

Schmittner (2005) has predicted that the biological productivity of the North Atlantic will decrease by 50% and ocean productivity worldwide by 20% due to climate change. Changes in productivity would greatly impact the availability of the small pelagics for fishmeal and oil. It is also possible that predicted changes in ocean circulation patterns will result in the occurrence of ENSO influences becoming more frequent, with impacts on the reliability of stocks of the small pelagics utilized for fishmeal and fish oil production.

Ocean acidification and aquaculture

In North America, ocean acidification is currently considered a serious near-term threat because of its potential to alter ocean food webs in a relatively short time period (De Silva & Soto 2009). For aquaculture, ocean acidification particularly influences shell formation and affects filter-feeding shellfish. Protecting vulnerable marine organisms grown in aquaculture facilities from the effects of ocean acidification may be possible in theory, but it presents practical challenges. Aquaculture is often conducted on land in tanks or ponds that are filled with coastal seawater or within coastal ocean pens. Adjusting seawater chemistry before supplying culture tanks on land would require equipment and monitoring that might increase the overhead of aquaculture operations, and aqua-cultured animals in nearshore operations cannot be shielded from ocean acidification (Cooley & Doney 2009).

Social impacts of climate change on aquaculture

Impacts of climate change on capture fisheries, such as damage to physical capital and impacts on transportation and marketing systems and channels, are likely to be mirrored in aquaculture (Cochrane et al. 2009). Likely the greatest social impact of climate change that must be dealt with by the aquaculture industry is on human health. Seafood consumption may have a number of health benefits, including improved cardiovascular function, reduced inflammatory disease, reduced macular degeneration, reduced mental depression, and higher IQ (Institute of Medicine 2006, FAO/WHO 2011). It is also clear that aquaculture will be necessary to supply the increased volume of seafood needed to support a growing global population. If climate change reduces wild harvest, then the production from aquaculture will have to be that much greater to meet the demand.

Offshore energy development

Oil and gas

Offshore oil and gas development has been increasing in recent years. The oil and gas industry and its consumers now face having to adapt to climate changes that they contributed to generating. Industry reaction will vary by type (international companies, small independents, or state owned); specific geographic location; local policies and regulations; the company's ethics; and combined industry performance on the national and international markets. Figure 11 illustrates the perspective of adaptive government regulations in Alaska, which have been changing with warmer temperatures (Arctic Climate Impact Assessment [ACIA] 2004). These issues need to be analyzed and understood in a broad context, given that the oil and gas industry not only delivers oil and gas, but also provides jobs.

The impacts of climate change on the marine sector of the oil and gas industry can be direct (e.g., on-site changes of environmental conditions) or indirect (e.g., pressures exerted by the public and governments). The financial impact of climate policies and restricted access to reserves, due to environmental protections or because companies are developing resources in less-accessible locations every year, is estimated to reduce shareholder value by between 1% and 7%, depending on the company (Austin & Sauer 2002).

Impact factors   Currently, five primary impacts of climate change have been associated with recent offshore oil and gas exploration (Acclimatise 2009a):

1. *Increased Pressure on Water Resources*: Changing rainfall amounts, the need for potable water, and droughts will all increase the demand for water, which is key in sustaining oil and gas production.

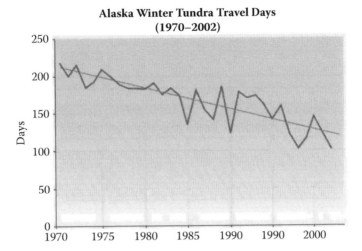

Figure 11 Adaptive government regulations in Alaska for oil exploration. The number of days in which oil exploration activities on the tundra are allowed under the Alaska Department of Natural Resources standards halved from 1970 to 2002 due to permafrost thaw, which is disrupting transportation, damaging buildings and assets (and in particular pipelines) and increasing the risk of pollution. Operational costs are increasing for oil and gas companies. (From Arctic Climate Impact Assessment [ACIA] 2004.)

2. *Physical Asset Failure*: Several types of existing equipment not only are old but also were designed to function under climate conditions typical of 20 to 40 years ago. Included in this category are energy supplies (e.g., generators and batteries), off-site utilities, and waste and water treatment technologies.

3. *Employee Health and Safety Risks*: As environmental conditions change, the oil and gas industry is exploring potential oil and gas reserves in areas (e.g., ultradeep waters, the Arctic Ocean) where ambient conditions are significantly more extreme and dangerous for industry workers than areas under current use, with consequent increases in insurance costs, salaries and other operational costs.

4. *Drop in Value of Financial Assets*: To meet the growing demand for energy, oil and gas companies need to continue securing investments for new exploration, production and manufacturing. Potential investors and stakeholders are placing greater importance on the business impacts of climate change as the risks have an impact on cost and revenue drivers. Beyond the safety-driven increases noted, insurance costs could potentially rise because of greater risk of physical plant damage due to extreme weather events.

5. *Damage to Corporate Reputation*: As knowledge and awareness of climate change grows, any failure to monitor and report the impacts of climate change on social and ecological resources is increasingly likely to harm oil and gas companies' reputations. Contractual relationships that do not adequately foresee and manage risks driven by climate change may damage a company's reputation with stakeholders, increasing the risk of parties turning to litigation.

Changes in regulations Governments will variably have an impact on the oil and gas industry based on their particular climate change policies and regulations. Fish and marine mammal species foreign to the Arctic Ocean just 5 years ago are starting to be sighted for the first time off the northern shore of Alaska (Acclimatise 2009b). These new inhabitants of the Chukchi and Beaufort Seas will likely trigger new protective measures by state and federal regulatory agencies.

At lower latitudes, in the Gulf of Mexico area, it is likely that future tropical weather events (such as typhoons and hurricanes) will become more intense (Ulbrich et al. 2008). However, these modelling projections carry uncertainties, thus making it difficult to anticipate the magnitude of such changes in the development of new regulations. It may be that industry is already responding to potential changes in the regulatory environment. For example, in 2006, following the extreme 2005 hurricane season, the American Petroleum Institute launched a process for reviewing design and safety standards for offshore oil platforms. Additional regulations will certainly increase operational costs; however, these risks will also be accompanied by opportunities. Reduced sea ice coverage, for instance, will lead to the opening of new shipping lanes, facilitating the transport of crude oil between the Atlantic and the Pacific Oceans. New or amended regulatory regimes may thus be required for these areas.

Drivers for change and projections   New regulations will require companies to invest in alternatives to fossil fuels and develop cleaner and more sustainable energy sources. It is hoped those changes will be reflected by an increase in profits in the mid- and long terms. There is consensus that the adoption of carbon sequestration procedures, combined with the inclusion of renewable energy production, will transform the current oil and gas industry (Lovell 2010). This new version will have the capacity to deliver energy products obtained from both renewable and non-renewable sources, simultaneously reducing emissions as required for compliance with national and international regulations.

The future of the industry   The great dilemma that society and oil and gas companies face today is based on balancing growing requirements to limit greenhouse gas emissions with desires for increased energy consumption and company profits (Van den Hove et al. 2002). Different companies have taken different approaches to this dilemma (Van den Hove et al. 2002). For instance, some companies argue that the risk of climate change is less than the risk of negative impacts to a profitable oil and gas industry that can boost the economy and technological development since reduced profits would presumably impact the entire economy (Button 1992, Van den Hove et al. 2002). Other companies weigh this question of profits and the well-being of the economy versus climate change impacts on society and ultimately the economy as well, providing their own weights and justifications. It is also possible that some or all of these companies might focus on lowering emissions, especially if they are convinced that this approach will maximize profits.

Renewable energy (wind, ocean waves and currents)

While coastal and marine environments do not currently host commercial facilities that generate electricity, several projects are proposed, and pilot projects are being tested. The possible types of renewable energy that may be developed in coastal and marine environments include wind, wave, ocean current, tidal, hydrogen generation and solar. The Bureau of Ocean Energy Management (BOEM, then the Minerals Management Service) prepared a Programmatic Environmental Impact Statement (PEIS) (BOEM 2007) which examined the potential environmental consequences of alternative energy development. The PEIS analysis determined that wind, wave and ocean current technologies were the most advanced and likely to be developed on the outer continental shelf. Along the Atlantic Coast, one wind facility is approved for development, and several others are proposed. Wave energy is most intense along the Pacific Coast, and technology testing is under way. The Gulf Stream along the south-eastern coast of Florida is the most favourable for ocean current development, and one pilot project is in development.

With respect to climate impacts, coastal and marine renewable energy projects are evaluated as mitigation measures because they do not directly result in emissions of greenhouse gases. The effects of climate change on the renewable energy industry have not been assessed, either along the US coast or elsewhere. However, as with the oil and gas industries, climate change is expected to affect the industry. Potential impacts include damage to infrastructure from increased storm intensity through larger waves, stronger currents, or sediment erosion and potential change in the resource being harnessed, including changes in wind speed, wave height, or ocean current intensity or direction. These changes could have either a positive or negative effect.

The offshore renewable energy sector is nascent and, unlike the financially self-sustaining oil and gas industry, requires investment from the public sector, at least in the United States, as construction of offshore facilities is more costly; specifically, offshore wind is more expensive than onshore. Uncertainty based on climate change projections could alter the evaluation of risk and potentially deter speculative investments in this emerging industry.

Tourism and recreation

Tourism is an important part of the US economy, contributing $1.8 trillion in economic output and supporting 14.1 million jobs in 2011 (US Travel Association 2012). Tourism is also one of the few sectors that has been growing during the current tentative economic recovery, with 101 million international tourist arrivals to North America in 2011, up 2.9% from 2010 (United Nations World Tourism Organization [UNWTO] 2012). Nationally, 2.8% of gross domestic product, 7.52 million jobs and $1.11 trillion in travel and tourism total sales are supported by tourism (Office of Travel and Tourism Industries [OTTI] 2011a,b).

Coastal tourism and recreation are used to describe all tourism, leisure, and recreationally oriented activities that take place on the coast and in coastal waters. Main activities involved in coastal tourism and recreation include visiting beaches, diving and snorkelling, cruises, boating and sailing, and bird and marine mammal watching. In addition, infrastructure, such as hotels, restaurants, vacation homes, marinas, dive shops, harbours and beaches, are present in coastal areas to support these tourism and recreational activities. Tourism statistics are difficult to disaggregate solely for coastal areas; however, the most recent data from the OTTI show that in 2009–2010, nine of the top ten states and US territories visited by overseas travellers were coastal (including the Great Lakes), and seven of the top ten cities were located on the coast (OTTI 2011a,b). An estimated 1.5 billion person-trips for leisure, based on overnight trips in paid accommodations or travel to destinations 50 or more miles from home, occurred in the United States in 2010 (US Travel Association 2012).

In the face of climate change, impacts to marine resource distribution, variable weather conditions, and extreme events such as typhoons and hurricanes are expected to pose the most significant impacts on the tourism industry. However, the projected impacts of climate change are expected to affect tourism and recreation industries and their associated infrastructure in a variety of ways: positively, negatively and mixed (Scott et al. 2004, Moreno & Becken 2009, Yu et al. 2009). Unfortunately, the science of assessing predicted impacts on these industries is still in its early stages.

Many coastal and marine tourism and recreational activities depend on favourable weather and climate. Activities such as diving and snorkelling rely on comfortable water and air temperatures and calm waters for boat travel to snorkelling and dive sites. Weather-related impacts, such as potential changes in wind patterns and wave height and direction, could affect activities such as sailing and surfing. Temperature is predicted to have an impact on biophysical events, such as marine mammal and seabird migrations, affecting recreation involving watching or interacting with these animals (Lambert et al. 2010). Arctic cruise tourism is expected to increase with increasing sea-surface temperatures and decreased sea ice during Northern Hemisphere summers (Stewart et al. 2007). However, researchers warn that climatic warming in the Arctic may change the distribution of sea ice, resulting in negative implications for tourist transits in the High Arctic and North-west Passage regions (Stewart et al. 2007).

Sea-level rise may also have an impact on coastal tourism and recreation in a variety of ways. Increased sea levels are likely to reduce the size of sandy beaches in some areas and possibly increase erosion rates (Yu et al. 2009). Hawai'i's Waikiki Beach is recognized as a major tourism destination and a popular recreational spot for both visitors and residents. In 2007, approximately 4 million tourists visited Waikiki Beach, and hotels sold 3.9 million room nights (State of Hawai'i Department of Business, Economic Development and Tourism [DBEDT] 2008). Given the popularity and economic importance of Waikiki, the issue of beach erosion has been an ongoing concern. It is estimated that nearly $2 billion in overall visitor expenditures could be lost per year due to complete erosion. Also, an estimated $66 million in tax revenue would be lost. Indirect effects could include hotel industry job loss of over 6000 jobs per year. In addition, higher sea levels could cause the landwards migration or flooding of coastal lagoons and other coastal habitats for species, such as seabirds, that are attractive for wildlife viewing (Bird 1994). Furthermore, sea-level rise threatens coastal infrastructure, such as marinas, boardwalks, hotels and houses, both directly and indirectly with increased inundation and erosion (Scott et al. 2004).

Finally, HAB events have been shown to significantly reduce reported business revenues for the lodging and restaurant sectors in affected coastal communities, with implications for local and state tax revenues (Hoagland et al. 2002). Morgan et al. (2010) estimated how and why participation in marine-based activities (e.g., beach-going, fishing, and coastal restaurant patronage) were affected during a red tide event. The authors found that recreational activities of 63% and 70% of south-western Florida residents who go saltwater fishing or go to beaches, respectively, were adversely affected (cancelling, delaying, cutting short their trip, or relocating their trip) by red tide events during the previous year. Given that the geographic and temporal scale of red tides have been shown to be affected by water temperature and potentially water quality, it is likely that climate change could increase their occurrence.

Human health

In addition to current and future climate change impacts on the biophysical and socioeconomic aspects of marine resources, there is a broader growth in knowledge of the human health dimensions of global climate change. Public health scientists with the US National Center for Environmental Health at the Centers for Disease Control and Prevention (CDC) have identified a number of primary areas in which climate change has an impact on human health and will likely exacerbate human vulnerability and sensitivity in the future (McGeehin 2007). These include, but are not limited to, extreme weather-related injuries, morbidity and mortality; decline in access to drinkable water; increased food insecurity and malnutrition; rising pollutant-related respiratory problems; and increased spread of infectious disease. This list of consequences illustrates the complex and varied ways in which the social impact of climate change in the United States extends beyond actuarial statistics and measures of economic loss and includes broad, critical aspects of human health, well-being and vulnerability (Brown 1999).

Health and vulnerability

Complex social and ecosystem conditions inform the reach and range of climate change effects on health, which is "not some absolute state of being but an elastic concept that must be evaluated in a larger socio-cultural context" (Baer et al. 2003, p. 5). While the environmental health effects of climate change on marine resource users in particular are not widely known, there is growing recognition of human vulnerability and sensitivity in the wake of global climate change. Vulnerability is a fundamental concept for assessing the role of climate change in determining health, especially because it merges theory and empirical findings from disaster studies in general and public health science, social and economic analysis and risk assessment in particular (Baer & Singer 2009). Disease vulnerability and environmental health risk may both become increasingly central issues in research exploring connections among climate change, marine resource contamination or decline, and poverty, especially since subjugated populations in coastal regions tend to be marine resource users.

Waterborne and foodborne diseases

The impact of warming oceans on waterborne pathogens that cause both seafood-related and direct contact wound infections has generated growing concern in a time of climate change. Much attention has been directed at pathogens in the *Vibrio* family, especially *V. cholera*, in global health research and intervention (Lipp et al. 2002). More recently, with regard to US cases, there has been increased concern about other *Vibrio* species, including *V. parahaemolyticus* and *V. vulnificus*, both known sources of seafood-related acute gastroenteritis. In 2011, the CDC, which maintains a voluntary surveillance system of culture-confirmed *Vibrio* infections in the Gulf of Mexico region, estimated 45,000 annual cases of *V. parahaemolyticus* and 207 cases of *V. vulnificus* in the United States (CDC 2011, Hlavsa et al. 2011). The highest concentrations of *Vibrio* infections were in the Mid-Atlantic states that surround Chesapeake Bay, where 305 cases were reported in 2009. Given this high concentration in Chesapeake Bay, there is a high likelihood that these numbers could increase significantly if water temperatures in the bay rise in the future.

There has been reported expansion of *V. parahaemolyticus* in the Pacific North-west and Alaska that closely corresponds with climate anomalies related to El Niño (CDC 1998, McLaughlin et al. 2005, Martinez-Urtaza et al. 2010). In addition, in the days after Hurricane Katrina, 22 *Vibrio* wound infections were recorded, of which 3 were caused by *V. parahaemolyticus*, 2 of which led to the deaths of the infected individuals (CDC 2005).

Vibrio vulnificus is perhaps the most threatening pathogenic *Vibrio* in the United States because of its highly invasive nature and high fatality rate following infection (Horseman & Surani 2011). In recent years, it has come to be recognized as the most virulent foodborne pathogen in the United States, with a fatality rate as high as 60% (Oliver 2005), and has been responsible for the overwhelming majority of reported US seafood-related deaths (Oliver & Kaper 2007). *V. vulnificus* can be transmitted to humans by way of consumption and dermal exposure. The bacterium is frequently isolated from oysters and other shellfish in warm coastal waters during the summer months. Since it is naturally found in warm marine waters, people with open wounds can be exposed to *V. vulnificus* through direct contact with seawater. A review of this bacterium found the death rate among domestically acquired foodborne illness associated with *V. vulnificus* (34.8%) was significantly higher than any of the other 31 foodborne pathogens assessed (ranging from 0 to 17.3%) (Scallan et al. 2011). In addition, *V. vulnificus* is becoming a significant and growing source of potentially fatal wound infections associated with recreational swimming, fishing-related cuts, and seafood handling (Weis et al. 2011). One study reported that almost 70% of infected individuals developed secondary lesions requiring tissue debridement or limb amputation (Oliver 1989). *V. vulnificus* is most frequently found in water with a temperature above 20°C, which is especially important given the significant changes in temperature that are anticipated in coastal waters over the coming decades. While there is no definitive link between the increasing incidence of *Vibrio* and climate change, sufficient data, such as these, are available to warrant closer research attention (Greer et al. 2008).

In addition to members of the *Vibrio* family, a number of other marine pathogens merit monitoring in a warming environment. *Aeromonas hydrophila* is a widely distributed inhabitant of both fresh and salt waters, as well as a common fish pathogen. In humans, it is known to cause gastroenteritis and a variety of extraintestinal infections, including endocarditis, pneumonia, conjunctivitis, and urinary tract infections, and is also capable of causing localized wound infections in individuals with intact immune systems (Collier 2002). *Myobacterium marinum* infections, of which approximately 200 are reported each year in the United States, have been described as an emerging necrotizing mycobacteria-caused disease involving both marine and freshwater exposure during a water-related injury (Dobos et al. 1999). *Erysipelothrix rhusiopathiae* is found in diverse animal species, including fish and shellfish. Successfully transferred to humans through cuts, this pathogen causes cutaneous eruptions on the hands or fingers. It is popularly referred to as 'shrimp picker's disease' and 'crab poisoning' in marine locations (Brooke & Riley 1999). Increased rates of infection have been documented for these emerging diseases. Like *V. vulnificus*, each of these pathogens has the potential for increased rates of infection as a result of warming ocean waters.

Harmful algal blooms and climate change

HABs, which occur worldwide and in all US states, have recently increased in duration and geographic range and have also involved new species and impacts (Moore et al. 2008, Hallegraeff 2010, Anderson 2012). Many HABs produce potent toxins that can kill or sicken humans, fish, birds, turtles, marine mammals, domestic animals and pets. Human health is threatened through exposure to toxin-contaminated shellfish and fish, drinking water, or aerosols. Monitoring and adaptations, such as shellfish harvesting closures, beach closures, and drinking water treatment minimize threats to human health. Control measures, while critical to protect public health, can reduce the availability of important sources of nutrition or income to communities that depend on the impacted resources. The economic consequence, based on a subset of HAB events that affected the United States, has been conservatively estimated at $82 million per year (Hoagland & Scatasta 2006), but only some of the impacts listed are included in this estimate.

HAB occurrence may be altered by climate change impacts, including increases in water temperature, stratification, and increased CO_2; alteration of currents or hydrology; and changes in nutrient availability due to upwelling or run-off. However, HABs are caused by a diverse group of organisms, and their growth and toxicity will respond very differently to changing environmental conditions. Increases in cyanobacterial blooms (CyanoHABs), many of which produce cyanotoxins linked to liver, digestive, skin, and neurological illness and even death, have already been well documented and attributed to a combination of increased nutrients and climate change (Paerl & Paul 2012). Massive toxic blooms threaten drinking water supplies and recreational use of water bodies, especially in areas experiencing droughts. For other HABs, change in climate may increase the period of time when environmental conditions are suitable for blooms to occur. For example, in Puget Sound (Washington State), as in many areas in the north-western and north-eastern United States, shellfish harvesting is often closed for a period in the late spring or summer. The closure period corresponds to the window of opportunity when environmental conditions are optimal for blooms of *Alexandrium* (Moore et al. 2008, 2009). This toxic dinoflagellate produces potent neurotoxins that can cause illness and death in humans when they eat contaminated shellfish. Climate change scenarios indicate that the window of opportunity will increase substantially, even for modest climate change projections (Moore et al. 2011).

Climate change may also alter conditions so that they become unfavourable for HAB growth. Ciguatera fish poisoning (CFP) is caused by ciguatoxin produced by a benthic dinoflagellate, *Gambierdiscus*, living on macroalgae on tropical hard substrates, especially coral reefs. The toxins accumulate in higher trophic level fish, which, when consumed by humans, causes CFP, a debilitating illness. CFP is the most common HAB-caused illness in the world, and it also deters fish consumption in many areas where protein is in short supply. A recent study (Tester et al. 2010) showed that CFP incidence in the Caribbean is highest where water temperatures are highest and postulated that climate change may be one factor in recent outbreaks of CFP from fish caught near oil platforms in the more temperate Gulf of Mexico (Villareal et al. 2007). However, data from the South Pacific suggest that waters may become too hot for the causative organism to grow (Llewellyn 2010); therefore, it is possible that the geographic distribution of CFP may change, but not the incidence.

Health risks related to climate impacts on marine zoonotic diseases

A global analysis of trends in infectious diseases found that emerging infectious disease events were increasing over time and were dominated by zoonotic diseases (transmitted between animals and humans), with the majority of those diseases (72%) originating in wildlife (Jones et al. 2008). Climate change may have an impact on infectious zoonotic diseases by prolonging periods when diseases may be transmissible and by changing geographic ranges of disease and animal reservoirs (Greer et al. 2008).

It has been difficult to make a definitive link between increases in marine zoonotic disease and climate change due to multiple contributing stressors (Wilcox & Gubler 2005, Burek et al. 2008), as well as lack of sufficient baseline data for some organisms (Burek et al. 2008), but there are examples of changes in latitudinal distributions of infectious organisms. For example, *Lacazia loboi* is a cutaneous fungus that has been reported to infect humans and dolphins in tropical and transitional tropical climates. The disease more recently has been diagnosed in dolphins off the coast of North Carolina, which represents a change in the latitudinal distribution of this fungus (Rotstein et al. 2009). To detect such changes, it is critical to continue monitoring and conducting assessments of disease in marine animals to establish baselines and identify trends. Furthermore, an integrated monitoring and surveillance system will be important to provide early warnings and better public information for any emerging diseases that are a threat to human health. Some coastal and tribal communities that depend on marine animals as traditional sources of nutrition are particularly vulnerable to outbreaks that have an impact on their already at risk food supply.

Health risks of extreme weather events

People living in coastal environments might also be at greater health risk when considering the increases in extreme weather events resulting from global climate change (Greenough et al. 2001). The IPCC noted that warming would vary by region but would overall be accompanied by changes in precipitation, in the variability of climate, and in the frequency and intensity of some extreme weather phenomena (IPCC 2007). The health risks of extreme weather are many, including, but not limited to, injury from storm wreckage, risks associated with poor drainage and impaired sanitation, heat exhaustion and other heat-related illnesses (Semenza et al. 1999, McGeehin & Mirabelli 2001, Bernard & McGeehin 2004, Luber & McGeehin 2008), mental health illnesses (CDC 2006, Norris et al. 2006, van Griensven et al. 2006), and vector-borne and zoonotic diseases (Parmenter et al. 1999, Glass et al. 2000, Enscore et al. 2002, Collinge et al. 2005, Eisen et al. 2007, Gage et al. 2008).

Globalized seafood and emerging health risks

Due to a rise in both the globalization of seafood and demand in the US market, the Food and Drug Administration (FDA) reports that the United States now imports more than 80% of its seafood supply, including wild-caught fish and aquaculture fish (US Food and Drug Administration [USDA] 2008). This seafood originates from over 13,000 suppliers in over 160 countries, with China being the largest exporter of seafood to the United States by volume (US Government Accountability Office [GAO] 2004). Imported seafood can be a source of health risk involving multiple agents, especially bacteria (e.g., *Salmonella*, *Campylobacter*, verotoxin-producing *Escherichia coli*, *Listeria*); parasites (*Toxoplasma gondii*, *Cyclospora cayetanensis*, *Trichinella*); and viruses (norovirus, hepatitis A virus), as well as rarer infectious agents and mycotoxins (Buisson et al. 2008). The critical question is whether climate change and the further globalization of seafood will contribute to additional increases in the prevalence of infected seafood available to consumers in the American market.

Climate change has the potential to adversely impact imported seafood in two ways. First, climate change is a risk to the degree that warming oceans and other changes in the marine environment increase rates of infection of various seafood stocks worldwide (including locations that export seafood to the United States). Second, rising temperatures and changing weather patterns may result in inadequate cooling of seafood at various points in the import process, allowing the growth of infectious agents. These scenarios, combined with the relatively low level of FDA testing of imported seafood, suggest the need for increased attention to this potential threat to US public health in a time of climate change.

Governance challenges

Many natural resource governance institutions have been built assuming stable environmental conditions that are similar to observed historical experience (Peloso 2010). In many instances of greater climate variability or climate change, these assumptions will be challenged or no longer valid. Climate change impacts on the distribution and accessibility of living natural resources and ecosystems will in some cases require changes in jurisdictional boundaries established by national or international management institutions. While governments can employ technologies to achieve better climate change preparedness and response, it is important to keep in mind that "technologies are only as effective as the social and political networks that use them for risk assessment, planning and responding to disasters" (Dowty & Allen 2011). In this regard, the National Academy of Sciences has repeatedly called for early, active, continuous, and transparent 'community' involvement in risk management decisions, and not just reliance on technology-based decisions (NRC 1996, 2000a,b) orchestrated by government scientists and experts (Fischer 2000). This broader, more inclusive governance approach is ever more important as marine resource users and coastal communities (1) increasingly adapt and respond to changing security conditions, the restructuring of transportation networks, and a warming climate; and (2) demand further involvement in climate change discussions, especially those resulting in changes in marine resource management decisions and policies.

Fisheries management in the United States

Federal fisheries management occurs mainly within the framework of the Magnuson-Stevens Fishery Conservation and Management Act (MSA). Within this framework, eight fishery management councils develop fishing regulations for specific regions and fisheries in cooperation with the US federal government, represented by the National Marine Fisheries Service. Management plans from the councils are designed to meet 10 national standards set by the MSA. Important among these standards is the requirement to prevent overfishing while achieving optimum yield. This optimum yield is the basis for caps on total harvest or annual catch limits (ACLs). ACLs are established by the Scientific and Statistical Committees (SSCs) of the Fishery Management Councils. The SSCs often establish uncertainty buffers to prevent overfishing. The SSCs are able to adjust annual harvest recommendations to fluctuations in stock size to prevent overfishing. For example, there is evidence that warming trends in the Bering Sea have caused increased overlap of pollock and salmon stocks, which has led to increased salmon by-catch rates in the pollock fishery (Stram & Evans 2009). As a result, the North Pacific Fishery Management Council (NPFMC) has taken steps to minimize salmon by-catch rates by limiting pollock fishing at certain times and in certain areas.

Climate change has an impact on stock levels, spatial distribution, and year-to-year variability in stock levels, resulting in increased uncertainty that can have an impact on decisions made by fisheries management, such as total allowable catch (TAC) limits and ACLs. For example, the survey area on which a stock assessment is based may no longer appropriately cover the range of the species, perhaps leading to an assessment suggesting a stock decline as opposed to a shift in location. This could result in setting a TAC lower than it should be for a stock that is otherwise healthy, costing fishers their livelihood. Uncertainties, such as accurate stock range, are likely to be built into the setting of more conservative TACs and ACLs. Changes in marine target populations mean managers must anticipate problems and build flexibility into management plans; many current regulations tie fishers to particular species in specific areas (Organization for Economic Cooperation and Development [OECD] 2010). Further, significant discussion needs to take place on societal value of fish stocks and those communities that depend on fishing to sustain their livelihoods. Transboundary and other jurisdictional issues will also emerge (Herrick et al. 2007).

Governance challenges associated with transboundary stocks are likely to increase, given that international agreements on fishing shared stocks are based on stable, historical abundance and spatial distribution patterns. Miller and Munro (2003) used a theoretical model to illustrate potential problems with transboundary Pacific salmon stocks in the United States and Canada. Their results highlight the need to update existing, or in some cases negotiate new, more flexible, international agreements. To adequately account for, and ultimately reduce, the amount of uncertainty related to climate change, enhanced monitoring and improved stock assessment methodology are needed.

Offshore energy development

Climate change will present new opportunities and new challenges to offshore energy development. New areas, especially in the Arctic, will become accessible to development and to marine energy transport, but this will also bring industry activities into contact and potential conflict with new environments and other uses, such as subsistence and tourism. Operations in new and traditional areas may face new climactic challenges with increasing storms, more severe operating conditions and more sensitive species and ecosystems. In addition, routine operations, such as oil transport, are likely to carry increased risk in Arctic waters, where lack of charts, extreme weather conditions and poor oil spill response capability hamper safe operations. Additional regulations will certainly increase operational costs, although in the Arctic, these risks will also be accompanied by opportunities as the reduced sea-ice coverage will lead to the opening of new shipping lanes, facilitating the transport of crude oil between the Atlantic and the Pacific Oceans. There is potential for restructuring of both current fisheries and offshore energy policy and management regimes through approaches such as the creation of protected areas for various marine resources and habitats.

Tourism and recreation

Tourism is seldom recognized as a single sector for policy and regulation. Policies in many different sectors, such as shipping (cruise ships), fisheries management (marine sanctuaries and protected areas), habitat protection (regulation of coastal development), and the business sector (departments and chambers of commerce), affect tourism and the tourism industry. As such, changes made by the governance structures in each of these sectors in response to climate change are likely to have an impact on tourism and recreation in the United States. An important consideration is the safety of life at sea, as with decreasing summer sea ice, tourists and cruise vessels venture into higher latitudes and uncharted waters far from established ports and rescue capabilities. The International Maritime Organization and the Arctic Council are already cooperating to increase governance measures in this area.

Human health

The consequences of climate change related to human health showcase the complex and variegated ways in which the social impact of climate change in the United States extends beyond measures of economic loss to broader and very critical aspects of human health, well-being, and vulnerability (Brown 1999). There are a large number of agencies in each country at the international, national, state and local levels that develop policy and regulations for these areas, including seafood safety, water quality, disaster response and disease outbreak avoidance and containment and that will need to develop adaptive means to respond to these changes.

International implications of climate change

Climate change and marine ecosystems neither begin nor end at the US border. Many marine organisms, such as fish, marine mammals, and seabirds, are highly migratory and do not remain in one jurisdictional boundary. As climate change increases, certain species will likely shift their ranges, expanding into countries where they were previously absent. Current protected area networks may not match critical sites needed in the future. The focus of much conservation work has historically been on critically endangered species. It is crucial that in light of climate change, attention is also given to ensuring that other species and populations remain robust and resilient to the changes that are projected to occur throughout the marine biome (Simmonds & Eliott 2009).

Multilateral regional fisheries management organizations (RFMOs) of which the US is a member are aware of the issue of climate change. Flexible management strategies will be needed that will allow these organizations to manage fisheries sustainably. Only half of the 12 RFMOs that include US fish species have taken or are taking actions to address climate change (Table 4). In addition, none of the six existing bilateral fisheries agreements between the United States and neighbouring countries have formally addressed climate change issues. Bilateral fishing regimes, which tend to be renegotiated periodically, will have to evolve over time in response to climate-related changes in abundance and spatial and temporal changes in fish stocks. The fishery governance process is much slower to adapt for an RFMO, for which the process is built on a stable decision environment (convention or treaty) and existing environmental conditions.

Security and transportation issues are at play in terms of expected climate change impacts to ocean services in the United States. Most notably, climate shifts in the Arctic (especially decreases in sea-ice coverage) are provoking discussion on the future of ocean governance, including marine resource and ecosystem-based management. Perhaps the most noteworthy issue in this arena is the increase in shipping accessibility in the Arctic. National security concerns and threats to national sovereignty have also been a recent focus of attention (Campbell et al. 2007, Borgerson 2008, Lackenbauer 2011). Ocean change will lead to an expanded geopolitical discussion involving the relationships among politics, territory, and state sovereignty on local, national, and international scales (Nuttall & Callaghan 2000).

International forums that deal with species conservation have a major role in providing coordination and direction. Therefore, international collaboration is both fundamental and foundational to understanding and managing climate change in the United States. Strengthening existing international partnerships, and developing new partnerships for knowledge sharing and strategy development, will be necessary to understand and address climate change impacts on marine ecosystems and communities around the world. Working closely with partner countries to enhance the level of understanding of climate change impacts is needed to build capacity and to effectively plan and implement adaptation actions.

Implications of climate change in international conventions and treaties

A number of international treaties and conventions have been developed to aid in addressing ocean issues that affect multiple jurisdictions and countries. Many of these either focus primarily on marine resources or involve them in some fashion. Exploring and strengthening synergies between these treaties and conventions would provide increased value, better coordination, improved focus and facilitation of the development of key priorities (Robinson et al. 2005). The following discussion includes only a subset of the larger body of international conventions and treaties that are considering climate change.

Convention on Migratory Species

The Convention on Migratory Species (CMS) of Wild Animals is the only global intergovernmental convention that is established exclusively for the conservation and management of migratory species (Robinson et al. 2005). The CMS recognizes that nations have a duty to protect migratory species that live within or pass through their jurisdictional boundaries, and if effective management of these species is to occur, concerted actions will be required from all nations (range states) in which a species spends any part of its life cycle (Robinson et al. 2005). Species are listed under two CMS appendices: Those species threatened with extinction are listed in Appendix I, and species that would benefit from internationally coordinated efforts are listed in Appendix II. The

Table 4   Primary regional fisheries management organizations (RFMOs) and arrangements that include US living marine resources, by organization/membership, mission, relevant species, and climate change actions, 2012

Organization/US membership status	Mission	Relevant species	Climate change actions
Commission for the Conservation of Antarctic Marine Living Resources (CCAMLR)/member	Protect and conserve the marine living resources in the waters surrounding Antarctica	Fish, molluscs, crustaceans and all other species of living organisms, including birds	CCAMLR includes climate change on the agenda of its Scientific Committee, which reports on this item to the commission. Climate is also a factor considered in the development of a proposal for a marine protected area in the Ross Sea.
North Atlantic Salmon Conservation Organization (NASCO)/member	Promote scientific research and the conservation, restoration, enhancement, and rational management of salmon stocks in the North Atlantic Ocean	Atlantic salmon (*Salmo salar*)	NASCO is concerned about the potential impacts of climate change on wild Atlantic salmon and requested the International Council for the Exploration of the Sea (ICES), which provides scientific advice to the organization, to advise it on the potential implications of climate change for salmon management at the 29th NASCO Annual Meeting held in Edinburgh, Scotland, on 5–8 June, 2012. NASCO has not published any studies directly addressing climate change and salmon to date.
North Pacific Anadromous Fish Commission (NPAFC)/member	Promote the conservation of anadromous stocks and ecologically related species in the high-seas areas of the North Pacific Ocean	Pacific salmon (chum, coho, pink, sockeye, chinook, cherry, and steelhead)	The Bering-Aleutian Salmon International Survey-II (BASIS-II) is NPAFC's coordinated program of cooperative research on Pacific salmon in the Bering Sea designed to clarify the mechanisms of biological response by salmon to the conditions caused by climate change. Climate change and its impact on salmon have been discussed in a symposium and two special publications: (1) a report on understanding impacts of future climate and ocean changes on the population dynamics of Pacific salmon (Beamish et al. 2009) and (2) a bibliography of literature associated with climate and ocean change impacts on Pacific salmon (Beamish et al. 2010). The overarching theme of the NPAFC 2011–2015 Science Plan is "Forecast of Pacific Salmon Production in the Ocean Ecosystems Under Changing Climate."
Northwest Atlantic Fisheries Organization (NAFO)/member	Study, conserve and manage fishery resources in the NAFO Regulatory Area in the North Atlantic Ocean beyond 200-mile zones of member states	Cod, flounders, redfish, capelin, hake, skates, shrimp	The NAFO Scientific Council Standing Committee on Fisheries Environment has been discussing change patterns, including climate change, for nearly 50 years. Beginning in 1964, NAFO conducted four symposia on decadal reviews (1950–1959, 1960–1969, 1970–1979, 1980s–1990s) of environmental conditions in the North-West Atlantic and their influence on fish stocks.

continued

Table 4 (continued) Primary regional fisheries management organizations (RFMOs) and arrangements that include US living marine resources, by organization/membership, mission, relevant species, and climate change actions, 2012

Organization/US membership status	Mission	Relevant species	Climate change actions
Western and Central Pacific Fisheries Commission (WCPFC)/ member	Ensure, through effective management, the long-term conservation and sustainable use of highly migratory fish stocks in the western and central Pacific Ocean in accordance with the 1982 U.N. Convention on the Law of the Sea and the 1995 U.N. Fish Stocks Agreement	All fish stocks of the species listed in Annex 1 of the 1982 Convention on the Law of the Sea occurring in the convention area and such other species of fish as the commission may determine	There is a growing awareness in the WCPFC Science Committee that the impact of oceanographic and climate variability is a key area of uncertainty, and that it should be integrated in future stock assessments (Commission for the Conservation and Management of Highly Migratory Fish Stocks in the Western and Central Pacific Ocean. Summary Report of the Seventh Regular Session of the Scientific Committee, 21 September 2011).
Western Central Atlantic Fishery Commission (WECAFC)/ member	Promote the effective conservation, management and development of the living marine resources of the area of competence of the commission and address common problems of fisheries management and development faced by members of the commission	All living marine resources, without prejudice to the management responsibilities and authority of other competent fisheries and other living marine resources management organizations or arrangements in the area	At the 13th session of the commission in 2008, parties addressed an agenda item titled "Climate Change Implications for Fisheries and Aquaculture: Contributions to Global Discussion From FAO." Parties agreed that there was a need for improved coordination and collaboration between countries in the region in improving disaster preparedness. In particular, there was a need to improve the collation and distribution of available information on climate change and its likely impacts. Fishers have said that little information is reaching them on climate change in relation to small-scale fisheries. Advance warning also needs to be improved in the region. Parties concluded that implementation of an ecosystem approach to fisheries management was an important mechanism for maximizing the resilience of marine ecosystems to climate change. The fifth session of WECAFC's Scientific Advisory Group recommended that the commission pay attention and provide appropriate response to the impact of climate change and climate variability on marine ecosystems, fishing communities and fisheries and aquaculture in general (WECAFC/XIV/2012/4).

United States is a range state for many of the marine species listed in these appendices, including whales, seabirds, turtles, and sharks (Table 5).

Cooperation between countries to tackle the impacts of climate change on specific migratory species is critical for effective management. The CMS provides an important opportunity to develop climate change strategies at the international level. A number of climate change actions have already been undertaken by CMS. One such action was an effort led by the Zoological Society of London to develop and test a climate change vulnerability assessment method for the United Nations Environmental Program CMS Secretariat on approximately half of the Appendix I species.

Table 5 Marine species with US ranges listed in the Convention on Migratory Species Appendices

Appendix	Taxa	Species
I	Mammals	Humpback whale (*Megaptera novaeangliae*), bowhead whale (*Balaena mysticetus*), blue whale (*Balaenoptera musculus*), northern Atlantic right whale (*Eubalaena glacialis*), North Pacific right whale (*Eubalaena japonica*)
I	Birds	Short-tailed albatross (*Phoebastria albatrus*), Bermuda petrel (*Pterodroma cahow*), Hawaiian petrel (*Pterodroma sandwichensis*), pink-footed shearwater (*Puffinus creatopus*)
I/II	Mammals	Sperm whale (*Physeter macrocephalus*), sei whale (*Balaenoptera borealis*), fin whale (*Balaenoptera physalus*), West Indian manatee (*Trichechus nanatus*)
I/II	Birds	Steller's eider (*Polysticta stelleri*)
I/II	Reptiles	Green turtle (*Chelonia mydas*), loggerhead turtle (*Caretta caretta*), hawksbill turtle (*Eretmochelys imbricata*), Kemp's Ridley turtle (*Lepidochelys kempii*), olive Ridley turtle (*Lepidochelys olivacea*), leatherback turtle (*Dermochelys coriacea*)
I/II	Fish	Basking shark (*Cetorhinus maximus*), great white shark (*Carcharodon carcharias*), manta ray (*Manta birostris*)
II	Mammals	Narwhal (*Monodon monoceros*), pantropical spotted dolphin (*Stenella attenuata*), spinner dolphin (*Stenella longirostris*), striped dolphin (*Stenella coeruleoalba*), killer whale (*Orcinus orca*), Baird's beaked whale (*Berardius bairdii*), Beluga whale (*Delphinapterus leucas*), northern bottlenose whale (*Hyperoodon ampullatus*), Bryde's whale (*Balaenoptera edeni*), dugong (*Dugong dugong*)
II	Birds	Black-footed albatross (*Phoebastria nigripes*), laysan albatross (*Phoebastria immutabilis*), black-browed albatross (*Thalassarche melanophris*), shy albatross (*Thalassarche cauta*), Salvin's albatross (*Thalassarche salvini*), white-chinned petrel (*Procellaria aequinoctialis*), spectacled petrel (*Procellaria conspicillata*), roseate tern (*Sterna dougallii*), Arctic tern (*Sterna paradisaea*), little tern (*Sterna albifrons*)
I/II	Fish	Whale shark (*Rhincodon typus*), shortfin mako shark (*Isurus oxyrinchus*), longfin mako shark (*Isurus paucus*), porbeagle (*Lamna nasus*), spiny dogfish (*Squalus acanthias*), green sturgeon (*Acipenser medirostris*)

Of these, approximately 50% are marine species. Results indicated that all of these species will be negatively impacted by climate change, including many species with ranges in the United States (Table 6). Continuing this work in a more quantitative way is considered a key priority for the future.

A technical workshop held in June 2011 included experts from academia, non-governmental organizations (NGOs), intergovernmental organizations (IGOs) and government agencies that work on issues associated with migratory species and climate change (CMS 2011). This workshop helped to charter a way forward and recommended key areas for future action and focus, including

- Predicting how future range shifts should be considered;
- Establishing long-term datasets and baselines of species listed under CMS, as well as their critical prey items;
- Developing historical and shifting baseline maps illustrating threats at spatiotemporal scales and zoning maps for use in planning of renewable energy projects;
- Focusing on populations that are resilient and adaptive to climate change;
- Integrating ecological networks into the design of protected areas;
- Highlighting that mitigation of climate change can be potentially more harmful to migratory species if sites are not carefully selected (e.g., wind farms may cause high mortality to bird and bat populations without proper siting and precautions);
- Recognizing that tertiary effects, such as new shipping routes in the Arctic, are increasing disturbance and exploitation of marine migrants, and biodiversity-related multilateral environmental agreements may be beneficial to coordinate responses;

Table 6 Marine species under the Convention on Migratory Species with US ranges that are vulnerable to climate change

Vulnerability	Taxa	Species
High	Reptiles	Green turtle (*Chelonia mydas*), hawksbill turtle (*Eretmochelys imbricata*), Kemp's Ridley turtle (*Lepidochelys kempii*), loggerhead turtle (*Caretta caretta*), olive Ridley turtle (*Lepidochelys olivacea*), leatherback turtle (*Dermochelys coriacea*)
High	Mammals	North Pacific right whale (*Eubalaena japonica*), Northern Atlantic right whale (*Eubalaena glacialis*), bowhead whale (*Balaena mysticetus*), blue whale (*Balaenoptera musculus*), narwhal (*Monodon monoceros*)
High	Birds	Short-tailed albatross (*Phoebastria albatrus*), Bermuda petrel (*Pterodroma cahow*), Steller's eider (*Polysticta stelleri*)
Medium	Mammals	Sperm whale (*Physeter macrocephalus*), sei whale (*Balaenoptera borealis*), humpback whale (*Megaptera novaeangliae*)
Medium	Fish	Basking shark (*Cetorhinus maximus*), great white shark (*Carcharodon carcharias*)

- Continuing to address research needs related to emerging issues such as disease, invasive species and ecosystem changes;
- Building capacity at the local level through climate change literacy training, participatory monitoring and creating incentives among communities for conservation; and
- Integrating climate change policies into additional multilateral agreements and strengthening collaboration with the Convention on Biological Diversity (CBD), Ramsar, the Bern Convention and United Nations Framework Convention on Climate Change (UNFCCC).

There is a need for multinational large-scale and long-term work to better understand risks associated with ocean and marine resources as a result of climate change. Focus must be given to ensuring that species and populations, in addition to those that are critically endangered, are able to combat the effects of climate change predicted to occur throughout the marine biome (Simmonds & Eliott 2009).

Convention on wetlands of international importance (Ramsar)

The United States is a contracting party to Ramsar, which is an intergovernmental treaty that provides the framework for voluntary national action and international cooperation for the conservation and wise use of wetlands and their resources. Parties of Ramsar

> work towards the wise use of all their wetlands through national land-use planning, appropriate policies and legislation, management actions, and public education … and ensure their effective management; and cooperate internationally concerning transboundary wetlands, shared wetland systems, shared species, and development projects that may affect wetlands. (http://www.ramsar.org)

There are nearly 2000 designated Ramsar sites around the world, over 30 of which are in the United States.

Many marine and coastal wetlands identified by Ramsar are particularly important habitats for ocean and marine species and include, among others, permanent shallow marine waters (e.g., sea bays and straits); marine subtidal aquatic beds, which include kelp beds, seagrass beds, tropical marine meadows; coral reefs; and rocky marine shores (e.g., rocky offshore islands, sea cliffs). The Ramsar Convention, similar to CMS, has explicitly recognized the threats caused by climate change and plays an important role in the future by understanding that additional work is required in understanding the relationship between wetlands and climate changes.

Convention on International Trade in Endangered Species of Wild Fauna and Flora

The United States is a party to the Convention on International Trade in Endangered Species (CITES) of Wild Fauna and Flora, which regulates the trade in species based on their conservation status. Appendix I species are those threatened with extinction and are, or may be, affected by trade. Parties are not allowed to trade in Appendix I species for commercial purposes. Appendix II species are not necessarily threatened with extinction but might be if trade is not regulated, and parties may trade in these species as long as trade is not detrimental to their survival. Appendix III species are listed unilaterally by parties seeking international cooperation in controlling trade.

Recently, CITES has begun to focus attention on the issue of climate change. In 2010, the scientific aspects of the provisions of CITES and the resolutions of the conference of the parties were assessed to determine which provisions were actually or likely to be affected by climate change and to make recommendations for further action. A working group was created to draft recommendations to be presented at the 62nd Standing Committee meeting in July 2012. The working group was tasked with addressing climate change in six CITES processes or mechanisms: species listing, non-detriment findings, periodic review of the appendices, management of nationally established export quotas, review of significant trade, and trade in alien invasive species. It was the general consensus that the decision-making framework developed within CITES is flexible enough to accommodate the consideration of climate change in each of its six processes or mechanisms.

Inter-American Convention for the protection and conservation of sea turtles

The Inter-American Convention (IAC) is focused on marine turtles, which are particularly vulnerable to climate change. It promotes the protection, conservation and recovery of the populations of marine turtles and those habitats on which they depend on the basis of the best-available data and taking into consideration the environmental, socioeconomic and cultural characteristics of the parties (Article II, Text of the Convention). These actions should cover both nesting beaches and the parties' territorial waters. International collaboration is essential to marine turtle conservation because turtles that breed on US beaches (or migrate through US waters) need continued protection once they leave US jurisdiction. IAC is an intergovernmental treaty that provides the legal framework for countries in the Western Hemisphere to take actions for the benefit of these species. Continued active involvement of the United States in international expert groups and initiatives provides key support to emerging collective adaptation action for marine turtles.

In 2009, the parties agreed to a number of actions specifically to address the impacts of climate change on marine turtles, including management actions as well as research and monitoring (http://www.iacseaturtle.org). According to annual reports submitted by the parties, overall performance against these goals has been fair; however, climate adaptation concerns are often overtaken by more immediate priorities, such as by-catch and non-climate impacts on nesting beaches.

Convention on Biological Diversity

Climate change is likely to become one of the most significant drivers of biodiversity loss by the end of the century (Millennium Ecosystem Assessment [MEA] 2005). Conserving and restoring marine ecosystems (including their genetic and species diversity) is essential for the overall goals of both the CBD and the UNFCCC. The objectives of the CBD are 3-fold: (1) the conservation of biological diversity, (2) the sustainable use of its components, and (3) the fair and equitable sharing of the benefits arising from the utilization of genetic resources. The CBD affirms that conservation of biodiversity is a common concern of humankind and reaffirms that nations have sovereign rights over their own biological resources. The convention covers both terrestrial and marine biota, and parties are explicitly required to implement the CBD consistent with the rights and obligations of states under the United Nations Convention on the Law of the Sea (UNCLOS). The convention was opened for signature at the United Nations Convention on Environment and Development in Rio de Janeiro, June 1992, and entered into force on 29 December 1993. The United States has signed the convention but has not yet ratified it.

The CBD Conference of the Parties has made over 40 decisions regarding biodiversity and climate change since the convention entered into force. Among those decisions, perhaps the most promising was adopted at the 10th meeting of the CBD in Nagoya, Japan, in 2010, which urged parties to "enhance the conservation, sustainable use and restoration of marine and coastal habitats that are vulnerable to the effects of climate change or which contribute to climate-change mitigation, such as mangroves, peatlands, tidal salt marshes, kelp forests and seagrass beds" (CBD 2010 COP 10 Decision X/33). The decision not only reaffirmed the importance of these habitats and the need to protect them but also played a role in achieving the objectives of the UNFCCC, the Ramsar Convention on Wetlands, and the CBD.

Climate change considerations in other international organizations

Agreement for the Conservation of Albatross and Petrels

The Agreement for the Conservation of Albatross and Petrels (ACAP) is an international multi-lateral agreement that seeks to reduce known threats to albatrosses and petrels through coordinating international activity. Twenty-two species of albatrosses and seven species of petrels are currently listed under ACAP. Since 2008, the ACAP Advisory Committee has had a standing agenda item "Impacts of Global Climate Change". Recently, ACAP parties have acknowledged that there is growing scientific evidence that present climate change is already affecting marine ecosystems at all levels of the food webs, and projection of future change suggests that these effects will increase considerably. For this reason, the parties recognized that it is important to review the potential impact of global climate variability and change on the conservation status of albatrosses and petrels. However, published studies to date are limited to a few species in the Indian Ocean. Therefore, the committee has recommended that parties and range states encourage further analyses on the combined impacts of environmental change and fisheries on albatross and petrel population trends.

International Whaling Commission

The International Whaling Commission (IWC) is the body charged with the proper conservation of the world's whale stocks, thus making possible the orderly development of the whaling industry. The main duty of the IWC is to review and revise as necessary the measures laid down in the Schedule to the Convention, which govern the conduct of whaling throughout the world. These measures, among other things, provide for the complete protection of certain species; designate specified areas as whale sanctuaries; set limits on the numbers and size of whales that may be taken; proscribe open and closed seasons and areas for whaling; and prohibit the capture of suckling calves and female whales accompanied by calves. The compilation of catch reports and other statistical and biological records is also required.

Climate change and its impacts on cetacean species have been highlighted in discussions at the IWC Scientific Committee, which has considerable expertise in understanding and modelling climate impacts. The IWC passed a resolution requesting contracting governments to incorporate climate change considerations into existing conservation and management plans, directing the Scientific Committee to continue its work on studies of climate change and the impacts of other environmental changes on cetaceans, as appropriate, and appealed to all contracting governments to take urgent action to reduce the rate and extent of climate change. The IWC has also hosted a number of workshops to enhance collaborations among various experts in cetacean biology, marine ecosystems, modelling and climate change, as well as improving the conservation outcomes for cetaceans under climate change scenarios.

Future climate change-related challenges facing whale stocks require innovative, large-scale, long-term and multinational response from scientists, conservation managers and decision makers. Moreover, the reactions to emerging developments and changes will need to be swift (Simmonds & Eliott 2009).

Commission for the Conservation of Antarctic Marine Living Resources

The Commission for the Conservation of Antarctic Marine Living Resources (CCAMLR) was established mainly in response to concerns that an increase in krill catches in the Southern Ocean could have a serious effect on populations of krill and other marine life, particularly on birds, marine mammals and fish, which depend on krill for food. Climate change is on the agenda of CCAMLR's Scientific Committee, which reports on this item to the commission. Climate change is also a factor the United States is considering in development of a proposal for a MPA in the Ross Sea. The United States would like to leave one area in the Ross Sea open to fishing and close an equivalent area so that the impacts of climate can be differentiated from the impacts of fishing.

North Pacific Marine Science Organization

The primary role of the North Pacific Marine Science Organization (PICES) is to promote and coordinate marine research undertaken by the parties (Canada, Japan, China, Korea, Russia, and the United States) in the temperate and sub-Arctic region of the North Pacific Ocean and its adjacent seas. It also endeavours to advance scientific knowledge about the ocean environment, global weather and climate change, living resources and their ecosystems, and the impacts of human activities as well as promote the collection and rapid exchange of scientific information on these issues. PICES provides an international forum to promote greater understanding of the biological and oceanographic processes of the North Pacific Ocean and its role in global environment.

PICES has published numerous scientific reports on the impacts of climate and climate change on fish species in the North Pacific, including reports forecasting climate impacts on future production of commercially exploited fish and shellfish (e.g., Hollowed et al. 2008). Since 2002, PICES has also hosted approximately 11 international symposia, 15 workshops, and 5 special sessions with climate change-related themes (see http://www.pices.int/publications/default.aspx).

Wider Caribbean Sea Turtle Conservation Network

The Wider Caribbean Sea Turtle Conservation Network (WIDECAST) is an active network of biologists, managers, community leaders and educators in more than 40 nations (including the United States) and territories committed to an integrated, regional capacity that ensures the recovery and sustainable management of depleted marine turtle populations. WIDECAST has conducted workshops geared towards marine turtle conservationists and MPA practitioners, covering climate-related topics, including monitoring, vulnerability assessment, selecting and prioritizing adaptation options and communicating climate change.

Climate change considerations by regional fisheries management organizations and living marine resource conservation organizations

In 2011, the United Nations General Assembly adopted a draft resolution (A/66/L.22) that expressed concern over the current and projected adverse effects of climate change on food security and the sustainability of fisheries. The resolution urged nations, either directly or through appropriate subregional, regional or global organizations or arrangements, to intensify efforts to assess and address, as appropriate, the impacts of global climate change on the sustainability of fish stocks and the habitats that support them.

Multilateral RFMOs in which the United States is a member are cognizant of the issue of climate change, but with few exceptions, most have done little to reduce the possible effects or develop contingency plans despite the growing body of fisheries research on climate change. None of the six existing bilateral fisheries agreements between the United States and neighbouring countries—five with Canada (albacore tuna, Pacific salmon, Pacific hake, Pacific halibut, and Great Lakes fisheries) and one with Russia (North Pacific and Bering Sea fisheries)—have formally addressed climate change issues. Bilateral fishing regimes undergo renegotiation periodically; the effects of climate change on fish stock range will have to be taken into consideration in these negotiations. The fishery governance process is much slower to adapt for an RFMO due to its consensus decision-making process, which is an additional complicating factor, especially if the RFMO has many member countries.

Straddling fish stocks

In some cases, changes in the location of straddling fish stocks may lead to challenges in international fisheries management. The United Nations defines straddling stocks as stocks of fish that migrate between, or occur in both, the EEZ of one or more states and the high seas (Figure 12). One example of such a challenge occurred within the US-Canada Pacific Salmon Commission (McIlgorm 2010). Pacific salmon are anadromous fish that cross state and international boundaries in their oceanic migrations. Fish spawned in the rivers of one jurisdiction are vulnerable to harvest in other jurisdictions. The turbulent history of US and Canadian cooperative management of their respective salmon harvests suggests that environmental variability may complicate the management of such shared resources. For 6 years, beginning in 1993, the United States and Canada were unable to agree on a full set of salmon-fishing regimes under the terms of the Treaty between the Government of Canada and the Government of the United States of America Concerning Pacific Salmon. The breakdown

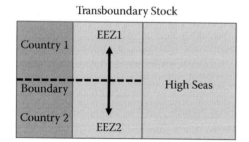

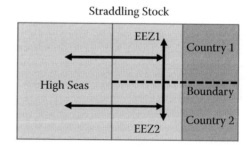

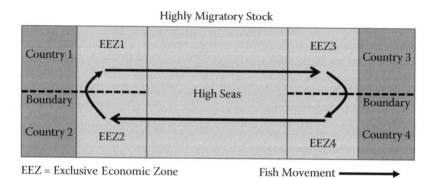

Figure 12 Types of fish stocks. Illustration of transboundary, straddling, and highly migratory species ranges.

in cooperation was fuelled by strongly divergent trends in Alaskan and southern salmon abundance and a consequent change in the balance of each nation's interceptions of salmon spawned in the other nation's rivers. A period of high productivity in Alaska contributed to increased Alaskan interceptions of British Columbia salmon at a time when Pacific North-west coho and Chinook salmon could least withstand additional actions by the Canadian salmon fleet. The mounting crisis led to a fundamental shift in the approach taken by the two nations to determine their respective salmon harvest shares. On 30 June 1999, Canada and the United States signed a 10-year agreement that laid the groundwork for a more sustainable and cooperative abundance-based management regime (Miller & Munro 2003). That agreement was renegotiated and extended for another 10 years in 2009. The latest agreement resulted in substantial investments in cooperative research programmes, improved modelling capabilities, and refined harvest rate calculations.

Future climate change will continue to have an impact on fisheries governance under the Pacific Salmon Treaty in the future as fisheries management science tries to assess the impacts of climate change on Pacific salmon (McIlgorm 2010).

Transboundary fish stocks

Climate change is also likely to have an impact on the spatial distribution of transboundary fish stocks. These are stocks that range in the EEZs of at least two countries (Figure 12), such as hake stocks off the US Atlantic and Pacific coasts, which both straddle the US-Canada border (Helser & Alade 2012). Pacific hake spawn off the coast of California and forage off the coasts of Oregon, Washington and Canada. Ocean conditions influence the latitudinal extent of Pacific hake and silver hake foraging migrations (Agostini et al. 2007, Nye et al. 2009). In addition, the age structure of the population also has a strong effect on the northerly distribution of Pacific hake; that is, older fish are found farther north in warm years. In 2003, the United States and Canada signed a treaty to establish national shares of the coastwide Pacific hake stock. Shares were initially determined by the percentage of the hake stock found in each country's waters during the summer fishing season. Shifts in spatial distribution of the foraging distribution will have an impact on the stock assessment and may necessitate changes to current management agreements for these stocks.

Another consideration is the lack of formal conservation and management agreements for transboundary fish stocks. Pacific sardines, which migrate across international boundaries, are a case in point. Currently, no international management agreement exists for Pacific sardines. However, scientists and members of industry from the United States, Mexico, and Canada informally meet at the annual Trinational Sardine Forum, where research results and ideas are exchanged. In view of the combined impact of fishing and ocean climate variability on the sardine stock, an important emerging issue is the need for stable transboundary management, given potential changes in the stock's availability within the affected countries' EEZs. The question becomes whether cooperative management of the stock will result in economic and biological gains to all three countries. If cooperative conservation and management involve a positive sum game, a related concern then becomes whether or not cooperative management will be stable in the face of climate change (Herrick et al. 2007).

Similar situations exist for a number of West Coast transboundary groundfish stocks, such as sablefish, petrale sole, and numerous rockfish species. These are currently managed separately by the US Pacific Fishery Management Council and Canada's Department of Fisheries and Oceans in British Columbia.

Highly migratory fish stocks

Some fish populations migrate over long distances, passing through multiple territorial waters (Figure 12). The term *highly migratory species* comes from Article 64 of UNCLOS. Although the convention does not provide an operational definition of the term, UNCLOS Annex 1 lists the species considered highly migratory by parties to the convention. The list includes tuna species (albacore, bluefin, bigeye, skipjack, yellowfin, blackfin, little tunny, southern bluefin and bullet) and tuna-like species (pomfret, marlin, sailfish, swordfish, saury and ocean-going sharks, dolphins and other cetaceans).

The stability and success of RFMOs that govern the harvests of straddling and highly migratory fish stocks will depend, in part, on how effectively they can maintain member nations' incentives to cooperate, despite the uncertainties and shifting opportunities that may result from climate-driven changes in productivity, migratory behaviour or catchability of the fish stocks governed by the RFMO (Miller 2007).

The reliance of island nations on tuna fisheries, and the potential adverse effects of climate change on this resource, emphasizes the need for precautionary approaches to management. The Western and Central Pacific are complex ocean fisheries with island nations and foreign fishers, including the United States, taking purse seine and longline tuna catches in areas to the north and east of Papua New Guinea. Independent states in the Pacific have several regional organizations, such as the South Pacific Forum Fisheries Agency (FFA) and the Western and Central Pacific Fishery Commission. The majority of fishery production is by distant water fishing nations under access agreements to FFA member states. Gaining economic benefits from domestic tuna fishery processing by island states is a priority to supplement income from access license fees.

Tuna are relatively sensitive to ocean temperature and move quickly to areas of preferred temperature. Consequently, the total stock of tuna does not necessarily change dramatically if temperature changes, but the spatial distribution may shift substantially (Aaheim & Sygna 2000). Climate change will have an impact on all industrial fishers and processors due to changes in the location of fishing sites, with vessels spending several months of the year inside a national EEZ as fish move to the high seas or to an adjacent EEZ. This movement has implications for licensing of foreign fishing vessels and tuna canneries using local suppliers. The increased risk of fish availability will have implications for future capital investment and labour requirements. Fishery governance systems need to be aware of potential climate change impacts on annual catches and the location of fish schools, increasing the variability in an already-complex system (McIlgorm 2010).

Arctic

Climate change is expected to have profound impacts in the Arctic; some of these changes are already being observed. The loss of sea ice is expected to have an impact on the timing of seasonal production and extend the growing season. Expansion or movement of some sub-Arctic species into the Arctic may occur over time, and the rates of expansion and movement will differ by species in relation to their vulnerability to climate change. There is a potential for expansion of fishing in response to periods of reduced ice cover in the Chukchi and Beaufort Seas. In recognition of this possibility, the NPFMC acted swiftly to close the US Arctic EEZ to commercial fishing until sufficient information is obtained to manage the stocks sustainably (Wilson & Ormseth 2009). While this action protects stocks within the US Arctic EEZ, the Arctic high-seas waters are international waters, and international fishing agreements may be needed for fisheries sustainability.

On a regional fisheries level, the potential for spatial displacement of aquatic resources and people as a result of climate change impacts will require existing regional structures and processes to be strengthened or enhanced. Agreements, both multilateral and bilateral, will need flexibility to adapt to changing circumstances, particularly unanticipated, climate-driven changes in stock levels or distribution across EEZs or high-seas areas.

Blue carbon

Blue carbon is a term to describe the carbon dioxide sequestered by seagrasses, tidal marshes, and mangroves and stored in large quantities in both the plants and in the sediment of coastal and marine ecosystems. Accounting for the carbon sequestration value of coastal marine systems has the potential to be a transformational tool in the implementation of improved coastal policy and management. Currently, no policy, financing, management or other systems specifically value the role of marine and coastal ecosystems in sequestering greenhouse gases or the potential emissions that result from degradation or conversion of these systems. There is, however, increasing attention on carbon sequestration in blue carbon systems both within the United States and internationally.

While not specifically dealing with blue carbon, a number of policy and financing mechanisms currently exist that support nature-based climate change mitigation solutions under UNFCCC, for example, Reducing Emissions from Deforestation and Forest Degradation (REDD+), National Appropriate Mitigation Actions (NAMAs), and the Clean Development Mechanisms (CDM) and Land-Use, Land-Use Change and Forestry (LULUCF) (Climate Focus 2011). These mechanisms provide incentives and financial support for national-level accounting and project-level activities, including conservation, restoration and sustainable use of natural systems such as forests and peat-lands. Coastal ecosystems can be integrated into these existing UNFCCC-supported mechanisms.

There is a recognized need for increasing decision-maker awareness of the importance of coastal carbon, and to that end, UNFCCC has invited the submission of information on emissions from coastal and marine ecosystems (UNFCCC 2011). The IPCC has also established an expert working group to update the 2006 IPCC guidelines for including wetlands in national greenhouse gas inventories. This revision will include a chapter on 'coastal wetlands', previously absent (IPCC 2011).

The Blue Carbon Initiative is a global agenda to maintain the blue carbon stored in coastal ecosystems and avoid emissions from their destruction. The initiative, coordinated by Conservation International (CI), the International Union for Conservation of Nature (IUCN), and the Intergovernmental Oceanic Commission (IOC) of UNESCO, has established international expert working groups to develop (1) the necessary scientific basis and tools and (2) the international- and national-level policy frameworks needed to support blue carbon-based policy, management, conservation and science globally. Field-based demonstration projects have been identified by the working groups as a current priority to demonstrating the viability of blue carbon projects, facilitating the development of practical, science-based methodologies and building capacity in target countries.

The Verified Carbon Standard, an international carbon registry, is updating its requirements to provide for the inclusion of eligible wetlands projects for carbon financing. A number of countries, including Indonesia, Costa Rica and Ecuador, have identified blue carbon as a priority issue and are currently developing strategies and approaches. A number of US federal agencies are currently investigating the integration of coastal blue carbon into their priority activities. These countries are in need of technical and resource support to complete this process and implement effective coastal-carbon-based management and policy.

Climate change and other international issues

Climate change impacts on ocean services in the United States are provoking discussion on the future of ocean governance, including marine resource and ecosystem-based management. Perhaps the most noteworthy issue in this arena is the increase in shipping accessibility in the Arctic. National security concerns and threats to national sovereignty have also been a recent focus of attention (Campbell et al. 2007, Borgerson 2008, Lackenbauer 2011). According to some researchers, the Arctic region could slide into a new era, featuring jurisdictional conflicts and increasingly severe clashes over the extraction of natural resources among the global powers (Berkman & Young 2009). Others are confident that existing bilateral and international arrangements are robust enough to handle the new challenges. In any case, given that the effects of climate change will vary across regions of the United States and the world, diverse and novel governance and security challenges will likely emerge.

Climate change adaptation and mitigation actions often extend beyond regional scales and regional governance and security concerns. In general, the warming of the ocean is leading to a redrawing of the biophysical map of Earth. This process will lead to an expanded geopolitical discussion involving the relationship among politics, territory, and state sovereignty on local, national and international scales (Nuttall & Callaghan 2000).

The topic of 'security' is a growing theme of global environmental change discussions, especially those focusing on climate change (Barnett 2006). Sea-level rise and extreme events could potentially lead to human migration, both within the United States and movement from other nations into the United States. In addition, climate-related impacts on international food security could lead to foreign conflict. The central role of the United States in emerging international discussions on the relationship between climate change and security response strategies centre around past precedent, with the United States showing immediate response to natural disasters both via the military and monetary aid (Campbell et al. 2007). In addition to energy security, global trade, terrorism, nuclear non-proliferation, and global poverty, global climate change may become a significant foreign policy and national security challenge as it complicates and exacerbates many more traditional security issues.

According to the Arctic Council's 2009 AMSA report, "The Arctic is now experiencing some of the most rapid and severe climate change on earth. … Of direct relevance to future Arctic marine activity, and to the AMSA, is that potentially accelerating Arctic sea ice retreat improves marine access throughout the Arctic Ocean" (AMSA 2009, p. 26). With sea ice receding in the Arctic as a result of warming temperatures, global shipping patterns are already changing and will continue to change considerably in the decades to come (Cressey 2007, Stewart et al. 2007, Berkman & Young 2009, Khon et al. 2010).

As the Northern Sea Route and North-west Passage routes become more passable by vessels because of melting Arctic sea ice, these regions are experiencing greater maritime travel (Khon et al. 2010), and sailors are witnessing "an age-old dream come true" (Kerr 2002, p. 1490). This regional transformation has global economic implications. The international shipping industry influences much of current world trade, suggesting that increasing the capacity of maritime transport in the Arctic may highly affect the import and export of goods throughout the global economy. Furthermore, world seaborne trade is increasing, which is tightening the linkages among economic growth, trade, and demand for maritime transport services (Kitagawa 2008). Given this state of affairs, it is likely that climate change impacts on marine resources and users will involve direct and indirect challenges and opportunities for the US seaborne transportation sector.

Ocean management challenges, adaptation approaches, and opportunities in a changing climate

Individuals, communities, resource managers, and governments across the United States are beginning to understand, plan for, and address the impacts of climate change on oceans. While the practice of climate adaptation is relatively nascent (particularly for marine systems), strategies and actions are emerging. Adaptation planning requires access to best-available science, including long-term monitoring and assessment of environmental and societal change to understand baselines, track changes through time, and evaluate the effectiveness of actions. Tools and services are currently being developed to meet the growing demand for user-friendly, science-based information that supports ocean adaptation efforts. Two-way communication between scientists and practitioners is critical for ensuring that information meets the needs of decision makers.

Climate change presents not just a challenge but also an opportunity to revisit and improve existing plans and management strategies to make them more robust and forward looking. Integration of climate change into management and stewardship efforts, such as fishery management plans (FMPs) and the design of protected areas, will enhance ocean resilience. A key strategy is reducing non-climatic stressors (e.g., pollution, habitat destruction) to enhance ecosystem function and resilience to climate change. Existing legal and regulatory frameworks can also be leveraged to promote ocean adaptation efforts. Future success will depend on flexible, adaptive management that can accommodate uncertainty.

Challenges and opportunities for adaptation in marine systems

Adaptation involves the processes of preparing for and building resilience to climate change, as well as responding to unavoidable impacts (IPCC 2007, NRC 2010a). The climate impacts observed today will increase in severity in the future, even if greenhouse gas emissions are substantially limited in the near term (NRC 2010a). Therefore, climate change adaptation is a critical component of society's effort to foster a more sustainable future through enhancing the social, economic, and ecological resilience of ocean systems.

While diverse adaptation actions exist, most processes use the following general approach, as articulated by the NRC (2010a): identify relevant current and future climate changes, assess risks and vulnerabilities; develop and implement adaptation options; and create an iterative process by which adaptation actions can be re-evaluated and redesigned as necessary (Figure 13). Tenets of adaptation for ecosystem managers include protecting adequate and appropriate space, reducing non-climatic stressors that interact with the impacts of climate change, and adopting adaptive management practices (e.g., Hansen et al. 2010, Glick et al. 2011).

Compared to terrestrial, aquatic, and coastal systems, relatively few adaptation actions have been designed and implemented for marine systems. To better understand perceived and real barriers to action, a survey was conducted of North American coastal and marine managers, who articulated the following as barriers (Gregg et al. 2011):

- Lack of economic resources and budgetary constraints;
- Lack of institutional support, governance, and mandates to take adaptation action;
- Lack of institutional capacity and guidance on how to take action;

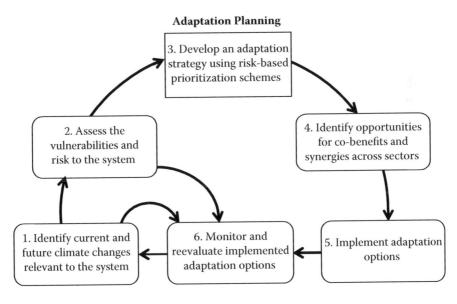

Figure 13 Conceptual framework for developing and implementing adaptation strategies. (From Stein et al. 2012.)

- Lack of key information on locally and regionally specific climate projections and tools to support assessments and monitoring;
- Uncertainty about risk and vulnerability; and
- Lack of awareness, stakeholder support, and engagement.

These constraints have been echoed by others (e.g., IPCC 2007, US Climate Change Science Program [CCSP] 2008, Glick et al. 2009).

Fortunately, solutions exist for overcoming many of these barriers, including enhanced provision of information, tools, and services that support ocean-related adaptation decisions and integration of climate change into existing policies, practices, and management.

Information, tools, and services to support ocean adaptation

Indicators, tools, and services are currently being developed to meet the growing demand for user-friendly, science-based information that supports ocean adaptation efforts. Incorporation of science is essential for successful adaptation planning, implementation, and evaluation. Decision makers rely on science to assess vulnerability and risk to a plausible range of climate change futures, understand potential impacts, inform adaptive actions, and evaluate the effectiveness of response options. Sustained interaction and feedback between scientists and information users (e.g., adaptation practitioners) can help to ensure that the information provided is accessible, understandable, and relevant. In addition, these interactions can grow the awareness of scientists regarding the key information needs of decision makers.

Importance of long-term observations and monitoring for adaptive management

The establishment of current baselines and trends is a core element of adaptation approaches (US CCSP 2008). Long-term data are essential for enhancing understanding of how ecosystems and human communities respond and adapt to climate change (Heinz Center 2008, Peterson & Baringer 2009). A range of observations on physical, ecological, social, and economic systems is needed to provide information on past and current trends, as well as to gain insight into future conditions. These observations not only detect changes and help communicate trends to managers and the public, but also are necessary for assessing risks, developing meaningful climate indices, and supporting adaptive management. Observations and monitoring data can provide critical insight into the relative contributions of anthropogenic change versus natural variability in ocean systems. In addition, long-term data can inform the development of more accurate and higher-resolution climate models that enhance the predictive capacity of managers and other decision makers.

Identification and accurate measurement of key variables can inform the development of ocean adaptation actions. Physical parameters (e.g., water temperature, water quality, salinity, pH and solar radiation; National Climate Assessment [NCA] 2010a); ecological parameters (e.g., phenology, species abundance and distribution, diversity and primary productivity; NCA 2010b); and socioeconomic parameters (e.g., demographics, food supplies, social and economic well-being and public health; NCA 2011) can help support assessments of vulnerability as well as the development of strategies for minimizing climate risks to ecosystems and communities. One opportunity is to leverage existing observation and monitoring systems, including those in MPAs, to establish 'sentinel sites' for understanding and managing for climate change (National Ocean Council 2012). For example, under the 1999 Marine Life Protection Act, and through a public-private partnership, California is implementing a statewide, 1100-mile network of MPAs to protect marine life, habitats, and ecosystems. The California Ocean Protection Council has invested over $20 million to conduct baseline characterizations of the ecosystem and to develop a novel approach for objective, scientifically rigorous, and cost-effective MPA monitoring. A statewide network of MPAs, in which other anthropogenic stressors are reduced, also provides a large-scale, natural laboratory to understand how climate changes manifest in ocean ecosystems. The innovative approaches to MPA monitoring being developed in the state also provide a framework that can be applied to inform the climate change management dialogue.

A challenge ahead is to ensure that coastal and ocean resource managers have access to high-quality information at resolutions commensurate with the scales at which decisions are made (Fauver 2008, National Ocean Council 2012). For example, oyster growers in the Pacific North-west depend on local observations to alert them to potentially harmful changes in ocean pH levels so that they can take proactive measures to protect young animals. Oceanographic models developed at the Woods Hole Oceanographic Institution and supported by long-term observations are now able to predict blooms of the toxic alga *Alexandrium fundyense*, providing local public officials and harvesters with an early-warning tool to minimize health risks and economic losses from tainted shellfish (Li et al. 2009). These examples illustrate the utility of accessible information at relevant decision scales.

Barriers remain in providing long-term information to support ocean adaptation decisions. There is a growing need to distill and synthesize large quantities of data into useful products that can provide practical information to inform management. The lack of a systematic approach to sustained, high-resolution climate observations is currently constraining the ability to develop informative and meaningful climate indices (NCA 2011). Efforts such as the Ocean Health Index offer a promising approach for synthesizing and improving accessibility of information on ocean stressors, as well as the valuable ecosystem services on which humans depend (Halpern et al. 2012). The index defines a healthy ocean as one that can sustainably deliver a range of benefits to people now and in the future and measures this health along 10 widely held public goals for the ocean and coasts, such as clean water, food provision, livelihoods, and cultural values. As such, the index provides assessments that are relevant to management objectives and mandates and allows systems that are sustainably used rather than simply protected to score highly. The index converts into a common measure the disparate ways in which climate impacts, fisheries production, pollution, species protection, coastal jobs, and other factors are assessed. The index can be used to assess the impacts of climate change on each of the 10 public goals, as well as the likely benefit to any given goal under management scenarios that target climate impact reduction versus improved fisheries management, land-based pollution regulations, or other measures (Halpern et al. 2012). User-friendly climate tools, services, and products based on long-term data will be essential for understanding changes and informing adaptation measures (Heinz Center 2008).

Tools and services for supporting ocean management in a changing climate

Efforts are under way to enhance the development and deployment of science in support of adaptation, to improve understanding and awareness of climate-related risks, and to enhance analytic capacity to translate understanding into planning and management activities (e.g., Moser & Luers 2008). For science to be useful to decision makers, the information provided must be timely, accessible, relevant, and credible. While critical knowledge gaps exist, there is a wealth of climate- and ocean-related science pertinent to adaptation. However, the majority of this information is currently 'inaccessible' to adaptation practitioners; it is unavailable, too technical to be understood and applied by non-scientists, or does not address the specific needs of decision makers. To address this challenge, diverse user-friendly tools and services are emerging, for example:

- The Sea Grant Climate Network is an online resource that includes adaptation-relevant information for coasts and oceans, a discussion forum, links to upcoming events, and social networking opportunities for the broader sea grant community, including extension agents. (http://sgcnetwork.ning.com/)
- Coral bleaching is a significant climate challenge for marine resource managers. Tools and services have been developed to help reef managers anticipate and respond to bleaching

events. For example, NOAA's Coral Reef Watch has developed several tools, including the Satellite Bleaching Alert system, an automated e-mail alert system that notifies subscribers when thermal conditions become conducive to bleaching at select reef sites. (http://coralfreewatch.noaa.gov/satellite/index.html)

- "Scanning the Conservation Horizon: A Guide to Climate Change Vulnerability Assessment" (Glick et al. 2011), produced by the National Wildlife Federation in partnership with the Department of Defense, the National Park Service, the US Fish and Wildlife Service (USFWS), the US Forest Service, NOAA, and the US Geological Survey, provides conservation practitioners and resource managers with methods and guidance for understanding and addressing the impacts of climate change on species and ecosystems. (http://www.nwf.org/~/media/PDFs/Global-Warming/Climate-Smart-Conservation/NWFScanningtheConservationHorizonFINAL92311.pdf?dmc=1&ts=20130317T1640584210)
- The Climate Adaptation Knowledge Exchange (CAKE), a joint project of Island Press and EcoAdapt, is aimed at building a shared knowledge base for managing natural systems, including oceans, in the face of climate change. (http://www.cakex.org/)
- The Coastal Climate Adaptation website of the NOAA Coastal Services Center provides diverse resources in support of coastal and ocean adaptation, such as state-level adaptation plans, case studies, climate communications information, risk and vulnerability assessments, guidance, and outreach materials. (http://collaborate.csc.noaa.gov/climateadaptation/default.aspx)

These tools and guidance documents help decision makers and managers navigate the complex landscape of information as they work to enhance preparedness and response efforts to safeguard ocean resources in a changing climate. The existing and emerging efforts to coordinate and provide timely, useful, and relevant climate information, tools, and services serve as a critical platform. However, a great deal of work remains in providing accessible information that meets the diverse set of adaptation planning, implementation, and evaluation challenges faced by marine resource managers and practitioners. Creative partnerships will be required in the near-term future to improve multidirectional communication and ensure that information provided by the scientific community meets the needs of decision makers.

Opportunities for integrating climate change into US ocean policy and management

While climate change presents challenges to marine resource managers and other ocean decision makers, solutions exist for incorporating climate change into ocean management. For example, because climatic and non-climatic stressors interact, reducing stressors over which there is more direct control (e.g., land-based pollution, habitat destruction) can enhance resilience to climate change and ocean acidification (Lubchenco & Petes 2010, Kelly et al. 2011). Next, we describe opportunities for reducing climate-related vulnerabilities of oceans by incorporating climate change considerations into marine spatial planning and MPA design, fisheries management, and application of existing legislative and regulatory frameworks.

Incorporating climate change into marine spatial planning and marine protected area design

Both coastal and marine spatial planning (CMSP) and MPAs spatially allocate human uses of the ocean as a means to better protect and sustainably use marine resources. CMSP focuses on all human uses of the oceans and seeks to allocate those uses across the ocean in a way that minimizes ecological, social, cultural, and economic impacts, while supporting and improving resource use and conservation goals (e.g., Ehler & Douvere 2009). MPAs instead focus primarily on limiting access to some or (in the case of 'no-take' marine reserves) all human uses within particular locations, typically for conservation or fisheries management purposes (e.g., Klein et al. 2008). Improving the enforcement and management of existing protected areas and refugia and increasing connectivity and the amount of protected space provide mechanisms for enhancing climate resilience (Glick et al. 2009). In addition, to enhance long-term effectiveness, it is critical that CMSP and MPA processes incorporate climate change into their planning, implementation, and evaluation efforts.

Accounting for the impacts of global climate change in CMSP and MPA planning may appear challenging as impacts can be diffuse and are often not under local control or management. However, there are at least three direct mechanisms for incorporating climate change into the design of management plans: (1) build resilience to climate impacts, (2) account for spatial patterns of climate impacts, and (3) anticipate future patterns of change.

Building climate resilience into spatial management remains the most commonly pursued approach (Halpern et al. 2008a, McLeod et al. 2009), in part because this mechanism is relatively straightforward and can leverage existing regulations and mandates. Targeted actions to limit or remove non-climatic stressors can help reduce the cumulative impact of total stressors, thus improving the ability of the ecosystem to cope with increases in climate-related stressors (Halpern et al. 2008b, 2010). For example, land-based pollution is the overwhelmingly dominant stressor to coastal areas of the Gulf of Mexico due to nutrient run-off from the human-dominated Mississippi River watershed, which drains into adjacent coastal areas (Halpern et al. 2009a). Efforts to reduce land-based stress would enhance resilience to climate impacts, such as sea-level rise, by improving the health of coastal salt marshes and wetlands. Furthermore, the size of protected areas or zones for limited use can be increased, and spacing between conservation patches decreased, as a means of buffering climate-related impacts (and other increasing or catastrophic stressors; e.g., Allison et al. 2003, McLeod et al. 2009). For example, the large areas encompassed by the MPAs recently established in the north-western Hawaiian Islands (Papahānaumokuākea Marine National Monument) and US Pacific holdings (Marianas Trench, Rose Atoll, and Pacific Remote Islands National Marine Monuments) are protected from many human activities. Therefore, these MPAs may be more resilient to climate impacts than are areas subject to higher levels of non-climatic stress from human activities.

Equally important are efforts to account for smaller-scale variation in climate impacts when designing spatially explicit resource management. Patterns of existing and projected impacts of changing sea-surface temperature, ocean acidification, and sea-level rise can be highly variable (Halpern et al. 2008b, Burrows et al. 2011). For management with a conservation goal, as in the case of MPAs, one potential strategy is to place protected areas in locations that exhibit high resilience to climate change. As assessments designed to inform CMSP and MPA planning processes engage and inform more sectors and more comprehensive planning, these efforts will be better able to address costs, benefits, and trade-offs across multiple sectors (e.g., White et al. 2012).

Anticipating future patterns of climate impacts and using these predictions in CMSP and MPA design are the most challenging, primarily because of the difficulty in predicting small-scale patterns of future climate impacts. Projected shifts in species ranges and ocean circulation patterns can be used to anticipate where species will exist in the future, as well as the potential for population connectivity through larval transport. MPA planning and CMSP processes can be designed to both anticipate and facilitate the transition to new geographies through strategies such as the creation of 'stepping stone' reserves that offer refuge or habitat to species as they migrate in response to warming waters (McLeod et al. 2009). In some cases, implementing networks of MPAs may help to diffuse climate risks by protecting multiple replicates of the full range of habitats and ecological communities within an ecosystem (US CCSP 2008).

Integrating climate change into fisheries management

Climate-related processes are affecting, and will continue to affect, the production of fisheries resources in marine ecosystems under US jurisdiction and beyond (Cochrane et al. 2009, Doney et al. 2012). Fish resources may respond to climate change in a variety of ways (e.g., changes in mortality, migration, distribution), and these changes can have important ramifications for fishery population dynamics, the ability to assess the status of fish populations, and the validity of future stock forecasts and rebuilding plans (Kraak et al. 2009).

The future sustainability and adaptation of fish resources in a changing climate depends on understanding past, current, and projected future climate impacts, and incorporating this information into the scientific bases of fishery management decisions, so that decision makers can effectively respond to impacts on existing fisheries and take advantage of new opportunities as conditions change (Link et al. 2010, Sumaila et al. 2011). Although some progress is being made, much work remains to ensure that fisheries management can effectively prepare for and adapt to the impacts of climate change on fish resources, as well as the communities and economies that depend on them (Hare et al. 2010, Link et al. 2010).

Most of the progress to date is in understanding climate impacts on fisheries. For example, in some US regions, oceanographic and fisheries observing systems are increasingly being mobilized to monitor and track the impacts of climate variability and change on fish and other living marine resources. These observing systems have been instrumental in producing the growing number of studies documenting persistent changes in spatial distribution of fishes attributable to large-scale changes in oceanographic processes (Nye et al. 2009, 2011, Link et al. 2010, Overholtz et al. 2011). The development of 'marine ecosystem status reports' in some regions is also providing a key mechanism for compiling and assessing marine ecosystem conditions as part of efforts to move towards ecosystem-based management (EAP 2009).

An increasing number of efforts are under way to understand and project the risks and impacts of climate variability and change on fish populations, as well as to advance modelling tools and their application to fisheries management (e.g., Hollowed et al. 2009, Stock et al. 2011). For example, Hollowed et al. (2009) developed and tested a framework for modelling fish and shellfish responses to future climate change. Mueter et al. (2011a) applied a variety of modelling techniques to project negative impacts of climate change (e.g., reduced recruitment) on walleye pollock populations in the Bering Sea. Another promising step is the development of ecosystem models to help explore the complex dynamics of marine ecosystems in a changing climate (Link et al. 2010). In addition, there is a growing body of literature and tools for assessing the vulnerability of natural resources in a changing climate (e.g., Glick et al. 2011). While most of this work has historically focused on terrestrial or freshwater environments, there has been some recent effort to develop and conduct vulnerability assessments for marine species (Johnson & Welch 2010).

There are relatively few examples of fishery management efforts that have explicitly incorporated climate-related information, but these efforts are expected to increase as more information and tools on climate impacts and vulnerabilities become available. The key questions remaining are: What are the adaptation options available to fisheries managers, and how and when should they be applied? At present, there appears to be little information and guidance to support fisheries management decisions along this path. Link et al. (2011) provided guidelines for incorporating distribution shifts into fisheries management, concluding that their approach is "feasible with existing information, and as such, fisheries managers should be able to begin addressing the role of changes in stock distribution in these fish stocks" (p. 461). Consideration of climate impacts on fishery resources will likely become more common with the development and application of ecosystem-based approaches, through mechanisms such as integration of changing environmental and ecological conditions into FMPs.

Efforts to integrate climate considerations into existing legislative and regulatory frameworks

Recent years have witnessed an increase in awareness that climate strategies must include efforts to both limit and adapt to climate change (e.g., Lazarus 2009, Ruhl 2010). Regulatory and management frameworks must be able to operate and remain effective in the face of increased and potentially significant uncertainty and change (Craig 2010, Gregg et al. 2011). While no single piece of existing US federal legislation directly targets climate change adaptation in the marine environment, there are several potential mechanisms, some of which are already being implemented, for incorporating climate change considerations into existing statutory and regulatory processes (e.g., Sussman et al. 2010, Gregg et al. 2011). As noted by the GAO, federal resource management agencies "are generally authorized ... to address changes in resource conditions resulting from climate change in their management activities" (US GAO 2007, p. 2). Whether agencies are required to address climate change depends on their delegated statutory and regulatory authority.

Broadly applicable policy initiatives may enable climate change adaptation in the ocean and marine environment. For example, the National Ocean Council has developed draft implementation plans for two relevant national priority objectives: (1) "resiliency and adaptation to climate change and ocean acidification" and (2) "changing conditions in the Arctic" (NOC 2012). The Council on Environmental Quality (CEQ) has also drafted guidance for federal agencies on incorporating consideration of greenhouse gas emissions and adaptation measures into environmental reviews conducted pursuant to the National Environmental Policy Act (CEQ 2010, NEPA 42 U.S.C. § 4321 et seq.).

In addition, the United States is undertaking specific efforts to address adaptation to climate change impacts in the marine environment through existing legislative and regulatory frameworks. The following discussion describes efforts related to incorporating climate change considerations into regulation and management.

Ocean acidification   The Clean Water Act (CWA) has been cited as a potential mechanism for managing climate change impacts in US waters (e.g., Craig 2009, Kelly et al. 2011). The purpose of the CWA is to restore and maintain the chemical, physical, and biological integrity of US waters (CWA 33 U.S.C. § 1251 et seq.). One of the statutory tools that may help manage ocean acidification is the designation of impaired waters, the designation for water bodies that fail to meet specified water quality standards. Pursuant to a settlement between the CBD and the EPA, the EPA solicited input on state approaches to determining whether waters are threatened or impaired by ocean acidification (*CBD v. EPA* 2009, *Federal Register* 2010a). The result was the agency's reassertion that states should seek scientific information on impacts and, when sufficient information is available, list as impaired those water bodies for which pH is below the recommended range (CWA § 304(a), US EPA 2010, Kelly et al. 2011). States can also list as impaired water bodies that fail to meet biological water quality standards because of ocean acidification, that is, water bodies that are failing to meet criteria established for coral reef ecosystems, bivalves, or other organisms protected under CWA aquatic life-designated uses (Bradley et al. 2010).

Threatened and endangered species   Public and private efforts have worked to ensure consideration of climate impacts through protected species management laws and listing decisions. In accordance with the Endangered Species Act (16 U.S.C. § 1531 et seq.), NOAA and the USFWS designate, protect, and recover threatened and endangered species. These agencies cited climate change impacts, such as increased sea-surface temperatures, sea-level rise, loss of sea ice, and ocean acidification, as habitat stressors in their decisions to list as threatened elkhorn and staghorn coral species (*Federal Register* 2006), polar bears (*Federal Register* 2008), and the southern distinct population segment of the spotted seal (*Federal Register* 2010b), as well as the finding that listing of the Pacific walrus as threatened or endangered is warranted (*Federal Register* 2011). NOAA also identified climate change as a key factor in finding that a threatened or endangered listing may be warranted for 82 coral species (*Federal Register* 2010c). In addition to listings, agencies could factor climate adaptation considerations into critical habitat designations, recovery plans, and consultations on proposed federal actions (Kostyack & Rohlf 2008, Craig 2009, Owen 2012).

Fisheries management   The MSA (16 U.S.C. §1801 et seq.) requires regional fishery management councils to develop FMPs, which must be approved by the secretary of commerce. While there are currently no requirements specifying consideration of climate change in FMPs, at least one regional council has begun to consider such impacts. In 2009, the NPFMC issued an FMP that closed Arctic fisheries to commercial harvesting (NPFMC 2009). Recognizing that loss of sea ice could remove barriers to previously inaccessible fisheries, the Arctic FMP established a prohibition against commercial fishing until adequate information is available to support sustainable management (NPFMC 2009). The decision reflects a precautionary approach to managing and developing fisheries in the face of climate change.

Enhancing the resilience of the nation's oceans to climate change will require action at all levels. The US federal statutory, regulatory, and policy efforts represent some of the existing approaches that agencies are taking to support adaptation to and management of climate change impacts in the marine environment.

Emerging frameworks and actions for ocean adaptation

Although the science and practice of marine adaptation are relatively nascent, many individuals, communities, ecosystem managers, and governments across the United States are developing strategies for enhancing ocean resilience in the face of a changing climate. Adaptation frameworks, including ocean-related adaptation efforts, are emerging across the US federal government (Center for Climate and Energy Solutions 2012). In addition, diverse frameworks for ocean adaptation action have been developed at national, regional, state, local, and non-governmental levels (Table 7). These efforts provide a platform to inspire and support the planning and implementation of on-the-ground actions.

Incorporation of climate change into forward-looking, adaptive management actions is beginning to occur for marine systems throughout the United States. For example, the North Pacific Climate Regimes and Ecosystem Productivity (NPCREP) project is working closely with resource managers to understand and address impacts of climate change on North Pacific and Bering Sea ecosystems. From 2000 to 2005, anomalously high sea-surface temperatures in the Bering Sea coincided with low pollock recruitment. NPCREP provided supplemental information that when considered in concert with the fisheries stock assessment corroborated the evidence that pollock recruitment was below average and added conservation was recommended. This information helped to inform the decisions of the NPFMC's Scientific and Statistical Committee. Based on recommendations from the committee, the NPFMC temporarily reduced pollock quotas between 2006 and 2010 until conditions became more favourable. This effort illustrates adaptive management based on changing environmental conditions.

To understand and address climatic and non-climatic threats to coral reefs, the US Geological Survey's Coral Reef Ecosystem Studies (CREST) project is investigating drivers and trends of coral reef ecosystem change. CREST is conducting monitoring and research efforts in national parks (Dry Tortugas, Virgin Islands, Biscayne) and areas of the Florida Keys National Marine Sanctuary. Projects include habitat mapping, assessment of calcification change related to ocean acidification, identification of diseases, and improving understanding of reef responses to sea-level change, among others. This work will improve understanding about coral health, advance the ability to forecast future change, and guide management decisions.

Oyster producers in the Pacific North-west have been facing persistently low seed survival, in part due to acidified waters. These failures are resulting in low harvest rates and economic impacts to shellfish hatcheries. In response, the Pacific Coast Shellfish Growers Association and partners launched the Emergency Oyster Seed Project in Washington State to establish monitoring programmes in key estuaries, develop solutions for enhancing hatchery production, and identify resilient oyster genotypes. Some hatcheries are already implementing adaptive management practices by coordinating water intake to avoid periods of high acidity.

Table 7 Examples of ocean-related climate adaptation frameworks in the United States

Adaptation project	Description
National/federal	
Interagency Climate Change Adaptation Task Force (ICCATF)	The ICCATF was initiated in spring 2009 to determine progress on federal agency actions in support of national adaptation and to develop recommendations for additional actions. The ICCATF is composed of more than 20 federal agencies and executive branch offices and has involved more than 300 federal employees. One outcome of this effort is a mandate (under Executive Order 13514) for all federal agencies, including those with ocean-related responsibilities, to develop adaptation plans.
National Fish, Wildlife & Plants Climate Adaptation Strategy	The National Fish, Wildlife, & Plants Climate Adaptation Strategy, initiated through Congressional directive in 2009, is currently under development. The strategy provides a nationwide blueprint for coordinated action among federal, state, tribal, and non-governmental entities to safeguard the nation's valuable natural resources, including marine resources, in a changing climate. The draft strategy was released in January 2012 for public review and input. The final version was released in Spring 2013.
National Ocean Policy	The National Ocean Policy was created in July 2010 under Executive Order 13547 to form a comprehensive, integrated framework for the stewardship of the ocean, coasts, and Great Lakes of the United States. In January 2012, a draft implementation strategy was released, including a set of interagency actions and milestones focused on enhancing the resiliency of oceans to climate change and ocean acidification.
Regional	
West Coast Governors Alliance on Ocean Health	The US West Coast Governors launched an Agreement on Ocean Health in 2006 to create a framework for regional collaboration on protection and management of ocean and coastal resources. The alliance includes a Climate Change Action Coordination Team, which is initially focusing on a West Coast assessment of shoreline change and anticipated impacts to coastal areas and communities due to climate change over the next several decades. This effort will inform adaptation to climate change and coastal hazards.
State	
State of California Climate Adaptation Strategy	In 2009, recognizing the need to prepare for climate change, the state of California released its Climate Adaptation Strategy. The Coastal and Ocean Resources Working Group is currently working to implement components of the strategy through assessing the risks of sea-level rise, mapping susceptible transportation areas, and conducting vulnerability assessments, among other actions.
Massachusetts Climate Change Adaptation Report	In response to the state's Global Warming Solutions Act of 2008, the secretary of energy and environmental affairs and the Massachusetts Climate Change Adaptation Advisory Committee produced the Massachusetts Climate Change Adaptation Report to describe state-level climate impacts, vulnerabilities, and adaptation strategies for key sectors, including coastal zone and oceans.
Non-governmental	
Alaskan Marine Arctic Conservation Action Plan for the Chukchi and Beaufort Seas	The Nature Conservancy's Alaska chapter developed the Conservation Action Plan to guide the organization's management and conservation efforts in the region. Climate change is identified in the plan as the primary threat to the region's natural resources. An expert panel helped guide the selection of primary conservation targets (bowhead whales, ice-dependent marine mammals, seabirds, boulder patch communities, benthic fauna, fish). Recommendations include promoting adaptation strategies and ecosystem-based management, investing in baseline and long-term data collection, and identifying and protecting climate refugia, among others.
A Climate Change Action Plan for the Florida Reef System (2010–2015)	The Florida Reef Resilience Program, a public-private partnership, released a Climate Change Action Plan in 2010. Florida reefs are subject to non-climatic and climatic stressors that are degrading overall ecosystem resilience. Priority actions identified in the plan include expanding disturbance response monitoring throughout the Florida Reef Tract, developing a marine zoning plan to address non-climate stressors, decreasing negative user impacts from fishing and diving, mapping areas of high and low resilience to prioritize protection, and restoring resistant reefs. The plan can be adopted by reef managers into existing management plans.

These efforts represent only a handful of the emerging ocean adaptation activities under way in the United States. They demonstrate that while much work remains to be done, actions are taking place across a spectrum of adaptation-related efforts, from risk, impact, and vulnerability assessments; to the development of guidance and tools; to on-the-ground implementation. In general, most ocean-related adaptation activities are still in the phases of improving understanding of risks, impacts, and vulnerabilities; some are in planning phases; and a few are in implementation. However, this also means that we can expect to see more actions in the future as marine managers and decision makers advance through the process of adaptation.

Sustaining the assessment of climate impacts on oceans and marine resources

Based on current understanding and projections, it is likely that marine ecosystems under US and other jurisdictions will continue to be affected by climate change through a suite of changes in ocean physical, chemical, biological, social, and economic systems (Hollowed et al. 2011, Sumaila et al. 2011, Doney et al. 2012). Despite this foundation of information, many uncertainties and gaps remain in understanding the current and future impacts of climate change and ocean acidification on marine ecosystems (Doney et al. 2012). Assessing what is known about past, current and future impacts is a challenging task, and as this review has shown, scientific knowledge is critical for gaining insight into the effects of climate change on the ocean environment and what those changes mean for the economy, social, cultural and personal well-being of the US population and to the health of the ocean itself.

The current body of literature documenting the effects of climate change on the ocean is growing rapidly, and it is critical to enhance integration of new knowledge into public and private-sector responses locally, nationally and internationally. One area of great need is interdisciplinary research that brings together scientists studying the biophysical effects of climate change with social scientists to develop a much fuller understanding of both the biophysical and human dimensions of climate change.

Improving understanding and projections of climate change impacts on natural and managed ecosystems remains a critical challenge. It is difficult to forecast how species, habitats, ecosystems, economies, and nations will respond when confronted by environmental conditions that are often outside the range of conditions experienced today. Communicating how climate change and ocean acidification are impacting the world's oceans and ocean resources, particularly in those areas where changes are expected to be seen in both the short and long term, is vital. Scientists and policy makers are now beginning to think about how to prepare the nation for climate change. The following are next steps that, if taken, would greatly aid in addressing this challenge; this is not intended to be a comprehensive list, and items are not listed in priority order.

Observations and monitoring

Monitoring by ships, buoys, satellites, scientists in the field, and even narwhals and seals carrying monitoring devices provides data that are the foundation for documenting the conditions and trends in US coastal and ocean ecosystems over time. Continuation, and in some areas expansion, of these monitoring efforts is essential to track trends and provide robust early warnings of future changes. Many critical information gaps remain with respect to understanding and documenting the impacts of climate change on ocean systems. Key steps to advance this area include

- Increase capacity and coordination of existing observing systems to collect, synthesize and deliver integrated information on physical, chemical, biological and social/economic impacts of climate change on US marine ecosystems.
- Monitor impacts of climate change on oceanic circulation, extreme weather events (including tropical cyclone activity), stratification, and sea-level rise.

- Improve detection and early warning of climate variability and change for ocean systems.
- Identify and implement a set of core physical, biological, and societal indicators of the condition of marine ecosystems that can be used to track and assess the impacts of climate change and ocean acidification, as well as the effectiveness of mitigation and adaptation efforts, over time at regional-to-national scales.
- Conduct integrated monitoring of ocean ecosystem and related socioeconomic change.
- Develop integrated ocean-related human health early-warning systems through deployment of marine sensors for monitoring, updating of public health surveillance systems, and developing of risk communication tools and strategies.

Research

A strategic, use-inspired and integrated science agenda is necessary for informing and supporting efforts to prepare for and respond to a changing climate. Advancing understanding and application of new knowledge will require sustained dialogue, mutual information exchange, and feedback among scientists, decision makers, and information users. Critical research gaps include

- Investigating the degree to which marine organisms can acclimate and populations can genetically adapt to rapid environmental change.
- Determining the cumulative impacts of multiple stressors (both climatic and non-climatic) on marine organisms within ecologically relevant contexts and understand the underlying mechanisms through which these impacts occur.
- Supporting research on climate- and ocean-related social, behavioural, and economic science needs (e.g., assessing adaptation options and trade-offs, determining costs of action vs. inaction, designing effective governance processes, identifying thresholds and tipping points in social systems, investigating decision making under uncertainty).
- Creating an interdisciplinary research effort and community-based risk assessment of the effects of ocean acidification.
- Improving understanding and valuation of climate-related impacts on ocean ecosystem services.
- Advancing scientific knowledge and application of the role of blue carbon in climate and ocean systems.
- Developing and implementing methods for evaluating the effectiveness of adaptation actions to enable a flexible and responsive management approach.

Modelling

Models can serve as valuable tools for improving understanding, prediction, and projection of climate-related changes in ocean systems. However, there are currently critical information needs associated with climate and ocean modelling that, if addressed, would improve the ability of decision makers to plan for the future. Therefore, there is a growing need to

- Advance modelling and projection of the integrated impacts of climate change on ocean physical, chemical, biological, and social systems at finer spatial (e.g., regional) and temporal (e.g., decadal) scales most relevant for decision makers.
- Improve understanding and projection of sea-level rise (including the potential for rapid ice melt), changes in ocean currents, and stratification.
- Improve prediction of environmental and ecological conditions that lead to non-linearities and tipping points in coastal and marine ecosystems.
- Enhance the development of spatially explicit predictions of ecosystem responses to climate change (particularly for local-to-regional scales), including estimates of uncertainty.

- Develop local-to-regional scale projections of climate impacts on ocean tourism and recreation.
- Develop models of fleet dynamics, vessel operator income, and community social and economic impacts linked to biophysical models of climate-induced changes.
- Develop dynamic models that incorporate risk and uncertainty to inform decisions on investment in adaptation and mitigation measures for ocean systems.

Communication

Climate change is a global problem with wide-ranging impacts, and it is essential that climate-related information, as well as the information needs of decision makers, be communicated effectively across many groups, including scientists, citizen scientists, partners, educators, resource managers, and elected officials, among others. Therefore, there is a growing need to

- Increase the capacity for integrating, synthesizing and sharing data acquired at different levels of space, time, taxonomic resolution, and so on into useful forms for diverse information users to support informed decision making.
- Coordinate and increase communication between decision makers and science providers to ensure that the most critical information needs are being met related to impacts, vulnerabilities, and adaptation of ocean systems in a changing climate.
- Build and support mechanisms with neighbouring countries and other international partners for assessing and addressing impacts of climate change and ocean acidification on marine ecosystems.

This review has shown the impacts of climate-related change on ocean physical, chemical, biological, and human social systems in the United States. This review has also elaborated on questions that continue to dominate climate change discussions, such as the following:

- What climate-related changes have already occurred in US marine ecosystems?
- Why are these changes happening?
- What effects are anticipated in the near future?
- What are the long-term trends of climate change?
- What can be done to prepare for and reduce impacts?
- How is science informing planning and policy?

Meeting the challenges of climate change in general and climate change impacts on the marine environment and ocean services requires interdisciplinary efforts and strengthened connections between science and decision making now and in the years to come (Østreng 2010).

References

Aaheim, A. & Sygna, L. 2000. Economic impacts of climate change on tuna fisheries in Fiji Islands and Kiribati. CICERO Report 2000(4), Oslo, Norway.

Abreu, M.H., Pereira, R., Sousa-Pinto, I. & Yarish, C. 2011. Ecophysiological studies of the non-indigenous species *Gracilaria vermiculophylla* (Rhodophyta) and its abundance patterns in Ria de Aveiro lagoon, Portugal. *European Journal of Phycology* **46**, 453–464.

Acclimatise. 2009a. Building business resilience to inevitable climate change. Carbon Disclosure Project Report 2008. Global oil and gas. Oxford, UK: Acclimatise and Climate Risk Management.

Acclimatise. 2009b. Understanding the investment implications of adapting to climate change—oil and gas. Oxford, UK: Acclimatise and Climate Risk Management.

Agostini, V.N., Francis, R.C., Hollowed, A.B., Pierce, S., Wilson, C. & Hendrix, A.N. 2006. The relationship between hake (*Merluccius productus*) distribution and poleward sub-surface flow in the California Current System. *Canadian Journal of Fisheries and Aquatic Sciences* **63**, 2648–2659.

Agostini, V.N., Hendrix, A.N., Hollowed, A.B., Wilson, C.D., Pierce, S. & Francis, R.C. 2007. Climate and Pacific hake: a geostatistical modeling approach. *Journal of Marine Systems* **71**, 237–248.

Alekseev, G., Danilov, A., Kattsov, V., Kuz'mina, S. & Ivanov, N., 2009. Changes in the climate and sea ice of the Northern Hemisphere in the 20th and 21st centuries from data of observations and modeling. *Izvestiya Atmospheric and Oceanic Physics* **43**, 675–686.

Alexander, R.B., Smith, R.A., Schwarz, G.E., Boyer, E.W., Nolan, J.V. & Brakebill, J.W. 2008. Differences in phosphorus and nitrogen delivery to the Gulf of Mexico from the Mississippi River Basin. *Environmental Science and Technology* **42**, 822–830.

Allison, G.W., Gaines, S.D., Lubchenco, J. & Possingham, H.P. 2003. Ensuring persistence of marine reserves: catastrophes require adopting an insurance factor. *Ecological Applications* **13**, S8–S24.

Altizer, S., Harvell, C.D. & Friedle, E. 2003. Rapid evolutionary dynamics and disease threats to biodiversity. *Trends in Ecology and Evolution* **18**, 589–596.

Alvarez-Filip, L., Dulvy, N.K., Gill, J.A., Côté, I.M. & Watkinson, A.R. 2009. Flattening of Caribbean coral reefs: region-wide declines in architectural complexity. *Proceedings of the Royal Society B* **276**, 3019–3025.

Andersen, N. & Malahoff, A. 1977. *The Fate of Fossil Fuel CO$_2$ in the Oceans*. New York: Plenum.

Anderson, D. 2012. Harmful algae. Online. http://www.whoi.edu/website/redtide/home (accessed 10 January 2012).

Anderson, D.M. 1989. Toxic algal blooms and red tides. In *Red Tides: Biology, Environmental Science, and Toxicology*, T. Okaichi et al. (eds). New York: Elsevier, 11–16.

Anderson, D.M. 2009. Approaches to monitoring, control and management of harmful algal blooms (HABs). *Ocean and Coastal Management* **52**, 342–347.

Anestis, A., Pörtner, H.O., Karagiannis, D., Angelidis, P., Staikou, A. & Michaelidis, B. 2010a. Response of *Mytilus galloprovincialis* (L.) to increasing seawater temperature and to marteliosis: metabolic and physiological parameters. *Comparative Biochemistry and Physiology Part A* **156**, 57–66.

Anestis, A., Pörtner, H.O. & Michaelidis, B. 2010b. Anaerobic metabolic patterns related to stress responses in hypoxia exposed mussels *Mytilus galloprovincialis*. *Journal of Experimental Marine Biology and Ecology* **394**, 123–133.

Anthony, K.R.N., Kline, D.I., Diaz-Pulido, G., Dove, S. & Hoegh-Guldberg, O. 2008. Ocean acidification causes bleaching and productivity loss in coral reef builders. *Proceedings of the National Academy of Sciences USA* **105**, 17442–17446.

Archer, C.L. & Caldeira, K. 2008. Historical trends in the jet streams. *Geophysical Research Letters* **35**, L08803.

Arctic Climate Impact Assessment (ACIA). 2004. Impacts of a warming Arctic. Arctic Climate Impact Assessment Overview Report. New York: Cambridge University Press.

Arctic Marine Shipping Assessment (AMSA). 2009. Protection of the Arctic Marine Environment. Akureyri, Iceland: Arctic Council. Online. http://arctic.gov/publications/AMSA_2009_2nd_print.pdf

Arctic Monitoring and Assessment Program (AMAP). 2008. Arctic oil and gas 2007. Oslo: Arctic Monitoring and Assessment Program, 40 pp.

Article II, Text of the Convention. Convention on International Trade in Endangered Species of Wild Fauna and Flora. Signed at Washington, D.C. on 3 March 1973. Amended at Bonn on 22 June 1979.

Arzel, O., England, M., de Verdière, A.C. & Huck, T. 2011. Abrupt millennial variability and interdecadal-interstadial oscillations in a global coupled model: sensitivity to the background climate state. *Climate Dynamics* (Online First), 1–17. doi:10.1007/s00382-011-1117-y

Auster, P.J. & Link, J.S. 2009. Compensation and recovery of feeding guilds in a northwest Atlantic shelf fish community. *Marine Ecology Progress Series* **382**, 163–172.

Austin, D. & Sauer, A. 2002. Changing oil: emerging environmental risks and shareholder value in the oil and gas industry. Washington, DC: World Resources Institute.

Badjeck, M., Allison, E., Halls, A. & Dulvy, N. 2010. Impacts of climate variability and change on fishery-based livelihoods. *Marine Policy* **34**, 375–383.

Baer, H. & Singer, M. 2009. *Global Warming and the Political Ecology of Health: Emerging Crises and Systemic Solutions*. Walnut Creek, California: Left Coast Press.

Baer, H., Singer, M. & Susser, I. 2003. *Medical Anthropology and the World System.* Westport, Connecticut: Praeger, 2nd edition, p. 5.

Baker, J.D., Littnan, C.L. & Johnston, D.W. 2006. Potential effects of sea level rise on the terrestrial habitats of endangered and endemic megafauna in the northwestern Hawaiian Islands. *Endangered Species Research* **2**, 21–30.

Barange, M. & Perry, R.I. 2009. Physical and ecological impacts of climate change relevant to marine and inland capture fisheries and aquaculture. In *Climate Change Implications for Fisheries and Aquaculture: Overview of Current Scientific Knowledge*, K. Cochrane et al. (eds). FAO Fisheries and Aquaculture Technical Paper No. 530. Rome: Food and Agricultural Organization of the United Nations, 7–106.

Barlas, M.E., Deutsch, C.J., de Wit, M. & Ward-Gieger, L.I. (eds) 2011. Florida manatee cold-related unusual mortality event, January-April 2010. Final report to USFWS (grant 40181AG037). St. Petersburg: Florida Fish and Wildlife Conservation Commission, 138 pp.

Barnett, J. 2006. Climate change, insecurity, and injustice. In *Fairness in Adaptation to Climate Change*, W.N. Adger et al. (eds). Cambridge, Massachusetts: MIT Press, 115–129.

Barnosky, A.D., Matzke, N., Tomiya, S., Wogan, G.O.U., Swartz, B. Quental, T.B., Marshall, C., McGuire, J.L., Lindsey, E.L., Maguire, K.C., Mersey, B. & Ferrer, E.A. 2011. Has the Earth's sixth mass extinction already arrived? *Nature* **470**, 51–57.

Barry, J.P., Baxter, C.H., Sagarin, R.D. & Gilman, S.E. 1995. Climate-related, long-term faunal changes in a California rocky intertidal community. *Science* **267**, 672–675.

Barth, J.A., Menge, B.A., Lubchenco, J., Chan, F., Bane, J.M., Kirincich, A.R., McManus, M.A., Nielsen, K.J., Pierce, S.D. & Washburn, L. 2007. Delayed upwelling alters nearshore coastal ocean ecosystems in the northern California current. *Proceedings of the National Academy of Sciences USA* **104**, 3719–3724.

Beamish, R.J., Lange, K.L., Riddell, B.E. & Urawa, S. 2010. Climate impacts on pacific salmon: bibliography. Special Publication No. 2. Vancouver, BC: North Pacific Anadromous Fish Commission.

Beamish, R.J., Riddell, B.E., Lange, K.L., Farley, E., Jr., Kang, S., Nagasawa, T., Radchenko, V., Temnykh, O. & Urawa, S. 2009. A long-term research and monitoring plan (LRMP) for Pacific salmon (*Oncorhynchus* spp.) in the North Pacific Ocean. Special Publication No. 1. Vancouver, BC: North Pacific Anadromous Fish Commission.

Beaufort, L, Probert, I., De Garidel-Thoron, T., Bendif, E.M., Ruiz-Pino, D., Metzl, N., Goyet, C., Buchet, N., Coupel, P., Greland, M., Rost, B., Rickaby, R.E.M. & de Vargas, C. 2011. Sensitivity of coccolithophores to carbonate chemistry and ocean acidification. *Nature* **476**, 80–83.

Beaugrand, G., Brander, K.M., Lindley, J.A., Souissi, S. & Reid, P.C. 2003. Plankton effect on cod recruitment in the North Sea. *Nature* **426**, 661–664.

Beaugrand, G., Reid, P.C., Ibañez, F., Lindley, J.A. & Edwards, M. 2002. Reorganization of North Atlantic marine copepod biodiversity and climate. *Science* **296**, 1692–1694.

Behrenfeld, M.J., O'Malley, R.T., Siegel, D.A., McClain, C.R., Sarmiento, J.L., Feldman, G.C., Milligan, A.J., Falkowski, P.G., Letelier, R.M. & Boss, E.S. 2006. Climate-driven trends in contemporary ocean productivity. *Nature* **444**, 752–755.

Bell, G. & Collins, S. 2008. Adaptation, extinction, and global change. *Evolutionary Applications* **1**, 3–16.

Bell, J.D., Johnson, J.E. & Hobday A.J. 2011. Vulnerability of tropical Pacific fisheries and aquaculture to climate change. Noumea, New Caledonia: Secretariat of the Pacific Community.

Beman, J.M., Chow, C.E., King, A.L., Feng, Y., Fuhrman, J.A., Andersson, A., Bates, N.R., Popp, B.N. & Hutchins, D.A. 2011. Global declines in oceanic nitrification rates as a consequence of ocean acidification. *Proceedings of the National Academy of Sciences USA* **108**, 208–213.

Bender, M.A., Knutson, T.R., Tuleya, R.E., Sirutis, J.J., Vecchi, G.A., Garner, S.T. & Held, I.M. 2010. Modeled impacts of anthropogenic warming on the frequency of intense Atlantic hurricanes. *Science* **327**, 454–458.

Berger, S.A., Diehl, S., Stibor, H., Trommer, G., Ruhenstroth, M., Wild, A., Weigert, A., Jäger, C.G. & Striebel, M. 2007 Water temperature and mixing depth affect timing and magnitude of events during spring succession of the plankton. *Oecologia* **150**, 643–654.

Berkman, P.A. & Young, O.R. 2009. Governance and environmental change in the Arctic Ocean. *Science* **324**, 239–240.

Bernard, S. & McGeehin, M. 2004. Municipal heat wave response plans. *American Journal of Public Health* **94**, 1520–1522.

Bertram, D.F. & Kaiser, G.W. 1993. Rhinoceros auklet (*Cerorhinea monocerata*) nestling diet may gauge Pacific sand lance (*Ammodytes hexapterus*) recruitment. *Canadian Journal of Fisheries and Aquatic Sciences* **50**, 1908–1915.

Beukema, J.J., Dekker, R. & Jansen, J.M. 2009. Some like it cold: populations of the tellinid bivalve *Macoma balthica* (L.) suffer in various ways from a warming climate. *Marine Ecology Progress Series* **384**, 135–145.

Biggs, R., Carpenter, S.R. & Brock, W.A. 2009. Turning back from the brink: detecting an impending regime shift in time to avert it. *Proceedings of the National Academy of Sciences USA* **106**, 826–831.

Bindoff, N.L., Willebrand, J., Artale, V., Cazenave, A., Gregory, J.M., Gulev, S., Hanawa, K., Le Quéré, C., Levitus, S., Nojiri, Y., Shum, C.K., Talley, L.D. & Unnikrishnan, A. 2007. Observations: oceanic climate change and sea level. In *Climate Change 2007: The Physical Science Basis: Contribution of Working Group I to the Fourth Assessment Report of the Intergovernmental Panel on Climate Change*, S. Soloman et al. (eds). Cambridge, UK: Cambridge University Press, 385–432.

Bird, E.C.F. 1994. Physical setting and geomorphology of coastal lagoons. In *Coastal Lagoon Processes*, B. Kjerfve (ed.). Amsterdam: Elsevier.

Bluhm, B.A. & Gradinger, R. 2008. Regional variability in food availability for Arctic marine mammals. *Ecological Applications* **18**, S77–S96.

Blunden, J., Arndt, D.S. & Baringer, M.O. 2011: State of the climate in 2010. *Bulletin of the American Meteorological Society* **92**, S1–S266.

Boesch, D.F., Coles, V.J., Kimmel, D.G. & Miller, W.D. 2007. Coastal dead zones and global climate change: ramifications of climate change for the Chesapeake Bay. In *Regional Impacts of Climate Change: Four Case Studies in the United States*. Arlington, VA: Pew Center for Global Climate Change, 54–70.

Bograd, S.J., Castro, C.G., Di Lorenzo, E., Palacios, D.M., Bailey, H., Gilly, W. & Chavez, F.P. 2008. Oxygen declines and the shoaling of the hypoxic boundary in the California Current. *Geophysical Research Letters* **35**, L12607.

Bograd, S.J., Schroeder, I., Sarkar, N., Qiu, X., Sydeman, W.J. & Schwing, F.B. 2009. Phenology of coastal upwelling in the California Current. *Geophysical Research Letters* **36**, L01602.

Bony, S., Colman, R., Kattsov, V.M., Allan, R.P., Bretherton, C.S., Dufresne, J.L., Hall, A., Hallegatte, S., Holland, M.M., Ingram, W., Randall, D.A., Soden, B.J., Tselioudis, G. & Webb, M.J. 2006. How well do we understand and evaluate climate change feedback processes? *Journal of Climate* **19**, 3445–3482.

Borgerson, S.G. 2008. Arctic meltdown: the economic and security implications of global warming. *Foreign Affairs* **87**, 63–77.

Bossart, G.D., Baden, D.G., Ewing, R.Y., Roberts, B. & Wright, S.D. 1998. Brevetoxicosis in manatees (*Trichechus manatus latirostris*) from the 1996 epizootic: gross, histologic, and immunohistochemical features. *Toxicologic Pathology* **26**, 276–282.

Bossart, G.D., Meisner, R.A., Rommel, S.A., Ghim, S. & Jenson, A.B. 2002. Pathological features of the Florida manatee cold stress syndrome. *Aquatic Mammals* **29**, 9–17.

Boyd, P.W., Doney, S.C., Strzepek, R., Dusenberry, J., Lindsay, K. & Fung, I. 2008. Climate-mediated changes to mixed-layer properties in the Southern Ocean: how will key phytoplankton species respond? *Biogeosciences* **5**, 847–864.

Braby, C.E. & Somero, G.N. 2006. Following the heart: temperature and salinity effects on heart rate in native and invasive species of blue mussels (genus *Mytilus*). *Journal of Experimental Biology* **209**, 2554–2566.

Bradley, P.A., Fore, L.S., Fisher, W.S. & Davis, W.S. 2010. Coral reef biological criteria: using the Clean Water Act to protect a national treasure. EPA/600/R-10/054, July 2010. Narragansett, RI: U.S. Environmental Protection Agency, Office of Research and Development.

Brander, K. 2010. Impacts of climate change on fisheries. *Journal of Marine Systems* **79**, 389–402.

Brewer, P.G. & Hester, K. 2009. Ocean acidification and the increasing transparency of the ocean to low-frequency sound. *Oceanography* **22**, 86–93.

Bricker, S., Longstaff, B., Dennison, W., Jones, A., Boicourt, K., Wicks, C. & Woerner, J. 2007. Effects of nutrient enrichment in the nation's estuaries: a decade of change. NOAA Coastal Ocean Program Decision Analysis Series No. 26. Silver Spring, MD: Nation Centers for Coastal Ocean Science.

Broecker, W.S. 1991. The great ocean conveyor. *Oceanography* **4**, 79–89.

Broecker, W.S. & Takahashi, T. 1966. Calcium carbonate precipitation on the Bahama Banks. *Journal of Geophysical Research* **71**, 1575–1602.

Broitman, B.R. & Kinlan, B.P. 2006. Spatial scales of benthic and pelagic producer biomass in a coastal upwelling ecosystem. *Marine Ecology Progress Series* **327**, 15–25.

Brook, B., Sodhi, N. & Bradshaw, C. 2008. Synergies among extinction drivers under global change. *Trends in Ecology & Evolution* **23**, 453–460.

Brooke, C. & Riley, T. 1999. *Erysipelothrix rhusiopathiae:* bacteriology, epidemiology and clinical manifestations of an occupational pathogen. *Journal of Medical Microbiology* **48**, 789–799.

Brown, K. 1999. Climate anthropology: taking global warming to the people. *Science* **283**, 1440–1441.

Buisson, Y., Marié, J. & Davoust, B. 2008. These infectious diseases imported with food. *Bulletin Société Pathologie Exotique* **101**, 343–347.

Bureau of Ocean Energy Management (BOEM). 2007. Final programmatic environmental impact statement for alternative energy development and production and alternate use of facilities on the outer continental shelf. OCS Report MMS 2007-046; USDOI, MMS, 2007. Online. http://ocsenergy.anl.gov/ (accessed 2 February 2012).

Burek, K., Gulland, F. & O'Hara, T. 2008. Effects of climate change on Arctic marine mammal health. *Ecological Applications* **18**, S126–S134.

Burke, L., Reytar, K., Spalding, M., Perry, A., et al. 2011. Reefs at risk revisited. Washington, DC: World Resources Institute.

Burrows, M.T., Schoeman, D.S., Buckley, L.B., Moore, P., Poloczanska, E.S., Brander, K.M., Brown, C., Bruno, J.F., Duarte, C.M., Halpern, B.S., Holding, J., Kappel, C.V., Kiessling, W., O'Connor, M.I., Pandolfi, J.M., Parmesan, C., Schwing, F.B., Sydeman, W.J. & Richardson, A.J. 2011. The pace of shifting climate in marine and terrestrial ecosystems. *Science* **334**, 652–655.

Bussell, J.A., Gidman, E.A., Causton, D.R., Gwynn-Jones, D., Malham, S.K., Jones, M.L.M., Reynolds, B. & Seed, R. 2008. Changes in the immune response and metabolic fingerprint of the mussel, *Mytilus edulis* (Linnaeus) in response to lowered salinity and physical stress. *Journal of Experimental Marine Biology and Ecology* **358**, 78–85.

Butchart, S.H.M., Walpole, M., Collen, B., van Strien, A., Scharlemann, J.P.W., Almond, R.E.A., Baillie, J.E.M., Bomhard, B., Brown, C., Bruno, J., Carpenter, K.E., Carr, G.M., Chanson, J., Chenery, A.M., Csirke, J., Davidson, N.C., Dentener, F., Foster, M., Galli, A., Galloway, J.N., Genovesi, P., Gregory, R.D., Hockings, M., Kapos, V., Lamarque, J.-F., Leverington, F., Loh, J., McGeoch, M.A., McRae, L., Minasyan, A., Hernández Morcillo, M., Oldfield, T.E.E., Pauly, D., Quader, S., Revenga, C., Sauer, J.R., Skolnik, B., Spear, D., Stanwell-Smith, D., Stuart, S.N, Symes, A., Tierney, M., Tyrrell, T.D., Vié, J.-C. & Watson, R. 2010. Global biodiversity: indicators of recent declines. *Science* **328**, 1164–1168.

Button, G. 1992. *Social conflict and the formation of emergent groups in a technological disaster: the Exxon-Valdez oil spill and the response of residents in the area of Homer, Alaska.* PhD dissertation, Brandeis University, Waltham, Massachusetts.

Byrd, V.G., Sydeman, W.J., Renner, H.M. & Minobe, S. 2008. Responses of piscivorous seabirds at the Pribilof Islands to ocean climate. *Deep-Sea Research Part II* **55**, 1856–1867.

Byrne, M. 2011. Impact of ocean warming and ocean acidification on marine invertebrate life history stages: vulnerabilities and potential for persistence in a changing ocean. *Oceanography and Marine Biology: an Annual Review* **49**, 1–42.

Byrnes, J.E., Reed, D.C., Cardinale, B.J., Cavanaugh, K.C., Holbrook, S.J. & Schmitt, R.J. 2011. Climate-driven increases in storm frequency simplify kelp forest food webs. *Global Change Biology* **17**, 2513–2524.

Byrnes, J.E., Reynolds, P.L. & Stachowicz, J.J. 2007. Invasions and extinctions reshape coastal marine food webs. *PLoS one* **2**, e295.

Caldeira, K. & Wickett, M.E. 2003. Oceanography: anthropogenic carbon and ocean pH. *Nature* **425**, 365.

Caldeira, K. & Wickett, M.E. 2005. Ocean model predictions of chemistry changes from carbon dioxide emissions to the atmosphere and ocean. *Journal of Geophysical Research* **110**, C09S04, 12 pp. doi:10.1029/2004JC002671.

Callaway, D., Eamer, J., Edwardsen, E., Jack, C., Marcy, S., Olrun, A., Patkotak, M., Rexford, D. & Whiting, A. 1999. Effects of climate change on subsistence communities in Alaska. In *Assessing the Consequences of Climate Change in Alaska and the Bering Sea Region,* G. Weller & P.A. Anderson (eds). Fairbanks, AK: Center for Global Change and Arctic System Research, the University of Alaska Fairbanks.

Campbell, K.M., Gulledge, J., McNeill, J.R., Podesta, J., Ogden, P., Fuerth, L., Woolsey, R.J., Lennon, A.T.J., Smith, J., Weitz, R. & Mix, D. 2007. *The Age of Consequences: The Foreign Policy and National Security Implications of Global Climate Change.* Washington, DC: Center for Strategic & International Studies and Center for a New American Security.

Canadell, J.G. & Raupach, M.R. 2008. Managing forests for climate change mitigation. *Science* **320**, 1456–1457.

Cardinale, B.J., Matulich, K.L., Hooper, D.U., Byrnes, J.E., Duffy, E., Gamfeldt, L., Balvanera, P., O'Connor, M.I. & Gonzalez, A. 2011. The functional role of producer diversity in ecosystems. *American Journal of Botany* **98**, 572–592.

Carlton, J.T. 1996. Pattern, process, and prediction in marine invasion ecology. *Biological Conservation* **78**, 97–106.

Carpenter, K.E., Abrar, M., Aeby, G., Aronson, R.B., Banks, S., Bruckner, A., Chiriboga, A., Cortés, J., Delbeek, J.C., DeVantier, L., Edgar, G.J., Edwards, A.J., Fenner, D., Guzmán, H.M., Hoeksema, B.W., Hodgson, G., Johan, O., Licuanan, W.Y., Livingstone, S.R., Lovell, E.R., Moore, J.A., Obura, D.O., Ochavillo, D., Polidoro, B.A., Precht, W.R., Quibilan, M.C., Reboton, C., Richards, Z.T., Rogers, A.D., Sanciango, J., Sheppard, A., Sheppard, C., Smith, J., Stuart, S., Turak, E., Veron, J.E.N., Wallace, C., Weil, E. & Wood, E. 2008. One-third of reef-building corals face elevated extinction risk from climate change and local impacts. *Science* **321**, 560–563.

Carson, H.S., Lopez-Duarte, P.C., Rasmussen, L., Wang, D. & Levin, L.A. 2010. Reproductive timing alters population connectivity in marine metapopulations. *Current Biology* **20**, 1926–1931.

Cazenave, A. & Llovel, W. 2010. Contemporary sea level rise. *Annual Review of Marine Science* **2**, 145–173.

Center for Biological Diversity v. Environmental Protection Agency. 2009. No. 2:09-cv-00670-JCC (W.D. Wash).

Center for Climate and Energy Solutions. 2012. Climate change adaptation: what Federal agencies are doing. February 2012 update. Online. http://www.c2es.org/docUploads/federal-agencies-adaptation.pdf (accessed 12 March 2013).

Centers for Disease Control and Prevention (CDC). 1998. Outbreak of *Vibrio parahaemolyticus* infections associated with eating raw oysters—Pacific Northwest, 1997. *Morbidity and Mortality Weekly Report* **47**, 457–462.

Centers for Disease Control and Prevention (CDC). 2005 *Vibrio* illnesses after Hurricane Katrina—multiple states, August–September 2005. *Morbidity and Mortality Weekly Report* **54**, 928–931.

Centers for Disease Control and Prevention (CDC). 2006. Morbidity surveillance after Hurricane Katrina— Arkansas, Louisiana, Mississippi, and Texas, September 2005. *Morbidity and Mortality Weekly Report* **55**, 727–731.

Centers for Disease Control and Prevention (CDC). 2011. Surveillance for waterborne disease outbreaks and other health events associated with recreational water—United States, 2007–2008. *Morbidity and Mortality Weekly Review Surveillance Summary* **60**, 1–32.

Chan, F., Barth, J.A., Lubchenco, J., Kirincich, A., Weeks, H., Peterson, W.T. & Menge, B.A. 2008. Emergence of anoxia in the California current large marine ecosystem. *Science* **319**, 920.

Chastel, O., Weimerskirch, H. & Jouventin, P. 1993. High annual variability in reproductive success and survival of an Antarctic seabird, the snow petrel *Pagodroma nivea*: a 27 year study. *Oecologia* **94**, 278–284.

Chavez, F.P., Messié, M. & Pennington, J.T. 2011. Marine primary production in relation to climate variability and change. *Annual Review of Marine Science* **3**, 227–260.

Chavez, F.P., Ryan, J., Lluch-Cota, S.E. & Niquen, M.C. 2003. From anchovies to sardines and back: multi-decadal change in the Pacific Ocean. *Science* **299**, 217–221.

Chernook, V.I. & Boltnev, A.I. 2008. Regular instrumental aerial surveys detect a sharp drop in the birth rates of the harp seal in the White Sea. In *Collection of scientific papers from the Marine Mammals of the Holarctic V Conference, Odessa, Ukraine, 14–18 October*, 100–104.

Cheung, W.W.L., Lam, V.W.Y., Sarmiento, J.L., Kearney, K., Watson, R. & Pauly, D. 2009. Projecting global marine biodiversity impacts under climate change scenarios. *Fish and Fisheries* **10**, 235–251.

Cheung, W.W.L., Lam, V.W.Y., Sarmiento, J.L., Kearney, K., Watson, R., Zeller, D. & Pauly, D. 2010. Large-scale redistribution of maximum fisheries catch potential in the global ocean under climate change. *Global Change Biology* **16**, 24–35.

Christensen, J.H., Hewitson, B., Busuioc, A., Chen, A., Gao, X., Held, I., Jones, R., Kolli, R.K., Kwon, W.-T., Laprise, R., Magaña Rueda, V., Mearns, L., Menéndez, C.G., Räisänen, J., Rinke, A., Sarr, A. & Whetton, P. 2007. Regional climate projections. In *Climate Change 2007: The Physical Science Basis. Contribution of Working Group I to the Fourth Assessment Report of the Intergovernmental Panel on Climate Change*, S. Soloman et al. (eds). Cambridge, UK: Cambridge University Press, 847–940.

Clay, P. & Olson, J. 2008. Defining 'fishing communities': vulnerability and the Magnuson-Stevens Fisheries Conservation and Management Act. *Human Ecology Review* 15, 143–160.

Climate Focus. 2011. Blue carbon: policy option assessment. Report prepared for the Linden Trust for Conservation by Climate Focus, Washington, DC. Online. http://www.climatefocus.com/documents/files/blue_carbon_.pdf (accessed 12 March 2013).

Cochrane, K., De Young, C., Soto, D. & Bahri, T. (eds) 2009. Climate change implications for fisheries and aquaculture: overview of current scientific knowledge. FAO Fisheries and Aquaculture Technical Paper No. 530. Rome: FAO.

Cohen, A.N. & Carlton, J.T. 1998. Accelerating invasion rate in a highly invaded estuary. *Science* 279, 555–558.

Colburn, L.L. & Jepson, M. 2012. Social indicators of gentrification pressure in fishing communities: a context for social impact assessment [Invited paper]. *Coastal Management Journal* 40, 289–300.

Colegrove, K.M., Lowenstine, L.J. & Gulland, F.M.D. 2005. Leptospirosis in northern elephant seals (*Mirounga angustirostris*) stranded along the California coast. *Journal of Wildlife Diseases* 41, 426–430.

Collier, D. 2002. Cutaneous infections from coastal and marine bacteria. *Dermatological Therapy* 15, 1–9.

Collinge, S.K., Johnson, W.C., Ray, C., Matchett, R., Grentsen, J., Cully, J.F., Jr., Gage, K.L., Kosoy, M.Y., Loye, J.E. & Martin, A.P. 2005. Testing the generality of a trophic-cascade model for plague. *Ecohealth* 2, 1–11.

Collins, M., An, S.I., Cai, W., Ganachaud, A., Guilyardi, E., Jin, F.F., Jochum, M., Lengaigne, M., Power, S., Timmermann, A., Vecchi, G. & Wittenberg, A. 2010. The impact of global warming on the tropical Pacific Ocean and El Niño. *Nature Geoscience* 3, 391–397.

Collins, S. & Bell, G. 2004. Phenotypic consequences of 1000 generations of selection at elevated CO_2 in a green alga. *Nature* 431, 566–569.

Colombi, B.J. 2009. Salmon nation: climate change and tribal sovereignty. In *Anthropology and Climate Change: From Encounters to Actions*, S. Crate & M. Nuttall (eds). Walnut Creek, California: Left Coast Press, 186–196.

Comiso, J. & Nishio, F. 2008. Trends in the sea ice cover using enhanced and compatible AMSR-E, SSM/I, and SMMR data. *Journal of Geophysical Research* 113, C02S07.

Comiso, J.C., Parkinson, C.L., Gersten, R. & Stock, L. 2008. Accelerated decline in the Arctic sea ice cover. *Geophysical Research Letters*, 35, L01703.

Commission for the Conservation and Management of Highly Migratory Fish Stocks in the Western and Central Pacific Ocean. 2011. Scientific Committee Seventh Regular Session Summary Report, Kolonia, Pohnpei, Federated States of Micronesia, 203 p.

Committee on Environment and Natural Resources (CERN). 2010. Scientific assessment of hypoxia in U.S. coastal waters. Washington, DC: Interagency Working Group on Harmful Algal Blooms, Hypoxia, and Human Health of the Joint Subcommittee on Ocean Science and Technology.

Connell, S.D. & Russell, B.D. 2010. The direct effects of increasing CO_2 and temperature on non-calcifying organisms: increasing the potential for phase shifts in kelp forests. *Proceedings of the Royal Society of London B* 277, 1409–1415.

Convention on Biological Diversity (CBD). 2010. Conference of the Parties (COP). COP 10 Decision X/33. Biodiversity and Climate Change. Online. www.cbd.int/doc/decisions/COP-10/cop-10-dec-33-en.pdf (accessed 12 March 2013).

Convention on Migratory Species (CMS). 2011. *Proceedings of the UNEP/CMS Technical Workshop on the Impact of Climate Change on Migratory Species: the Current Status and Avenues for Action. Tour du Valat, Camargue, France*. Bonn, Germany: CMS.

Cook, T., Folli, M., Klinck, J., Ford, S. & Miller, J. 1998. The relationship between increasing sea-surface temperature and the northward spread of *Perkinsus marinus* (Dermo) disease epizootics in oysters. *Estuarine, Coastal and Shelf Science* 46, 587–597.

Cooley, S.R. & Doney, S.C. 2009. Anticipating ocean acidification's economic consequences for commercial fisheries. *Environmental Research Letters* 4, 024007.

Cooley, S.R., Kite-Powell, H.L. & Doney, S.C. 2009. Ocean acidification's potential to alter global marine ecosystem services. *Oceanography* **22**, 172–180.

Cooper, L.W., Ashjian, C.J., Smith, S.L., Codispoti, L.A., Grebmeier, J.M., Campbell, R.G. & Sherr, E.B. 2006. Rapid seasonal sea-ice retreat in the Arctic could be affecting Pacific walrus (*Odobenus rosmarus divergens*) recruitment. *Aquatic Mammals* **32**, 98–102.

Cooper, S.R. & Brush, G.S., 1991. Long-term history of Chesapeake Bay anoxia. *Science* **254**, 992–996.

Cooper, T.F., De 'Ath, G., Fabricius, K.E. & Lough, J.M. 2008. Declining coral calcification in massive *Porites* in two nearshore regions of the northern Great Barrier Reef. *Global Change Biology* **14**, 529–538.

Coulthard, S. 2008. Adapting to environmental change in artisanal fisheries—insights from a South Indian Lagoon. *Global Environmental Change* **18**, 479–489.

Council on Environmental Quality (CEQ). 2010. Memorandum from draft NEPA guidance on consideration of the effects of climate change and greenhouse gas emissions, from Nancy H. Sutley, Chair, to Heads of Federal Departments and Agencies. February 18. Online. http://ceq.hss.doe.gov/nepa/regs/Consideration_of_Effects_of_ghg_draft_NEPA_Guidance_FINAL_02182010.pdf.

Craig, R.K. 2009. The Clean Water Act on the cutting edge: climate change regulation and water-quality regulation. *Natural Resources and Environment* **24**, 14–18.

Craig, R.K. 2010. Stationarity is dead—long live transformation: five principles for climate change adaptation law. *Harvard Environmental Law Review* **34**, 9–75.

Crain, C.M., Kroeker, K. & Halpern, B.S. 2008. Interactive and cumulative effects of multiple human stressors in marine systems. *Ecology Letters* **11**, 1304–1315.

Cravatte, S., Delcroix, T., Zhang, D., McPhaden, M. & Leloup, J. 2009. Observed freshening and warming of the western Pacific Warm Pool. *Climate Dynamics* **33**, 565–589.

Cressey, D. 2007. Arctic melt opens Northwest Passage. *Nature* **449**, 267.

Crook, E.D., Potts, D., Rebolledo-Vieyra, M., Hernandez, L. & Paytan, A. 2011. Calcifying coral abundance near low-pH springs: implications for future ocean acidification. *Coral Reefs* **31**, 239–245.

Crooks, S., Findsen, J., Igusky, K., Orr, M.K. & Brew, D. 2009. Greenhouse gas mitigation typology issues paper: tidal wetlands restoration (2009). Report prepared for California Climate Action Registry. Online. http://www.climateactionreserve.org/wp-content/uploads/2009/03/future-protocol-development_tidal-wetlands.pdf (accessed 12 March 2013).

Crooks, S., Herr, D., Tamelander, J., Laffoley, D. & Vandever, J. 2011. Mitigating climate change through restoration and management of coastal wetlands and near-shore marine ecosystems: challenges and opportunities. Environment Department Paper 121. Washington, DC: World Bank.

Cubillos, J.C., Wright, S.W., Nash, G., de Salas, M.F., Griffiths, B., Tilbrook, B., Poisson, A. & Hallegraeff, G.M. 2007. Calcification morphotypes of the coccolithophorid *Emiliania huxleyi* in the Southern Ocean: changes in 2001 to 2006 compared to historical data. *Marine Ecology Progress Series* **348**, 47–54.

Cushing, D.H. 1996. *Towards a Science of Recruitment in Fish Populations*. Oldendorf/Luhe, Germany: Ecology Institute.

Czaja, A., Robertson, A. & T. Huck, 2003. The role of coupled processes in producing NAO variability. Geophysical Monograph 134. Washington, DC: American Geophysical Union, 147–172.

Dahlhoff, E.P. 2004. Biochemical indicators of stress and metabolism: applications for marine ecological studies. *Annual Review of Physiology* **66**, 183–207.

Dalton, M. 2001. El Niño, expectations, and fishing effort in Monterey Bay, California. *Journal of Environmental Economics and Management* **42**, 336–359.

Dayton, P.K. 1985. Ecology of kelp communities. *Annual Review of Ecology and Systematics* **16**, 215–245.

Dayton, P.K., Tegner, M.J., Edwards, P.B. & Riser, K.L. 1999. Temporal and spatial scales of kelp demography: the role of oceanographic climate. *Ecological Monographs* **69**, 219–250.

Deacutis, C.F., Murray, D.P., Prell, W., Saarman, E. & Korhun, L. 2006. Hypoxia in the upper half of Narragansett Bay, RI, during August 2001 and 2002. *Northeastern Naturalist* **13**, 173–198.

Denman, K., Christian, J.R., Steiner, N., Pörtner, H.O. & Nojiri, Y. 2011. Potential impacts of future ocean acidification on marine ecosystems and fisheries: current knowledge and recommendations for future research. *ICES Journal of Marine Science* **68**, 1019–1029.

Denny, M.W., Hunt, L.J.H., Miller, L.P. & Harley, C.D.G. 2009. On the prediction of extreme ecological events. *Ecological Monographs* **79**, 397–421.

de Rivera, C.E., Steves, B.P., Fofonoff, P.W., Hines, A.H. & Ruiz, G.M. 2011. Potential for high-latitude marine invasions along western North America. *Diversity and Distributions* **17**, 1198–1209.

Deser, C., Alexander, M.A., Xie, S.P., Phillips, A.S. 2010. Sea surface temperature variability: patterns and mechanisms. *Annual Review of Marine Science* **2**, 115–143.

Deser, C. & Teng, H., 2008. Evolution of Arctic sea ice concentration trends and the role of atmospheric circulation forcing, 1979–2007. *Geophysical Research Letters* **35**, L02504.

De Silva, S.S. & Soto, D. 2009. Climate change and aquaculture: potential impacts, adaptation and mitigation. FAO Fisheries and Aquaculture Technical Paper 537. Rome, Italy: Food and Agricultural Organization of the United Nations, pp. 137–215.

Dessler, A.E. 2011. Cloud variations and the Earth's energy budget. *Geophysical Research Letters* **38**, L19701.

Deverel, S.J. & Leighton, D.A. 2010. Historic, Recent and Future Subsidence, Sacramento–San Joaquin Delta, California, USA. *San Francisco Estuary and Watershed Science*, 8(2). Online. http://escholarship.org/uc/item/7xd4x0xw

Diaz, R.J. & Rosenberg, R. 1995. Marine benthic hypoxia: a review of its ecological effects and the behavioural responses of benthic macrofauna. *Oceanography and Marine Biology: an Annual Review* **33**, 245–303.

Diaz, R.J. & Rosenberg, R. 2008. Spreading dead zones and consequences for marine ecosystems. *Science* **321**, 926–929.

Diaz-Pulido, G., Anthony, K.R.N., Kline, D.I., Dove, S. & Hoegh-Guldberg, O. 2012. Interactions between ocean acidification and warming on the mortality and dissolution of coralline algae. *Journal of Phycology* **48**, 32–39.

Diaz-Pulido, G., Gouezo, M., Tilbrook, B., Dove, S. & Anthony, K.R. 2011. High CO_2 enhances the competitive strength of seaweeds over corals. *Ecology Letters* **14**, 156–162.

Diaz-Pulido, G., McCook, L.J., Dove, S., Berkelmans, R., Roff, G., Kline, D.I., Weeks, S., Evans, R.D., Williamson, D.H. & Hoegh-Guldberg, O. 2009. Doom and boom on a resilient reef: climate change, algal overgrowth and coral recovery. *PLoS one* **4**, e5239.

Dijkstra, J.A., Boudreau, J. & Dionne, M. 2012. Species-specific mediation of temperature and community interactions by multiple foundation species. *Oikos* **121**(5), 646–654.

Di Lorenzo, E., Miller, A.J., Scheider, N. & McWilliams, J.C. 2005. The warming of the California Current system: dynamics and ecosystem implications. *Journal of Physical Oceanography* **35**, 336–362.

DIPNET (Disease Interactions and Pathogen exchange between farmed and wild aquatic animal populations—a European Network). 2007. Review of disease interactions and pathogen exchange between farmed and wild finfish and shellfish in Europe (DIPNET), R. Raynard, T. Wahli, I. Vatsos & S. Mortensen (eds). Oslo: VESO. Online. http://www.revistaaquatic.com/DIPNET/docs/doc.asp?id=48

Dirzo, R. & Raven, P.H. 2003. Global state of biodiversity and loss. *Annual Review of Environment and Resources* **28**, 137–167.

Dobos, K., Quinn, F., Ashford, D., Horsburgh, C. & King, C. 1999. Emergence of a unique group of necrotizing mycobacterial diseases. *Emerging Infectious Disease* **5**, 367–378.

Donato, D.C., Kauffman, J.B., Murdiyarso, D., Kurnianto, S., Stidman, M. & Kanninen, M. 2011. Mangroves among the most carbon-rich forests in the tropics. *Nature Geoscience* **4**, 293–297.

Doney, S.C., Fabry, V.J., Feely, R.A. & Kleypas, J.A. 2009. Ocean acidification: the other CO2 problem. *Annual Review of Marine Science* **1**, 169–192.

Doney, S.C., Ruckelshaus, M., Duffy, J.E., Barry, J.P., Chan, F., English, C.A., Galindo, H.M., Grebmeier, J.M., Hollowed, A.B., Knowlton, N., Polovina, J., Rabalais, N.N., Sydeman, W.J. & Talley, L.D. 2012. Climate change impacts on marine ecosystems. *Annual Review of Marine Science* **4**, 11–37.

Dowty, R.A. & Allen, B.L. 2011. Afterward. In *Dynamics of Disaster: Lessons on Risk, Response and Recovery*, R.A. Dowty & B.L. Allen (eds). London: Earthscan, 203–207.

Dudgeon, S.R., Aronson, R.B., Bruno, J.F. & Precht, W.F. 2010. Phase shifts and stable states on coral reefs. *Marine Ecology Progress Series* **413**, 201–216.

Durack, P.J. & Wijffels, S.E. 2010. Fifty-year trends in global ocean salinities and their relationship to broad-scale warming. *Journal of Climate* **23**, 4342–4362.

Durant, J.M., Hjermann, D.O., Ottersen, G. & Stenseth, N.C. 2007. Climate and the match or mismatch between predator requirements and resource availability. *Climate Research* **33**, 271–283.

Ebeling, A.W., Laur, D.R. & Rowley, R.J. 1985. Severe storm disturbances and reversal of community structure in a southern California kelp forest. *Marine Biology* **84**, 287–294.

Ecosystem Assessment Program (EAP). 2009. Ecosystem assessment report for the northeast U.S. Continental Shelf Large Marine Ecosystem. Woods Hole, Massachussetts: U.S. Department of Commerce, Northeast Fisheries Science Center Reference Document 09-11, 61 pp. Available from National Marine Fisheries Service, 166 Water Street, Woods Hole, Massachusetts 02543-1026 or online at http://www.nefsc.noaa.gov/nefsc/publications/.

Ecosystem Assessment Program (EAP). 2012. Ecosystem status report for the Northeast Shelf Large Marine Ecosystem–2011. U.S. Department of Commerce, Northeast Fisheries Science Center Reference Document 12-07; 32 p. Available from: National Marine Fisheries Service, 166 Water Street, Woods Hole, MA 02543-1026, or online at http://www.nefsc.noaa.gov/nefsc/publications/.

Edwards, M. & Richardson, A.J. 2004. Impact of climate change on marine pelagic phenology and trophic mismatch. *Nature* **430**, 881–884.

Ehler, C. & Douvere, F. 2009. Marine spatial planning: a step-by-step approach toward ecosystem-based management. IOC Manual and Guides No. 53, IOCAM Dosier No. 6. Paris: Intergovernmental Oceanographic Commission and Man and the Biosphere Programme.

Eisen, R.J., Glass, G.E., Eisen, L., Cheek, J., Enscore, R.E., Ettestad, P. & Gage, K.L. 2007. A spatial model of shared risk for plague and hantavirus pulmonary syndrome in the southwestern United States. *American Journal of Tropical Medicine and Hygiene* **77**, 999–1004.

Enscore, R.E., Biggerstaff, B.J., Brown, T.L., Fulgham, R.F., Reynolds, P.J., Engelthaler, D.M., Levy, C.E., Parmenter, R.R., Montenieri, J.A., Cheek, J.E., Grinnell, R.K., Ettestad, P.J. & Gage, K.L. 2002. Modeling relationships between climate and the frequency of human plague cases in the southwestern United States, 1960–1997. *American Journal of Tropical Medicine and Hygiene* **66**, 186–196.

Erlandsson, J. & McQuaid, C.D. 2004. Spatial structure of recruitment in the mussel *Perna perna* at local scales: effects of adults, algae and recruit size. *Marine Ecology Progress Series* **267**, 173–185.

Fabricius, K.A., Langdon, C., Uthicke, S., Humphrey, C., Noonan, S., De'ath, G., Okazaki, R., Muehllehner, N., Glass, M.S. & Lough, J.M. 2011. Losers and winners in coral reefs acclimatized to elevated carbon dioxide concentrations. *Nature Climate Change* **1**, 165–169.

Fabry, V.J. 2008. Marine calcifiers in a high-CO_2 ocean. *Science* **320**, 1020–1022.

Fabry, V.J., McClintock, J.B., Mathis, J.T. & Grebmeier, J.M. 2009. Ocean acidification at high latitudes: the bellwether. *Oceanography* **22**, 160–171.

Fabry, V.J., Seibel, B.A., Feely, R.A. & Orr, J.C. 2008. Impacts of ocean acidification on marine fauna and ecosystem processes. ICES *Journal of Marine Science* **65**, 414–432.

Fautin, D., Dalton, P., Incze, L.S., Leong, J.A.C., Pautzke, C., Rosenberg, A., Sandifer, P., Sedberry, G., Tunnell, J.W., Jr., Abbott, I., Brainard, R.E., Brodeur, M., Eldredge, L.G., Feldman, M., Moretzsohn, F., Vroom, P.S., Wainstein, M. & Wolff, N. 2010. An overview of marine biodiversity in United States waters. *PLoS one* **5**, e11914.

Fauver, S. 2008. Climate-related needs assessment synthesis for coastal management. Charleston, SC: National Oceanic and Atmospheric Administration, Coastal Services Center.

Federal Register. 2006. Department of Commerce: endangered and threatened species: final listing determinations for elkhorn coral and staghorn coral. 71 Federal Register 26852, May 9.

Federal Register. 2008. Department of the Interior: endangered and threatened wildlife and plants; determination of threatened status for the polar bear (*Ursus maritimus*) throughout its range. 73 Federal Register 28212, May 15.

Federal Register. 2010a. Environmental Protection Agency: Clean Water Act section 303(d): notice of call for public comment on 303(d) program and ocean acidification. 75 Federal Register 13537, March 22.

Federal Register. 2010b. National Oceanic and Atmospheric Administration: endangered and threatened wildlife and plants; threatened status for the southern distinct population segment of the spotted seal. 75 Federal Register 65, 239, October 22.

Federal Register. 2010c. National Oceanic and Atmospheric Administration: endangered and threatened wildlife; notice of 90-day finding on a petition to list 83 species of corals as threatened or endangered under the Endangered Species Act (ESA). 75 Federal Register 6616, February 10.

Federal Register. 2011. Department of the Interior: endangered and threatened wildlife and plants; 12-month finding on a petition to list the Pacific walrus as endangered or threatened. 76 Federal Register 7634, February 10.

Fedler, A.J. & Ditton, R.B. 1994. Understanding angler motivations in fisheries management. *Fisheries* **19**, 6–13.

Feely, R.A., Doney, S.C. & Cooley, S.R. 2009. Ocean acidification: present conditions and future changes in a high-CO_2 world. *Oceanography*, **22**, 36–47.

Feely, R.A., Fabry, V.J., Dickson, A., Gattuso, J.P., Bijma, J., Riebesell, U., Doney, S., Turley, C., Saino, T., Lee, K., Anthony, K. & Kelypas, J. 2010. An international observational network for ocean acidification. In *Proceedings of Ocean Observations '09: Sustained Ocean Observations and Information for Society (Vol. 2), Venice, Italy, 21–25 September 2009*, J. Hall et al. (eds). ESA Publication WPP-306. doi:10.5270/OceanObs09.

Feely, R.A., Sabine, C.L., Hernandez-Ayon, J.M., Ianson, D. & Hales, B. 2008. Evidence for upwelling of corrosive 'acidified' water along the continental shelf. *Science* **320**,1490–1492.

Feely, R.A., Sabine, C.L., Lee, K., Berelson, W., Kleypas, J., Fabry, V.J. & Millero, F.J. 2004. Impact of anthropogenic CO_2 on the $CaCO_3$ system in the oceans. *Science* **305**, 362–366.

Feng, Y., Warner, M.E., Zhang, Y., Sun, J., Fu, F.-X., Rose, J.M. & Hutchins, D.A. 2008. Interactive effects of increased pCO_2, temperature and irradiance on the marine coccolithophore *Emiliania huxleyi* (Prymnesiophyceae). *European Journal of Phycology* **43**, 87–98.

Ferguson, S.H., Stirling, I. & McLoughlin, P. 2005. Climate change and ringed seal (*Phoca hispida*) recruitment in western Hudson Bay. *Marine Mammal Science* **21**, 121–135.

Findlay, H.S., Burrows, M.T., Kendall, M.A., Spicer, J.I. & Widdicombe, S. 2010. Can ocean acidification affect population dynamics of the barnacle *Semibalanus balanoides* at its southern range edge? *Ecology* **91**, 2931–2940.

Fire, S.E., Fauquier, D., Flewelling, L.J., Henry, M., Naar, J., Pierce, R. & Wells, R.S. 2007. Brevetoxin exposure in bottlenose dolphins (*Tursiops truncatus*) associated with *Karenia brevis* blooms in Sarasota Bay, Florida. *Marine Biology* **152**(4), 827–834.

Firth, L.B., Knights, A.M. & Bell, S.S. 2011. Air temperature and winter mortality: implications for the persistence of the invasive mussel, *Perna viridis* in the intertidal zone of the southeastern United States. *Journal of Experimental Marine Biology and Ecology* **400**, 250–256.

Fischbach, A.S., Amstrup, S.C. & Douglas, D.C. 2007. Landward and eastward shift of Alaskan polar bear denning associated with recent sea ice changes. *Polar Biology* **30**, 1395–1405.

Fischer, F. 2000. *Citizens, Experts, and the Environment: The Politics of Local Knowledge.* Durham, North Carolina: Duke University Press.

Fletcher, C. 2010. Hawai'i's changing climate. Briefing Sheet, 2010 UH Sea Grant Program. Honolulu, Hawaii: Center for Island Climate Adaptation and Policy (ICAP).

Flewelling, L.J., Naar, J.P., Abbott, J.P., Baden, D.G., Barros, N.B., Bossart, G.D., Dottein, M.-Y.D., Hammond, D.G., Haubold, E.M., Heil, C.A., Henry, M.S., Jacocks, H.M., Leighfield, T.A., Pierce, R.H., Pithford, T.D., Rommel, S.A., Scott, P.S., Steiginger, K.A., Truby, E.W., Van Dolah, F.M. & Landsberg, J.H. 2005. Brevetoxicosis: red tides and marine mammal mortalities. *Nature* **435**, 755–756.

Fodrie, F.J., Heck, K.L., Powers, S.P., Graham, W.M. & Robinson, K.L. 2010. Climate-related, decadal-scale assemblage changes of seagrass-associated fishes in the northern Gulf of Mexico. *Global Change Biology* **16**, 48–59.

Fogarty, M.J., Incze, L., Hayhoe, K., Mountain, D.G. & Manning, J. 2008. Potential climate change impacts on Atlantic cod (*Gadus morhua*) off the northeastern USA. *Mitigation and Adaptation Strategies for Global Change* **13**, 453–466.

Foley, A.M., Singel, K.E., Dutton, P.H., Summers, T.M., Redlow, A.E. & Lessman, J. 2007. Characteristics of a green turtle (*Chelonia mydas*) assemblage in northwestern Florida determined during a hypothermic stunning event. *Gulf of Mexico Science* **25**, 131–143.

Folke, C., Carpenter, S., Walker, B., Scheffer, M., Elmqvist, T., Gunderson, L. & Holling, C.S. 2004. Regime shifts, resilience, and biodiversity in ecosystem management. *Annual Review of Ecology, Evolution, and Systematics* **35**, 557–581.

Food and Agriculture Organization of the United Nations & World Health Organization (FAO/WHO). 2011. Report of the joint FAO/WHO expert consultation on the risks and benefits of fish consumption. Rome: Food and Agriculture Organization of the United Nations; Geneva: World Health Organization.

Ford, S.E. 1996. Range extension by the oyster parasite *Perkinsus marinus* into the northeastern United States: response to climate change? *Journal of Shellfish Research* **15**, 45–56.

Fourqurean, J.W., Duarte, C.M., Kennedy, H., Marbà, N., Holmer, M., Mateo, M.A., Apostolaki, E.T., Kendrick, G.A., Krause-Jensen, D., McGlathery, K.J. & Serrano, O. 2012. Seagrass ecosystems as a globally significant carbon stock. *Nature Geoscience* **5**, 505–509.

Friedlingstein, P., Bopp, L. & Ciais, P. 2001. Positive feedback between future climate change and the carbon cycle. *Geophysical Research Letters*, **28**, 1543–1546.

Frumhoff, P.C., McCarthy, J.J., Melillo, J.M., Moser, S.C. & Wuebbles, D.J. 2007. Confronting climate change in the U.S. northeast: science, impacts, and solutions. Synthesis report of the Northeast Climate Impacts Assessment (NECIA). Cambridge, MA: Union of Concerned Scientists (UCS).

Fu, F.X., Mulholland, M.R., Garcia, N., Beck, A., Bernhardt, P.W., Warner, M.E., Sañudo-Wilhelmy, S.A. & Hutchins, D.A. 2008. Interactions between changing pCO_2, N_2 fixation, and Fe limitation in the marine unicellular cyanobacterium *Crocosphaera*. *Limnology and Oceanography* **53**, 2472–2484.

Fu, F.X., Place, A.R., Garcia, N.S. & Hutchins, D.A. 2010. Effects of changing pCO_2 and phosphate availability on toxin production and physiology of the harmful bloom dinoflagellate *Karlodinium veneficum*. *Aquatic Microbial Ecology* **59**, 55–65.

Gage, K., Burkot, T., Eisen, R. & Hayes, E. 2008. Climate and vectorborne diseases. *American Journal of Preventive Medicine* **35**, 436–450.

Gale, R.P. 1991. Gentrification of America's coasts: impacts of the growth machine on commercial fishermen. *Society and Natural Resources* **4**, 101–121.

Galloway, J.N., Townsend, A.R., Erisman, J.W., Bekunda, M., Cai, Z., Freney, J.R., Martinelli, L.A., Seitzinger, S.P. & Sutton, M.A. 2008. Transformation of the nitrogen cycle: recent trends, questions, and potential solutions. *Science* **320**, 889–892.

Garcia-Reyes, M. & Largier, J.L. 2010. Observations of increased wind-driven coastal upwelling off central California. *Journal of Geophysical Research* **115**, CO4011.

Gardner, T.A., Cote, I.M., Gill, J.A., Grant, A. & Watkinson, A.R. 2003. Long-term region-wide declines in Caribbean corals. *Science* **301**, 958–960.

Garlich-Miller, J., MacCracken, J.G., Snyder, J., Meehan, R., Myers, M., Wilder, J.M., Lance, E. & Matz, A. 2011. Status review of the Pacific walrus (*Odobenus rosmarus divergens*). Anchorage, AK: U.S. Fish and Wildlife Service, Marine Mammals Management.

Gaston, A.J., Gilchrist, H.G., Mallory, M.L. & Smith, P.A. 2009. Changes in seasonal events, peak food availability, and consequent breeding adjustment in a marine bird: a case of progressive mismatching. *The Condor* **111**, 111–119.

Gaston, A.J. & Woo, K. 2008. Razorbills (*Alca torda*) follow subarctic prey into the Canadian Arctic: colonization results from climate change? *The Auk* **125**, 939–942.

Gattuso, J.P. & Hansson, L. (eds) 2011. *Ocean Acidification*. Oxford, UK: Oxford University Press.

Gaynor, K., Katz, A.R., Park, S.Y., Nakata, M., Clark, T.A. & Effler, P.V. 2007. Leptospirosis on Oahu: an outbreak associated with flooding of a university campus. *American Journal of Tropical Medicine and Hygiene* **76**, 882–885.

Gazeau, F., Gattuso, J.P., Dawber, C., Pronker, A.E., Peene, F., Peene, J., Heip, C.H. & Middelburg, J.J. 2010. Effect of ocean acidification on the early life stages of the blue mussel (*Mytilus edulis*). *Biogeosciences* **7**, 2051–2060.

Gazeau, F., Quiblier, C., Jansen, J.M., Gattuso, J.P., Middelburg, J.J. & Heip, C.H.R. 2007. Impact of elevated CO2 on shellfish calcification. *Geophysical Research Letters* **34**, LO7603.

Gearheard, S., Matumeak, W., Angutikjuaq, I., Maslanik, J., Huntington, H.P., Leavitt, J., Kagak, D.M., Tigullaraq, G. & Barry, R.G. 2006. "It's not that simple": a collaborative comparison of sea ice environments, their uses, observed changes, and adaptations in Barrow, Alaska, USA, and Clyde River, Nunavut, Canada. *Ambio* **35**, 203–211.

Gedan, K.B. & Bertness, M.D. 2010. How will warming affect the salt marsh foundation species *Spartina patens* and its ecological role? *Oecologia* **164**, 479–487.

Giles, K., Laxon, S. & Ridout, A. 2008. Circumpolar thinning of Arctic sea ice following the 2007 record ice extent minimum. *Geophysical Research Letters* **35**, L22502.

Giri, C., Ochieng, E., Tieszen, L., Zhu, Z., Singh, A., Loveland, T., Masek, J. & Duke, N. 2010. Status and distribution of mangrove forests of the world using earth observation satellite data. *Global Ecology and Biogeography* **20**(1), 154–159.

Gjøen, H.M. & Bentsen, H.B. 1997. Past, present, and future of genetic improvement in salmon aquaculture. ICES *Journal of Marine Science* **54**, 1009–1014.

Glass, G.E., Cheek, J.E., Patz, J.A., Shields, T.M., Doyle, T.J., Thoroughman, D.A., Hunt, D.K., Enscore, R.E., Gage, K.L., Irland, C. & Bryan, R. 2000. Anticipating risk areas for Hantavirus pulmonary syndrome with remotely sensed data: re-examination of the 1993 outbreak. *Emerging Infectious Diseases* **6**, 238–247.

Gleeson, M.W. & Strong, A.E. 1995. Applying MCSST to coral reef bleaching. *Advances in Space Research* **16**, 151–154.

Glenn, R.P. & Pugh, T.L. 2006. Epizootic shell disease in American lobster (*Homarus americanus*) in Massachusetts coastal waters: interactions of temperature, maturity, and intermolt duration. *Journal of Crustacean Biology* **26**, 639–645.

Glenn, S., Arnone, R., Bergmann, T., Bissett, W.P., Crowley, M., Cullen, J., Gryzmski, J., Haidvogel, D., Kohut, J., Moline, M., Oliver, M., Orrico, C., Sherrell, R., Song, T., Weidemann, A., Chant, R. & Schofield, O. 2004. Biogeochemical impact of summertime coastal upwelling on the New Jersey shelf. *Journal of Geophysical Research* **109**, C12S02.

Glenn, S.M., Crowley, M.F., Haidvogel, D.B. & Song, Y.T. 1996. Underwater observatory captures coastal upwelling off New Jersey. *EOS* **77**, 233–236.

Glick, P., Staudt, A. & Stein, B. 2009. *A New Era for Conservation: Review of Climate Change Adaptation Literature.* Washington, DC: National Wildlife Federation.

Glick, P., Stein, B.A. & Edelson, N.A. (eds) 2011. Scanning the conservation horizon: a guide to climate change vulnerability assessment. Washington, DC: National Wildlife Federation. Online. http://www.nwf.org/~/media/PDFs/Global-Warming/Climate-Smart-Conservation/NWFScanningtheConservationHorizonFINAL92311.pdf?dmc=1&ts=20130317T1724145272 (accessed 17 March 2013).

Global Carbon Project. 2011. Carbon budget 2010—highlights. December 5. Online. http://www.globalcarbonproject.org/carbonbudget/ (accessed 27 March 2011).

Gooding, B., Harley, C.D.G. & Tang, E. 2009. Multiple climate variables increase growth and feeding rates of a keystone predator. *Proceedings of the National Academy of Sciences USA* **106**, 9316–9321.

Goodkin, N.F., Hughen, K.A., Doney, S. & Curry, W.B. 2008. Increased multidecadal variability of the North Atlantic Oscillation since 1781. *Nature Geoscience* **1**, 844–848.

Gouhier, T.C., Guichard, F. & Menge, B.A. 2010. Ecological processes can synchronize marine population dynamics over continental scales. *Proceedings of the National Academy of Sciences USA* **107**, 8281–8286.

Grafton, R.Q. 2010. Adaptation to climate change in marine capture fisheries. *Marine Policy* **34**, 606–615.

Graham, M.H. 2004. Effects of local deforestation on the diversity and structure of Southern California giant kelp forest food webs. *Ecosystems* **7**, 341–357.

Graham, M.H., Halpern, B.S. & Carr, M.H. 2008. Diversity and dynamics of Californian subtidal kelp forests. In *Food Webs and the Dynamics of Marine Benthic Ecosystems,* T.R. McClanahan & G.R. Branch (eds). Oxford, UK: Oxford University Press, 103–134.

Graham, M.H., Kinlan, B.P. & Grosburg, R.K. 2010. Post-glacial redistribution and shifts in productivity of giant kelp forests. *Proceedings of the Royal Society Series B* **277**, 399–406.

Graham, M.H., Vásquez, J.A. & Buschmann, A.H. 2007. Global ecology of the giant kelp *Macrocystis*: from ecotypes to ecosystems. *Oceanography and Marine Biology: an Annual Review* **45**, 39–88.

Graham, N.A.J., Wilson, S.K., Jennings, S., Polunin, N.V.C., Bijoux, J.P. & Robinson, J. 2006. Dynamic fragility of oceanic coral reef ecosystems. *Proceedings of the National Academy of Sciences USA* **103**, 8425–8429.

Grantham, B.A., Chan, F., Nielsen, K.J., Fox, D.S., Barth, J.A., Huyer, A., Lubchenco, J. & Menge, B.A. 2004. Upwelling-driven nearshore hypoxia signals ecosystem and oceanographic changes in the northeast Pacific. *Nature* **429**, 749–754.

Greene, C.H., Pershing, A.J., Cronin, T.M. & Ceci, N. 2008. Arctic climate change and its impacts on the ecology of the North Atlantic. *Ecology* **89**, S24–S38.

Greene, R.M., Lehrter, J.C. & Hagy, J.D., III. 2009. Multiple regression models for hindcasting and forecasting midsummer hypoxia in the Gulf of Mexico. *Ecological Applications* **19**, 1161–1175.

Greenough, G., McGeehin, M., Bernard, S. & Trtanj, J. 2001. The potential impacts of climate change on health impacts of extreme weather events in the United States. *Environmental Health Perspectives* **109**,185–189.

Greer, A., Ng, V. & Fisman, D. 2008. Climate change and infectious diseases in North America: the road ahead. *Canadian Medical Association Journal* **178**, 715–722.

Gregg, R.M., Hansen, L.J., Feifel, K.M., Hitt, J.L., Kershner, J.M., Score, A. & Hoffman, J.R. 2011. *The State of Marine and Coastal Adaptation in North America: A Synthesis of Emerging Ideas.* Bainbridge Island, Washington: EcoAdapt.

Grémillet, D. & Boulinier, T. 2009. Spatial ecology and conservation of seabirds facing global climate change: a review. *Marine Ecology Progress Series* **391**, 121–137.

Gulland, F.M.D., Koski, M., Lowenstine, L.J., Colagross, A., Morgan, L. & Spraker, T. 1996. Leptospirosis in California sea lions (*Zalophus californianus*) stranded along the central California coast, 1981–1994. *Journal of Wildlife Diseases* **32**, 572–580.

Gunter, G., Williams, R.H., Davis, C.C. & Smith, F.G.W. 1948. Catastrophic mass mortality of marine animals and coincident phytoplankton bloom on the west coast of Florida, November 1946 to August 1947. *Ecological Monographs* **18**, 309–324.

Hagy, J.D., III & Murrell, M.C. 2007. Susceptibility of a northern Gulf of Mexico estuary to hypoxia: an analysis using box models. *Estuarine, Coastal and Shelf Science* **74**, 239–253.

Hallegraeff, G.M. 1993. A review of harmful algal blooms and their apparent global increase. *Phycologia* **32**, 79–99.

Hallegraeff, G.M. 2010. Ocean climate change, phytoplankton community responses, and harmful algal blooms: a formidable predictive challenge. *Journal of Phycology* **46**, 220–235.

Hall-Spencer, J.M., Rodolfo-Metalpa, R., Martin, S., Ransome, E., Fine, M., Turner, S.M., Rowley, S.J., Tedesco, D. & Buia, M.C. 2008. Volcanic carbon dioxide vents show ecosystem effects of ocean acidification. *Nature* **454**, 96–99.

Halpern, B.S., Ebert, C.M., Kappel, C.V., Madin, E.M.P., Micheli, F., Perry, M., Selkoe, K.A. & Walbridge, S. 2009a. Global priority areas for incorporating land-sea connections in marine conservation. *Conservation Letters* **2**, 189–196.

Halpern, B.S., Kappel, C.V., Selkoe, K.A., Micheli, F., Ebert, C., Kontgis, C., Crain, C.M., Martone, R., Shearer, C. & Teck, S.J. 2009b. Mapping cumulative human impacts to California Current marine ecosystems. *Conservation Letters* **2**, 138–148.

Halpern, B.S., Lester, S.E. & McLeod, K.L. 2010. Placing marine protected areas onto the ecosystem-based management seascape. *Proceedings of the National Academy of Sciences USA* **107**, 18312–18317.

Halpern, B.S., Longo, C., Hardy, D., McLeod, K.L., Samhouri, J.F., Katona, S.K., Kleisner, K., Lester, S.E., O'Leary, J., Ranelletti, M., Rosenberg, A.A., Scarborough, C., Selig, E.R., Best, B.D., Brumbaugh, D.R., Chapin, F.S., Crowder, L.B., Daly, K.L., Doney, S.C., Elfes, C., Fogarty, M.J., Gaines, S.D., Jacobsen, K.I., Karrer, L.B., Leslie, H.M., Neeley, E., Pauly, D., Polasky, S., Ris, B., St. Martin, K., Stone, G.S., Sumaila, U.R. & Zeller, D. 2012. An index to assess the health and benefits of the global ocean. *Nature* **488**, 615–620.

Halpern, B.S., McLeod, K.L., Rosenberg, A.A. & Crowder, L.B. 2008a. Managing for cumulative impacts in ecosystem-based management through ocean zoning. *Ocean & Coastal Management* **51**, 203–211.

Halpern, B.S., Walbridge, S., Selkoe, K.A., Kappel, C.V., Micheli, F., D'Agrosa, C., Bruno, J., Casey, K.S., Ebert, C., Fox, H.E., Fujita, R., Heinemann, D., Lenihan, H.S., Madin, E.M.P., Perry, M., Selig, E., Spalding, M., Steneck, R. & Watson, R. 2008b. A global map of human impact on marine ecosystems. *Science* **319**, 948–952.

Hansen, I.S., Keul, N., Sorensen, J.T., Erichsen, A. & Anderson, J.H. 2007. Oxygen map for the Baltic Sea. BALANCE Interim Report No. 17. Copenhagen, Denmark: HELCOM project.

Hansen, J., Nazarenko, L., Ruedy, R., Sato, M., Willis, J., Del Genio, A., Koch, D., Lacis, A., Lo, K., Menon, S., Novakov, T., Perlwitz, J., Russell, G., Schmidt, G.A. & Tausnev, N. 2005. Earth's energy imbalance: confirmation and implications. *Science* **308**, 1431–1435.

Hansen, L., Hoffman, J., Drews, C. & Mielbrecht, E. 2010. Designing climate-smart conservation: guidance and case studies. *Conservation Biology* **24**, 63–69.

Hare, C.E., Leblanc, K., Ditullio, G.R., Kudela, R.M., Zhang, Y., Lee, P.A., Riseman, S. & Hutchins, D.A. 2007. Consequences of increased temperature and CO_2 for phytoplankton community structure in the Bering Sea. *Marine Ecology Progress Series* **352**, 9–16.

Hare, J.A. & Able, K.W. 2007. Mechanistic links between climate and fisheries along the east coast of the United States: explaining population outbursts of Atlantic croaker (*Micropogonias undulatus*). *Fisheries Oceanography* **16**, 31–45.

Hare, J.A., Alexander, M.A., Fogarty, M.J., Williams, E.H. & Scott, J.D. 2010. Forecasting the dynamics of a coastal fishery species using a coupled climate–population model. *Ecological Applications* **20**, 452–464.

Hare, J.A. & Cowen, R.K. 1997. Size, growth, development, and survival of the planktonic larvae of *Pomatomus saltatrix* (Pisces: Pomatomidae). *Ecology* **78**, 2415–2431.

Hare, S.R. & Mantua, N.J. 2000. Empirical evidence for North Pacific regime shifts in 1977 and 1989. *Progress in Oceanography* **47**, 103–145.

Harley, C.D.G. 2003. Abiotic stress and herbivory interact to set range limits across a two-dimensional stress gradient. *Ecology* **84**, 1477–1488.

Harley, C.D.G. 2008. Tidal dynamics, topographic orientation, and temperature-mediated mass mortalities on rocky shores. *Marine Ecology Progress Series* **371**, 37–46.

Harley, C.D.G. 2011. Climate change, keystone predation, and biodiversity loss. *Science* **334**, 1124–1127.

Harley, C.D.G., Hughes, A.R., Hultgren, K.M., Miner, B.G., Sorte, C.J.B., Thornber, C.S., Rodriguez, L.F., Tomanek, L. & Williams, S.L. 2006. The impacts of climate change in coastal marine systems. *Ecology Letters* **9**, 228–241.

Harley, C.D.G. & Paine, R.T. 2009. Contingencies and compounded rare perturbations dictate sudden distributional shifts during periods of gradual climate change. *Proceedings of the National Academy of Sciences USA* **106**, 11172–11176.

Harrison, G.G., Sharp, M., Manalo-LeClair, G., Ramirez, A. & McGarvey, N. 2007. Food security among California's low-income adults improves, but most severely affected do not share in improvement. Policy Brief (PB2007–6), Los Angeles, California: UCLA Center for Health Policy Research, 1–11.

Harrold, C. & Reed, D.C. 1985. Food availability, sea urchin grazing, and kelp forest community structure. *Ecology* **66**, 1160–1169.

Harvell, C.D., Kim, K., Burkholder, J.M., Colwell, R.R., Epstein, P.R., Grimes, D.J., Hofmann, E.E., Lipp, E.K., Osterhaus, A.D.M.E., Overstreet, R.M., Porter, J.W., Smith, G.W. & Vasta, G.R. 1999. Emerging marine diseases—climate links and anthropogenic factors. *Science* **285**, 1505–1510.

Harvell, D., Altizer, S., Cattadori, I.M., Harrington, L. & Weil, E. 2009. Climate change and wildlife diseases: when does the host matter the most? *Ecology* **90**, 912–920.

Hatala, J.A., Detto, M., Sonnentag, O., Deverel, S.J., Verfaillie, J. & Baldocchi, D.D. 2012. Greenhouse gas (CO_2, CH_4, H_2O) fluxes from drained and flooded agricultural peatlands in the Sacramento San Joaquin Delta. *Agriculture, Ecosystems and Environment* **150**, 1–18.

Hawkes, L.A., Broderick, A.C., Coyne, M.S. & Godfrey, M.H. 2006. Phenotypically linked dichotomy in sea turtle foraging requires multiple conservation approaches. *Current Biology* **16**, 990–995.

Hawkes, L.A., Broderick, A.C., Godfrey, M.H. & Godley, B.J. 2007. Investigating the potential impacts of climate change on a marine turtle population. *Global Change Biology* **13**, 923–932.

Hawkes, L.A., Broderick, A.C., Godfrey, M.H. & Godley, B.J. 2009. Climate change and marine turtles. *Endangered Species Research* **7**, 137–154.

Haynie, A.C. & Pfeiffer, L. 2012. Why economics matters for understanding the effects of climate change on fisheries. *ICES Journal of Marine Science: Journal du Conseil* doi:10.1093/icesjms/fss021.

Hays, G.C., Richardson, A.J. & Robinson, C. 2005. Climate change and marine plankton. *Trends in Ecology and Evolution* **20**, 337–344.

Heide-Jørgensen, M.P., Laidre, K.L., Quackenbush, L.T. & Citta, J.J. 2011. The Northwest Passage opens for bowhead whales. *Biology Letters* **8**, 270–273.

Heinz Center. 2008. *Strategies for Managing the Effects of Climate Change on Wildlife and Ecosystems.* Washington, DC: Heinz Center.

Heisler, J., Gilbert, P.M., Burkholder, J.M., Anderson, D.M., Cochlan, W., Dennison, W.C., Dortch, Q., Gobler, C.J., Heil, C.A., Humphries, E., Lewitus, A., Magnien, R., Marshall, H.G., Sellner, K., Stockwell, D.A., Stoecker, D.K. & Suddleson, M. 2008. Eutrophication and harmful algal blooms: a scientific consensus. *Harmful Algae* **8**, 3–13.

Held, I.M. & Soden, B.J. 2006. Robust responses of the hydrological cycle to global warming *Journal of Climate* **19**, 5686–5699.

Hellmann, J.J., Byers, J.E., Bierwagen, B.G. & Dukes, J.S. 2008. Five potential consequences of climate change for invasive species. *Conservation Biology* **22**, 534–543.

Helly, J.J. & Levin, L.A. 2004. Global distribution of naturally occurring marine hypoxia on continental margins. *Deep-Sea Research Part I: Oceanographic Research Papers* **51**, 1159–1168.

Helm, K.P., Bindoff, N.L. & Church, J.A., 2010. Changes in the global hydrological-cycle inferred from ocean salinity. *Geophysical Research Letters* **37**, L18701.

Helmuth, B. 2009. From cells to coastlines: how can we use physiology to forecast the impacts of climate change? *The Journal of Experimental Biology* **212**, 753–760.

Helmuth, B., Broitman, B.R., Blanchette, C.A., Gilman, S., Halpin, P., Harley, C.D.G., O'Donnell, M.J., Hofmann, G.E., Menge, B. & Strickland, D. 2006a. Mosaic patterns of thermal stress in the rocky intertidal zone: implications for climate change. *Ecological Monographs* **76**, 461–479.

Helmuth, B., Mieszkowska, N., Moore, P. & Hawkins, S.J. 2006b. Living on the edge of two changing worlds: forecasting the responses of rocky intertidal ecosystems to climate change. *Annual Review of Ecology, Evolution, and Systematics* **37**, 373–404.

Helser, T.E. & Alade, L. 2012. A retrospective of the hake stocks off the Atlantic and Pacific coasts of the United States: uncertainties and challenges facing assessment and management in a complex environment. *Fisheries Research* **114**, 2–18.

Herbert, R.J.H., Southward, A.J., Clarke, R.T., Sheader, M. & Hawkins, S.J. 2009. Persistent border: an analysis of the geographic boundary of an intertidal species. *Marine Ecology Progress Series* **379**, 135–150.

Herrick, S. F., Jr., Norton, J.G., Mason, J.E. & Bessey, C. 2007. Management application of an empirical model of sardine-climate regime shifts. *Marine Policy* **31**, 71–80.

Hester, K.C., Peltzer, E.T., Kirkwood, W.J. & Brewer, P.G. 2008. Unanticipated consequences of ocean acidification: A noisier ocean at lower pH. *Geophysical Research Letters* **35**, L19601.

Hiatt, T., Dalton, M., Felthoven, R., Fissel, B., Garber-Yonts, B., Haynie, A., Himes-Cornell, A., Kasperski, S., Lee, J., Lew, D., Pfeiffer, L., Sepez, J. & Seung, C. 2010. Stock assessment and fishery evaluation report for the groundfish fisheries of the Gulf of Alaska and Bering Sea/Aleutian Islands area: economic status of the groundfish fisheries off Alaska. Anchorage, Alaska: North Pacific Fishery Management Council.

Higdon, J.W. & Ferguson, S.H. 2009. Loss of Arctic sea ice causing punctuated change in sightings of killer whales (*Orcinus orca*) over the past century. *Ecological Applications* **19**, 1365–1375.

Hilborn, R., Quinn, T.P., Schindler, D.E. & Rogers, D.E. 2003. Biocomplexity and fisheries sustainability. *Proceedings of the National Academy of Sciences USA* **100**, 6564–6568.

Hlavsa, M.C., Roberts, V.A., Anderson, A.R., Hill, V.R., Kahler, A.M., Orr, M., Garrison, L.E., Hicks, L.A., Newton, A., Hilborn, E.D., Wade, T.J., Beach, M.J. & Yoder, J.S. 2011. Surveillance for waterborne disease outbreaks and other health events associated with recreational water—United States, 2007–2008. *Morbidity and Mortality Weekly Report* **60**, 1–32.

Hoagland, P., Anderson, D.M., Kaoru, Y. & White, A.W. 2002. The economic effects of harmful algal blooms in the United States: estimates, assessment issues, and information needs. *Estuaries* **25**, 819–837.

Hoagland, P. & Scatasta, S. 2006. The economic effects of harmful algal blooms. In *Ecology of Harmful Algae*, E. Granéli & J.T. Turner (eds). Ecological Studies **189**. Dordrecht, The Netherlands: Springer-Verlag, 391–402.

Hobbs, R.J., Arico, S., Aronson, J., Baron, J.S., Bridgewater, P., Cramer, V.A., Epstein, P.R., Ewel, J.J., Klink, C.A., Lugo, A.E., Norton, D., Ojima, D., Richardson, D.M., Sanderson, E.W., Valladares, F., Vilà, M., Zamora, R. & Zobel, M. 2006. Novel ecosystems: theoretical and management aspects of the new ecological world order. *Global Ecology and Biogeography* **15**, 1–7.

Hochachka, P.W. & Somero, G.N. 2002. *Biochemical Adaptation: Mechanism and Process in Physiological Evolution.* New York: Oxford University Press.

Hoegh-Guldberg, O. & Bruno, J.F. 2010. The impact of climate change on the world's marine ecosystems. *Science* **328**, 1523–1528.

Hoegh-Guldberg, O., Mumby, P.J., Hooten, A.J., Steneck, R.S., Greenfield, P., Gomez, E., Harvell, C.D., Sale, P.J., Edwards, A.J., Caldeira, K., Knowlton, N., Eakin, C.M., Iglesias-Prieto, R., Muthiga, N., Bradbury, R.H., Dubi, A. & Hatziolos, M.E. 2007. Coral reefs under rapid climate change and ocean acidification. *Science* **318**, 1737–1742.

Hoegh-Guldberg, O., Ortiz, J.C. & Dove, S. 2011. The future of coral reefs. *Science* **334**, 1494–1495.

Hofmann, A.F., Peltzer, E.T., Walz, P.M. & Brewer, P.G. 2011. Hypoxia by degrees: establishing definitions for a changing ocean. *Deep-Sea Research Part I-Oceanographic Research Papers* **58**, 1212–1226.

Hofmann, G.E., Barry, J.P., Edmunds, P.J., Gates, R.D., Hutchins, D.A., Klinger, T. & Sewell, M.A. 2010. The effect of ocean acidification on calcifying organisms in marine ecosystems: an organism to ecosystem perspective. *Annual Review of Ecology, Evolution, and Systematics* **41**, 127–147.

Hofmann, G.E. & Place, S.P. 2007. Genomics-enabled research in marine ecology: challenges, risks and pay-offs. *Marine Ecology Progress Series* **332**, 244–259.

Hofmann, G.E. & Todgham, A.E. 2010. Living in the now: physiological mechanisms to tolerate a rapidly changing environment. *Annual Review of Physiology* **72**, 172–145.

Holbrook, S.J., Schmitt, R.J. & Stephens, J.S., Jr. 1997. Changes in an assemblage of temperate reef fishes associated with a climate shift. *Ecological Applications* **7**, 1299–1310.

Holland, D.S., Pinto da Silva, P. & Wiersma, J. 2010. A survey of social capital and attitudes toward management in the New England groundfish fishery. Washington, DC: U.S. Department of Commerce.

Holland, M.M., Bitz, C.M., Eby, M. & Weaver, A.J., 2001. The role of ice, ocean interactions in the variability of the North Atlantic Thermohaline Circulation. *Journal of Climate* **14**, 656–675.

Hollowed, A.B., Beamish, R.J., Okey, T.A. & Schirripa, M.J. 2008. Sidney, British Columbia, Canada: PICES scientific report, North Pacific Marine Science Organization (PICES). Reports of PICES/NPRB workshops on forecasting climate impacts on future production of commercially exploited fish and shellfish. Sidney, British Columbia: North Pacific Marine Science Organization.

Hollowed, A.B., Bond, N.A., Wilderbuer, T.K., Stockhausen, W.T., A'mar, Z.T., Beamish, R.J., Overland, J.E. & Schirripa, M.J. 2009. A framework for modelling fish and shellfish responses to future climate change. *ICES Journal of Marine Science* **66**, 1584–1594.

Hollowed, A., Barange, M., Ito, S., Kim, S., Loeng, H. & Peck, M. 2011. Climate change effects on fish and fisheries: forecasting impacts, assessing ecosystem responses, and evaluating management strategies. *ICES Journal of Marine Science* **68**, 984–985.

Hönisch, B., Ridgwell, A., Schmidt, D.N., Thomas, E., Gibbs, S.J., Sluijs, A., Zeebe, R., Kump, L., Martindale, R.C., Greene, S.E., Kiessling, W., Ries, J., Zachos, J.C., Royer, D.L., Barker, S., Marchitto, T.M., Jr., Moyer, R., Pelejero, C., Ziveri, P., Foster, G.L. & Williams, B. 2012. The geological record of ocean acidification. *Science* **335**, 1058–1063.

Hooff, R.C. & Peterson, W.T. 2006 Recent increases in copepod biodiversity as an indicator of changes in ocean and climate conditions in the northern California current ecosystem. *Limnology and Oceanography* **51**, 2042–2051.

Horseman, M. & Surani, S. 2011. A comprehensive review of *Vibrio vulnificus*: an important cause of severe sepsis and skin and soft-tissue infection. *International Journal of Infectious Diseases* **15**, e157–e166.

Hsieh, C.H., Reiss, C.S., Hewitt, R.P. & Sugihara, G. 2008. Spatial analysis shows that fishing enhances the climatic sensitivity of marine fishes. *Canadian Journal of Fisheries and Aquatic Sciences* **6**, 947–961.

Hu, A., Meehl, G.A., Han, W. & Yin, J. 2011. Effect of the potential melting of the Greenland ice sheet on the meridional overturning circulation and global climate in the future. *Deep-Sea Research Part II: Topical Studies in Oceanography* **58**, 1914–1926.

Hughes, T.P., Graham, N.A.J., Jackson, J.B.C., Mumby, P.J. & Steneck, R.S. 2010. Rising to the challenge of sustaining coral reef resilience. *Trends in Ecology and Evolution* **25**, 633–642.

Hughes, T.P., Rodrigues, M.J., Bellwood, D.R., Ceccarelli, D., Hoegh-Guldberg, O., McCook, L., Moltschaniwskyj, N., Pratchett, M.S., Steneck, R.S. & Willis, B. 2007. Phase shifts, herbivory, and the resilience of coral reefs to climate change. *Current Biology* **17**, 360–365.

Huisman, J., Pham Thi, N.N., Karl, D.M. & Sommeijer, B. 2006. Reduced mixing generates oscillations and chaos in the oceanic deep chlorophyll maximum. *Nature* **439**, 322–325.

Huisman, J., van Oostveen, P. & Weissing, F.J. 1999. Critical depth and critical turbulence: two different mechanisms for the development of phytoplankton blooms. *Limnology and Oceanography* **44**, 1781-1787.

Humston, R., Olson, D.B. & Ault, J.S. 2004. Behavioral assumptions in models of fish movement and their influence on population dynamics. *Transactions of the American Fisheries Society* **133**, 1304–1328.

Hunt, G.L., Jr., Coyle, K.O., Eisner, L.B., Farley, E.V., Heintz, R.A., Mueter, F., Napp, J.M., Overland, J.E., Ressler, P.H., Salo, S. & Stabeno, P.J. 2011. Climate impacts on eastern Bering Sea foodwebs: a synthesis of new data and an assessment of the oscillating control hypothesis. *ICES Journal of Marine Science* **68**, 1230–1243.

Hurrell, J.W. & Deser, C. 2010. North Atlantic climate variability: the role of the North Atlantic Oscillation. *Journal of Marine Systems* **79**, 231–244.

Hutchins, D.A. 2011. Oceanography: forecasting the rain ratio. *Nature* **476**, 41–42.

Hutchins, D.A., Fu, F.X., Zhang, Y., Warner, M.E., Feng, Y., Portune, K., Bernhardt, P.W. & Mulholland, M.R. 2007. CO_2 control of *Trichodesmium* N_2 fixation, photosynthesis, growth rates, and elemental ratios: implications for past, present, and future ocean biogeochemistry. *Limnology and Oceanography* **52**, 1293–1304.

Hutchins, D.A., Mulholland, M.R. & Fu, F. 2009. Nutrient cycles and marine microbes in a CO_2-enriched ocean. *Oceanography* **22**,128–145.

Hyrenbach, K.D. & Veit, R.R. 2003. Ocean warming and seabird communities of the southern California Current System (1987–98): responses at multiple temporal scales. *Deep-Sea Research II* **50**, 2537–2565.

Idjadi, J.A. & Edmunds, P.J. 2006. Scleractinian corals as facilitators for other invertebrates on a Caribbean reef. *Marine Ecology Progress Series* **319**, 117–127.

Ingles, P. & McIlvaine-Newsad, H. 2007. Any port in the storm: the effects of Hurricane Katrina on two fishing communities in Louisiana. NAPA Bulletin 28: Anthropology and Fisheries Management in the United States: Methodology for Research Issue **28**, Hoboken, New Jersey: Blackwell Publishing. 69–86.

Inoue, J. & Hori, M.E., 2011. Arctic cyclogenesis at the marginal ice zone: a contributory mechanism for the temperature amplification? *Geophysical Research Letters* **38**, L12502.

Institute of Medicine. 2006. Seafood choices. Online. http://www.iom.edu/~/media/Files/Report%20 Files/2006/Seafood-Choices-Balancing-Benefits-and-Risks/11762_SeafoodChoicesReportBrief.ashx (accessed 17 March 2013).

Intergovernmental Panel on Climate Change (IPCC). 2007. *Climate Change 2007: Impacts, Adaptation and Vulnerability. Contribution of Working Group II to the Fourth Assessment Report of the Intergovernmental Panel on Climate Change*, M.L. Parry et al. (eds). Cambridge, UK: Cambridge University Press.

Intergovernmental Panel on Climatic Change (IPCC). 2011. 33rd Session Abu Dhabi, 10–13 May. Activities of the Task Force on National Greenhouse Gas Inventories. Online. http://www.ipcc.ch/meetings/session33/ doc07_p33_tfi_activities.pdf (accessed 17 March 2013).

Irwin, A.J. & Oliver, M.J. 2009. Are ocean deserts getting larger? *Geophysical Research Letters* **36**, L18609.

Iwasaki, S., Razafindrabe, B.H.N. & Shaw, R. 2009. Fishery livelihoods and adaptation to climate change: a case study of Chilika lagoon, India. *Mitigation and Adaptation Strategies for Global Change* **14**, 339–355.

Jallow, B.P., Toure, S., Barrow, M.M.K. & Mathieu, A.A. 1999. Coastal zone of the Gambia and the Abidjan region in Côte d'Ivoire: sea level rise vulnerability, response strategies, and adaptation options. *Climate Research* **12**, 129–136.

Jansen, J.M., Pronker, A.E., Kube, S., Sokolowski, A., Sola, J.C., Marquiegui, M.A., Schiedek, D., Wendelaar Bonga, S., Wolowicz, M. & Hummel, H. 2007. Geographic and seasonal patterns and limits on the adaptive response to temperature of European *Mytilus* spp. and *Macoma balthica* populations. *Oecologia* **154**, 23–34.

Jevrejeva, S., Moore, J.C., Grinsted, A., Woodworth, P.L. 2008. Recent global sea level acceleration started over 200 years ago? *Geophysical. Research Letters* **35**, 8–11.

Jiang, L. & Pu, Z. 2009. Different effects of species diversity on temporal stability in single trophic and multi-trophic communities. *The American Naturalist* **174**, 651–659.

Johansson, J.O.R. 2002. Historical and current observations on macroalgae in the Hillsborough Bay Estuary (Tampa Bay), Florida. In *Understanding the Role of Macroalgae in Shallow Estuaries,* M. McGinty et al. (eds). Linthicum: Maryland Department of Natural Resources Maritime Institute, 26–28.

Johnson, J.E. & Welch, D.J. 2010. Marine fisheries management in a changing climate: a review of vulnerability and future options. *Reviews in Fisheries Science* **18**, 106–124.

Jones, K., Patel, N.G., Levy, M.A., Storeygard, A., Balk, D., Gittleman, J.L. & Daszak, P. 2008. Global trends in emerging infectious diseases. *Nature* **451**(7181), 990–993.

Jones, S.J., Mieszkowska, N. & Wethey, D.S. 2009. Linking thermal tolerances and biogeography: *Mytilus edulis* (L.) at its southern limit on the East Coast of the United States. *Biological Bulletin* **217**, 73–85.

Joseph, J.E. & Chiu, C.S., 2010. A computational assessment of the sensitivity of ambient noise level to ocean acidification. *The Journal of the Acoustical Society of America* **128**, EL144–EL149.

Justić, D., Bierman, V.J., Jr., Scavia, D. & Hetland, R.D. 2007. Forecasting gulf's hypoxia: the next 50 years? *Estuaries and Coasts* **30**, 791–801.

Kanzow, T., Cunningham, S.A., Johns, W.E., Hirschi, J.J.M., Marotzke, J., Baringer, M.O., Meinen, C.S., Chidichimo, M.P., Atkinson, C., Beal, L.M., Bryden, H.L. & Collins, J. 2010. Seasonal variability of the Atlantic meridional overturning circulation at 26.5° N. *Journal of Climate* **23**, 5678–5698.

Karl, T.R., Melillo, J.M. & Peterson, T.C. (eds) 2009. *Global Climate Change Impacts in the United States.* New York: Cambridge University Press.

Karlson, K., Rosenberg, R. & Bonsdorff, E. 2002. Temporal and spatial large-scale effects on eutrophication and oxygen deficiency on the benthic fauna in Scandinavian and Baltic waters—a review. *Oceanography and Marine Biology: an Annual Review* **40**, 427–489.

Kaschner, K., Tittensor, D.P., Ready, J., Gerrodette, T. & Worm, B. 2011. Current and future patterns of global marine mammal biodiversity. *PLoS one* **6**, e19653.

Katz, A.R., Buchholz, A.E., Hinson, K., Park, S.Y. & Effler, P.V. 2011. Leptospirosis in Hawaii, USA, 1999– 2008. *Emerging Infectious Diseases* **17**, 221–226.

Kavry, V.I., Boltunov, A.N. & Nikiforov, V.V. 2008. New coastal haulouts of walruses (*Odobenus rosmarus*)— response to the climate changes. In *Collection of Scientific Papers from the Marine Mammals of the Holarctic V Conference, Odessa, Ukraine*, 248–251.

Kelly, K.A., Singh, S. & Huang, R.X. 1999. Seasonal variations of the sea surface height in the Gulf Stream region. *Journal of Geophysical Research* **29**, 313–327.

Kelly, M.W., Sanford, E. & Grosberg, R.K. 2012. Limited potential for adaptation to climate change in a broadly distributed marine crustacean. *Proceedings of the Royal Society B* **279**, 349–356.

Kelly, R.P., Foley, M.M., Fisher, W.S., Feely, R.A., Halpern, B.S., Waldbusser, G.G. & Caldwell, M.R. 2011. Mitigating local causes of ocean acidification with existing laws. *Science* **332**, 1036–1037.

Kerr, R.A. 2002. A warmer Arctic means change for all. *Science* **297**, 1490–1492.

Khon, V.C., Mokhov, I.I., Latif, M., Semenov, V.A. & Park, W. 2010. Perspectives of northern sea route and Northwest Passage in the twenty-first century. *Climate Change* **100**, 757–768.

Kik, M.J.L., Goris, M.G., Bos, J.H., Hartskeerl, R.A. & Dorrestein, G.M. 2006. An outbreak of leptospirosis in seals (*Phoca vitulina*) in captivity. *Veterinary Quarterly* **38**, 33–39.

Kildow, J.T., Colgan, C.S. & Scourse, J.S. 2009. State of the U.S. ocean and coastal economies. National Ocean Economics Program. Online. www.oceaneconomics.org/nationalreport/ (accessed 17 March 2013).

King, J.R., Agostini, V.N., Harvey, C.J., McFarlane, G.A., Foreman, M.G.G., Overland, J.E., Di Lorenzo, E., Bond, N.A. & Aydin, K.Y. 2011. Climate forcing and the California Current ecosystem. *ICES Journal of Marine Science* **68**, 1199–1216.

Kitagawa, H. 2008. Arctic routing: challenges and opportunities. *WMU Journal of Maritime Affairs* **7**, 485–503.

Kitts, A., Pinto da Silva, P. & Rountree, B. 2007. The evolution of collaborative management in the northeast USA tilefish fishery. *Marine Policy* **31**, 192–200.

Klein, C.J., Chan, A., Kircher, L., Cundiff, A.J., Gardner, N., Hrovat, Y., Scholz, A., Kendall, B.E. & Airamé, S. 2008. Striking a balance between biodiversity conservation and socioeconomic viability in the design of marine protected areas. *Conservation Biology* **22**, 691–700.

Kleypas, J.A. & Langdon, C. 2006. Coral reefs and changing seawater chemistry. In Coral Reefs and Climate Change: Science and Management. *AGU Monograph Series, Coastal and Estuarine Studies* **61**, J. Phinney et al. (eds). Washington, DC: American Geophysical Union, 691–700.

Knutson, T.R., McBride, J.L., Chan, J., Emanuel, K., Holland, G., Landsea, C., Held, I., Kossin, J.P., Srivastava, A.K. & Sugi, M., 2010. Tropical cyclones and climate change. *Nature Geoscience* **3**, 157–163.

Kordas, R.L., Harley, C.D.G. & O'Connor, M.I. 2011. Community ecology in a warming world: the influence of temperature on interspecific interactions. *Journal of Experimental Marine Biology and Ecology* **400**, 218–226.

Kostyack, J. & Rohlf, D. 2008. Conserving endangered species in an era of global warming. *Environmental Law Reporter News and Analysis* **38**, 10203–10213.

Kovacs, K.M., Lydersen, C., Overland, J.E. & Moore, S.E. 2010. Impacts of changing sea-ice conditions on Arctic marine mammals. *Marine Biodiversity* **41**(1), 181–194.

Kraak, S.B.M., Daan, N. & Pastoors, M.A. 2009. Biased stock assessment when using multiple, hardly over- lapping, tuning series if fishing trends vary spatially. *ICES Journal of Marine Science* **66**, 2272–2277.

Kreuder, C., Miller, M.A., Lowenstine, L.J., Conrad, P.A., Carpenter, T.E., Jessup, D.A. & Mazet, J.A.K. 2005. Evaluation of cardiac lesions and risk factors associated with myocarditis and dilated cardiomyopathy in southern sea otters. *American Journal of Veterinary Research* **66**, 289–299.

Kristiansen, T., Drinkwater, K.F., Lough, R.G. & Sundby, S. 2011. Recruitment variability in North Atlantic cod and match-mismatch dynamics. *PLoS one* **6**, e17456.

Kroeker, K., Kordas, R.L., Crim, R.N. & Singh, G.G. 2010. Meta-analysis reveals negative yet variable effects of ocean acidification on marine organisms. *Ecology Letters* **13**, 1419–1434.

Kroeker, K.J., Micheli, F., Gambi, M.C. & Martz, T.R. 2011. Divergent ecosystem responses within a benthic marine community to ocean acidification. *Proceedings of the National Academy of Sciences USA* **108**, 14515–14520.

Krupnik, I., Aporta, C., Gearheard, S., Laidler, G. & Kielse Holm, L. (eds) 2010. *SIKU: Knowing Our Ice. Documenting Inuit Sea Ice Knowledge and Use.* New York: Springer.

Kuhlbrodt, T., Rahmstorf, S., Zickfeld, K., Vikebø, F., Sundby, S., Hofmann, M., Link, P., Bondeau, A., Cramer, W. & Jaeger, C. 2009. An integrated assessment of changes in the thermohaline circulation. *Climatic Change* **96**, 489–537.

Kuo, E.S.L. & Sanford, E. 2009. Geographic variation in the upper thermal limits of an intertidal snail: implications for climate envelope models. *Marine Ecology Progress Series* **388**, 137–146.

Kushnir, Y., Seager, R., Ting, M., Naik, N. & Nakamura, J. 2010. Mechanisms of tropical Atlantic SST influence on North American precipitation variability. *Journal of Climate* **23**, 5610–5628.

Kwok, R. & Untersteiner, N. 2011. The thinning of Arctic sea ice. *Physics Today* **64**, 36.

Lackenbauer, P.W. (ed.) 2011. *Canadian Arctic Sovereignty and Security: Historical Perspectives.* Calgary Papers in Military and Strategic Studies. Calgary, Alberta, Canada: Centre for Military and Strategic Studies and University of Calgary Press.

Lafferty, K.D. 2009. The ecology of climate change and infectious diseases. *Ecology* **90**, 888–900.

Lambert, E., Hunter, C., Pierce, G.J. & MacLeod, C.D. 2010. Sustainable whale-watching tourism and climate change: towards a framework of resilience. *Journal of Sustainable Tourism* **18**, 409–427.

Landsberg, J.H., Flewelling, L.J. & Naar, J. 2009. *Karenia brevis* red tides, brevetoxins in the food web, and impacts on natural resources: decadal advancements. *Harmful Algae* **8**, 598–607.

Lapointe, B.E., Barile, P.J., Littler, M.M., Littler, D.S., Bedford, B.J. & Gasque, C. 2005. Macroalgal blooms on southeast Florida coral reefs I. Nutrient stoichiometry of the invasive green alga *Codium isthmocladum* in the wider Caribbean indicates nutrient enrichment. *Harmful Algae* **4**, 1092–1105.

Large, S.I., Smee, D.L. & Trussell, G.C. 2011. Environmental conditions influence the frequency of prey responses to predation risk. *Marine Ecology Progress Series* **422**, 41–49.

Last, P.R., White, W.T., Gledhill, D.C., Hobday, A.J., Brown, R., Edgar, G.J. & Pecl, G. 2011. Long-term shifts in abundance and distribution of a temperate fish fauna: a response to climate change and fishing practices. *Global Ecology and Biogeography* **20**, 58–72.

Lau, C.L., Smythe, L.D., Craig, S.B. & Weinstein, P. 2010. Climate change, flooding, urbanisation and leptospirosis: fuelling the fire? *Transactions of the Royal Society of Tropical Medicine and Hygiene* **104**, 631–638.

Laxon, S.W., Giles, K.A., Ridout, A.L., Winghm, D.J., Willatt, R., Cullen, R., Kwok, R., Schweiger, A., Zhang, J., Haas, C., Hendricks, S., Krishfield, R., Kurtz, N., Farrell, S. & Davidson, M. 2013. CryoSat-2 estimates of Arctic sea ice thickness and volume, *Geophysical Research Letters* DOI: 10.1002/grl.50193.

Lazarus, R.J. 2009. Super wicked problems and climate change: restraining the present to liberate the future. *Cornell Law Review* **94**, 1153–1232.

Lehodey, P., Chai, F. & Hampton, J. 2003. Modelling climate-related variability of tuna populations from a coupled ocean-biogeochemical populations dynamics model. *Fisheries Oceanography* **12**, 483–494.

Lehodey, P., Senina, I., Sibert, J., Bopp, L., Calmettes, B., Hampton, J. & Murtgudde, R. 2010. Preliminary forecasts of Pacific bigeye tuna population trends under the A2 IPCC scenario. *Progress in Oceanography* **86**, 302–315.

Lesser, M.P., Bailey, M.A., Merselis, D.G. & Morrison, J.R. 2010. Physiological response of the blue mussel *Mytilus edulis* to differences in food and temperature in the Gulf of Maine. *Comparative Biochemistry and Physiology Part A* **156**, 541–551.

Levett, P.N. 2001. Leptospirosis. *Clinical Microbiology Reviews* **14**, 296–326.

Levitus, S., Antonov, J.I., Boyer, T.P., Locarnini, R.A., Garcia, H.E. & Mishonov, A.V. 2009. Global ocean heat content 1955–2008 in light of recently revealed instrumentation problems. *Geophysical Research Letters* **36**, L07608, doi:10.1029/2008GL037155.

Li, Y., He, R., McGillicuddy, D.J., Jr., Anderson, D.M. & Keafer, B.A. 2009. Investigation of the 2006 *Alexandrium fundyense* bloom in the Gulf of Maine: *in situ* observations and numerical modeling. *Continental Shelf Research* **29**, 2069–2082.

Ling, S.D., Johnson, C.R., Frusher, S.D. & Ridgway, K.R. 2009. Overfishing reduces resilience of kelp beds to climate-driven catastrophic phase shift. *Proceedings of the National Academy USA* **106**, 22341–22345.

Link, J.S., Fulton, E.A. & Gamble, R.J, 2010. The northeast U.S. application of ATLANTIS: a full system model exploring marine ecosystem dynamics in a living marine resource management context. *Progress in Oceanography* **87**, 214–234.

Link, J.S., Nye, J.A. & Hare, J.A. 2011. Guidelines for incorporating fish distribution shifts into a fisheries management context. *Fish and Fisheries* **12**, 461–469.

Lipp, E., Huq, A. & Colwell, R. 2002. Effects of global climate on infectious disease: the cholera model. *Clinical Microbiology Reviews* **15**, 757–770.

Llewellyn, L.E. 2010. Revisiting the association between sea surface temperature and the epidemiology of fish poisoning in the South Pacific: reassessing the link between ciguatera and climate change. *Toxicon* **56**, 691–697.

Lloyd-Smith, J.O., Greig, D.J., Hietala, S., Ghneim, G.S., Palmer, L., St. Leger, J., Grenfell, B.T. & Gulland, F.M. 2007. Cyclical changes in seroprevalence of leptospirosis in California sea lions: endemic and epidemic disease in one host species? *BMC Infectious Diseases* **7**, 125–136.

Lockwood, B.L. & Somero, G.N. 2011a. Transcriptomic responses to salinity stress in invasive and native blue mussels (genus *Mytilus*). *Molecular Ecology* **20**, 517–529.

Lockwood, B.L. & Somero, G.N. 2011b. Invasive and native blue mussels (genus *Mytilus*) on the California coast: the role of physiology in a biological invasion. *Journal of Experimental Marine Biology and Ecology* **400**, 167–174.

Lombard, F., da Rocha, R.E., Bijma, J. & Gattuso, J.P. 2010. Effect of carbonate ion concentration and irradiance on calcification in planktonic Foraminifera. *Biogeosciences* **7**, 247–255.

Lopez, C.B., Dortch, Q., Jewett, E.B. & Garrison, D. 2008. *Scientific Assessment of Marine Harmful Algal Blooms*. Washington, DC: Interagency Working Group on Harmful Algal Blooms, Hypoxia, and Human Health of the Joint Subcommittee on Ocean Science and Technology.

Loring, P.A. & Gerlach, S.C. 2009. Food, culture, and human health in Alaska: an integrative health approach to food security. *Environmental Science and Policy* **12**, 466–478.

Loring, P.A. & Gerlach, S.C. 2010. Food security and conservation of Yukon River salmon: are we asking too much of the Yukon River? *Sustainability* **2**, 2965–2987.

Loring, P.A., Gerlach, S.C., Atkinson, D.E. & Murray, M.S. 2011. Ways to help and ways to hinder: governance for successful livelihoods in a changing climate. *Arctic* **64**, 73–88.

Lovell, B. 2010 *Challenged by Carbon: the Oil Industry and Climate Change*. Cambridge, UK: Cambridge University Press.

Low, L. 2008. United States of America—Alaska region. In *Impacts of Climate and Climate Change on Key Species in the Fisheries of the North Pacific*, R. Beamish (ed.). PICES Scientific Report No. 35. Sidney, British Columbia: North Pacific Marine Science Organization (PICES), 163–205.

Lubchenco, J. & Petes, L.E. 2010. The interconnected biosphere: Science at the ocean's tipping points. *Oceanography* **23**, 115–129.

Luber, G. & McGeehin, M. 2008. Climate change and extreme heat events. *American Journal of Preventive Medicine* **35**, 429–435.

Luettich, R.A., Jr., Carr, S.D., Reynolds-Fleming, J.V., Fulcher, C.W. & McNinch, J.E. 2002. Semi-diurnal seiching in a shallow, micro-tidal lagoonal estuary. *Continental Shelf Research* **22**, 1669–1681.

Lumpkin, R. & Speer, K., 2007. Global ocean meridional overturning. *Journal of Physical Oceanography* **37**, 2550–2562.

Lüthi, D., Le Floch, M., Bereiter, B., Blunier, T., Barnola, J., Urs, S., Raynaud, D., Jouzel, J., Fisher, H., Kawamura, K. & Stocker, T. 2008. High-resolution carbon dioxide concentration record 650,000–800,000 years before present. *Nature* **453**, 379–382.

Macdonald, R.W., Harner, T. & Fyfe, J. 2005. Recent climate change in the Arctic and its impact on contaminant pathways and interpretation of temporal trend data. *Science of the Total Environment* **342**, 5–86.

Macey, B.M., Achilihu, I.O., Burnett, K.G. & Burnett, L.E. 2008. Effects of hypercapnic hypoxia on inactivation and elimination of *Vibrio campbellii* in the eastern oyster, *Crassostrea virginica*. *Applied and Environmental Microbiology* **74**, 6077–6084.

Mahon, R. 2002. Adaptation of fisheries and fishing communities to the impacts of climate change in the CARICOM region: issue paper-draft. Mainstreaming adaptation to climate change (MACC) of the Caribbean Center for Climate Change (CCCC). Washington, DC: Organization of American States.

Mainelli, M., DeMaria, M., Shay, L.K. & Goni, G. 2008. Oceanic heat content estimation to operational forecasting of recent Atlantic category 5 hurricanes. *Weather and Forecasting* **23**, 3–16.

Mann, K.H. 1973. Seaweeds: their productivity and strategy for growth. *Science* **182**, 975–981.

Mantua, N.J., Hare, S.R., Zhang, Y., Wallace, J.M. & Francis, R.C. 1997. A Pacific interdecadal climate oscillation with impacts on salmon production. *Bulletin of the American Meteorological Society* **78**, 1069–1079.

Mantua, N., Tohver, I. & Hamlet, A. 2010. Climate change impacts on streamflow extremes and summertime stream temperature and their possible consequences for freshwater salmon habitat in Washington State. *Climatic Change* **102**, 187–223.

Markowski, M., Knapp, A., Neumann, J.E. & Gates, J. 1999. The economic impact of climate change on the U.S. commercial fishing industry. In *The Impact of Climate Change on the United States Economy*, R. Mendelsohn & J.E. Neumann (eds). New York: Cambridge University Press, 237–264.

Martin, S. & Gattuso, J.P. 2009. Response of Mediterranean coralline algae to ocean acidification and elevated temperature. *Global Change Biology* **15**, 2089–2100.

Martínez, E.A., Cárdenas, L. & Pinto, R. 2003. Recovery and genetic diversity of the intertidal kelp *Lessonia nigrescens* (Phaeophyceae) 20 years after El Niño 1982/83. *Journal of Phycology* **39**, 504–508.

Martinez-Urtaza, J., Bowers, J.C., Trinanes, J. & DePaola, A. 2010. Climate anomalies and the increasing risk of *Vibrio parahaemolyticus* and *Vibrio vulnificus* illnesses. *Food Research International* **43**, 1780–1790.

Maslanik, J., Fowler, C., Stroeve, J., Drobot, S., Zwally, J., Yi, D. & Emery, W., 2007. A younger, thinner Arctic ice cover: increased potential for rapid, extensive sea-ice loss. *Geophysical Research Letters*, **34**, L24501.

Matson, P.G. & Edwards, M.S. 2007. Effects of ocean temperature on the southern range limits of two understory kelps, *Pterygophora californica* and *Eisenia arborea*, at multiple life-stages. *Marine Biology* **151**, 1941–1949.

Maurstad, A. 2000. To fish or not to fish: small-scale fishing and changing regulations of the cod fishery in northern Norway. *Human Organization* **59**, 37–47.

Mayfield, A.B. & Gates, R.D. 2007. Osmoregulation in anthozoan–dinoflagellate symbiosis. *Comparative Biochemistry and Physiology—Part A: Molecular & Integrative Physiology* **147**, 1–10.

McCay, B.J., Brandt, S. & Creed, C.F. 2011. Human dimensions of climate change and fisheries in a coupled system: the Atlantic surfclam case. *ICES Journal of Marine Science* **68**,1354.

McClain, C.R., Signorini, S.R. & Christian, J.R. 2004. Subtropical gyre variability observed by ocean-color satellites *Deep-Sea Research II* **51**, 281–301.

McGeehin, M. 2007. CDC's role in addressing the health effects of climate change. *CDC Conference: Safe Healthier People*, May 4–5, Atlanta, Georgia.

McGeehin, M. & Mirabelli, M. 2001. The potential impacts of climate variability and change on temperature-related morbidity and mortality in the United States. *Environmental Health Perspectives* **109**, 191–198.

McIlgorm, A. 2010. Economic impacts of climate change on sustainable tuna and billfish management: insights from the Western Pacific. *Progress in Oceanography* **86**, 187–191.

McLaughlin, A., DePaola, C., Bopp, K., Martinek, N., Napolilli, B.S. & Allison, C. 2005. Outbreak of *Vibrio parahaemolyticus* gastroenteritis associated with Alaskan oysters. *New England Journal of Medicine* **353**,1463–1470.

McLeod, E., Chmura, G.L., Bouillon, S., Salm, R., Björk, M., Duarte, C.M., Lovelock, C.E., Schlesinger, W.H. & Silliman, B.R. 2011. A blueprint for blue carbon: toward an improved understanding of the role of vegetated coastal habitats in sequestering CO_2. *Frontiers in Ecology and the Environment* **9**, 552–560.

McLeod, E., Salm, R., Green, A. & Almany, J. 2009. Designing marine protected area networks to address the impacts of climate change. *Frontiers in Ecology and the Environment* **7**, 362–370.

McNeeley, S. 2009. *Seasons out of balance: vulnerability and sustainable adaptation in Alaska.* PhD dissertation, University of Alaska Fairbanks.

Meehl, G.A., Stocker, T.F., Collins, W.D., Friedlingstein, P., Gaye, A.T., Gregory, J.M., Kitoh, A., Knutti, R., Murphy, J.M., Noda, A., Raper, S.P.B., Watterson, I.G., Weaver, A.J. & Zhao, Z.C. 2007. Global climate projections. In *Climate Change 2007: The Physical Science Basis. Contribution of Working Group I to the Fourth Assessment Report of the Intergovernmental Panel on Climate Change*, S. Solomon et al. (eds). Cambridge, UK: Cambridge University Press, 589–662.

Meites, E., Jay, M.T., Deresinski, S., Shieh, W.J., Zaki, S.R., Tompkins, L. & Smith, D.S. 2004. Reemerging leptospirosis, California. *Emerging Infectious Diseases* **10**, 406–412.

Menge, B.A., Chan, F. & Lubchenco, J. 2008. Response of a rocky intertidal ecosystem engineer and community dominant to climate change. *Ecology Letters* **11**, 151–162.

Merico, A., Tyrrell, T., Lessard, E.J., Oguz, T., Stabeno, P.J., Zeeman, S.I. & Whitledge, T.E. 2004. Modelling phytoplankton succession on the Bering Sea shelf: role of climate influences and trophic interactions in generating *Emiliania huxleyi* blooms 1997–2000. *Deep-Sea Research Part I: Oceanographic Research Papers* **51**, 1803–1826.

Merzouk, A. & Johnson, L.E. 2011. Kelp distribution in the northwest Atlantic Ocean under a changing climate. *Journal of Experimental Marine Biology and Ecology* **400**, 90–98.

Mikulski, C.M., Burnett, L.E. & Burnett, K.G. 2000. The effects of hypercapnic hypoxia on the survival of shrimp challenged with *Vibrio parahaemolyticus*. *Journal of Shellfish Research* **19**, 301–311.

Millennium Ecosystem Assessment (MEA). 2005. Ecosystems and human well-being: biodiversity synthesis. Washington, DC: World Resources Institute.

Miller, G., Alley, R., Brigham-Grette, J., Fitzpatrick, J., Polyak, L., Serreze, M. & White, J. 2010. Arctic amplification: can the past constrain the future? *Quaternary Science Reviews* **29**, 1779–1790.

Miller, J., Muller, E., Rogers, C., Waara, R., Atkinson, A., Whelan, K.R.T., Patterson, M. & Witcher, B. 2009. Coral disease following massive bleaching in 2005 causes 60 percent decline in coral cover on reefs in the U.S. Virgin Islands. *Coral Reefs* **28**, 925–937.

Miller, J.R. & Russell, G.L. 1992. The impact of global warming on river runoff. *Journal of Geophysical Research* **97**, 2757–2764.

Miller, K.A. 2007. Climate variability and tropical tuna: management challenges for highly migratory fish stocks. *Marine Policy* **31**, 56–70.

Miller, K.A. & Munro, G.R. 2003. Climate and cooperation: a new perspective on the management of shared fish stocks. *Marine Resource Economics* **19**, 367–393.

Miller, M.A., Conrad, P.A., Harris, M., Hatfield, B., Langlois, G., Jessup, D.A., Magargal, S.L., Packham, A.E., Toy-Choutka, S., Melli, A.C., Murray, M.A., Gulland, F.M. & Grigg, M.E. 2010. A protozoal-associated epizootic impacting marine wildlife: mass-mortality of southern sea otters (*Enhydra lutris nereis*) due to *Sarcocystis neurona* infection. *Veterinary Parasitology* **172**, 183–194.

Mills, J., Gage, K. & Khan, A.S. 2010. Potential influence of climate change on vector-borne and zoonotic diseases: a review and proposed research plan. *Environmental Health Perspectives* **118**, 1507–1514.

Mislan, K.A.S. & Wethey, D.S. 2011. Gridded meteorological data as a resource for mechanistic macroecology in coastal environments. *Ecological Applications* **21**, 2678–2690.

Moeser, G.M. & Carrington, E. 2006. Seasonal variation in mussel byssal thread mechanics. *Journal of Experimental Biology* **209**, 1996–2003.

Moffitt, C.M., Stewart, B., Lapatra, S., Brunson, R., Bartholomew, J., Peterson, J. & Amos, K.H. 1998. Pathogens and diseases of fish in aquatic ecosystems: implications in fisheries management. *Journal of Aquatic Animal Health* **10**, 95–100.

Monaco, C.J. & Helmuth, B. 2011. Tipping points, thresholds, and the keystone role of physiology in marine climate change research. *Advances in Marine Biology* **60**, 123–160.

Monahan, A.M., Miller, I.S. & Nally, J.E. 2009. Leptospirosis: risks during recreational activities. *Journal of Applied Microbiology* **107**, 707–716.

Montevecchi, W.A. & Myers, R.A. 1997. Centurial and decadal oceanographic influences on changes in northern gannet populations and diets in the north-west Atlantic: implications for climate change. *ICES Journal of Marine Science* **54**, 608–614.

Moore, C. 2011. Welfare impacts of ocean acidification: an integrated assessment model of the U.S. mollusk fishery, Working Paper 11–06. Washington, DC: National Center for Environmental Economics, U.S. Environmental Protection Agency.

Moore, K.A. & Jarvis, J.C. 2008. Environmental factors affecting recent summertime eelgrass diebacks in the lower Chesapeake Bay: implications for long-term persistence. *Journal of Coastal Research* **55**, 135–147.

Moore, S. & Gill, M. 2011. Marine ecosystems summary. In *Arctic Report Card 2011*, J. Richter-Menge, M.O. Jeffries & J.E. Overland (eds), 63–64. Online. http://www.arctic.noaa.gov/report11/ArcticReportCard_full_report.pdf

Moore, S.E. 2008. Marine mammals as ecosystem sentinels. *Journal of Mammalogy* **89**, 534–540.

Moore, S.E., Grebmeier, J.M. & Davies, J.R. 2003. Gray whale distribution relative to forage habitat in the northern Bering Sea: current conditions and retrospective summary. *Canadian Journal of Zoology* **81**, 734–742.

Moore, S.K., Mantua, N.J., Hickey, B.M. & Trainer, V.L. 2009. Recent trends in paralytic shellfish toxins in Puget Sound, relationships to climate, and capacity for prediction of toxic events. *Harmful Algae* **8**, 463–477.

Moore, S.K., Mantua, N.J. & Salathé, E.P., Jr. 2011. Past trends and future scenarios for environmental conditions favoring the accumulation of paralytic shellfish toxins in Puget Sound shellfish. *Harmful Algae* **10**, 521–529.

Moore, S.K., Trainer, V.L., Mantua, N.J., Parker, M.S., Laws, E.A., Backer, L.C. & Fleming, L.E. 2008. Impacts of climate variability and future climate change on harmful algal blooms and human health. *Environmental Health* **7**, S4.

Morán, X.A.G., López-Urrutia, Á., Calvo-Díaz, A. & Li, W.K.W. 2010. Increasing importance of small phytoplankton in a warmer ocean. *Global Change Biology* **16**, 1137–1144.

Moreno, A. & Becken, S. 2009. A climate change vulnerability assessment methodology for coastal tourism. *Journal of Sustainable Tourism* **17**, 473–488.

Morgan, K.L., Larkin, S.L. & Adams, C.M. 2010. Red tides and participation in marine-based activities: estimating the response of Southwest Florida residents. *Harmful Algae* **9**, 333–341.

Morreale, S.J., Meylan, A.B., Sadove, S.B. & Standora, E.A. 1992. Annual occurrence and winter mortality of marine turtles in New York waters. *Journal of Herpetology* **26**, 301–308.

Moser, S. & Luers, A.L. 2008. Managing climate risks in California: the need to engage resource managers for successful adaptation to change. *Climatic Change* **87**, 309–322.

Moy, A.D., Howard, W.R., Bray, S.G. & Trull, T.W. 2009. Reduced calcification in modern Southern Ocean planktonic Foraminifera. *Nature Geoscience* **2**, 276–280.

Mueter, F.J., Bond, N.A., Lanelli, J.N. & Hollowed, A.B. 2011a. Expected declines in recruitment of walleye pollock (*Theragra chalcogramma*) in the eastern Bering Sea under future climate change. *ICES Journal of Marine Science* **68**, 1284–1296.

Mueter, F.J., Siddon, E.C. & Hunt, G.L., Jr. 2011b. Climate change brings uncertain future for subarctic marine ecosystems and fishes. In *North by 2020: Perspectives on Alaska's Changing Social-Ecological Systems*, A.L. Lovecraft & H. Eicken (eds). Fairbanks, Alaska: University of Alaska Press, 329–357.

Mumby, P.J., Elliott, I.A., Eakin, C.M., Skirving, W., Paris, C.B., Edwards, H.J., Enríquez, S., Iglesias-Prieto, R., Cherubin, L.M. & Stevens, J.R. 2011a. Reserve design for uncertain responses of coral reefs to climate change. *Ecology Letters* **14**, 132–140.

Mumby, P.J., Iglesias-Prieto, R., Hooten, A.J., Sale, P.F., Hoegh-Guldberg, O., Edwards, A.J., Harvell, C.D., Gomez, E.D., Knowlton, N., Hatziolos, M.E., Kyewalyanga, M.S. & Muthiga, N. 2011b. Revisiting climate thresholds and ecosystem collapse. *Frontiers in Ecology and the Environment* **9**, 94–95.

Mumby, P.J. & Steneck, R.S. 2011. The resilience of coral reefs and its implications for reef management. In *Coral Reefs: An Ecosystem in Transition*, Z. Dubinsky & N. Stambler (eds). Dordrecht, the Netherlands: Springer, 509–519.

Musick, J.A. & Limpus, C.J. 1997. Habitat utilization and migration in juvenile sea turtles. In *The Biology of Sea Turtles,* Vol. 1, P.L. Lutz & J.A. Musick (eds). Boca Raton, FL: CRC Press, 137–163.

Nam, S., Kim, H.J. & Send, U. 2011. Amplification of hypoxic and acidic events by La Niña conditions on the continental shelf off California. *Geophysical Research Letters* **38**, L22602.

National Climate Assessment (NCA) Report Series. 2010a. Volume 5b. Monitoring climate changes and its impacts: physical climate indicators. Washington, DC: U.S. Global Change Research Program.

National Climate Assessment (NCA) Report Series. 2010b. Volume 5a. Ecosystem responses to climate change: selecting indicators and integrating observation networks. Washington, DC: U.S. Global Change Research Program.

National Climate Assessment (NCA) Report Series. 2011. Volume 5c. Climate change impacts and responses: societal indicators for the National Climate Assessment. Washington, DC: U.S. Global Change Research Program.

National Climate Data Center (NCDC). 2010. 2009/2010 cold season. Online. http://www.ncdc.noaa.gov/special-reports/2009-2010-cold-season.html (accessed 12 March 2012).

National Marine Fisheries Service (NOAA Fisheries). 2009. Fishing communities of the United States, 2006. U.S. Department of Commerce, NOAA Technical Memo. NOAA Fisheries-F/SPO-98.

National Marine Fisheries Service (NOAA Fisheries). 2010. Fisheries economics of the United States, 2009. U.S. Department of Commerce, NOAA Technical Memo. NOAA Fisheries-F/SPO-118, 172.

National Ocean Council (NOC). 2012. Draft national ocean policy implementation plan. Washington, DC: NOC. Online. http://www.whitehouse.gov/administration/eop/oceans/implementationplan (accessed 17 March 2013).

National Oceanic and Atmospheric Administration (NOAA). 1997a. Wetlands, fisheries, and economics in the New England coastal states. *Habitat Connections* **1**, 3.

National Oceanic and Atmospheric Administration (NOAA). 1997b. Wetlands, fisheries, and economics in the Mid-Atlantic coastal states. *Habitat Connections* **1**, 5.

National Research Council (NRC). 1996. *Understanding Risk: Informing Decisions in a Democratic Society.* Washington, DC: National Academies Press.

National Research Council (NRC). 2000a. *Long-Term Institutional Management of U.S. Department of Energy Legacy Waste Sites.* Washington, DC: National Academies Press.

National Research Council (NRC). 2000b. *A Risk Management Strategy for PCB-Contaminated Sediments.* Washington, DC: National Academies Press.

National Research Council (NRC). 2010a. *Adapting to the Impacts of Climate Change. America's Climate Choices.* Washington, DC: National Academies Press.

National Research Council (NRC). 2010b. *Ocean Acidification: a National Strategy to Meet the Challenges of a Changing Ocean.* Washington, DC: National Academies Press.

Neuwald, J.L. & Valenzuela, N. 2011. The lesser known challenge of climate change: thermal variance and sex-reversal in vertebrates with temperature-dependent sex determination. *PLoS one* **6**, 18117.

Nghiem, S., Rigor, I., Perovich, D., Clemente-Colon, P., Weatherly, J. & Neumann, G. 2007. Rapid reduction of Arctic perennial sea ice. *Geophysical Research Letters* **34**, L19504.

Norris, F.H., Speier, A., Henderson, A.K., Davis, S.I., Purcell, D.W., Stratford, B.D., Baker, C.K., Reissman, D.B. & Daley, W.R. 2006. Assessment of health-related needs after Hurricanes Katrina and Rita—Orleans and Jefferson Parishes, New Orleans area, Louisiana, October 17–22, 2005. *Morbidity and Mortality Weekly Report* **55**, 38–41.

North Pacific Fishery Management Council (NPFMC). 2009. Fishery management plan for fish resources of the Arctic Management Area. Online. http://alaskafisheries.noaa.gov/npfmc/PDFdocuments/fmp/Arctic/ArcticFMP.pdf (accessed 17 March 2013).

Nuttall, M. & T.V. Callaghan (eds) 2000. *The Arctic: Environment, People, Policy.* New York: Taylor and Francis.

Nye, J., Link, J.S., Hare, J.A. & Overholtz, W.J. 2009. Changing spatial distributions of Northwest Atlantic Fish stocks in relation to temperature and stock size. *Marine Ecology Progress Series* **393**, 111–129.

Nye, J.A., Joyce, T.M., Kwon, Y. & Link, J.A. 2011. Silver hake tracks changes in Northwest Atlantic circulation. *Nature Communications* **2**, 412.

O'Connor, M.I. 2009. Warming strengthens an herbivore-plant interaction. *Ecology* **90**, 388–398.

O'Connor, M.I., Bruno, J.F., Gaines, S.D., Halpern, B.S., Lester, S.E., Kinlan, B.P. & Weiss, J.M. 2007. Temperature control of larval dispersal and the implications for marine ecology, evolution, and conservation. *Proceedings of the National Academy of Sciences USA* **104**, 1266–1271.

O'Connor, M.I., Piehler, M.F., Leech, D.M., Anton, A. & Bruno, J.F. 2009. Warming and resource availability shift food web structure and metabolism. *PLoS Biology* **7**, e1000178.

Office of Travel and Tourism Industries (OTTI). 2011a. *Fast Facts: United States Travel and Tourism Industry 2010.* June. Washington, DC: U.S. Department of Commerce, ITA, Office of Travel and Tourism Industries. Online. http://tinet.ita.doc.gov/outreachpages/inbound.general_information.inbound_overview.html (accessed 6 February 2012).

Office of Travel and Tourism Industries (OTTI). 2011b. *Overseas Visitation Estimates for U.S. States, Cities, and Census Regions: 2010.* May. Washington, DC: U.S. Department of Commerce, ITA, Office of Travel and Tourism Industries. Online. http://tinet.ita.doc.gov/outreachpages/inbound.general_information. inbound_overview.html (accessed 6 February 2012).

Oliver, J. 1989. *Vibrio vulnificus.* In *Foodborne Bacterial Pathogens,* M.P. Doyle (ed.). New York: Dekker, 569–599.

Oliver, J. 2005. Wound infections caused by *Vibrio vulnificus* and other marine bacteria. *Epidemiology and Infection* **122**, 383–391.

Oliver, J. & Kaper, J. 2007. *Vibrio* species. In *Food Microbiology: Fundamentals and Frontiers,* M.P. Doyle & L.R. Beuchat (eds). Washington, DC: ASM Press, 3rd edition, 343–379.

Organization for Economic Cooperation and Development (OECD). 2010. *The Economics of Adapting Fisheries to Climate Change*. Paris: OECD. Online. http://www.oecd-ilibrary.org/agriculture-and-food/the-economics-of-adapting-fisheries-to-climate-change_9789264090415-en (accessed 17 January 2012).

Orr, J.C., Fabry, V.J., Aumont, O., Bopp, L., Doney, S.C., Feely, R.A., Gnanadesikan, A., Gruber, N., Ishida, A., Joos, F., Key, R.M., Lindsay, K., Maier-Reimer, E., Matear, R., Monfray, P., Mouchet, A., Najjar, R.G., Plattner, G.K., Rodgers, K.B., Sabine, C.L., Sarmiento, J.L., Schlitzer, R., Slater, R.D., Totterdell, I.J., Weirig, M.F., Yamanaka, Y. & Yool, A. 2005. Anthropogenic ocean acidification over the twenty-first century and its impact on calcifying organisms. *Nature* **437**, 681–686.

Osgood, K.E. (ed.) 2008. Climate impacts on U.S. living marine resources: National Marine Fisheries Service concerns, activities and needs. U.S. Department of Commerce, NOAA Technical Memo. NMFS-F/SPO-89.

Østreng, W. 2010. *Science without Boundaries: Interdisciplinarity in Research, Society, and Politics*. New York: University Press of America.

Ottersen, G., Kim, S., Huse, G., Polovina, J.J. & Stenseth, N.C. 2010. Major pathways by which climate may force marine fish populations. *Journal of Marine Systems* **79**, 343–360.

Overholtz, W.J., Hare, J.A. & Keith, C.M. 2011. Impacts of interannual environmental forcing and climate change on the distribution of Atlantic mackerel in the U.S. Northeast continental shelf. *Marine and Coastal Fisheries: Dynamics, Management and Ecosystem Science* **3**, 219–252.

Overland, J., Wood, K. & Wang, M. 2011. Warm Arctic–cold continents: climate impacts of the newly open Arctic Sea. *Polar Research* **30**, 15787.

Overland, J.E. & Wang, M. 2007. Future climate of the North Pacific Ocean. *Eos, Transactions, American Geophysical Union* **88**, 178–182.

Overland, J.E., Alheit, J., Bakun, A., Hurrell, J.W., Mackas, D.L. & Miller, A.J. 2010. Climate controls on marine ecosystems and fish populations. *Journal of Marine Systems* **79**, 305–315.

Owen, D. 2012. Critical habitat and the challenge of regulating small harms. *Florida Law Review* **64**, 141–200.

Paerl, H.W. & Paul, V.J. 2012 Climate change: links to global expansion of harmful cyanobacteria. *Water Research* **46**, 1349–1363.

Paine, R.T. 1992. Food web analysis through field measurements of per capita interaction strength. *Nature* **355**, 73–75.

Palacios, D.M., Bograd, S.J., Mendelsshon, R. & Schwing, F.B. 2004. Long-term and seasonal trends in stratification in the California Current, 1950–1993. *Journal of Geophysical Research* **109**, C10016.

Palm, H.W. 2011. Fish parasites as biological indicators in a changing world: can we monitor environmental impact and climate change? In *Progress in Parasitology*, H. Mehlhorn (ed.). Berlin: Springer-Verlag, 223–250.

Palmer, C.T. 1990. Telling the truth (up to a point): radio communication among Maine lobstermen. *Human Organization* **49**, 157–163.

Pandolfi, J.M., Connolly, S.R., Marshall, D.J. & Cohen, A.L. 2011. Projecting coral reef futures under global warming and ocean acidification. *Science* **333**, 418–422.

Park, K., Kim, C.K. & Schroeder, W.W. 2007. Temporal variability in summertime bottom hypoxia in shallow areas of Mobile Bay, Alabama. *Estuaries and Coasts* **30**, 54–65.

Parmenter, R.R., Yadav, E.P., Parmenter, C.A., Ettestad, P. & Gage, K.L. 1999. Incidence of plague associated with increased winter-spring precipitation in New Mexico, USA. *American Journal of Tropical Medicine and Hygiene* **61**, 814–821.

Parmesan, C. & Yohe, G. 2003. A globally coherent fingerprint of climate change impacts across natural systems. *Nature* **421**, 37–42.

Peeters, F., Straile, D., Lorke, A. & Livingstone, D.M. 2007. Earlier onset of the spring phytoplankton bloom in lakes of the temperate zone in a warmer climate. *Global Change Biology* **13**, 1898–1909.

Peloso, M. 2010. *Adapting to rising sea levels*. PhD dissertation, Duke University, Durham, NC.

Peperzak, L. 2003. Climate change and harmful algal blooms in the North Sea. *Acta Oecologia* **24**, 139–144.

Pereira, H.M., Leadley, P.W., Proença, V., Alkemade, R., Scharlemann, J.P.W., Fernandez-Manjarrés, J.F., Araújo, M.B., Balvanera, P., Biggs, R., Cheung, W.W.L., Chini, L., Cooper, H.D., Gilman, E.L., Guénette, S., Hurtt, G.C., Huntington, H.P., Mace, G.M., Oberdorff, T., Revenga, C., Rodrigues, P., Scholes, R.J., Sumaila, U.R. & Walpole, M. 2010. Scenarios for global biodiversity in the 21st century. *Science* **330**, 1496–1501.

Perovich, D., Meier, W., Maslanik, J. & Richter-Menge, J. 2011. *Sea Ice. Marine Ecosystems Study. Arctic Report Card: Update for 2011.* Online. http://www.arctic.noaa.gov/reportcard/sea_ice.html (accessed 6 December 2011).

Perry, A., Low, P., Ellis, J. & Reynolds, J. 2005. Climate change and distribution shifts in marine fishes. *Science* **308**, 1912–1915.

Peterson, T.C. & Baringer, M.O. (eds) 2009. State of the climate in 2008. *Special Supplement to the Bulletin of the American Meteorological Society* **90**, S1–S196.

Peterson, W.T. & Keister, J.E. 2003. Interannual variability in copepod community composition at a coastal station in the northern California Current: a multivariate approach. *Deep-Sea Research Part II: Topical Studies in Oceanography* **50**, 2499–2517.

Petes, L.E., Menge, B.A. & Harris, A.L. 2008. Intertidal mussels exhibit energetic trade-offs between reproduction and stress resistance. *Ecological Monographs* **78**, 387–402.

Petes, L.E., Menge, B.A. & Murphy, G.D. 2007. Environmental stress reduces survival, growth, and reproduction in New Zealand intertidal mussels. *Journal of Experimental Marine Biology and Ecology* **351**, 83–91.

Petit, J.R., Jouzel, J., Raynaud, D., Barkov, N.I., Barnola, J.-M., Basile, I., Bender, M., Chappellaz, J., Davis, M., Delaygue, G., Delmotte, M., Kotlyakov, V.M., Legrand, M., Lipenkov, V.Y., Lorius, C., Pépin, L., Ritz, C., Saltzman, E. & Stievenard, M. 1999. Climate and atmospheric history of the past 420,000 years from the Vostok ice core, Antarctica. *Nature* **399**, 429–436.

Pfeiffer, L. & Haynie A.C. 2012. The effect of decreasing seasonal sea-ice cover on the winter Bering Sea pollock fishery. *ICES Journal of Marine Science: Journal du Conseil* June 8, doi:10.1093/icesjms/fss097.

Philippart, C.J.M., van Aken, H.M., Beukema, J.J., Bos, O.G., Cadée, G.C. & Dekker, R. 2003. Climate-related changes in recruitment of the bivalve *Macoma balthica*. *Limnology and Oceanography* **48**, 2171–2185.

Phillips, P. & Morrow, B.H. 2007. Social science research needs: focus on vulnerable populations, forecasting, and warnings. *Natural Hazards Review* **8**, 61–68.

Pimm, S. 2008. Biodiversity: climate change or habitat loss—which will kill more species? *Current Biology* **18**, R117–R119.

Pimm, S., Russell, G., Gittleman, J. & Brooks, T. 1995. The future of biodiversity. *Science* **269**, 347–350.

Pincebourde, S., Sanford, E., Casas, J. & Helmuth, B. 2012. Temporal coincidence of environmental stress events modulates predation rates. *Ecology Letters* **15**, 680–688.

Pincebourde, S., Sanford, E. & Helmuth, B. 2008. Body temperature during low tide alters the feeding performance of a top intertidal predator. *Limnology and Oceanography* **53**, 1562–1573.

Place, S.P., Menge, B.A. & Hofmann, G.E. 2012. Transcriptome profiles link environmental variation and physiological response of *Mytilus californianus* between Pacific tides. *Functional Ecology* **2**, 144–155.

Place, S.P., O'Donnell, M.J. & Hofmann, G.E. 2008. Gene expression in the intertidal mussel *Mytilus californianus*: physiological response to environmental factors on a biogeographic scale. *Marine Ecology Progress Series* **356**, 1–14.

Poloczanska, E.S., Hawkins, S.J., Southward, A.J. & Burrows, M.T. 2008. Modeling the response of populations of competing species to climate change. *Ecology* **89**, 3138–3149.

Poloczanska, E.S., Limpus, C.J. & Hays, G.C. 2009. Vulnerability of marine turtles to climate change. *Advances in Marine Biology* **56**, 151–211.

Polovina, J., Uchida, I., Balazs, G., Howell, E.A., Parker, D. & Dutton, P. 2006. The Kuroshio Extension Bifurcation Region: a pelagic hotspot for juvenile loggerhead sea turtles. *Deep-Sea Research II* **53**, 326–339.

Polovina, J.J. 2005. Climate variation, regime shifts, and implications for sustainable fisheries. *Bulletin of Marine Science* **76**, 233–244.

Polovina, J.J. 2007. Decadal variation in the trans-Pacific migration of northern bluefin tuna (*Thunnus thynnus*) coherent with climate-induced change in prey abundance. *Fisheries Oceanography* **5**, 114–119.

Polovina, J.J., Dunne, J.P., Woodworth, P.A. & Howell, E.A. 2011. Projected expansion of the subtropical biome and contraction of the temperate and equatorial upwelling biomes in the North Pacific under global warming. *ICES Journal of Marine Science* **68**, 986–995.

Polovina, J.J. & Haight, W.R. 1999. Climate variation, ecosystem dynamics, and fisheries management in the Northwestern Hawaiian Islands. In *Ecosystem Approaches for Fisheries Management*. Fairbanks, Alaska: Alaska Sea Grant College Program AK-SG-99-01, 23–32.

Polovina, J.J., Howell, E.A. & Abecassis, M. 2008. Ocean's least productive waters are expanding. *Geophysical Research Letters* **35**, L03618.

Pörtner, H.O. 2008. Ecosystem effects of ocean acidification in times of ocean warming: a physiologist's view. *Marine Ecology Progress Series* **373**, 203–217.

Pörtner, H.O. 2010. Oxygen- and capacity-limitation of thermal tolerance: a matrix for integrating climate-related stressor effects in marine ecosystems. *The Journal of Experimental Biology* **213**, 881–893.

Pörtner, H.O., Bock, C., Knust, R., Lannig, G., Lucassen, M., Mark, F.C. & Sartoris, F.J. 2008. Cod and climate in a latitudinal cline: physiological analyses of climate effects in marine fishes. *Climate Research* **37**, 253–270.

Pörtner, H.O. & Farrell, A.P. 2008. Physiology and climate change. *Science* **322**, 690–692.

Porzio, L., Buia, M.C. & Hall-Spencer, J.M. 2011. Effects of ocean acidification on macroalgal communities. *Journal of Experimental Marine Biology and Ecology* **400**, 278–287.

Powers, S.P., Peterson, C.H., Grabowski, J.H. & Lenihan, H.S. 2009. Success of constructed oyster reefs in no-harvest sanctuaries: implications for restoration. *Marine Ecology Progress Series* **389**, 159–170.

Pratchett, M.S., Munday, P.L., Graham, N.A.J., Kronen, M., Pinca, S., Friedman, K., Brewer, T.D., Bell, J.D., Wilson, S.K., Cinner, J.E., Kinch, J.P., Lawton, R.J., Williams, A.J., Chapman, L., Magron F. & Webb, A. 2011. Vulnerability of coastal fisheries in the tropical Pacific to climate change. In *Vulnerability of Tropical Pacific Fisheries and Aquaculture to Climate Change*, J.D. Bell et al. (eds). Noumea, New Caledonia: Secretariat of the Pacific Community.

Pratchett, M.S., Munday, M.S., Wilson, S.K., Graham, N.A.J., Cinner, J.E., Bellwood, D.R., Jones, G.P., Polunin, N.V.C. & McClanahan, T.R. 2008. Effects of climate-induced coral bleaching on coral reef fishes: ecological and economic consequences. *Oceanography and Marine Biology: an Annual Review* **46**, 251–296.

Purvis, A., Jones, K.E. & Mace, G.M. 2000 extinction. *BioEssays* **22**, 1123–1133.

Pyke, C.R., Thomas, R., Porter, R.D., Hellmann, J.J., Dukes, J.S., Lodge, D.M. & Chavarria, G. 2008. Current practices and future opportunities for policy on climate change and invasive species. *Conservation Biology* **22**, 585–592.

Rabalais, N.N., Turner, R.E., Diaz, R.J. & Justić, D. 2009. Global change and eutrophication of coastal waters. *ICES Journal of Marine Science* **66**, 1528–1537.

Rabalais, N.N., Turner, R.E., Sen Gupta, B.K., Boesch, D.F., Chapman, P. & Murrell, M.C. 2007. Hypoxia in the northern Gulf of Mexico: does the science support the plan to reduce, mitigate and control hypoxia? *Estuaries and Coasts* **30**, 753–772.

Rahel, F.J. & Olden, J.D. 2008. Assessing the effects of climate change on aquatic invasive species. *Conservation Biology* **22**, 521–533.

Randall, D.A., Wood, R.A., Bony, S., Colman, R., Fichefet, T., Fyfe, J., Kattsov, V., Pitman, A., Shukla, J., Srinivasan, J., Stouffer, R.J., Sumi, A. & Taylor, K.E. 2007. Climate models and their evaluation. In *Climate Change 2007: The Physical Science Basis. Contribution of Working Group I to the Fourth Assessment Report of the Intergovernmental Panel on Climate Change,* S. Solomon et al. (eds). Cambridge, UK: Cambridge University Press, 589–662.

Rattenbury, K., Kielland, K., Finstad, G. & Schneider, W. 2009. A reindeer herder's perspective on caribou, weather and socio-economic change on the Seward Peninsula, Alaska. *Polar Research* **28**, 71–88.

Rayner, N.A., Brohan, P., Parker, D.E., Folland, C.K., Kennedy, J.J., Vanicek, M., Ansell, T.J. & Tett, S.F.B. 2006. Improved analyses of changes and uncertainties in sea surface temperature measured *in situ* since the mid-nineteenth century: the HadSST2 dataset. *Journal of Climate* **19**, 446–469.

Reed, D.C., Rassweiler, A. & Arkema, K.K. 2008. Biomass rather than growth rate determines variation in net primary production by giant kelp. *Ecology* **89**, 2493–2505.

Reeder, D.B. & Chiu, C.S. 2010. Ocean acidification and its impact on ocean noise: phenomenology and analysis. *The Journal of the Acoustical Society of America* **128**, EL137–EL143.

Regehr, E.V., Amstrup, S.C. & Stirling, I. 2006. Polar bear population status in the southern Beaufort Sea. Open-File Report 2006-1337. Reston, Virginia: U.S. Geological Survey.

Regehr, E.V., Hunter, C.M., Caswell, H., Armstrup, S.C. & Stirling, I. 2010. Survival and breeding of polar bears in the southern Beaufort Sea in relation to sea ice. *Journal of Animal Ecology* **79**, 117–127.

Reid, P.C., Johns, D.G., Edwards, M., Starr, M., Poulin, M. & Snoeijs, P. 2007. A biological consequence of reducing Arctic ice cover: arrival of the Pacific diatom *Neodenticula seminae* in the North Atlantic for the first time in 800,000 years. *Global Change Biology* **13**, 1910–1921.

Revelle, R. & Seuss, H.E. 1957. Carbon dioxide exchange between atmosphere and ocean and the question of an increase of atmospheric CO_2 during past decades. *Tellus* **9**, 18–27.

Reynolds-Fleming, J.V. & Luettich, R.A., Jr. 2004. Wind-driven lateral variability in a partially mixed estuary. *Estuarine, Coastal and Shelf Science* **60**, 395–407.

Riebesell, U., Schulz, K.G., Bellerby, R.G.J., Botros, M., Fritsche, P., Meyerhöfer, M., Neill, C., Nondal, G., Oschlies, A., Wohlers, J. & Zöllner, E. 2007. Enhanced biological carbon consumption in a high CO_2 ocean. *Nature* **450**, 545–548.

Riebesell, U., Zondervan, I., Rost, B., Tortell, P.D., Zeebe, R.E. & Morel, F.M.M. 2000. Reduced calcification of marine plankton in response to increased atmospheric CO_2. *Nature* **407**, 364–367.

Ries, J., Cohen, A. & McCorkle, D. 2009. Marine calcifiers exhibit mixed responses to CO_2-induced ocean acidification. *Geology* **37**(12), 1131–1134.

Ries, J.B. 2010. Review: geological and experimental evidence for secular variation in seawater Mg/Ca (calcite-aragonite seas) and its effects on marine biological calcification. *Biogeosciences* **7**, 2795–2849.

Rind, D., Perlwitz, J., Lonergan, P. & Lerner, J. 2005. AO/NAO response to climate change: 2. Relative importance of low- and high-latitude temperature changes. *Journal of Geophysical Research* **110**, D12108.

Robinson, R.A., Learmouth, J.A., Hutson, A.M., Macleod, C.D., Sparks, T.H., Leech, D.I., Pierce, G.J., Rehfisch, M.M. & Crick, H.Q.P. 2005. Climate change and migratory species. BTO Research Report 414. Thetford, Norfolk, UK: British Trust for Ornithology, The Nunnery.

Robinson, S. & the Gloucester Community Panel. 2003. A study of Gloucester's commercial fishing infrastructure. Community Panels Project. Online. http://seagrant.mit.edu/cmss/comm_mtgs/commmtgs.html (accessed 17 January 2012).

Robinson, S. & the Gloucester Community Panel. 2005. Commercial fishing industry needs on Gloucester Harbor, now and in the future. Community Panels Project. Online. http://seagrant.mit.edu/cmss/comm_mtgs/commmtgs.html (accessed 17 January 2012).

Roff, D.A. 1992. *Evolution of Life Histories: Theory and Analysis*. New York, NY, USA: Chapman and Hall.

Roleda, M.Y., Morris, J.N., McGraw, C.M. & Hurd, C.L. 2012. Ocean acidification and seaweed reproduction: increased CO_2 ameliorates the negative effect of lowered pH on meiospore germination in the giant kelp *Macrocystis pyrifera* (Laminariales, Phaeophyceae). *Global Change Biology* **18**, 854–864.

Rotstein, D.S., Burdett, L.G., McLellan, W., Schwacke, L., Rowles, T., Terio, K.A., Schultz, S. & Pabst, A. 2009. Lobomycosis in offshore bottlenose dolphins (*Tursiops truncatus*), North Carolina. *Emerging Infectious Diseases* **15**, 588–590.

Ruhl, J.B. 2010. Climate change adaptation and the structural transformation of environmental law. *Environmental Law* **40**, 363–431.

Ruiz, G.M., Fofonoff, P., Hines, A.H. & Grosholz, E.D. 1999. Non-indigenous species as stressors in estuarine and marine communities: assessing invasion impacts and interactions. *Limnology and Oceanography* **44**, 950–972.

Ruiz, G.M., Fofonoff, P.W., Carlton, J.T., Wonham, M.J. & Hines, A.H. 2000. Invasion of coastal marine communities in North America: apparent patterns, processes, and biases. *Annual Review of Ecology and Systematics* **31**, 481–531.

Runge, J.A., Kovach, A.I., Churchill, J.H., Kerr, L.A., Morrison, J.R., Beardsley, R.C., Berlinsky, D.L., Chen, C., Cadrin, S.X., Davis, C.S., Ford, K.H., Grabowski, J.H., Howell, W.H., Ji, R., Jones, R.J., Pershing, A.J., Record, N.R., Thomas, A.C., Sherwood, G.D., Tallack, S.M.L. & Townsend, D.W. 2010. Understanding climate impacts on recruitment and spatial dynamics of Atlantic cod in the Gulf of Maine: integration of observations and modeling. *Progress in Oceanography* **87**, 251–263.

Russell, B.D., Harley, C.D.G., Wernberg, T., Mieszkowska, N., Widdicombe, S., Hall-Spencer, J.M. & Connell, S.D. 2011. Predicting ecosystem shifts requires new approaches that integrate the effects of climate change across entire systems. *Biology Letters* doi: 10.1098/rsbl.2011.0779.

Sabine, C.L., Feely, R.A., Gruber, N., Key, R.M., Lee, K., Bullister, J.L., Wanninkhof, R., Wong, C.S., Wallace, D.W.R., Tilbrook, B., Millero, F.J., Peng, T.S., Kozyr, A., Ono, T. & Rios, A.F. 2004. The oceanic sink for anthropogenic CO_2. *Science* **305**, 367–371.

Sagarin, R.D. & Gaines, S.D. 2002. The 'abundant centre' distribution: to what extent is it a biogeographical rule? *Ecology Letters* **5**, 137–147.

St. Martin, K. & Hall-Arber, M. 2008. The missing layer: geo-technologies, communities, and implications for marine spatial planning. *Marine Policy* **32**, 779–786.

Sala, E. & Knowlton, N. 2006. Global marine biodiversity trends. *Annual Review of Environment and Resources* **31**, 93–122.

Salisbury, J., Green, M., Hunt, C. & Campbell, J. 2008. Coastal acidification by rivers: a new threat to shellfish? *Eos Transactions of the American Geophysical Union* **89**, 513.

Sanford, E. 1999. Regulation of keystone predation by small changes in ocean temperature. *Science* **283**, 2095–2097.

Sanford, E. & Kelly, M.W. 2011. Local adaptation in marine invertebrates. *Annual Review of Marine Science* **3**, 509–535.

Sarch, M.T. & Allison, E.H. 2000. Fluctuating fisheries in Africa's inland waters: well adapted livelihoods, maladapted management. In *Proceedings of the 10th International Conference of the Institute of Fisheries Economics and Trade, Corvallis, Oregon, July 9–14, 2000*. R.J. Johnston & A.L. Shriver (eds). Corvallis, Oregon: International Institute of Fisheries Economics and Trade (IIFET), 1-11.

Sarmiento, J.L., Gruber, N., Brzezinski, M.A. & Dunne, J.P. 2004. High-latitude controls of thermocline nutrients and low latitude biological productivity. *Nature*. **427**, 56–60.

Sartoris, F.J., Bock, C., Serendero, I., Lannig, G. & Pörtner, H.O. 2003. Temperature-dependent changes in energy metabolism, intracellular pH and blood oxygen tension in the Atlantic cod. *Journal of Fish Biology* **62**, 1239–1253.

Scallan, E., Hoekstra, R., Angulo, F., Tauxe, R., Widdowson, M.A., Roy, S., Jones, J. & Griffin, P. 2011. Foodborne illness acquired in the United States—major pathogens. *Emerging Infectious Diseases* **17**, 7–15.

Scheffer, M., Bascompte, J., Brock, W.A., Brovkin, V., Carpenter, S.R., Dakos, V., Held, H., van Nes, E.H., Rietkerk, M. & Sugihara, G. 2009. Early-warning signals for critical transitions. *Nature* **461**, 53–59.

Scheffer, M. & Carpenter, S.R. 2003. Catastrophic regime shifts in ecosystems: linking theory to observation. *Trends in Ecology and Evolution* **18**, 648–656.

Scheffer, M., Carpenter, S., Foley, J., Folke, C. & Walker, B. 2001. Catastrophic shifts in ecosystems. *Nature* **413**, 591–596.

Schewe, J., Levermann, A. & Meinshausen, M. 2010 Climate change under a scenario near 1.5C of global warming: monsoon intensification, ocean warming and steric sea level rise. *Earth System Dynamics Discussions*, **1**, 297–324.

Schiedek, D., Sundelin, B., Readman, J.W. & Macdonald, R.W. 2007. Interactions between climate change and contaminants. *Marine Pollution Bulletin* **54**, 1845–1856.

Schiel, D.R., Steinbeck, J.R. & Foster, M.S. 2004. Ten years of induced ocean warming causes comprehensive changes in marine benthic communities. *Ecology* **85**, 1833–1839.

Schindler, D.E., Hilborn, R., Chasco, B., Boatright, C.P., Quinn, T.P., Rogers, L.A. & Webster, M.S. 2010. Population diversity and the portfolio effect in an exploited species. *Nature* **465**, 609–613.

Schmidt, P.S., Serrao, E.A., Pearson, G.A., Riginos, C., Rawson, P.D., Hilbish, T.J., Brawley, S.H., Trussell, G.C., Carrington, E., Wethey, D.S., Grahame, J.W., Bonhomme, F. & Rand, D.M. 2008. Ecological genetics in the North Atlantic: environmental gradients and adaptation at specific loci. *Ecology* **89**, S91–S107.

Schmittner, A. 2005. Decline of the marine ecosystem caused by a reduction in the Atlantic overturning circulation. *Nature* **434**, 628–633.

Schneider, K.R., Van Thiel, L.E. & Helmuth, B. 2010. Interactive effects of food availability and aerial body temperature on the survival of two intertidal *Mytilus* species. *Journal of Thermal Biology* **35**, 161–166.

Scott, D., McBoyle, G. & Schwartzentruber, M. 2004. Climate change and the distribution of climatic resources for tourism in North America. *Climate Research* **27**, 105–117.

Screen, J.A. & Simmonds, I. 2010. The central role of diminishing sea ice in recent Arctic temperature amplification. *Nature* **464**, 1334–1337.

Seager, R., Ting, M., Held, I., Kushnir, Y., Lu, J., Vecchi, G., Huang, H., Harnik, N., Leetmaa, A., Lau, N., Li, C., Velez, J. & Naik, N. 2007. Model projections of an imminent transition to a more arid climate in southwestern North America. *Science* **316**, 1181–1184.

Segner, H. 2011. Moving beyond a descriptive aquatic toxicology: the value of biological process and trait information. *Aquatic Toxicology* **105**, 50–55.

Selkoe, K.A., Halpern, B.S., Ebert, C.M., Franklin, E.C., Selig, E.R., Casey, K.S., Bruno, J. & Toonen, R.J. 2009. A map of human impacts to a 'pristine' coral reef ecosystem, the Papahānaumokuākea Marine National Monument. *Coral Reefs* **28**, 635–650.

Selkoe, K.A., Halpern, B.S. & Toonen, R.J. 2008. Evaluating anthropogenic threats to the northwestern Hawaiian Islands. *Aquatic Conservation* **18**, 1149–1165.

Semenza, J., McCullough, J., Flanders, W. & McGeehin, M. 1999. Excess hospital admissions during the July 1995 heat wave in Chicago. *American Journal of Preventive Medicine* **16**, 269–277.

Serreze, M. & Barry, R. 2011. Processes and impacts of Arctic amplification. *Global and Planetary Change* **77**, 85–96.

Serreze, M. & Francis, J. 2006. The Arctic amplification debate. *Climatic Change* **76**, 241–264.

Shillinger, G., Palacios, D.L., Bailey, H. & Bograd, S. 2008. Persistent leatherback turtle migrations present opportunities for conservation. *PLoS Biology* **6**, e171.

Sigler, M.F., Renner, M., Danielson, S.L., Eisner, L.B., Lauth, R.R., Kuletz, K.J., Logerwell, E.A. & Hunt, G.L., Jr. 2011. Fluxes, fins, and feathers: relationships among the Bering, Chukchi, and Beaufort Seas in a time of climate change. *Oceanography* **24**, 250–265.

Silverman, J., Lazar, B., Cao, L., Caldiera, K. & Erez, J. 2009. Coral reefs may start dissolving when atmospheric CO_2 doubles. *Geophysical Research Letters* **36**, L05606.

Simmonds, M. & Eliott, W.J. 2009. Climate change and cetaceans: concerns and recent developments. *Journal of the Marine Biological Association of the UK* **89**, 203–210.

Simmonds, M.P. & Isaac, S.J. 2007 The impacts of climate change on marine mammals: early signs of significant problems. *Oryx* **41**, 19–26.

Skirrow, G. & Whitfield, M. 1975. The effect of increases in the atmospheric carbon dioxide content on the carbonate ion concentration of surface ocean water at 25°C. *Limnology and Oceanography* **20**, 103–108.

Slenning, B.D. 2010. Global climate change and implications for disease emergence. *Veterinary Pathology Online* **47**, 28–33.

Smith, C.R., Grange, L.J., Honig, D.L., Naudts, L., Huber, B., Guidi, L. & Domack, E. 2011. A large population of king crabs in Palmer Deep on the west Antarctic Peninsula shelf and potential invasive impacts. *Proceedings of the Royal Society B* **279**, 1017–1026.

Smith, J.R., Fong, P. & Ambrose, R.F. 2006. Dramatic declines in mussel bed community diversity. *Ecology* **87**, 1153–1161.

Soden, B.J. & Held, I.M. 2006. An assessment of climate feedbacks in coupled ocean atmosphere models. *Journal of Climate* **19**, 3354–3360.

Sokolova, I.M. & Lannig, G. 2008. Interactive effects of metal pollution and temperature on metabolism in aquatic ectotherms: implications of global climate change. *Climate Research* **37**, 181–201.

Somero, G.N. 2011. Comparative physiology: a 'crystal ball' for predicting consequences of global change. *American Journal of Physiology—Regulatory, Integrative and Comparative Physiology* **301**, R1–R14.

Sorte, C.J.B., Jones, S.J. & Miller, L.P. 2011. Geographic variation in temperature tolerance as an indicator of potential population responses to climate change. *Journal of Experimental Marine Biology and Ecology* **400**, 209–217.

Sorte, C.J.B., Williams, S.L. & Carlton, J.T. 2010a. Marine range shifts and species introductions: comparative spread rates and community impacts. *Global Ecology and Biogeography* **19**, 303–316.

Sorte, C.J.B., Williams, S.L. & Zerebecki, R.A. 2010b. Ocean warming increases threat of invasive species in a marine fouling community. *Ecology* **91**, 2198–2204.

Southward, A.J., Langmead, O., Hardman-Mountford, N.J., Aitken, J., Boalch, G.T., Dando, P.R., Genner, M.J., Joint, I., Kendall, M.A., Halliday, N.C., Harris, R.P., Leaper, R., Mieszkowska, N., Pingree, R.D., Richardson, A.J., Sims, D.W., Smith, T., Waine, A.W. & Hawkins, S.J. 2005. Long-term oceanographic and ecological research in the western English Channel. *Advances in Marine Biology* **47**, 1–105.

Spalding, M.D., Kainuma, M. & Collins, L. 2010. *World Atlas of Mangroves*. London: Earthscan.

Springer, A.M., Byrd, G.V. & Iverson, S.J. 2007. Hot oceanography: planktivorous seabirds reveal ecosystem responses to warming of the Bering Sea. *Marine Ecology Progress Series* **352**, 289–297.

Stachowicz, J., Bruno, J. & Duffy, J. 2007. Understanding the effects of marine biodiversity on communities and ecosystems. *Annual Review of Ecology, Evolution, and Systematics* **38**, 739–766.

Stachowicz, J.J., Fried, H., Osman, R.W. & Whitlatch, R.B. 2002a. Biodiversity, invasion resistance, and marine ecosystem function: reconciling pattern and process. *Ecology* **83**, 2575–2590.

Stachowicz, J.J., Terwin, J.R., Whitlatch, R.B. & Osman, R.W. 2002b. Linking climate change and biological invasions: ocean warming facilitates nonindigenous species invasions. *Proceedings of the National Academy of Sciences USA* **99**, 15497–15500.

Stafford, K.M., Moore, S.E. & Spillane, M. 2007. Gray whale calls recorded near Barrow, Alaska, throughout the winter of 2003–04. *Arctic* **60**, 167–172.

State of Hawai'i Department of Business, Economic Development and Tourism (DBEDT). 2008. Visitor statistics. Online. http://hawaii.gov/dbedt/info/visitor-stats/tourism/ (accessed 17 March 2013).

Stearns, S.C. 1992. *The Evolution of Life Histories*. Oxford, UK: Oxford University Press.

Stein, B.A., Staudt, A., Cross, M.S., Dubois, N., Enquist, C., Griffis, R., Hansen, L., Hellman, J., Lawler, J., Nelson, E., & Pairis, A. 2012. Adaptation to impacts of climate change on biodiversity, ecosystems, and ecosystem services. in Impacts of Climate Change on Biodiversity, Ecosystems, and Ecosystem Services: Technical Input to the 2013 National Climate Assessment. Cooperative Report to the 2013 National Climate Assessment. 296 p.

Steinacher, M., Joos, F., Frölicher, T.L., Bopp, L., Cadule, P., Cocco, V., Doney, S.C., Gehlen, M., Lindsay, K., Moore, J.K., Schneider, B. & Segschneider, J. 2010. Projected 21st century decrease in marine productivity: a multi-model analysis. *Biogeosciences* **7**, 979–1005.

Steinacher, M., Joos, F., Frölicher, T.L., Plattner, G.K. & Doney, S.C. 2009. Imminent ocean acidification in the Arctic projected with the NCAR global coupled carbon cycle-climate model. *Biogeosciences* **6**, 515–533.

Steinback, S., Wallmo, K. & Clay, P.M. 2007. Assessing subsistence on a regional scale for the Northeast U.S. Presented at the American Anthropological Association annual meeting, November, Washington, DC.

Steinback, S., Wallmo, K. & Clay, P.M. 2009. Saltwater sport fishing for food or income in the northeastern U.S.: statistical estimates and policy implications. *Marine Policy* **33**, 49–57.

Stenseth, N.C., Mysterud, A., Ottersen, G., Hurrell, J.W., Chan, K.-S. & Lima, M. 2002. Ecological effects of climate fluctuations. *Science* **297**, 1292–1296.

Stephens, G.L., Vane, D.G., Tanelli, S., Im, E., Durden, S., Rokey, M., Reinke, D., Partain, P., Mace, G.G., Austin, R., L'Ecuyer, T., Haynes, J., Lebsock, M., Suzuki, K., Waliser, D., Wu, D., Kay, J., Gettelman, A., Wang, Z. & Marchands, R. 2008. CloudSat mission: performance and early science after the first year of operation. *Journal of Geophysical Research* **113**, D00A18.

Stewart, E.J., Howell, S.E.L., Draper, D., Yackel, J. & Tivy, A. 2007. Sea ice in Canada's Arctic: implications for cruise tourism. *Arctic* **60**, 370–380.

Stirling, I., Lunn, N.J. & Iacozza, J. 1999. Long-term trends in the population ecology of polar bears in western Hudson Bay in relation to climate change. *Arctic* **52**, 294–306.

Stirling, I. & Parkinson, C.L. 2006. Possible effects of climate warming on selected populations of polar bears (*Ursus maritimus*) in the Canadian Arctic. *Arctic* **59**, 261–275.

Stock, C.A., Alexander, M.A., Bond, N.A., Brander, K.M., Cheung, W.W.L., Curchitser, E.N., Delworth, T.L., Dunne, J.P., Griffies, S.M., Haltuch, M.A., Hare, J.A., Hollowed, A.B., Lehodey, P., Levin, S.A., Link, J.S., Rose, K.A., Rykaczewski, R.R., Sarmiento, J.L., Stouffer, R.J., Schwing, F.B., Vecchi, G.A. & Werner, F.E. 2011. On the use of IPCC-class models to assess the impact of climate on living marine resources. *Progress in Oceanography* **88**, 1–27.

Storck, C.H., Postic, D., Lamaury, I. & Perez, J.M. 2008. Changes in epidemiology of leptospirosis in 2003–2004, a two El Niño Southern Oscillation period, Guadeloupe archipelago, French West Indies. *Epidemiology and Infection* **136**, 1407–1415.

Stouffer, R.J., Yin, J., Gregory, J.M., Dixon, K.W., Spelman, M.J., Hurlin, W., Weaver, A.J., Eby, M., Flato, G.M., Hasumi, H., Hu, A., Jungclaus, J.H., Kamenkovich, I.V., Levermann, A., Motoya, M., Murakami, S., Nawrath, S., Oka, A., Peltier, W.R., Robitaille, D.Y., Sokolov, A., Vettoretta, G. & Weber, S.L. 2006. Investigating the causes of the response of the thermohaline circulation to past and future climate changes. *Journal of Climate* **19**, 1365–1387.

Stram, D.L. & Evans, D.C.K. 2009. Fishery management responses to climate change in the North Pacific. *ICES Journal of Marine Science* **66**, 1633–1639.

Stroeve, J., Serreze, M., Drobot, S., Gearheard, S., Holland, M., Maslanik, J., Meier, W. & Scambos, T. 2008. Arctic Sea ice extent plummets in 2007. *Eos Transactions of the American Geophysical Union* **89**, 13.

Stroeve, J.C., Serreze, M.C., Holland, M.M., Kay, J.E., Meier, W. & Barrett, A.P. 2011. The Arctic's rapidly shrinking sea ice cover: a research synthesis. *Climatic Change* doi:10.1007/s10584-011-0101-1.

Stumm, W. & Morgan, J.J. (1970). *Aquatic Chemistry*. New York: Wiley.

Sturges, W. 1974. Sea level slope along continental boundaries. *Journal of Geophysical Research* **79**, 825–830.

Su, N.J., Sun, C.L., Punt, A.E., Yeh, S.Z. & DiNardo, G. 2011. Modeling the impacts of environmental variation on the distribution of blue marlin, *Makaira nigricans*, in the Pacific Ocean. *ICES Journal of Marine Science* **68**, 1072–1080.

Sumaila, U.R., Cheung, W., Lam, V.W.Y., Pauly, D. & Herrick, S. 2011. Climate change impacts on the biophysics and economics of world fisheries. *Nature Climate Change* **1**, 449–456.

Sun, J., Hutchins, D.A., Feng, Y., Seubert, E.L., Caron, D.A. & Fu, F.X. 2011. Effects of changing pCO2 and phosphate availability on domoic acid production and physiology of the marine harmful bloom diatom Pseudo-nitzschia multiseries. *Limnology and Oceanography* **56**, 829–840.

Sussman, E., Major, D.C., Deming, R., Esterman, P.R., Fadil, A., Fisher, A., Fucci, F., Gordon, R., Harris, C., Healy, J.K., Howe, C., Robb, K. & Smith, J. 2010. Climate change adaptation: fostering progress through law and regulation. *New York University Environmental Law Journal* **18**, 55–155.

Sydeman, W. & Thompson, S.A. 2010. The California Current integrated ecosystem assessment (IEA) module II: trends and variability in climate-ecosystem state. Final Report to NOAA/NMFS/Environmental Research Division. Petaluma, California: Farallon Institute for Advanced Ecosystem Research.

Sydeman, W.J., Mills, K., Santora, J. & Thompson, S.A. 2009. Seabirds and climate in the California Current—a synthesis of change. CalCOFI Report Vol. 50. Petaluma, California: Farallon Institute for Advanced Ecosystem Research.

Tacon, A.G.J., Hasan, M.R. & Metian, M. 2011. Demand and supply of feed ingredients for farmed fish and crustaceans: trends and prospects. FAO Fisheries and Aquaculture Technical Paper No. 564. Rome, Italy: Food and Agriculture Organization of the United Nations.

Takasuka, A., Oozeki, Y. & Aoki, I. 2007. Optimal growth temperature hypothesis: why do anchovy flourish and sardine collapse or vice versa under the same ocean regime? *Canadian Journal of Fisheries and Aquatic Sciences* **64**, 768–776.

Talmage, S.C. & Gobler, C.J. 2011. Effects of elevated temperature and carbon dioxide on the growth and survival of larvae and juveniles of three species of Northwest Atlantic bivalves. *PLoS one* **6**, e26941.

Taylor, D.I., Nixon, S.W., Granger, S.L., Buckley, B.A., McMahon, J.P. & Lin, H.J. 1995. Responses of coastal lagoon plant communities to different forms of nutrient enrichment: a mesocosm experiment. *Aquatic Botany* **52**, 19–34.

Teichberg, M., Fox, S.E., Aguila, C., Olsen, Y.S. & Valiela, I. 2008. Macroalgal responses to experimental nutrient enrichment in shallow coastal waters: growth, internal nutrient pools, and isotopic signatures. *Marine Ecology Progress Series* **368**, 117–126.

Tester, P.A., Feldman, R.L., Nau, A.W., Kibler, S.R. & Litaker, W.R. 2010. Ciguatera fish poisoning and sea surface temperatures in the Caribbean Sea and the West Indies. *Toxicon* **56**, 698–710.

Thomas, P.O. & Laidre, K.L. 2011. Biodiversity—cetaceans and pinnipeds (whales and seals). Online. http://www.arctic.noaa.gov/report11/biodiv_whales_walrus.html (accessed 17 March 2013).

Thornber, C.S., DiMilla, P., Nixon, S.W. & McKinney, R.A. 2008. Natural and anthropogenic nitrogen uptake by bloom-forming macroalgae. *Marine Pollution Bulletin* **56**, 261–269.

Tomanek, L. 2011. Environmental proteomics: changes in the proteome of marine organisms in response to environmental stress, pollutants, infection, symbiosis, and development. *Annual Review of Marine Sciences* **3**, 373–399.

Tomanek, L. & Zuzow, M.J. 2010. The proteomic response of the mussel congeners *Mytilus galloprovincialis* and *M. trossulus* to acute heat stress: implications for thermal tolerance limits and metabolic costs of thermal stress. *The Journal of Experimental Biology* **213**, 3559–3574.

Torrissen, O., Olsen, R.E., Toresen, R., Hemre, G.I., Tacon, A.G.J., Asche, F., Hardy, R.W. & Lall, S. 2011. Atlantic salmon (*Salmo salar*): the 'super-chicken' of the sea? *Reviews in Fisheries Science* **19**, 257–278.

Trenberth, K.E. 2011. Changes in precipitation with climate change. *Climate Research* **47**, 123–138.

Trenberth, K.E., Dai, A., Rasmussen, R.M. & Parsons, D.B. 2003 The changing character of precipitation. *Bulletin of the American Meteorological Society* **84**, 1205–1217.

Trenberth, K.E. & Fasullo, J. 2007. Water and energy budgets of hurricanes and implications for climate change. *Journal of Geophysical Research* **112**, D23107.

Trenberth, K.E., Fasullo, J. & Kiehl, J. 2009. Earth's global energy budget. *Bulletin of the American Meteorological Society* **90**, 311–323.

Trenberth, K.E., Jones, P.D., Ambenje, P., Bojariu, R., Easterling, D., Klein Tank, A., Parker, D., Rahimzadeh, F., Renwick, J.A., Rusticucci, M., Soden B. & Zhai, P. 2007. Observations: surface and atmospheric climate change. In *Climate Change 2007: The Physical Science Basis. Contribution of Working Group I to the Fourth Assessment Report of the Intergovernmental Panel on Climate Change*, S. Solomon et al. (eds). Cambridge, UK: Cambridge University Press, 235–336.

Trussell, G.C. & Etter, R.J. 2001. Integrating genetic and environmental forces that shape the evolution of geographic variation in a marine snail. *Genetica* **112–113**, 321–337.

Udovydchenkov, I.A., Duda, T.F., Doney, S.C. & Lima, I.D., 2010. Modeling deep ocean shipping noise in varying acidity conditions. *The Journal of the Acoustical Society of America* **128**, EL130–EL136.

Ulbrich, U., Pinto, J., Kupfer, H., Leckebusch, G., Spangehl, T. & Reyers, M. 2008. Changing Northern Hemisphere storm tracks in an ensemble of IPCC climate change simulations. *Journal of Climate* **21**, 1669–1679.

United Nations Framework Convention on Climate Change (UNFCCC). 2011. SBSTA report on research and observation. FCCC/SBSTA/2011/L.27. Geneva, Switzerland: United Nations Office at Geneva.

United Nations World Tourism Organization (UNWTO). 2012. U.N. world tourism barometer. January 2012— Statistical Annex, Volume 10, version 20/01/12. Madrid: Publications Unit, World Tourism Organization.

U.S. Climate Change Science Program (CCSP). 2008. Preliminary review of adaptation options for climate-sensitive ecosystems and resources. In *A Report by the U.S. Climate Change Science Program and the Subcommittee on Global Change Research*, S.H. Julius et al. (eds). Washington, DC: U.S. Environmental Protection Agency.

U.S. Commission on Ocean Policy (USCOP). 2004. An ocean blueprint for the 21st century. Final report. Washington, DC.

U.S. Environmental Protection Agency (EPA). 2008. *Effects of Climate Change for Aquatic Invasive Species and Implications for Management and Research*. Washington, DC: National Center for Environmental Assessment.

U.S. Environmental Protection Agency (EPA). 2010. Memorandum on integrated reporting and listing decisions related to ocean acidification, from Denise Keehner, director, to Water Division directors, Regions 1–10. 15 November.

U.S. Food and Drug Administration (USDA). 2008. Enhanced aquaculture and seafood inspection—report to Congress. Washington, DC: U.S. Food and Drug Administration.

U.S. Government Accountability Office (GAO). 2004. Food safety: FDA's imported seafood safety program shows some progress, but further improvements are needed. Report GAO 246. Washington, DC: GAO.

U.S. Government Accountability Office (GAO). 2007. Climate change: agencies should develop guidance for addressing the effects on federal land and water resources. Report to Congressional requesters. GAO-07-863.

U.S. Travel Association. 2012. Travel industry facts: In *Advance of the President's Speech at Walt Disney World*. January 18, 2012. Washington, DC: U.S. Travel Association. Online. http://www.ustravel.org/news/press-releases/travel-industry-facts-advance-presidentpercentE2percent80percent99s-speech-walt--disney-world (accessed 18 January 2012).

Urian, A.G., Hatle, J.D. & Gilg, M.R.. 2011. Thermal constraints for range expansion of the invasive green mussel, *Perna viridis*, in the southeastern United States. *Journal of Experimental Zoology* **315**, 12–21.

Valiela, I., McClelland, J., Hauxwell, J., Behr, P.J., Hersch, D. & Foreman, K. 1997. Macroalgal blooms in shallow estuaries: controls and ecophysiological and ecosystem consequences. *Limnology and Oceanography* **42**, 1105–1118.

Van den Hove, S., Menestrel, M.L. & de Bettignies, H.C. 2002. The oil industry and climate change: strategies and ethical dilemmas. *Climate Policy* **2**, 3–18.

van Griensven, F., Chakkraband, M.L., Thienkrua, W., Pengjuntr, W., Lopes Cardozo, B., Tantipiwatanaskul, P., Mock, P.A., Ekassawin, S., Varangrat, A., Gotway, C., Sabin, M. & Tappero, J.W. 2006. Thailand post-tsunami mental health study group. *Journal of the American Medical Association* **296**, 537–548.

Van Houtan, K.S. & Halley, J.M. 2011. Long-term climate forcing in loggerhead sea turtle nesting. *PLoS one* **6**, e19043.

Vaquer-Sunyer, R. & Duarte, C.M. 2008. Thresholds of hypoxia for marine biodiversity. *Proceedings of the National Academy of Sciences USA* **105**, 15452–15457.

Vecchi, G.A. & Soden, B.J. 2007. Effect of remote sea surface temperature change on tropical cyclone potential intensity. *Nature* **450**, 1066–1070.

Vellinga, M. & Wu, P. 2004. Low-latitude fresh-water influence on centennial variability of the Atlantic thermohaline circulation. *Journal of Climate* **17**, 4498–4511.

Vermeij, G.J. & Roopnarine, P.D. 2008. The coming Arctic invasion. *Science* **321**, 780–781.

Villareal, T.A., Hanson, S., Qualia, S., Jester, E.L.E., Granade, H.R. & Dickey, R.W. 2007. Petroleum production platforms as sites for the expansion of ciguatera in the northwestern Gulf of Mexico. *Harmful Algae* **6**, 253–259.

Wagner, J.D.M., Cole, J.E., Beck, J.W., Patchett, P.J., Henderson, G.M. & Barnett, H.R. 2010. Moisture variability in the southwestern United States linked to abrupt glacial climate change. *Nature Geoscience* **3**, 110–113.

Wake, C., Burakowski, L., Lines, G., McKenzie, K. & Huntington, T. 2006. Cross border indicators of climate change over the past century: northeastern United States and Canadian Maritime Region. The Climate Change Task Force of the Gulf of Maine Council on the Marine Environment in cooperation with Environment Canada and Clean Air-Cool Planet. Online. http://www.gulfofmaine.org/council/publications/cross-border-indicators-of-climate-change.pdf (accessed 17 March 2013).

Walker, N., Leben, R.R. & Balasubramanian, S. 2005. Hurricane forced upwelling and chlorophyll a enhancement within cold core cyclones in the Gulf of Mexico. *Geophysical Research Letters* **32**, L18610.

Walsh, J.E. & Chapman, W.L. 2001. 20th-century sea-ice variations from observational data. *Annals of Glaciology* **33**, 444–448.

Walther, K., Anger, K. & Pörtner, H.O. 2010. Effects of ocean acidification and warming on larval development of the spider crab *Hyas araneus* from different latitudes (54° vs. 79° N). *Marine Ecology Progress Series* **417**, 159–170.

Walther, K., Sartoris, F.J., Bock, C. & Pörtner, H.O. 2009. Impact of anthropogenic ocean acidification on thermal tolerance of the spider crab *Hyas araneus*. *Biogeosciences* **6**, 2207–2215.

Wang, M., Overland, J.E. & Bond, N.A., 2010. Climate projections for selected large marine ecosystems. *Journal of Marine Systems* **79**, 258–266.

Ward, J.R. & Lafferty, K.D. 2004. The elusive baseline of marine disease: are diseases in ocean ecosystems increasing? *Plos Biology* **2**, 542–547.

Warzybok, P.M. & Bradley, R.W. 2010. Status of seabirds on Southeast Farallon Island during the 2010 breeding season. Unpublished report to the U.S. Fish and Wildlife Service. PRBO Contribution Number 1769. Petaluma, California: PRBO Conservation Science.

Wassmann, P. 2011. Arctic marine ecosystems an era of rapid change. *Progress in Oceanography* **90**, 1–17.

Wassmann, P., Duarte, C.M., Agustí, S. & Sejr, M.K. 2011. Footprints of climate change in the Arctic marine ecosystem. *Global Change Biology* **17**, 1235–1249.

Weinberg, J.R. 2005. Bathymetric shift in the distribution of Atlantic surfclams: response to warmer ocean temperature. *ICES Journal of Marine Science: Journal du Conseil* **62**, 1444–1453.

Weis, K., Hammond, R., Hutchinson, R. & Blackmore, C. 2011. *Vibrio* illness in Florida, 1998–2007. *Epidemiology and Infection* **139**, 591–598.

Wernberg, T., Russell, B.D., Moore, P.J., Ling, S.D., Smale, D.A., Campbell, A., Coleman, M.A., Steinberg, P.D., Kendrick, G.A. & Connell, S.D. 2011a. Impacts of climate change in a global hotspot for temperate marine biodiversity and ocean warming. *Journal of Experimental Marine Biology and Ecology* **400**, 7–16.

Wernberg, T., Thomsen, M.S., Tuya, F. & Kendrick, G.A. 2011b. Biogenic habitat structure of seaweeds change along a latitudinal gradient in ocean temperature. *Journal of Experimental Marine Biology and Ecology* **400**, 264–271.

Wernberg, T., Thomsen, M.S., Tuya, F., Kendrick, G.A., Staehr, P.A. & Toohey, B.D. 2010. Decreasing resilience of kelp beds along a latitudinal temperature gradient: potential implications for a warmer future. *Ecology Letters* **13**, 685–694.

Westlund, L., Poulain, F., Bage, H. & van Anrooy, R. 2007. *Disaster Response and Risk Management in the Fisheries Sector.* Rome: FAO.

Wethey, D.S. 2002. Biogeography, competition, and microclimate: the barnacle *Chthamalus fragilis* in New England. *Integrative and Comparative Biology* **42**, 872–880.

Wethey, D.S. & Woodin, S.A. 2008. Ecological hindcasting of biogeographic responses to climate change in the European intertidal zone. *Hydrobiologia* **606**, 139–151.

Wethey, D.S., Woodin, S.A., Hilbish, T.J., Jones, S.J., Lima, F.P. & Brannock, P.M. 2011. Response of intertidal populations to climate: effects of extreme events versus long term change. *Journal of Experimental Marine Biology and Ecology* **400**, 132–144.

White, C., Halpern, B.S. & Kappel, C.V. 2012. Ecosystem service tradeoff analysis reveals the value of marine spatial planning for multiple ocean uses. *Proceedings of the National Academy of Sciences USA* **109**, 4696–4701.

White, D.M., Gerlach, S.C., Loring, P.A. & Tidwell, A. 2007. Food and water security in a changing arctic climate. *Environmental Research Letters* **2**, 4.

Widdicombe, S. & Spicer, J.I. 2008. Predicting the impact of ocean acidification on benthic biodiversity: What can animal physiology tell us? *Journal of Experimental Marine Biology and Ecology* **366**, 187–197.

Wilcox, B.A. & Gubler, D.J. 2005. Disease ecology and the global emergence of zoonotic pathogens. *Environmental Health and Preventive Medicine* **10**, 263–272.

Williams, J. & Jackson, S. 2007. Novel climates, no-analog communities, and ecological surprises. *Frontiers in Ecology and the Environment* **5**, 475–482.

Wilson, W.J. & Ormseth, O.A. 2009. A new management plan for the Arctic waters of the United States. *Fisheries* **34**, 555–558.

Winder, M. & Schindler, D.E., 2004. Climate change uncouples trophic interactions in an aquatic ecosystem. *Ecology* **85**, 2100–2106.

Wingfield, J.C. & Sapolsky, R.M. 2003. Reproduction and resistance to stress: when and how. *Journal of Neuroendocrinology* **15**, 711–724.

Witherington, B.E. & Ehrhart, L.M. 1989. Hypothermic stunning and mortality of marine turtles in the Indian River Lagoon System, Florida. *Copeia* **1989**, 696–703.

Wolf, S.G., Snyder, M.A., Sydeman, W.J., Doak, D.F. & Croll, D.A. 2010. Predicting population consequences of ocean climate change for an ecosystem sentinel, the seabird Cassin's auklet. *Global Change Biology* **16**, 1923–1935.

Wolf, S.G., Sydeman, W.J., Hipfner, J.M., Abraham, C.L., Tershy, B.R. & Croll, D.A. 2009. Range-wide reproductive consequences of ocean climate variability for the seabird Cassin's auklet. *Ecology* **90**, 742–753.

Wood, R.A., Collins, M., Gregory, J., Harris, G. & Vellinga, M. 2006. Towards a risk assessment for shutdown of the Atlantic thermohaline circulation. In *Avoiding Dangerous Climate Change*, H.-J. Schellnhuber et al. (eds). Cambridge, UK: Cambridge University Press, 49–55.

Wood, R.A., Vellinga, M. & Thorpe, R.B. 2003. Global warming and thermohaline circulation stability. *Philosophical Transactions of the Royal Society A* **361**, 1961–1975.

Wootton, J.T., Pfister, C.A. & Forester, J.D. 2008. Dynamic patterns and ecological impacts of declining ocean pH in a high-resolution multi-year dataset. *Proceedings of the National Academy of Sciences USA* **105**, 18848–18853.

Worm, B., Barbier, E.B., Beaumont, N., Duffy, J.E., Folke, C., Halpern, B.S., Jackson, J.B.C., Lotze, H.K., Micheli, F., Palumbi, S.R., Sala, E., Selkoe, K.A., Stachowicz, J.J. & Watson, R. 2006. Impacts of biodiversity loss on ocean ecosystem services. *Science* **314**, 787–790.

Yamane, L. & Gilman, S.E. 2009. Opposite responses by an intertidal predator to increasing aquatic and aerial temperatures. *Marine Ecology Progress Series* **393**, 27–36.

Yin, J., Griffies, S.M. & Stouffer, R.J. 2010. Spatial variability of sea level rise in twenty-first century projections. *Journal of Climate* **23**, 4585–4607.

Yu, G., Schwartz, Z. & Walsh, J.E. 2009. A weather-resolving index for assessing the impact of climate change on tourism related climate resources. *Climatic Change* **95**, 551–573.

Zacherl, D., Gaines, S.D. & Lonhart, S.I. 2003. The limits to biogeographical distributions: insights from the northward range extension of the marine snail, *Kelletia kelletii* (Forbes, 1852). *Journal of Biogeography* **30**, 913–924.

Zachos, J.C., Röhl, U., Schellenberg, S.A., Sluijs, A., Hodell, D.A., Kelly, D.C., Thomas, E., Nicolo, M., Raffi, I., Lourens, L.J., McCarren, H. & Krron, D. 2005. Rapid acidification of the ocean during the Paleocene-Eocene thermal maximum. *Science* **308**, 1611–1615.

Zardi, G., Nicastro, K., McQuaid, C.D., Hancke, J. & Helmuth, B. 2011. The combination of selection and dispersal helps explain genetic structure in intertidal mussels. *Oecologia* **165**, 947–958.

Zippay, M.L. & Hofmann, G.E. 2010. Effect of pH on gene expression and thermal tolerance of early life history stages of red abalone (*Haliotis rufescens*). *Journal of Shellfish Research* **29**, 429–439.

Zuerner, R.L., Cameron, C.E., Raverty, S., Robinson, J., Colegrove, K.M., Norman, K.M., Norman, S.A., Lambourn, D., Jeffries, S., Alt, D.P. & Gulland, F. 2009. Geographical dissemination of *Leptospira interrogans* serovar Pomona during seasonal migration of California sea lions. *Veterinary Microbiology* **137**, 105–110.

Oceanography and Marine Biology: An Annual Review, 2013, **51**, 193-280
© Roger N. Hughes, David Hughes, and I. Philip Smith, Editors
Taylor & Francis

THE SHORT-BEAKED COMMON DOLPHIN (*DELPHINUS DELPHIS*) IN THE NORTH-EAST ATLANTIC: DISTRIBUTION, ECOLOGY, MANAGEMENT AND CONSERVATION STATUS

SINÉAD MURPHY[1,2], EUNICE H. PINN[3] & PAUL D. JEPSON[1]

[1]Institute of Zoology, Zoological Society of London,
Regent's Park, London, NW1 4RY, United Kingdom
[2]C-MRG, Institute of Natural Sciences, Massey University,
Private Bag 102 904, Auckland 0745, New Zealand
E-mail: sinead.murphy@ioz.ac.nz (corresponding author)
[3]Joint Nature Conservation Committee, Inverdee House,
Baxter Street, Aberdeen, AB11 9QA, United Kingdom

The common dolphin is the second most abundant cetacean species in the North-east Atlantic, with a wide-ranging distribution and is, potentially, impacted by a wide variety of pressures and threats. To assess the conservation status of common dolphins in this region, it is essential to understand population structure, key drivers of population dynamics, key resources and the effects of stressors. In recent years, a number of studies have assessed population structure, distribution and abundance, life-history parameters, dietary requirements and the effect of stressors—especially those caused by anthropogenic interactions, such as incidental capture (i.e., by-catch) and pollutants. A full review of this work is presented, with particular focus on current and potential pressures and threats. Notwithstanding the recent research, due to the lack of baseline data (i.e., prior to human influence) on abundance and pregnancy rate and on historical direct and incidental capture rates, the actual conservation status of the North-east Atlantic common dolphin population is unknown. Current assessments of conservation status of the species are therefore reliant on recent data. However, these assessments are hindered by the lack of data on contemporary incidental capture rates in some fisheries and limited sampling in other fisheries, as well as large data gaps for other stressors. In addition, the numerous potential ways in which multiple and diverse stressors can interact remain poorly understood. This chapter provides an outline of a management framework and describes methods for future evaluation of conservation status through development of indicators focusing on not only population size and distribution but also mortality and condition. Recommendations for research and conservation actions are described.

Introduction

Common dolphins (*Delphinus delphis*) are one of the most abundant cetaceans in the North-east (NE) Atlantic and are potentially impacted by a wide variety of threats and pressures. A large number of studies since the year 2000 have focused on their biology, ecology, population structure, abundance, health status, foraging behaviour, interactions with fisheries, and pollutant levels, among other aspects. No baseline data (prior to anthropogenic impacts) are available for this species in the region, making comparisons to the modern situation impossible. Current assessments of conservation status of the species therefore rely on recent data. Such assessments require information on genetic diversity, evidence of density-dependent compensatory responses in reproductive parameters, and, most importantly, trend analysis of abundance estimates—though for the last only primary (preliminary) survey data are available for most of the NE Atlantic. As these parameters respond to a change after it has occurred, most likely years later, monitoring of anthropogenic (and environmental) impacts is extremely important to limit their effects. Under European Union legislation, Member States now have a legal obligation to undertake such monitoring in their national waters. The present review encompasses a comprehensive assessment of knowledge on the common dolphin in the NE Atlantic, with the aim of providing both a current evaluation of conservation status and valuable information for the future development of a conservation management plan in these waters.

Species identification

The common dolphin was first identified by Artedi in 1738 and later described by Linnaeus in 1758. Much confusion has arisen about whether the 'common dolphin' comprises one or more species due to the cosmopolitan distribution and high variability in morphological characters and pigmentation patterns of otherwise-similar forms. Almost two dozen nominal species have been described (Hershkovitz 1966, Heyning & Perrin 1994).

During the 1990s, studies focused on clarifying taxonomic status and proposed that only two distinct species of common dolphin existed: the short-beaked common dolphin *Delphinus delphis* Linnaeus, 1758 and the long-beaked common dolphin *D. capensis* Gray, 1828 (Heyning & Perrin 1994, Rosel et al. 1994). Contemporary taxonomic classifications further recognize an endemic subspecies of the short-beaked form, *D. delphis ponticus* (Barabash-Nikiforov, 1935), which is restricted to the Black Sea, and two endemic subspecies of the long-beaked form found in the Indo-Pacific region: *D. capensis capensis* Gray, 1828, and the extremely long-beaked *D. c. tropicalis* (van Bree, 1971) (Amaha 1994, Heyning & Perrin 1994, Rosel et al. 1994, Jefferson & Van Waerebeek 2002, Natoli et al. 2006, Perrin 2009).

Delphinus delphis and *D. capensis* were differentiated on the basis of an assessment of morphological data (including measurements of skeletal characters, such as overall body size, length of the rostrum, rostrum length/zygomatic width ratio [RL/ZGW], and tooth counts) and colouration patterns in two sympatric populations inhabiting coastal waters off California, United States (Heyning & Perrin 1994). An assessment of samples and data from outside this region led the authors to propose that *D. delphis* and *D. capensis* existed globally as two separate species. This view was further supported by a parallel genetic analysis of tissue samples from dolphins inhabiting Californian waters and from *D. delphis* in the eastern tropical Pacific (ETP) and Black Sea. No shared mitochondrial DNA (mtDNA) haplotypes or cytochrome *b* sequences were observed between the two morphotypes, and both gene regions exhibited nucleotide frequency differences and fixed nucleotide substitutions between the two morphotypes (Rosel et al. 1994). The estimated genetic divergence in the mitochondrial DNA control region between the two forms was 1.1% (Rosel et al. 1994).

However, the classification of individuals into the two species was not as clear cut as originally perceived, and questions have arisen over the use of morphology-based taxonomy in common dolphins. Additional morphological studies have shown large variability in body and skull size and RL/ZGW ratio in *D. delphis* inhabiting the North Atlantic and waters around New Zealand and southern Australia compared with *D. delphis* off the coast of California (Bell et al. 2002, Murphy et al. 2006, Westgate 2007, Jordan 2012). Studies using mtDNA identified that *D. delphis* and *D. capensis* do not show reciprocal monophyly (LeDuc et al. 1999, Amaral et al. 2007), and an assessment of nuclear divergence using amplified fragment length polymorphism markers suggested that *D. delphis* and *D. capensis* diverged only recently (Kingston & Rosel 2004). Overall, the short-beaked form, despite having high morphological variability, exhibits low genetic differentiation between populations, with evidence of gene flow across oceans, reflecting high mobility and a fluid social structure in this species (Natoli et al. 2006; assessed nine microsatellite and 369 base pairs [bp] of the mtDNA control region). The opposite is true for the long-beaked form: populations are highly differentiated, suggesting that separate populations may have evolved from independent founder events, then converged on the same morphotype (Natoli et al. 2006). A more recent international collaborative genetics study, which analysed sequences of the mitochondrial DNA cytochrome *b* gene, further revealed that the distribution of mitochondrial lineages does not correspond to the geographical distribution of the long-beaked morphotype, thus suggesting some ambiguity in the phylogenetic relationships and taxonomy within this species (Amaral et al. 2009, 2012a,b). Further analysis is required to clarify the species-level taxonomy of common dolphins (International Whaling Commission [IWC] 2009).

North-east Atlantic Ocean

Only the short-beaked form has been recorded in the North Atlantic (Murphy et al. 2006, Westgate 2007). Based on tooth counts, absolute length of rostrum and RL/GZW ratio, Murphy et al. (2006) proposed that common dolphins in the NE Atlantic are most similar to *Delphinus delphis* described by Heyning and Perrin (1994) but are larger than *D. delphis* in waters off California. The ranges of RL/GZW ratio defined by Heyning and Perrin (1994) for *D. delphis* and *D. capensis* were 1.21–1.47 and 1.55–1.77, respectively. Skulls of mature common dolphins from the NE Atlantic had RL/GZW ratios in the range of 1.31 to 1.57 (mean = 1.41, *n* = 111), with 95% of ratio values less than 1.52 (Murphy et al. 2006). Upper alveolar count and rostrum lengths had ranges of 41–56 and 233.6–299.8 mm, respectively, compared with 42–54 and 218–275 mm for California *D. delphis* (Heyning & Perrin 1994). In Mauritanian waters, morphological analysis indicates that both short- and long-beaked forms were present (Van Waerebeek 1997, Pinela et al. 2011), although stable isotope analysis of diet indicated that skull variability in this region (e.g., RL/GZW ratio range of 1.27–1.76) may be due to niche segregation rather than speciation (Pinela et al. 2011).

Species identification in the NE Atlantic has also been confirmed by genetic analysis. Using mtDNA, Natoli et al. (2006) identified only *D. delphis* in waters from the Canary Islands to Scotland (*n* = 100), with the nearest *D. capensis* population reported off Mauritania (*n* = 6), which was highly differentiated from long-beaked dolphins off South Africa. Amaral et al. (2007) also identified only *D. delphis* in a sample of 69 individuals stranded in Portugal, northern Spain and Scotland; the identification was based on analysis of two mitochondrial regions (control region and cytochrome *b* gene). Interestingly, that study documented a group of highly divergent individuals, 'Clade X' (five females throughout the sampled area). Genetic divergence between Clade X and *D. delphis* in the NE Atlantic was 1.59%, considerably higher than the divergence reported by Rosel et al. (1994) for *D. delphis* and *D. capensis*. Genetic divergence between Clade X and *D. capensis* was also high at 1.76% (Amaral et al. 2007). In a broader study that included populations from the Atlantic, Indian and Pacific Oceans, it became clear that Clade X was present not only in the North Atlantic short-beaked common dolphin, but also in *D. capensis tropicalis* of the Indian Ocean (Amaral et al. 2012b). The most likely explanation for such a divergent clade is a scenario of variance and secondary contact during the Pleistocene, a period when the Atlantic and Indo-Pacific basins were intermittently isolated. This explanation has been proposed for phylogeographic patterns in other large migratory marine animals (Amaral et al. 2012b).

Individuals resembling long-beaked common dolphins in colour pattern and rostrum length have been sighted off the Azores (Quérouil et al. 2010) but have not been genetically characterized.

Population structure in the North Atlantic Ocean

Common dolphins are widely distributed throughout the North Atlantic, occurring in many juris-dictions enforcing different environmental legislation. Knowledge of population status and range is essential for effective conservation and management of this species in the region, and this has provided the impetus for numerous population studies in recent years.

Geometric analysis of cranial variables revealed significant differences between the NE Atlantic and North-west (NW) Atlantic for both male and female short-beaked common dolphins inhabit-ing continental shelf and slope waters ($n = 149$ males; $n = 96$ females) (Westgate 2007). Rostral width dimensions were important discriminating variables, suggesting differing feeding strategies between the populations (Westgate 2007).

Using genetic data, Natoli et al. (2006) also found significant but low differentiation between *Delphinus delphis* from the NW ($n = 13$) and NE Atlantic ($n = 119$). Genetic differentiation was more marked in maternally inherited mtDNA markers than nuclear markers, which did not always indicate significant differentiation, suggesting a greater dispersal of males. However, analysis of mtDNA data using the MIGRATE program (Beerli 2012) indicated a possible bias in the long-term direction of migration of females in the North Atlantic, from west to east (Natoli et al. 2006). Following this study, Mirimin et al. (2009b) assessed genetic variability at the mtDNA control region (360 bp) and 14 microsatellite loci using a larger sample size of stranded common dolphins, and individuals incidentally caught in fishing gear, from continental shelf and slope waters of the NE ($n = 205$) and NW ($n = 219$) Atlantic. Results confirmed those of Natoli et al. (2006), and sig-nificant genetic differentiation was observed, which was more pronounced in mitochondrial than nuclear markers, suggesting the existence of at least two genetically distinct populations (Mirimin et al. 2009b). However, the low levels of genetic differentiation measured in both studies may arise from a recent population split or a high level of gene flow in the North Atlantic (Mirimin et al. 2009b). Natoli et al. (2006), Amaral et al. (2007) and Mirimin et al. (2009a) all reported highly significant negative values of a genetic index based on the distribution of alleles or haplotypes, Fu's F_S, indicating population expansion within the NE Atlantic. This index quantifies departures from the pattern of DNA polymorphism expected under a neutral model of evolution, caused, for example, by rapid population increase. A separate common dolphin population is recognized within the Mediterranean Sea based on mtDNA markers (Natoli et al. 2008).

Even in light of these recent studies, a full assessment of population structure in the North Atlantic has been hindered by sampling limitations as all genetic samples were obtained from con-tinental shelf and contiguous oceanic waters. Further investigations into whether common dolphins from the NE and NW Atlantic belong to the same population or to two separate, but highly con-nected, populations requires sampling from the entire range of this species in the North Atlantic (International Council for the Exploration of the Sea [ICES] Working Group on Marine Mammal Ecology [WGMME] 2009, Murphy et al. 2009a).

North-east Atlantic Ocean

There appears to be a slight latitudinal cline in size of male common dolphins in the NE Atlantic, with males in higher latitudes slightly larger in total body length, absolute skull width, mandible length and depth compared to individuals off the NW coast of Spain (Murphy et al. 2006). Analysis of cranial morphometrics has also revealed evidence of population differentiation, with female *Delphinus delphis* off Portugal showing segregation from more northerly sampled areas (Murphy et al. 2006). Although samples of mature individuals from southern regions were small in the study by Murphy et al. (2006), there were indications that females off the Portuguese coast (the most southerly sampled region) may be mixing with individuals in the Mediterranean Sea population or individuals from waters further south of the sampled region.

Morphological traits are influenced heavily by environmental factors and should be interpreted differently from genetic traits. However, the results of the cranial morphometric study were partially supported by genetic analysis as a directional movement of females from the western Mediterranean Sea (Alborán Sea) into the North Atlantic was reported (Natoli et al. 2008). Based on genetic analysis of nine microsatellites, no significant genetic differentiation was detected in common dolphins sampled in the Alborán Sea ($n = 34$) and the contiguous Atlantic, though mtDNA analysis (426 bp) indicated significant differentiation (Galicia, $n = 30$; Portugal, $n = 17$).

Viricel (2006) did not observe any significant genetic variation using mtDNA (933 bp) and (7–11) microsatellite nuclear loci markers in samples obtained from oceanic ($n = 14$) and neritic ($n = 106$) waters of the Bay of Biscay and the English Channel ($n = 48$). Both Natoli et al. (2006) and Amaral et al. (2007) extended the sampling region in the NE Atlantic for assessing population structure, but using smaller sample sizes. Natoli et al. (2006) reported low but significant genetic differentiation between common dolphins in Scottish waters ($n = 26$) and those from the Celtic Sea ($n = 41$) and Galicia (NW Spain, $n = 39$) using nine microsatellites, but not sequences from the maternally inherited mtDNA control region. In contrast, Amaral et al. (2007) detected no significant genetic structure using mtDNA cytochrome b sequence data in samples from Scotland, Galicia and Portugal, except when the sexes were analysed separately, but sample sizes were too small for further assessment.

Following these initial investigations, Mirimin et al. (2009a) undertook a large-scale study, incorporating samples and data produced by Viricel (2006) and Amaral et al. (2007). In this study, 25 microsatellites and 556 bp of the control region were analysed in common dolphins from Scotland ($n = 13$), Ireland ($n = 102$), Celtic Sea ($n = 75$), English Channel ($n = 2$), France ($n = 46$), and Portugal ($n = 39$). Results indicated strong gene flow and the presence of a large 'coastal' panmictic (random mating) population in the NE Atlantic (Mirimin et al. 2009a). The Scottish sample demonstrated a unimodal distribution of the observed number of differences between pairs of haplotypes, suggesting the population has passed through a recent demographic expansion, but not a significant, negative value of Fu's F_S, which suggests that the marginal position in the geographic range may have led to a lower genetic exchange rate with neighbouring aggregations (Mirimin et al. 2009a). Conversely, this was not the case for the Portuguese sample.

Moura (2010) assessed whether genetic differentiation within the NE Atlantic common dolphin population increases with geographic distance, that is, isolation by distance, which has been reported in the harbour porpoise, *Phocoena phocoena*, population in this region (Fontaine et al. 2007). Analysis of 15 microsatellite DNA markers in 466 samples from *Delphinus delphis*, inhabiting waters off Scotland to Madeira, showed a lack of population genetic structure along the European Atlantic coastline and a lack of evidence of isolation by distance. In addition, analysis of a large number of (Atlantic) Iberian common dolphins by Moura (2010) did not support the previous suggestion by Amaral et al. (2007) of fine-scale population structure in that region.

In more southern waters, analysis using mtDNA and 14 microsatellite loci did not detect evidence of genetic structure among common dolphins inhabiting waters off Madeira ($n = 56$) and the archipelago of the Azores ($n = 91$), which are 900 km apart (Quérouil et al. 2010). Further evidence of gene flow over large geographic distances was provided by a lack of significant genetic differentiation amongst pairwise comparisons using microsatellites and samples from the central-east (CE) Atlantic (Azores, Madeira and Canary Islands, $n = 13$) and NE Atlantic (Galicia, Celtic Sea and Scotland, $n = 106$; Natoli et al. 2006). Shared haplotypes were also common between both regions, though sample sizes were small. Although F_{ST} values indicated no significant difference between the CE and NE Atlantic, ϕ_{ST} values indicated a significant difference between samples from the CE Atlantic and those from Scotland and Galicia. Amaral et al. (2012a) reported genetic differentiation between putative populations in the NE Atlantic (Portugal, northern Spain, Ireland and Scotland, $n = 75$) and the CE Atlantic (Madeira, $n = 29$) (and also the NW Atlantic, $n = 38$) using 14 microsatellites. As both studies reported low and hence marginally significant levels of divergence, this suggests high levels of gene flow between the NE and CE Atlantic or a recent population split—a similar situation to that reported between the NE and NW Atlantic populations (Mirimin et al. 2009b).

Interestingly, unlike in the North Atlantic, fine-scale (≤1000 km) population genetic structure has been reported in *D. delphis* inhabiting the NE Pacific and waters around Australia and New Zealand (Möller et al. 2011, Amaral et al. 2012a). Oceanographic variables (such as ocean currents, chlorophyll concentrations, temperature and salinity) have been proposed as factors limiting the movements of common dolphins in these regions through their effects on local prey availability (Bilgmann et al. 2008, Möller et al. 2011, Amaral et al. 2012a). This was supported by a sea-scape genetics study that assessed total averages of chlorophyll concentration, water turbidity and sea-surface temperature (SST) for an 8-year-period and reported that marine productivity and SST were correlated with genetic structure in common dolphins at medium spatial scales, that is, within ocean basins, such as the North Atlantic and South Indo-Pacific (Amaral et al. 2012a).

In summary, morphometric and genetic assessments indicate that there is only one short-beaked common dolphin population in the NE Atlantic, ranging from Scotland to Portugal (Murphy et al. 2009a). Future studies should focus on sampling dolphins from offshore waters using remote biopsy darting systems and strategic sampling approaches (e.g., temporal [including seasonal] and spatial; ICES WGMME 2009, Mirimin et al. 2009a, Murphy et al. 2009a).

Ecological stocks

As a consequence of the low genetic differentiation in this species as a whole within the North Atlantic Ocean, it was proposed by the ASCOBANS/HELCOM[*] Small Cetacean Population Structure Workshop (Murphy et al. 2009a) and by the ICES WGMME (2009) that common dolphins in the NE Atlantic should be managed on an ecological timescale (i.e., managing 'ecological stocks'). Ecological stocks can be defined as groups of individuals of the same species that co-occur in space and time and have an equal opportunity to interact with each other (Waples & Gaggiotti 2006). To take this approach, it was recommended that directed studies using ecological markers should be carried out to identify the existence of ecological stocks in common dolphins through sampling a large number of animals throughout the region, including all age-sex-maturity classes (ICES WGMME 2009, Murphy et al. 2009a). In addition, statistical power analysis should be undertaken to determine appropriate sample sizes required to detect the existence of ecological stocks within the NE Atlantic population.

The ASCOBANS/HELCOM Small Cetacean Population Structure Workshop discussed how to integrate different lines of evidence, including both genetic and ecological markers, to obtain the best possible indication of relevant stock structure. A few generations was proposed as the appropriate time frame for defining a management unit (MU) (Evans & Teilmann 2009). An MU was identified as a group of individuals for which there are different lines of complementary evidence suggesting reduced exchange (migration/dispersal) rates (Evans & Teilmann 2009). It was recommended that quantitative parameters (e.g., a maximum of 10% migration per generation) should be set, though in most cases this information is not available and the theoretical framework for integration of different evidence bases has not been fully developed (Evans & Teilmann 2009).

'Elemental profiles', determined, for example, by stable isotopes, fatty acids and pollutants, can be used as 'ecological tracers' of resource and habitat use (Caurant et al. 2009). However, these profiles can be sensitive to the physiological and health status of the animal and decomposition state of dead animals (Aguilar et al. 1999, Das et al. 2004, Pierce et al. 2008, ICES WGMME 2009),

---

[*] ASCOBANS: 'Agreement on the Conservation of Small Cetaceans of the Baltic, North East Atlantic, Irish and North Seas'; HELCOM: the Helsinki Commission, the governing body of the Convention on the Protection of the Marine Environment of the Baltic Sea Area.

as well as biotic factors such as metabolism, age, sex, and reproductive status (Evans & Teilmann 2009). For example, females transfer some of their lipophilic pollutants, such as polychlorinated biphenyls (PCBs), to their offspring during gestation and lactation, which confounds the use of these markers for assessing ecological stocks. Accumulation of certain xenobiotics is greater in females, possibly due to their lower capacity for detoxification compared to males (Aguilar et al. 1999 and references therein). This has been attributed to differing sex hormones and their effect on the activity of enzymes responsible for pollutant degradation (Aguilar et al. 1999). Stable isotope ratios indicated that female *Delphinus delphis* in the Southern California Bight consumed higher trophic level prey (higher $\delta^{15}N$ value) than males (Berman-Kowalewski & Newsome 2009, S.D. Newsome, personal communication, December 2012). However, $\delta^{15}N$ values will be dependent not only on an individual's food source and trophic level but also on its nutritional and physiological state (Cherel et al. 2005, Habran et al. 2010). All these confounding effects have to be considered prior to proposing ecological stocks within a population on the basis of elemental profiles (ICES WGMME 2009). In addition, toxicokinetics (toxicant absorption, distribution, metabolism, and excretion) and toxicodynamics (molecular, biochemical, and physiological effects of toxicants or their metabolites in biological systems) of some elements are only well known in humans and small laboratory mammals (Evans & Teilmann 2009).

Differing ecological tracers provide information on different timescales. For example, stomach contents provide information on dietary intake over a few days, fatty acids in blubber provide information on dietary intake a few weeks earlier (Nordstrom et al. 2008), stable isotopes in muscle provide information over a timescale of several months (Mèndez-Fernandez et al. 2012), whereas stable isotopes in hard tissues, such as teeth (Graham et al. 2010), and cadmium in kidney (Lahaye et al. 2005) reflect dietary intake over years (Caurant et al. 2009). Again, studies based on these approaches have to take into account biotic factors, health status and condition, as well as temporal and seasonal variations in dietary intake and habitat use. Although one approach would be to assess numerous tracers to identify ecological stocks within the NE Atlantic population, the appropriate timeframe for defining an MU should be specified at the outset.

To date, the presence of ecological stocks has not been verified in the NE Atlantic common dolphin population owing to a lack of data (ICES WGMME 2009, 2012). Limited studies have been undertaken, with predominantly small sample sizes, thus increasing the possibility of sampling biases.

Das et al. (2003) reported that mean muscle and liver $\delta^{13}C$ were significantly more negative for short-beaked common dolphins off the Irish coast (muscle $\delta^{13}C = -17.1$, $n = 14$) compared to animals off the north coast of France (-16.5, $n = 8$), suggesting a more oceanic/offshore diet for the former group. Similar muscle $\delta^{13}C$ values were obtained in common dolphins from Galicia, Spain (-17.0, $n = 114$), though a slightly more negative value was obtained for common dolphins stranded in the Bay of Biscay, France (-17.4, $n = 26$) (Chouvelon et al. 2012, Mèndez-Fernandez et al. 2012). However, parameters such as age, sex, season, health and decomposition states were not assessed in these studies.

The half-life of lead (Pb) ranges from 5 to 20 years in hard tissues. As no geographic differences were found in lead concentrations or isotopic composition ($^{206}Pb/^{207}Pb$) in bone or teeth of stranded common dolphins from Brittany (NW France, $n = 17$; from a single mass stranding event [MSE] in 2002) and Galicia (NW Spain, $n = 16$), movements of animals between these continental shelf areas was proposed (Caurant et al. 2006). There were no sex-related differences in lead concentrations, though only 22% of the sample comprised males, and age was taken into account within the analysis.

Lahaye et al. (2005) proposed that there are two ecological stocks of *D. delphis* in the neritic and oceanic waters of the Bay of Biscay, based on higher renal cadmium levels in dolphins caught in the French summer tuna drift net fishery in the mid-1990s, compared to by-caught and stranded animals from French neritic waters ($n = 48$) sampled between 2001 and 2005, predominantly during winter and spring. Cadmium has a biological half-life of more than 10 years in mammals, and higher cadmium levels in the oceanic group were attributed to the consumption of oceanic cephalopods (Lahaye et al. 2005). Cadmium tissue concentrations did not differ between males and females, in contrast to other studies (Aguilar et al. 1999). Concerns over the sample size of the 'oceanic' stock (five males, five females) proposed by Lahaye et al. (2005) were highlighted by the ICES WGMME (2009) and Murphy et al. (2009b), and they thus did not recommend a two-stock approach in the Bay of Biscay.

Murphy et al. (2007a) estimated the mean generation time for this population as 12.94 years. Consequently, taking on board recommendations by Evans & Teilmann (2009) and ICES WGMME (2012), ecological tracers showing an integration of tens of years (i.e., a few generations) should be explored. Further analysis is needed to verify the existence of ecological stocks within the Bay of Biscay, including increasing the sample size of carcasses from the proposed oceanic stock (ICES WGMME 2009, Murphy et al. 2009a). An alternative approach is the further development and application of non-lethal biomarkers from cetacean skin biopsies. Biopsy sampling from cetaceans has become a valuable approach, providing data on genetics, prey preferences, foraging ecology, contaminant loads, and physiological processes (ICES WGMME 2012, Noren & Mocklin 2012). In addition, these samples may be more representative of the population than samples collected from dead or live-stranded animals that may be ill or emaciated (ICES WGMME 2012). However, it should be noted that trace elements generally have no affinity for lipids, and few studies to date have assessed population segregation through trace element analysis of skin tissue (Kunito et al. 2002, Evans & Teilmann 2009).

Distribution and abundance

In the NE Atlantic, short-beaked common dolphins are distributed, at least during summer, from coastal waters to the mid-Atlantic ridge and from south of the Azores and the Strait of Gibraltar to around 70°N, west of Norway, but are mainly found south of 60°N (Figure 1; Murphy 2004, Cañadas et al. 2009, Murphy et al. 2009a). In the NW Atlantic, the SNESSA (Southern New England to Scotian Shelf Abundance) survey undertaken in 2007 sighted individuals as far north as 56.9°N, with higher concentrations reported on the Scotian shelf than off southern Newfoundland (Lawson et al. 2009). Short-beaked common dolphins may in fact be distributed across the whole North Atlantic Ocean, between 35°N and 55°N (partially covering a region strongly influenced by the Gulf Stream/North Atlantic Drift); however, due to a lack of observer effort west of the mid-Atlantic ridge (approximately 30–40°W), the contemporary range is unknown (Figure 1). Furthermore, the distributional boundary of the NE Atlantic population has not been determined. As outlined previously, the sampling of individuals for genetic and cranial morphometric analysis has been confined to continental shelf and slope waters and oceanic waters of the Bay of Biscay.

Cañadas et al. (2009) assessed sightings made during summer from (1) the North Atlantic Sightings Surveys (NASS), undertaken throughout most of the central and eastern North Atlantic (north of about 40°N) in 1987, 1989, 1995 and 2001; (2) the MICA93 programme (Goujon et al. 1993b); and (3) the NE Atlantic segment of the Small Cetacean Abundance in the North Sea and adjacent waters (SCANS) survey from 1994 (Hammond et al. 2002). Analyses suggested that common dolphins were most commonly sighted in water temperatures above 15°C, depths of 400–1000 m (an association with shelf features was noted), and within the area bounded by latitudes 49–55°N and longitudes 20–30°W. It should be noted that SST data were only available for some of the waters surveyed. Average group size was 15 ± 2.2 individuals (± standard error, SE; range 1–239) and showed a significant increasing trend with depth from 8.0 ± 1.44 animals in waters less than 400 m to 18.6 ± 2.76 animals in water depths more than 2000 m (Cañadas et al. 2009). The most northerly sighting was at 56°45′ N, substantially further south than the most northerly observation for this species in the NE Atlantic at 73°34′ N 11°04′ E, made in August at an SST of 10.7°C (McBrearty et al.

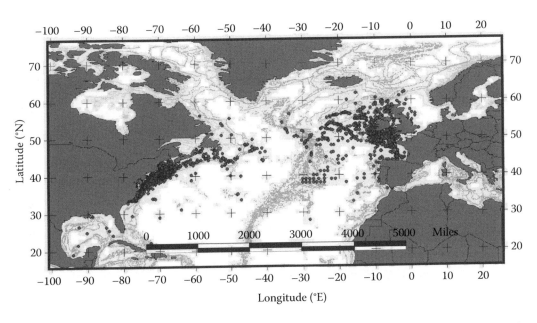

Figure 1 (See also colour figure in the insert) Distribution of common dolphin sightings in the North Atlantic (Murphy et al. 2009a). Data obtained between 1963 and 2007 by a large number of observer schemes, though the majority of sightings were obtained after 1980 and primarily during summer. Sightings data that contributed to this map were provided by Northeast Fisheries Science Center, NOAA; University of North Carolina at Wilmington; University of Rhode Island for access to BLM data; Canadian Wildlife Service for access to PIROP data; North Atlantic Marine Mammal Commission (NAMMCO), Museum of Natural History, Faroe Islands, and Marine Research Institute, Iceland, for access to NASS and T-NASS survey data; Joint Nature Conservation Committee; Sea Watch Foundation; Sea Mammal Research Unit (SMRU), St. Andrews University, for access to SCANS I data; Institut Français de Recherche pour l'Exploitation de la Mer (IFREMER) for access to MICA 1993 survey data; University College Cork for access to SIAR survey data; MAR-ECO survey data; Whale Watch Azores; and OBIS SEAMAP (Read et al. 2007).

1986, Cañadas et al. 2009). Interestingly, there was a gap in the distribution of the common dolphin in offshore waters in a rectangular area from 42°N 18°W to 48°N 12°W, waters that were surveyed in 1993 by the MICA programme (Cañadas et al. 2009).

Reid et al. (2003) mapped the distribution of common dolphins in western European waters using sightings data obtained primarily during summer, between 1978 and 1998. Highest numbers were reported in the Celtic Sea, St. George's Channel, western approaches of the English Channel, and off southern and western Ireland (Figure 2). The species was not observed in the eastern English Channel and only occasionally in the North Sea, mainly from June to September.

Further development of this data resource, including collation and inclusion of new datasets and development of statistical modelling techniques, has been initiated in recent years. Paxton & Thomas (2010) assessed common dolphin abundance and distribution in the Irish Sea between 1980 and 2009, with highest numbers occurring in St. George's Channel. They reported an increasing trend in the abundance of animals over time (Figure 3). To test the utility of such combined datasets for monitoring cetacean populations, Paxton & Thomas (2010) assessed the minimum population change that could be detected with a reasonable degree of certainty through power analysis of an index based on the ratio of population density estimates for different years (a ratio of 1 indicating no change). For common dolphin in the latest time period in the dataset (2003 to 2008), the minimum population density ratio, expressed on an annual basis, detectable with a statistical power of 0.8 was 0.978, equivalent to a 2.2% decline per year. Following from this work, the statistical techniques were further developed and data from the neighbouring Celtic Sea and the Greater Minch Area

Short-Beaked Common Dolphin *(Delphinus delphis)*

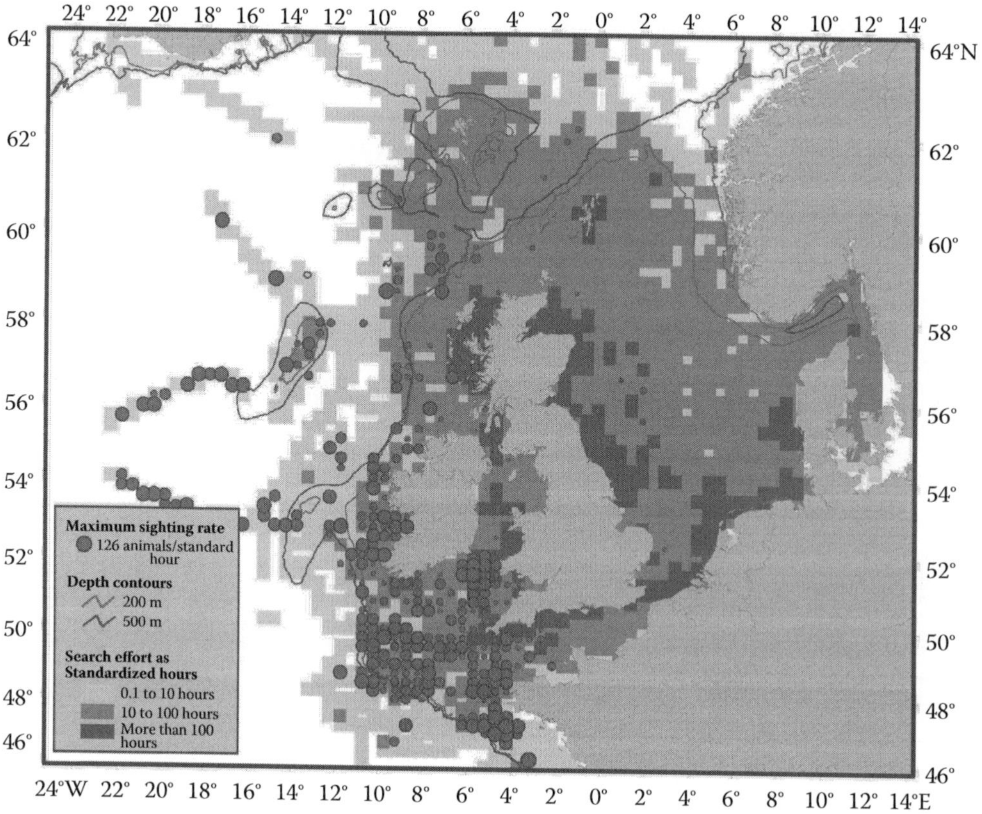

Figure 2 (See also colour figure in the insert) Distribution of common dolphins in western European waters. (Data obtained from 1978 to 1998; Reid et al. 2003. With permission.)

(western Scotland) were included (1982–2010). High population density of common dolphin was predicted throughout much of the Celtic Sea, but with high uncertainty owing to low observer effort in this region (Paxton et al. 2011). Paxton et al. (2011) noted that common dolphin abundance in the Celtic Sea, the Irish Sea, and off the west coast of Scotland generally peaked in the autumn but was also high in May and June. Numbers in the autumn peak varied between 50,200 (confidence interval [CI] = 30,800–113,600, coefficient of variation [CV] = 0.30) in 1995 and 180,900 (CI = 108,600–399,600, CV = 0.34) in 2008.

In more southern waters of the NE Atlantic, not assessed by these studies, common dolphins are one of the most frequently sighted cetaceans off the coasts of France, northern Spain, mainland Portugal and the islands of Madeira and the Azores (Silva et al. 2003, López et al. 2004, Marcos-Ipiña et al. 2005, Certain et al. 2008, 2011, Brito et al. 2009, Vieira et al. 2009, ICES WGMME 2010, Marcos et al. 2010, Pierce et al. 2010, Spyrakos et al. 2011, Moura et al. 2012). High densities in the Bay of Biscay are associated with the shelf break, though during spring common dolphins are more abundant closer to the coast, especially in areas of river plumes (Certain et al. 2008, 2011). The French PELGAS (pelagic acoustic spring) surveys between 2003 and 2008 indicated that common dolphins are distributed throughout this region in spring, with greatest abundance between the upper Gironde river plume to waters off the Vendée coast, around canyons in the south of the Bay (Cap Ferret and around), and in coastal waters off Brittany (ICES WGMME 2010, Certain et al. 2011).

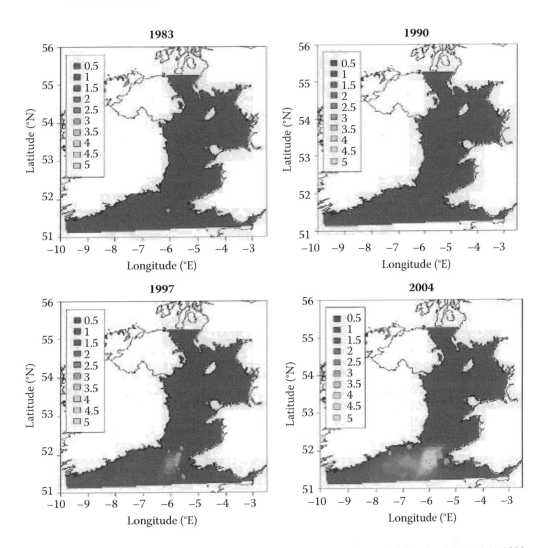

Figure 3 (See also colour figure in the insert) Predictions of common dolphin density for July 1983, 1990, 1997 and 2004 according to a two-stage modelling process. Green circles are proportional in area to estimated density of common dolphin associated with that segment locality. Numbers indicate upper bound of colour-coded densities (animals km^{-2}). (From Paxton & Thomas 2010. With permission.)

In the waters off Gipuzkoan, Basque, northern Spain, common dolphins were sighted year round, predominantly in water depths of 200–1200 m, and in the vicinity of the Cap Breton canyon (Marcos et al. 2010). Densities were higher in winter and decreased significantly in spring. Data were collected from ship-based surveys carried out between April 2003 and October 2008 (Marcos et al. 2010). Off Galicia (NW Spain), short-beaked common dolphins were most commonly sighted in deeper waters (>200 m), with the highest number of sightings in the second quarter of the year (López et al. 2004). Coastal sightings of *D. delphis* were most commonly where the continental shelf was narrowest, consistent with the dolphins occupying deeper waters (Pierce et al. 2010). In surveys undertaken onboard vessels fishing on the continental shelf in 2001 and 2003 (between February and September), *D. delphis* were primarily sighted from May to August, and mean group size was 25.4 individuals (Spyrakos et al. 2011). Once survey effort was taken into account, there was no relationship between sighting frequency and water depth, and although larger group sizes were observed in the north, off Galicia, there were fewer sightings in this area compared to the south (Spyrakos et al. 2011).

Individuals were sighted year-round off mainland Portugal, primarily along oceanic features such as canyons (Brito et al. 2009, Vieira et al. 2009). Moura et al. (2012) suggested that *D. delphis* has a patchy distribution, varying on an annual basis, associated primarily with chlorophyll concentrations, which possibly reflects a higher incidence of pelagic schooling fish in those areas. The highest encounter rates were recorded off Peniche, central Portugal (Moura et al. 2012). Off the Azores and Madeira archipelagos, the species is more frequently seen in coastal waters than off shore (Silva et al. 2003). While common dolphins are only seasonal visitors to the waters around Madeira, animals off the Azores exhibit some degree of site fidelity (Quérouil et al. 2010). In some years, there is a significant reduction in relative abundance off the Azores during summer and autumn (June to October) (ICES WGMME 2010).

Contemporary seasonal movements

Common dolphins are extremely mobile, and swimming speeds of 0.77 to 3.20 nautical miles per hour (1.43–5.93 km h^{-1}) have been estimated from radio tracking (Evans 1975, 1982), though animals can travel at 15–20 km h^{-1} and sometimes twice as fast (Murphy et al. 2008). Maximum burst swimming speeds of 8 m s^{-1} have been reported in captive *Delphinus delphis* (Rohr et al. 2002). Radio tracking showed that a female dolphin in the eastern Pacific Ocean travelled approximately 270 nautical miles (500 km) from the point of release over a 10-day period (Evans 1975, 1982). Two rehabilitated common dolphins were tagged with satellite-linked radio transmitters in 1994 and 1995, and one of the individuals was tracked for 31 days off the coast of California. The dolphin immediately moved off shore into deep water and began moving north. It travelled about 400 km within 5 days of release, then covered another 250 km, approximately, after which it continued travelling north until radio contact was lost (Zagzebski et al. 2006). The only other report of long-distance movement by a common dolphin was of a naturally marked individual that travelled a minimum of 1000 km across the Ionian and the Adriatic Seas (Genov et al. 2012).

In the NE Atlantic, there are no data on the habitat range of individual common dolphins, but distributional data suggest large-scale seasonal movements (ICES WGMME 2005). Analysis of sightings data collated by Reid et al. (2003) indicated that common dolphins are more widely dispersed along the continental shelf edge and in deep offshore waters, as well as off the west coast of Scotland and Ireland and in St. George's Channel, during the summer than in winter, when there are pronounced concentrations in shelf waters of the western English Channel, St. George's Channel, off shore in the Celtic Sea, and also off the coasts of NW France (western Brittany), south-western (SW) and NW Ireland and NW Scotland (ICES WGMME 2005). These seasonal movements may be related to prey availability and distribution (ICES WGMME 2005).

Brereton et al. (2005) also reported winter inshore movements of short-beaked common dolphins based on sightings data collected between 1995 and 2002 from 'platforms of opportunity', that is, ferries operating from Portsmouth, England, to Bilbao, Spain. From summer to winter, there was a 10-fold increase in the number of dolphins observed in the western English Channel. In July, highest densities and largest group sizes were seen along the northern shelf slope of the Bay of Biscay, coinciding with the mating/calving period (Murphy et al. 2005a, 2009b). Macleod et al. (2009) extended the sampling period to incorporate data collected up to 2006 and noted a similar pattern, in addition to a 5-fold increase (0.02 in 1996 to 0.11 in 2006) in wintertime occupancy in the western English Channel during the study period (though this increasing trend was not statistically significant). This increased wintertime occurrence coincided with a peak in reported strandings along the SW coast of the United Kingdom in the early 2000s (Jepson 2005) (see 'Stranding patterns' section).

Kiszka et al. (2007) also analysed sightings data collected opportunistically onboard ferries operating in the English Channel and Bay of Biscay between 1998 and 2002, predominantly from July to October. As in previous studies, summertime aggregations were larger in the northern Bay of Biscay, primarily along the shelf slope, than in the western English Channel. This may be related to the distribution of their preferred prey species in this area, *Sardina pilchardus* and *Trachurus trachurus* (Meynier 2004, Kiszka et al. 2007, Certain et al. 2011). Common dolphins also occurred seasonally off the French Channel coast, primarily around Ile d'Ouessant (western Brittany) and north of the Channel Islands, with only a few reported sightings in the eastern Channel— assessed using year-round sightings data collected between 1980 and 2000 from a variety of French sources (Kiszka et al. 2004).

Common dolphins are present in the Irish Sea at low abundance from late spring to late summer, and the distribution appears to shift southwards out of the Irish Sea during autumn and winter, though in the Celtic Deep, where higher densities exist, they remain at least until November (Evans et al. 2007, Wall et al. 2011, Baines & Evans 2012). Goold (1998) noted that the marked decrease in numbers of common dolphins off the western Wales coast between September and October suggests an autumn migration. SST distribution across the entire region was visualized using infrared satellite imagery, and it was hypothesized from these observations that the migration coincides with a break-up of the Celtic Sea Front (Goold 1998).

In the NW Atlantic, common dolphins also undertake seasonal migrations. The species is distributed from Cape Hatteras, North Carolina, north-east to Georges Bank (35–42°N) between mid-January and May, following which dolphins move on to Georges Bank and the Scotian Shelf from midsummer to autumn (Waring et al. 2011). Migrations on to the Scotian Shelf and continental shelf off Newfoundland occur when SST exceeds 11°C (Waring et al. 2011).

Long-term distribution patterns

There is evidence of changes in the distribution of the NE Atlantic population within the last century, with both an increased occurrence in more northern waters and movements into the North Sea. These shifts were observed between the 1920s and 1960s and largely since the 1990s. Increased stranding rates of common dolphins were documented in the southern North Sea during the early to mid-twentieth century along the Dutch (1920s–1960s; Bakker & Smeenk 1987) and Danish (1937–1952; including the inner Danish waters, Kinze 1995) coastlines and the eastern coast of England (1930s–1940s; Murphy et al. 2006). The increase in strandings in the North Sea partially coincided with a decline in strandings along the Irish and SW English coasts between the late 1930s and mid-1970s (Evans & Scanlan 1989, Murphy 2004, Murphy et al. 2006). This may indicate a shift in the general distribution of the species at that time (Fraser 1934, 1946, Sheldrick 1976, Evans & Scanlan 1989, Murphy 2004, Murphy et al. 2006).

Movements of common dolphins into the northern North Sea from the Atlantic occurred in the 1930s and led to an unusually large number of reported strandings along the Scottish eastern coast. Notably, this took place during an influx of the European flying squid, *Todarodes sagittatus*, a prey species of the common dolphin, into the North Sea (Fraser 1937, 1946, Evans & Scanlan 1989, Murphy et al. 2006). There was also an increase in strandings in the southern North Sea in that period, but it is unknown if this was due to some of these dolphins migrating further south or to individuals that entered the North Sea through the English Channel (Murphy et al. 2006). Based on cranial morphometric analysis, *Delphinus delphis* skulls collected in the Netherlands primarily between 1926 and 1953 were more similar to dolphin skulls collected during the last two decades from the south-west of the United Kingdom than to skulls collected in Scotland over the same time period. However, it was noted that the Scottish cranial sample may not be representative of animals inhabiting those waters in the early to mid-1900s as the distribution and abundance of the species in Scottish waters has fluctuated over the last century (MacLeod et al. 2005, Murphy et al. 2006).

There were infrequent sightings of common dolphins in the North Sea between 1978 and 1998, the majority of which were in northern British waters (Reid et al. 2003). Common dolphins were not sighted in the North Sea during either the SCANS-I or SCANS-II surveys undertaken in July 1994 and 2005, respectively. Conversely, during the 1990s and 2000s, common dolphins were documented (sightings and strandings) in both the North and Baltic Seas in Danish, German, Polish, Finnish, Swedish and Norwegian waters (ICES WGMME 2005). Six common dolphins were stranded along the Danish coastline between 2001 and 2003 and schools of up to 10 individuals were sighted (ICES WGMME 2005). Prior to this, the last reported stranding of a common dolphin in Danish waters was in 1978 (Kinze 1995), while sightings of the species were recorded in 1979, 1982, 1990 and 1996 (Kinze 1995, ICES WGMME 2005).

Øien and Hartvedt (2009) collated sightings of common dolphins in Norwegian waters recorded between 1968 and 2008. The species was sighted every year since 1976 as far north as 72°N and in almost every month apart from February and November. Sighting rates were highest between June and October, with a peak in June. There was an unusually high number of sightings (20) in 1998, the reasons for which are unknown (Øien & Hartvedt 2009). In addition to these records, common dolphins have been recorded each summer in the Moray Firth (Scottish North Sea) from 2006 to 2009, with up to 13 encounters and group sizes ranging from 2 to 450+ individuals (Robinson et al. 2010). This summertime presence has continued since 2009 (K.P. Robinson personal communication, June 2006). Macleod et al. (2005) also reported an increase in the abundance (sightings and strandings) of common dolphins off the NW Scottish coast during the period 1992 to 2003. Taken together, these data suggest that the distribution of common dolphins is once again expanding into more northern waters, including the North Sea, as apparently occurred in the early to mid-twentieth century.

Population abundance

There is no population trend information available for common dolphins in the NE Atlantic, and a lack of knowledge on the actual status of the population, based on sightings data.

Continental shelf waters

SCANS-II estimated that 56,221 (CV = 0.23) *Delphinus delphis* occupied continental shelf and slope waters of the NE Atlantic during July 2005 (Hammond et al. in press; Table 1; Figures 4 and 5). Highest densities were reported west of Ireland and Scotland, in the Celtic Sea and extending into St. George's Channel/southern Irish Sea, along the continental shelf off SW Ireland and west of Brittany, in the western English Channel, and waters off northern Spain and Portugal (ICES WGMME 2010). Earlier large-scale sightings studies, such as MICA (Goujon et al. 1993b) and SCANS-I (Hammond et al. 2002), also collected sightings data for common dolphins inhabiting oceanic and continental shelf waters (see Cañadas et al. 2009 for distribution of survey effort). However, the earlier studies are not comparable with SCANS-II since they used only a single-platform method, and they did not correct for animals missed on the survey track or for responsive movements (i.e., attraction of dolphins to the vessel).

Offshore waters

The Cetacean Offshore Distribution and Abundance in the European Atlantic (CODA) survey, undertaken in July 2007, estimated that there were 116,709 (CV = 0.34) *Delphinus delphis* in European offshore waters (beyond the continental shelf, from off NW Scotland to NW Spain) (CODA 2009;

Table 1 Estimates of group abundance, mean group size, animal abundance and animal density (individuals km^{-2}) of *Delphinus delphis* from the SCANS-II survey, July 2005

Block	Group abundance	Mean group size	Animal abundance	Animal density
B	378 (0.73)	13.0 (0.36)	4,919 (0.82)	0.040 (0.82)
N	1256 (0.58)	1.75 (0.14)	2,199 (0.60)	0.072 (0.60)
O	375 (0.69)	2.20 (0.36)	826 (0.78)	0.018 (0.78)
P	1058 (0.33)	11.6 (0.30)	15,957 (0.31)	0.081 (0.31)
Q	558 (0.98)	3.08 (0.32)	2,230 (0.87)	0.015 (0.87)
R	1266 (0.70)	9.2 (0.19)	11,661 (0.73)	0.302 (0.73)
W	1470 (0.29)	12.3 (0.27)	18,039 (0.23)	0.130 (0.23)
Z	314 (0.84)	1.25 (0.20)	392 (0.86)	0.012 (0.86)
Total	6675 (0.27)		56,221 (0.234)	
	[3969–11,230]		[35,748–88,419]	

Source: Hammond et al. in press.

Note: Coefficients of variation are given in parentheses. Figures in square brackets are 95% confidence intervals. There were no sightings of *Delphinus delphis* in blocks H, J, L, M, S, T, U, V and Y. See Figure 4 for SCANS-II survey map.

Table 2, Figure 4). Highest densities were observed in more southern areas of the surveyed region, with most sightings along the continental slope off western France and northern Spain (Figure 6).

The small numbers sighted off the western coast of Ireland during CODA were comparable to an earlier survey undertaken in July 2000; see Figure 4 for realized survey effort during the CODA survey and note low survey effort off the SW coast of Ireland. SIAR (Survey in Western Irish Waters and the Rockall Trough), which surveyed waters over the shelf break to the north and west of Ireland, estimated only 4496 individuals in this region (Ó Cadhla et al. 2003); see ICES WGMME (2005) for a map of summer abundance surveys in the NE Atlantic. The study area was about 120,000 km^2, covering the western Irish continental shelf, central and eastern Rockall Trough and from the Porcupine Bank to the Outer Hebrides. Results from both SIAR and CODA are in contrast to the large numbers of short-beaked common dolphins sighted off the west of Ireland during the 1990s. Faroese scientists who participated in the NASS 1995 survey covered two large areas (NASS-east and NASS-west) to the west of Ireland and Scotland. The estimated abundance of *D. delphis* in NASS-west, an area that extended beyond the CODA survey region, was 273,159 (CV = 0.26; 95% CI = 153,392–435,104) (this estimate was corrected for animals missed on the survey track and for responsive movement) (Cañadas et al. 2009). An abundance of 77,547 *D. delphis* was estimated for NASS-east, but due to limitations of the survey, this estimate was not considered reliable (Cañadas et al. 2009). Nevertheless, even allowing for the uncertainty in this estimate, it suggests a considerably greater number may have been present than during the SIAR survey.

The Trans North Atlantic Sightings Survey (T-NASS) was carried out at the same time as CODA but surveyed waters further to the north and off shore (Lawson et al. 2009). Few short-beaked common dolphins were sighted in areas where animals were seen in high abundance during the NASS 1995 survey. In 2009, the IWC Sub-Committee on Small Cetaceans identified several potential reasons for the observed changes in density/distribution, including (1) differences in sighting conditions (e.g., sea state), (2) uncertain species identification (as other dolphin species were sighted), (3) a true reduction in common dolphin density, (4) ship effect and (5) interannual distributional shifts. In addition, due to poor weather conditions during T-NASS, some of the planned survey tracks were not covered (IWC 2009).

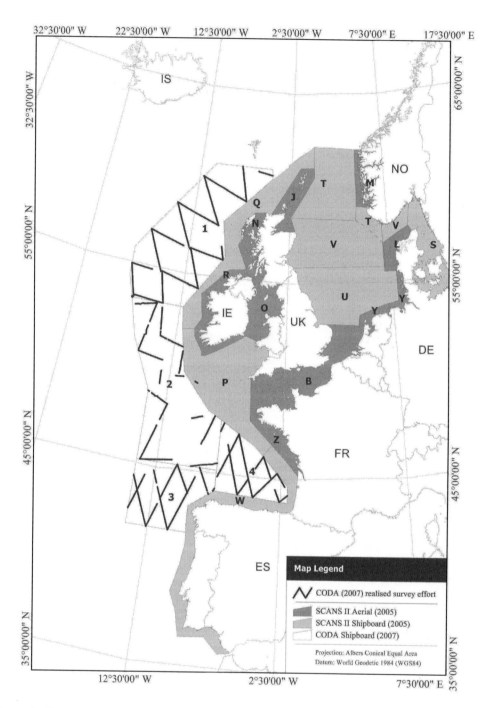

Figure 4   Survey blocks defined for the SCANS-II (undertaken in 2005) and CODA (undertaken in 2007) surveys. SCANS-II Blocks S, T, V, U, Q, P and W were surveyed by ship. The remaining blocks were surveyed from aircraft (SCANS-II 2008). CODA survey region divided into the survey blocks 1–4 and realized survey effort route (in black) (CODA 2009). (Map produced by Rene Swift from a projection of Albers conical equal area and datum from World Geodetic System 1984.)

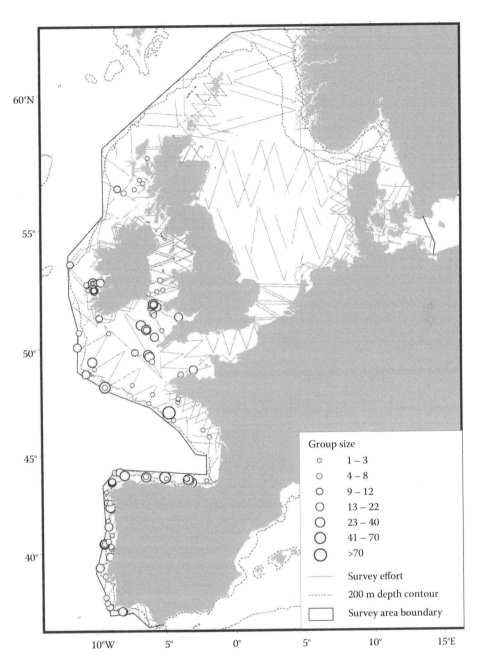

Figure 5 Sightings of common dolphins during SCANS-II in July 2005. (From Small Cetaceans in the European Atlantic and North Sea (SCANS-II), Final report to the European Commission under contract LIFE04NAT/GB/000245. St. Andrews, UK: Sea Mammal Research Unit, 2008; Hammond et al. in press, with permission from Elsevier.)

Table 2 Estimates of model-based (density surface modelling) animal abundance estimates, with coefficients of variation (CV) in brackets, and 95% confidence intervals, of *Delphinus delphis* from the CODA survey, July 2007

Block	Animal abundance (CV)	95% Confidence interval
1	4,216 (0.57)	1,478–12,027
2	52,749 (0.39)	25,054–111,059
3	21,071 (0.51)	8,270–53,689
4	38,673 (0.46)	16,464–90,839
Total	116,709 (0.34)	61,397–221,849

Sources: Cetacean Offshore Distribution and Abundance in the European Atlantic (CODA), 2009; ICES WGMME (2010).

Note: See Figure 4 for CODA survey blocks.

Life-history parameters

Extensive studies have been undertaken since 2000 assessing life-history parameters in the NE Atlantic *Delphinus delphis* population using samples from stranded and by-caught dolphins, funded by both national governments and the European Union's Fifth and Sixth Framework Programmes for Research, including large-scale projects such as BIOCET (Bioaccumulation of Persistent Organic Pollutants in Small Cetaceans in European Waters: Transport Pathways and Impact on Reproduction) and NECESSITY (Nephrops and Cetacean Species Selection Information and Technology). These have resulted in preliminary data on a large number of biological parameters, such as the population pregnancy rate and average age at sexual maturity, essential for effective conservation management. However, these could not be considered baseline data as there may have been anthropogenic impacts prior to these studies.

Size and morphology

In the early to mid-1900s, maximum body lengths of 250 to 270 cm were recorded for short-beaked common dolphins in the NE Atlantic (Harmer 1927, Fraser 1934, 1946, 1974). If these earlier studies identified the species correctly, there may be indications that maximum length is declining within the region. During the last 30 years, maximum lengths of 250 and 239 cm have been reported for males and females, respectively, though the majority of individuals were less than 230 cm (Collet 1981, Silva & Sequeira 2003, Murphy 2004, Murphy et al. 2006, 2009b, Murphy & Rogan 2006). Average length of newborn calves is 93 cm (range 89–110 cm) (Collet 1981, Murphy & Rogan 2006).

Short-beaked common dolphins in the NE Atlantic exhibit sexual size dimorphism, with males significantly larger than females in total body length and 19 of 23 other morphometric characters (Murphy & Rogan 2006). Although there is a statistically significant difference in size, it is only moderate: Murphy & Rogan (2006) calculated a sexual size dimorphism ratio of 1.06 using average adult body lengths of 201.2 cm for females and 212.9 cm for males. Sexual shape dimorphism (relative size) was not detected in body characters, apart from the presence of prominent postanal humps in mature males (Murphy & Rogan 2006). Interestingly, unlike in spinner (*Stenella longirostris*) and spotted dolphins (*S. attenuata*), the postanal hump in the common dolphin is composed of muscle and not connective tissue, suggesting different functions (Murphy et al. 2005a). It has been proposed that the postanal hump in the common dolphin may serve in female choice, allowing the identification of the healthiest male, that is, the male that can produce the largest quantity of sperm (Lewis 1991, Neumann et al. 2002, Murphy et al. 2005a) since preliminary investigations suggested that the size of the postanal hump is positively correlated with testis size (Lewis 1991).

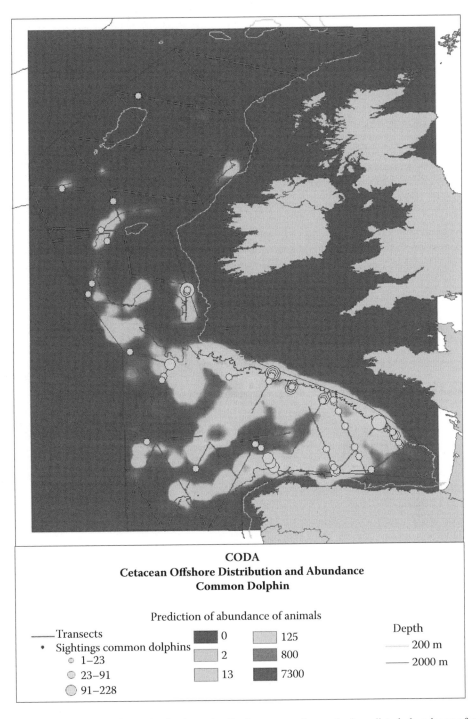

Figure 6 (See also colour figure in the insert)   Surface maps of smoothed predicted abundance of common dolphins in offshore waters, including the distribution of sightings (circles proportional to group size). (Cetacean Offshore Distribution and Abundance in the European Atlantic (CODA) 2009.)

Male short-beaked common dolphins are significantly larger than females in condylobasal length (CBL) and 21 other cranial morphometric characters (Murphy 2006). Sexual dimorphism in cranial shape was identified in seven characters, mainly related to the width of the skull. Mature males had (in relation to CBL) significantly greater rostrum width at base, zygomatic width, post-orbital width, width between temporal fossae, maximum width of external nares, orbital length, and mandible depth.

Population parameters

Average age at the attainment of sexual maturity was estimated at 11.9 years in males, based on examination of common dolphins sampled by the Irish and French stranding and by-catch observer programmes between 1991 and 2003 (Murphy et al. 2005a). Sexually mature males ranged from 195 to 233 cm in length and 8 to 28 years in age. Applying a single Gompertz growth curve to the male age data produced an asymptotic length of 206 cm (Murphy et al. 2005a). Murphy et al. (2009b) analysed a much larger sample size of stranded and by-caught female common dolphins collected throughout the NE Atlantic (Scotland to Portugal) between 1990 and 2006. Female body lengths ranged from 91 to 239 cm, and the maximum estimated age was 30 years (Murphy et al. 2009b, 2010). The asymptotic length, estimated using Richard's model, was 202 cm (Murphy et al. 2009b). Average age and length at sexual maturity were 8.2 years and 188 cm, respectively (Murphy et al. 2009b).

Based on mortality data from 248 mature females, an annual pregnancy rate of 26% and extended calving interval (gestation, lactation and resting periods) of 3.8 years were determined for the NE Atlantic population (Murphy et al. 2009b). There was no significant difference in the proportion of pregnant females between different geographical areas (Ireland, United Kingdom, France, NW Spain) of the NE Atlantic. The pregnancy rate was also estimated using a control group of 'healthy' individuals, that is, individuals not suffering from any infectious or non-infectious disease that may inhibit reproduction. As no significant difference was found in proportion of pregnant females between the control group and the whole sample, it appears that the sampling of stranded and by-caught short-beaked common dolphins is adequate for estimating population reproductive parameters. This is in contrast with other cetacean species, such as the harbour porpoise, with a large number of stranded individuals dying due to ill health or poor condition, including starvation, disease, and bacterial and parasitic infections (Deaville & Jepson 2011). In contrast, the majority of stranded common dolphins in this region were killed in fishing gear (by-catch) and subsequently washed ashore (Murphy et al. 2009b, Deaville & Jepson 2011).

The reproductive rate identified in the NE Atlantic population is relatively low compared to other *Delphinus* sp. populations, in which pregnancy rates are higher, such as South Africa, 40.2% (Mendolia 1989, Murphy et al. 2009b); or ETP, 47% (Danil & Chivers 2007). Murphy et al. (2009b) found no evidence of compensatory density-dependent responses in reproductive parameters. No significant differences were observed in the proportion of pregnant females, proportion of mature females simultaneously pregnant and lactating, average age attained at sexual maturity, or nutritional condition of females between two different time periods (1991–1999 and 2000–2006).

In the NE Atlantic, short-beaked common dolphins exhibit reproductive seasonality. A unimodal calving/mating period extends from April to September, possibly with a more active period in July and August (Murphy et al. 2005a, 2009b). Estimated individual conception dates of sampled fetuses ranged from 5 April to 2 October, though the average date of conception was 19 July, and 40% of individuals were conceived during this month (Murphy et al. 2009b). The gestation period in the population was estimated at 0.99 years (Murphy et al. 2009b). Even though the sample size was small, sexually mature and pubertal females were reported ovulating only during May to September (6 of 45 individuals examined) (Murphy 2004). Such an extended mating period in the NE Atlantic population allows females to undergo numerous ovulations, with some individuals possibly completing up to five reproductive cycles during this period (Murphy et al. 2010). This provides a substantial buffer for individuals that may not conceive during their first oestrous within the mating period.

Male gonadal tissue in this region also exhibits seasonality, evidenced by reduced testis weights and testicular cellular activity outside the mating period (Murphy et al. 2005a, Murphy & Rogan 2006). In the NE Atlantic, mature male common dolphins developed large testes, relative to their body size, with combined testes weight ranging from 415.9 to 5000 g. Macleod (2010) ranked common dolphins comparatively high among 31 cetacean species in their relative investment in testicular tissue, based on an assessment of both percentage testes (3.2% of body mass in the species) and residual testes mass (deviation from the mean testes mass-body mass relationship of 31 cetacean species). The presence of enlarged testes and the existence of moderate sexual dimorphism in the species suggest postmating competition among males (i.e., sperm competition), resulting from a promiscuous mating system (Murphy et al. 2005a). To date, there have been no studies of mating strategies in wild common dolphins in this region to verify this hypothesis. Males continue to produce sperm outside the mating period (Murphy et al. 2005a), which possibly enables them to partake in 'recreational' (non-reproductive) mating. Recreational mating has been noted in New Zealand waters (Neumann 2001, Murphy et al. 2005a), but the function of this behaviour is not known.

Age and sex segregation

Common dolphins are found in a wide range of group sizes, up to 1000 to 5000 individuals (Murphy 2004 and references therein). There is evidence that smaller groups are segregated by age and sex, especially during winter (i.e., outside the mating period). Three mass live strandings (three or more individuals) of 'nursery groups' have been reported along the Irish, French, and UK coastlines. In February 2001, there were 15 common dolphins stranded alive on the Mullet Peninsula, western coast of Ireland. Eleven dolphins were refloated, and five died, including one male yearling, three sexually mature females aged between 14 and 17 years and one pregnant 17-year-old female (Murphy 2004). Further evidence of nursery groups within this region arose when a mass live-stranding event involving approximately 100 individuals took place at Pleubian, France, in 2002. The animals that died comprised one male calf and 52 females aged between 0.5 to 2.8 years and 6 to 26 years (Dabin et al. 2008, Viricel et al. 2008). This suggests that weaned juveniles, subadult males and mature males segregate, at times, from nursery groups. Interestingly, genetic analysis of this nursery group revealed that variability within the mass-stranded group was similar to variability observed in single strandings of common dolphins along the French coastline; that is, mature females within the nursery group were not necessarily genetically related (Viricel et al. 2008). In June 2008, 26 *Delphinus delphis* died during a mass stranding in Cornwall, SW England (Jepson & Deaville 2009). The group comprised five lactating females and sexually immature individuals of both sexes, ranging in age from 2 to 9 years (Jepson et al. in press). Finally, a group of seven common dolphins live stranded in May 2002 on the western coast of Ireland resulted in the deaths of four individuals (two males and two females), ranging in age from 1 to 8.5 years (Glanville et al. 2003, Murphy 2004), which provides further evidence that juveniles and subadults may segregate, at times, from other social groups.

Examination of individuals incidentally caught during spring and summer in Portuguese gill nets, beach seine nets and trawls revealed that sexually mature females only associated with young calves, and sexually immature males either formed separate groups (sometimes with small numbers of immature females) or joined mature male groups (Silva & Sequeira 2003). There was almost a complete absence of sexually immature females in the by-catch (Silva & Sequeira 2003). In contrast, an assessment of dolphin by-catch in the Irish and French albacore tuna (*Thunnus alalunga*) drift net fishery suggested mixing of nursery and mature male bachelor groups, though this was during the mating period (Murphy & Rogan 2006). The NE Atlantic albacore tuna drift-net fishery operated over a large area, primarily beyond the continental shelf (see 'Fisheries interactions' section). The lack of juveniles and subadults (3 to 8 years) incidentally caught in this type of gear, which has low species selectivity above a minimum body size (Northridge 2009), suggests a significant summertime segregation of juveniles and subadults from other groups over a large geographic area, though this requires further investigation.

Mirimin et al. (2012) assessed group composition of *Delphinus delphis* that mass-stranded on the western coast of Ireland or were by-caught in the Irish albacore tuna and UK bass (*Dicentrarchus labrax*) fisheries; these authors used 14 microsatellite loci and 360 bp of the mtDNA control region. Parentage and kinship analyses revealed that dolphins caught in the same net tended to be unrelated to each other, with the exception of mother-offspring pairs (Mirimin et al. 2012). Individuals from the same group rarely shared the same mtDNA haplotype, apart from mother-offspring pairs, indicating the presence of multiple maternal lineages.

Feeding ecology

In the wild, several distinct feeding strategies have been described for individual common dolphins, including high-speed pursuits, 'fish whacking' (striking with the tail) and 'kerplunking' (rapid tail movement on the surface) (Neumann & Orams 2003, Burgess 2006), while cooperative feeding allows dolphins to exploit shoaling prey in an energetically advantageous way (Young & Cockcroft 1994, Brophy 2003). Coordinated feeding strategies includes 'carouselling', line abreast, wall formation, synchronous diving and bubble cloud production (Neumann & Orams 2003, Peschak 2005, Burgess 2006). Results from studies of feeding behaviour in New Zealand suggest that variation in prey distribution and productivity, possibly as a result of differing habitats (i.e., shallow waters vs. open ocean), may affect strategy selection (Neumann & Orams 2003, Burgess 2006, Stockin & Orams 2009).

Common dolphins in the NE Atlantic have been observed in mixed feeding aggregations comprising other cetaceans (e.g., *Stenella frontalis*, and *Tursiops truncatus*), large tunas and seabirds (Clua & Grosvalet 2001). In this region, common dolphins have shown associations with albacore tuna, though it is not known how long these associations last. The stomach contents of *Delphinus delphis* caught in drift nets set for albacore tuna included all the prey species (fish) found in the stomachs of tuna (Hassani et al. 1997). In the ETP, the strong associations between yellowfin tuna *Thunnus albacares* and pantropical spotted dolphins *Stenella attenuata* were attributed to the risk of predation, resulting in these species forming large, mixed-species groups, and not due to feeding advantages (Scott et al. 2012). At dawn and dusk off the Azores, feeding aggregations initiated and sustained by common dolphins feeding at the periphery of bait balls and actively herding fish towards the surface can be broken down by the arrival of large tuna (*Thunnus thynnus, T. albacares*) as they swim straight into the bait ball, foraging on fish (Clua & Grosvalet 2001). This suggests that the tunas benefit from these aggregations by accessing prey more easily. Most of the information on diet of common dolphins in the NE Atlantic arises from studies of stomach contents of stranded and by-caught individuals. While these studies have been highly informative, they are limited by sampling biases. They provide information on dietary preferences of individuals inhabiting inshore waters primarily during winter, when most strandings occur, and of by-caught dolphins that were either feeding on target prey species of a particular fishery or opportunistically exploiting enhanced prey availability around fishing operations, including non-target species of those fisheries. For example, common dolphins have been reported to feed both on discards and directly from the cod end of trawls and inside trawl nets on small non-target prey species (see 'Fisheries interactions' section). In addition to seasonal and age-sex biases in sampling, differences in the gut passage times and rates of digestion of different prey species may create additional limitations and biases (Pusineri et al. 2007). Recently, fatty acid and stable isotope signatures have been used to discriminate spatial and temporal differences in diet in the NE Atlantic population of *Delphinus delphis*.

Analysis of non-lethal biopsies of tissues such as blubber and muscle may provide a good general overview of the diet of this species. Variables such as age, sex and health status should be taken into account when interpreting results from these studies (see 'Ecological stocks' section).

Temporal, geographic and seasonal variations in diet

Common dolphins are opportunistic feeders (Young & Cockcroft 1994), though more recently it has been suggested that they select prey based on energy densities (Santos et al. 2004, Brophy et al. 2009, Spitz et al. 2010). In the NE Atlantic, the diet of common dolphins includes a wide variety of fish and squid species, though it is predominantly composed of a few main species that vary with season and region (Murphy et al. 2008). In areas where preferred prey species are in high abundance, common dolphins tend to select those species. Consequently, diet displays strong interannual and seasonal variation (Murphy et al. 2008). During winter, common dolphins in inshore waters prey mainly on shoaling pelagic fishes, whereas in summer, *Delphinus delphis* caught in tuna drift nets set at night beyond the continental shelf edge had fed predominantly on squid and mesopelagic fishes, such as lanternfish (Myctophidae). Other small delphinids in the NE Atlantic show similar diet plasticity. For example, striped dolphins, *Stenella coeruleoalba*, switch between migrating mesopelagic prey to neritic or coastal prey types (Spitz et al. 2006). It is not known whether common dolphins in the NE Atlantic follow the migratory patterns of their preferred prey, as dolphins have not been fitted with animal-borne tracking devices in this region. However, inshore movements of common dolphins into the Celtic Sea and western English Channel in winter have been attributed to feeding opportunities (see 'Distribution and abundance' section).

Offshore waters

Common dolphins caught in Irish and French albacore tuna drift nets during the 1990s (see 'Fisheries interactions' section) were predominantly feeding at night, when the migrating deep scattering layer approaches the surface (Hassani et al. 1997, Pusineri et al. 2007, Brophy et al. 2009). The dietary preferences of these by-caught common dolphins were assessed by two different studies. Pusineri et al. (2007) examined stomach contents from 63 *Delphinus delphis* caught in the French tuna drift nets from 39°N to 50°N and 10°W to 21°W. Animals sampled were biased towards younger dolphins, and age-sex differences in dietary preferences were not assessed in the study. Brophy et al. (2009) analysed dietary remains from 58 *D. delphis* caught in Irish drift nets, predominantly along and off the continental slope, SW of Ireland. The main prey species of animals caught in French drift nets were the myctophids *Notoscopelus kroyeri*, *Benthosema glaciale*, and *Myctophum punctatum*; the sternoptychid *Maurolicus muelleri*; and cephalopods *Ancistroteuthis lichtensteinii*, *Gonatus steenstrupi*, *Brachioteuthis riisei*, and *Teuthowenia megalops* (Pusineri et al. 2007). A number of these are small schooling species, a similar prey profile to that consumed in inshore waters of the Bay of Biscay (Pusineri et al. 2007). *Notoscopelus kroyeri* dominated the diet, occurring in 84% of stomachs, and accounted for 65% of abundance by number. Similar prey types were observed in the stomachs of *Delphinus delphis* caught in Irish drift nets. Myctophids were the most common family of fish and comprised *Myctophum punctatum* (29% of all prey items) and *Notoscopelus kroyeri* (22%). Of lesser importance was the main cephalopod species consumed, *Brachioteuthis riisei*. Other species included *Arctozenus risso*, *Maurolicus muelleri*, *Benthosema glaciale* and *Gonatus steenstrupi*. Interestingly, the horse mackerel (*Trachurus trachurus*, Carangidae) was the numerically dominant species (38%), with a high index of importance, though was listed only fourth in order of significance expressed as percentage frequency of occurrence. A few prey species were common to the stomachs of dolphins caught off shore and animals stranded along the Irish coast, a situation not observed in the Bay of Biscay (Meynier 2004, Pusineri et al. 2007). These prey species included horse mackerel, blue whiting (*Micromesistius poutassou*), *Benthosema glaciale* and *Brachioteuthis riisei* (Brophy et al. 2009).

Along the Mid-Atlantic Ridge, Doksæter et al. (2008) assessed spatial correlations between dolphin occurrence and candidate prey organisms that were recorded acoustically and sampled by midwater trawling. This indicated that mesopelagic fishes such as *Lampanyctus macdonaldi*, *Stomias boa ferox* and *Chauliodus sloani* were important prey for common dolphins.

Inshore waters

Ireland The diet of 76 stranded common dolphins (including by-caught dolphins that subsequently stranded; 21%) and individuals retrieved directly from fishing gear (9%), sampled over a 15-year period (1990–2004), was primarily composed of fish (97% by number, cephalopods 3% by number) represented by at least nine families and 14 species (Brophy et al. 2009). Gadoids were the best-represented family, with *Trisopterus* spp. (*T. esmarkii* and *T. minutus*) the most important species. *Trisopterus* spp. were the most common prey during the summer (April to September) and winter (October to March). Although gobies appeared to dominate the diet during the winter, this was largely due to the occurrence of 1822 otoliths in one stomach. Other fish species of importance were blue whiting, whiting (*Merlangius merlangus*), *Argentina* sp., Atlantic herring (*Clupea harengus*), and the European sprat (*Sprattus sprattus*). The average prey size present in stranded common dolphins (9.7 cm, SD = 6.45 cm) was considerably larger than that found in the stomachs of dolphins caught in Irish tuna drift nets (4.2 cm, SD = 2.25 cm). For whiting, 40% of the fish present in the diet were above the commercial minimum landing size (Brophy et al. 2009).

United Kingdom Common dolphins off the SW coast of the United Kingdom consume a wide variety of fish, but primarily sardine *Sardina pilchardus,*[*] mackerel, *Scomber scombrus*, horse mackerel, Norway pout, *Trisopterus esmarkii*, other clupeids and various squid species (Pascoe 1986, Kuiken et al. 1994, Natural History Museum 1995, Gosselin 2001). Gosselin (2001) assessed non-empty stomachs from 18 by-caught dolphins that were stranded along the SW coast of the United Kingdom between December 2000 and April 2001. Sardine and horse mackerel comprised 40% and 37% of the stomach contents, respectively, with mackerel and Norway pout found to a lesser extent. Diets were similar to those of by-caught common dolphins that were mass stranded along the SW coast of the United Kingdom during the first quarter of 1992 (Kuiken et al. 1994). Mackerel and pilchards dominated the diet; the size of the latter was remarkably large, ranging from 14 to 30 cm in length. In Scottish waters, 14 fish taxa and 2 cephalopod taxa (*Alloteuthis subulata*, unidentified Sepiolidae) were identified in the stomachs of nine common dolphins that were stranded between 2000 and 2003. Mackerel, followed by whiting, were the main prey consumed, together making up more than 40% of the estimated prey weight (Learmonth et al. 2004b).

France Stomach contents analysis of 26 common dolphins that were stranded along the Normandy coast (English Channel) revealed they consumed mainly gadoid fish (*Trisopterus luscus*), gobies and mackerel (De Pierrepont et al. 2005). Cephalopods occurred in small numbers in the diet. In inshore waters of the Bay of Biscay, four taxa contributed to the majority of the dietary remains of 71 common dolphins that were stranded between 1999 and 2002. These included sardine, anchovy *Engraulis encrasicolus*, sprat and horse mackerel, which represented 44.9%, 22.6%, 8.0% and 5.0% by mass of the fresh diet, respectively (Meynier et al. 2008). Cephalopods only constituted 5.6% of the diet by mass, comprising *Alloteuthis* spp., *Loligo* spp., and unidentified Loliginidae. Meynier et al. (2008) reported that the diet displayed strong seasonal and interannual variations in terms of both prey species composition and prey size distributions, reflecting prey availability in the area. However, estimated daily food intakes changed relatively little, as all diets included a high

---

[*] The UK Sea Fish Industry Authority classifies sardines as young pilchards. One criterion suggests fish shorter in length than 15 cm are sardines, and larger ones are pilchards.

proportion of lipid-rich fish (sardine, sprat, anchovy and mackerel; 73 to 93% by mass). Sardines were the dominant prey by mass in summer/autumn and winter, whereas the quantity of sprats was highest in spring. Horse mackerel was absent from the diet during the summer, and anchovy did not show any marked seasonal variation, though female common dolphins did consume more of this prey type during the summer. As noted in Irish waters, gobies were the most important prey by number during the winter (to summer), though it is not known if these small demersal fishes were primary prey or prey of other fish secondarily eaten by the dolphins (Meynier et al. 2008).

Spain Off Galicia, blue whiting and sardine together comprised 56.5% of reconstructed prey weight present in the stomachs of 414 stranded individuals (one-third of which showed signs of net entanglement), sampled predominantly during winter from 1991 to 2003 (Santos et al. 2004). Gobies were the most numerous prey group. Other species included horse mackerel, sand smelt *Atherina* sp., and cephalopods such as *Loligo* sp. In total, 25 fish species and 15 cephalopod species were identified in the dietary remains. Although the diet was primarily composed of only two species, there was evidence that common dolphins off Galicia were opportunistic feeders (Santos et al. 2004). For example, higher numbers of sardines were consumed in years of higher sardine abundance and lower recruitment of blue whiting. This was possibly due to sardines having a higher calorific value than blue whiting (Santos et al. 2004). Strong seasonal variations were also observed in dietary preferences, with higher numbers of gobies, *Atherina* sp. and small squid (*Alloteuthis* sp.) consumed during the first quarter. On the whole, dolphins were exploiting different size classes of prey compared to those targeted by fisheries, apart from sardines (15.5- to 22.5-cm long), which were well above the minimum landing size of 11 cm (Santos et al. 2004). A recent stable isotope study has confirmed the importance of blue whiting and sardines in the diet of common dolphins off Galicia (Mèndez-Fernández et al. 2012).

Portugal Silva (1999) analysed stomach contents from 50 stranded and by-caught common dolphins sampled between 1987 and 1997. Even though 27 different fish species and 8 cephalopod species were identified, the diet was mainly composed of a small number of taxa (Silva 1999). Six fish species (sardine, blue whiting, *Atherina* sp., horse mackerel and scombrid species) composed 84% of the total estimated weight, and sardines were the most important prey item, occurring in 81% of stomachs (27% of prey by number and 43% of estimated weight). Common dolphins stranded along the Portuguese coast appear to have a higher proportion of sardines in their diet than animals stranded along the Galician coastline (Santos et al. 2004). Spring acoustic surveys carried out by both countries since 1986 showed that sardines are more common in Portuguese waters than off Galicia (Carrera & Porteiro 2003, Santos et al. 2004).

Age- and sex-related dietary requirements

Dietary studies indicate that weaning can commence between 3 and 6 months after birth (Brophy et al. 2009). Studies of reproductive parameters, however, suggest that females may lactate for up to 10 months after parturition (Murphy 2004), and the length of the lactation period may increase with maternal age, as noted in other delphinid populations, including common dolphins elsewhere (e.g., Danil & Chivers 2007).

Few studies have assessed differences in dietary preferences between age-sex maturity classes. Meynier et al. (2008) reported that there were significant differences between age-sex maturity groups of common dolphins stranded in the Bay of Biscay, in terms of both prey species composition and prey size distributions. Prey length was weakly correlated to dolphin length, and prey composition of mature males was less diverse than that of mature females and immature dolphins, with mature males predominantly consuming sardines. Adult males also fed on larger sardines than the other age-sex groups. Off the Portuguese coast, limited variation in dietary preferences was observed between different age-sex maturity groups. Sardine was the dominant prey species in the diet of all groups, with the exception of immature males, which ate more blue whiting (Silva 1999). Interestingly, both Silva (1999) and Meynier et al. (2008) reported that cephalopods were a minor component of the diet of mature males. Diets of *Delphinus delphis* in both offshore (summer) and inshore (predominantly winter) habitats in Irish waters revealed no significant difference in total prey numbers, prey species number, or proportion of cephalopods in the diet between different age-sex maturity groups, except that the stomachs of stranded females contained a significantly higher number of prey items than stranded males (Brophy et al. 2009). In addition, there were positive relationships between dolphin body length and both total prey numbers and number of prey species in the offshore group.

The difference in diets between inshore and offshore areas recorded by Brophy et al. (2009) provides evidence of seasonal offshore-onshore movements of common dolphins in the NE Atlantic. The energy requirement of (pregnant and) lactating *D. delphis* and their calves may contribute to the offshore movement of some mature individuals (and calves) during the spring and summer to take advantage of nutrient-rich prey at times when neritic prey are nutrient poor (or have dispersed to/from spawning grounds) (Brophy et al. 2009). Many species of pelagic fish, such as *Trisopterus* spp., whiting, sprat, and Atlantic horse mackerel, spawn to the south and SW of Ireland during the spring and summer and then migrate north (Jákupsstovu 2002, Dransfeld et al. 2004, Brophy et al. 2009). During spawning, the lipid content of these fish species falls, reducing their calorific value as prey (Brophy et al. 2009). Interestingly, the offshore dietary sample of horse mackerel comprised fish less than 1 year of age (i.e., prespawning stage). Myctophids, which are consumed in offshore waters during the summer, are reported to have higher lipid content than other marine fish species (Saito & Murata 1998, Lea et al. 2002, Brophy et al. 2009), though the size range (age class) of the myctophids that are consumed is unknown. As noted previously, the general absence of juvenile/subadult *Delphinus delphis* in the by-catch of Irish tuna drift nets suggests that they were not present in the area where this fishery operated (Murphy & Rogan 2006), which may suggest a different feeding strategy in summer. Off Portugal, immature males were found to consume blue whiting and showed a tendency to be caught in pelagic trawls targeting that species (Fernández-Contreras et al. 2010) (see 'Fisheries interactions' section).

A subsequent study assessing energy requirements of common dolphins in the Bay of Biscay has confirmed the selection of high-quality foods during summer. As noted, the diet of individuals sampled during the 1990s in this region was dominated by the myctophid *Notoscopelus kroyeri*, a high-energy prey species (Spitz et al. 2010). Surveys of the epi- to mesopelagic oceanic fish community off the Bay of Biscay in October 2002, 2003 and 2008 revealed that the alepocephalid *Xenodermichtys copei*, a low-energy prey that was not consumed by common dolphins, was the most abundant species (Spitz et al. 2010).

Health status and cause of death

National marine mammal stranding networks have for many years collected basic data (date, location, species, etc.) on single and mass stranding events in many countries. Some of these datasets are now nearly 100 years old. However, it is only in recent years that more-detailed investigations have been conducted on stranded animals through systematic necropsies of stranded carcasses. These new studies have provided new insights into health status, causes of mortality and causes of MSEs that could not be revealed by any other method.

Stranding patterns

Strandings or beaching of carcasses—dead or alive, single or mass (involving two or more individuals)—can be influenced by many factors, some known and some more speculative. These include variation in cetacean population density, interannual and seasonal variations in climatic factors like prevailing onshore winds, and coastline length. Drivers of cetacean mortality can also influence stranding patterns, including behaviour and location of commercial fishery operations, high-intensity acoustic activity and disease epizootic events (Murphy 2004, Geraci & Lounsbury 2005, Deaville & Jepson 2011). There are a number of speculative 'natural' causes for cetaceans becoming stranded alive, including behavioural tendencies and group cohesion, disease in one or more individuals in a social group leading to some or all of the remainder of the group being stranded alive; unusual environmental conditions, such as electrical storms and other meteorological events; becoming trapped by a receding tide; geomagnetic disturbances and errors in navigation while following geomagnetic contours; confused navigation arising from 'bathymetric conditions' (i.e., misleading depth contours); disturbance of echolocation by multiple reflections in bays; pursuing prey too close to shore; and earthquakes (reviewed in Geraci & Lounsbury 2005, Sundarama et al. 2006). However, few MSEs have been forensically investigated. More recently, anthropogenic factors, such as high-intensity acoustic transmission, as used in naval operations, have been increasingly implicated in cetacean MSEs (Jepson et al. 2003, Fernandez et al. 2005, Southall et al. 2006, Weilgart 2007a), including an MSE of short-beaked common dolphins in the UK in 2008 (Jepson et al. in press).

Common dolphins frequently become stranded in the NE Atlantic, especially along the coastlines of Ireland (Berrow & Rogan 1997, Murphy 2004); Britain (Sabin et al. 2002, Jepson 2005, Deaville & Jepson 2011); France (Tregenza & Collet 1998, Van Canneyt et al. 2011); Spain (López et al. 2002); and Portugal (Silva & Sequeira 2003). Recorded strandings of common dolphins have increased since 1990, possibly as a result of increased coastal vigilance, a change in the distribution and abundance of common dolphins, an increase in adverse anthropogenic activities, or a combination of these factors. Annual numbers of strandings have fluctuated in recent years in more northern waters, with similar trends apparent in France and the United Kingdom (see Figure 7; Deaville & Jepson 2011). Strandings have shown a consistent spatial and seasonal pattern, with pronounced winter peaks (ICES WGMME 2005), and for most countries, a high proportion of the common dolphins that became stranded during these winter peaks exhibited external evidence of

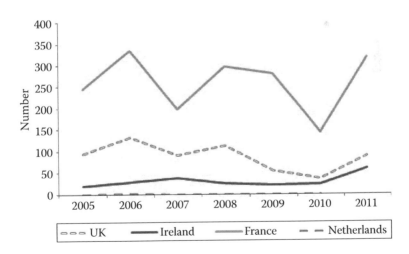

Figure 7 Interannual variation in strandings of short-beaked common dolphins in North-west Europe (2005–2011). One common dolphin became stranded on the Dutch coastline in 2006. (Adapted from Deaville & Jepson 2011. Data provided by the UK Cetacean Strandings Investigation Programme; Irish Whale and Dolphin Group; Centre de Recherche sur les Mammifères Marins, Université de La Rochelle, France; and Naturalis [National Museum of Natural History] in the Netherlands.)

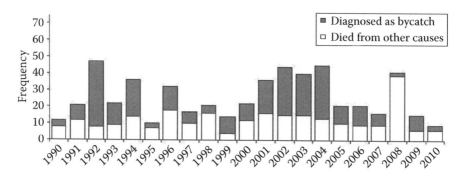

Figure 8 Cause of death for common dolphins necropsied by the UK Cetacean Strandings Investigation Programme (*n* = 542, 1990–2010). Data for 2008 include 26 common dolphins that died during a mass stranding event in Cornwall.

by-catch or pathological evidence of by-catch as the most likely cause of death (Kuiken et al. 1994, Tregenza & Collet 1998, Murphy 2004, Jepson 2005, Deaville & Jepson 2011, Pikesley et al. 2011, Castège et al. 2012, Peltier et al. 2012). Analysis of the age of by-caught dolphins that subsequently became stranded along the Irish, UK, and French coastlines between 1990 and 2006 showed an increased mortality of juveniles, with a peak in mortality of 3-year-olds (Murphy et al. 2007b). It is not established what fisheries or type of fishing nets were involved in these incidental mortalities.

On the Irish coast, the majority of recorded common dolphin strandings have been along the western and southern shores, corresponding with areas of highest sighting rates and direction of prevailing westerly winds (Murphy 2004). Between 1990 and 2003, of all strandings, 49% occurred during the first quarter of the year, and at least 25% of these dolphins were identified as by-catch (Murphy 2004). There was a peak in strandings of 37 common dolphins in 2003 (Brophy et al. 2006); prior to this, the average stranding rate was 11 dolphins per year (Murphy 2004). Peaks in annual numbers stranded (>15 dolphins) resulted from both fisheries interactions and live stranding events (Murphy 2004).

A large number and proportion of common dolphins that became stranded along the UK coastline were diagnosed as by-catch in most years since 1991, mostly between January and April along the SW coast of England (Cornwall and Devon) (Deaville & Jepson 2011). The annual number and by-catch proportion (see Figure 8) of stranded common dolphins in SW England increased in the late 1990s to a peak in 2004 (annual stranding numbers not shown) and then gradually declined thereafter (Jepson 2005, Deaville & Jepson 2011). The reasons for the increase and then reduction in numbers (and proportion) of stranded common dolphins diagnosed as by-catch in SW England around 2004 are not fully understood (Jepson 2005, Deaville & Jepson 2011). In 2011, however, UK stranding numbers were more than twice that of 2010 (see Figure 7).

Since 1989, there were several years with high stranding rates of common dolphin (>250 dolphins per annum) along the French Atlantic coastline, with up to 508 *Delphinus delphis* reported stranded in the year 2000; common dolphins also became stranded at lower frequencies along the French western channel coast (Tregenza & Collet 1998, Van Canneyt et al. 2011). Although there was considerable interannual variability, years of high stranding levels corresponded to events of multiple strandings (ICES WGMME 2005). These events typically occur over periods of 10 to 30 days and involve stranding of predominantly common dolphins, of which about two-thirds show amputation of tail flukes, pectoral flippers or dorsal fin; broken rostrum; or opening of the abdominal cavity—features typical of by-caught animals that have been returned to the sea by fishers (ICES WGMME 2005, Murphy et al. 2007b). The majority of strandings during these peaks are in the southern Bay of Biscay between January and March, and the sex ratio is skewed towards (juvenile) males (ICES WGMME 2005, Murphy et al. 2007b). The fact that juvenile males are more heavily exposed to adverse interactions with fishing gear may result from differential utilization of space and food (ICES WGMME 2005).

In the southern Bay of Biscay between 1980 and 2002, there was strong seasonality in the at-sea encounter rate of common dolphins (sightings data), strandings, and fishing activity (measured by landings of French fleets), with peaks in all three during the first quarter (Castège et al. 2012). However, the relatively high at-sea encounter rate during summer—a period of fewer fish landings—was not reflected in strandings (Castège et al. 2012). Analysis of drift patterns of cetacean carcasses suggested that as many as 57% and 87% of all stranded common dolphins recorded along the French Atlantic coast originated from the continental shelf of the southern Bay of Biscay within the 100-m and 500-m isobaths, respectively (Peltier et al. 2012). Interestingly, of the tagged dead dolphins released by fisheries in the region, only a very small percentage were actually recovered ashore, and results suggested that approximately 84% of dead cetaceans would sink (Peltier et al. 2012). Peltier et al. (2012) estimated that the numbers of stranded common dolphins reported in winter between 2004 and 2009 (207 individuals on average, of which a minimum of 50% were by-caught) would be about one order of magnitude below the true numbers dying at sea (approximately 2000 individuals per winter, of which at least 1000 would have been by-caught). In recent years, however, as in the UK, the stranding numbers have declined, with 136 common dolphins reported stranded along the Atlantic coast of France (and 8 along the English Channel) in 2010. Of these, at least 60 dolphins showed signs of by-catch in fishing gear (Van Canneyt et al. 2011). Stranding numbers increased again in 2011, though they were still lower than records from the 1990s and early 2000s (Van Canneyt et al. 2012; see Figure 7).

The common dolphin is the most frequently recorded cetacean species in strandings along the Galician coast (López et al. 2002). The annual stranding numbers increased from 1990 to 1999, and as in other areas, there was a peak in strandings during the first quarter (March) (López et al. 2002). This peak coincided broadly with the time of the year when the upwelling index is lowest and winds from the west predominate. Interestingly, a secondary peak was observed in August. Overall, the number of strandings was low in autumn, when easterly winds prevail (López et al. 2002). Between 1990 and 2007, there were 1747 common dolphins stranded along the Galician coastline, of which 606 were 'fresh' enough to assess for evidence of fisheries interactions (Read et al. 2009). There were 146 common dolphins diagnosed as by-catch, with more juveniles than adults recorded and more males than females (López et al. 2002, Read et al. 2009). López et al. (2002) reported significant seasonal differences in average size, with the smallest average body length observed in the first quarter and the largest in the third quarter of the year. The percentage of stranded common dolphins showing signs of fishery interactions has increased over time, from 23% during the 1990s (López et al. 2002) to around 41% in the late 2000s (Read et al. 2009).

Between 1975 and 1998, there were 431 common dolphins stranded along the coast of Portugal (Silva & Sequeira 2003). Interannual variations and pronounced peaks in strandings were reported in some years, and as noted in other studies, annual numbers generally increased during the study period. Highest numbers of strandings occurred from December to April, with a peak in March, primarily along the northern and central coasts. The geographic distribution of strandings was attributed to differences in the distribution or abundance of animals, oceanographic conditions, or bathymetric conditions of the region (Silva & Sequeira 2003). The abundance and spatial distribution of the main prey species (sardines and blue whiting) of the common dolphin off the Portuguese coast corresponded with the distribution of strandings. Significantly more immature individuals were found stranded, and the sex ratio was also biased towards males. Fisheries interactions may have contributed to at least 47% of mortalities in the strandings data (Silva & Sequeira 2003).

Overall, stranding recording programmes in the NE Atlantic show a peak in strandings during the first quarter of the year, stranded individuals are biased towards immature males, and a large proportion of animals show evidence of incidental capture.

Mass mortality events

Mass mortality events have been documented in common dolphins in the NE Atlantic. For example, a common dolphin mass mortality event resulting from by-catch occurred in the United Kingdom in 1992 (Kuiken et al. 1994), though it should be noted that high incidences of by-caught dolphins along the SW coast of England during the winter months are annual events (Jepson 2005, Deaville & Jepson 2011). In 1994, a mass mortality occurred in the Black Sea due to distemper caused by cetacean morbillivirus (CMV) infection (Birkun et al. 1999). This CMV epizootic event in Black Sea common dolphins followed a similar CMV epizootic event of striped dolphins in the Mediterranean Sea between 1990 and 1992 (Kennedy 1998).

Cetacean MSEs are often rather loosely described as two or more individuals of the same species (excluding a cow-calf pair) coming ashore, usually alive, at the same time and place (Geraci & Lounsbury 2005). Common dolphin MSEs have occasionally been recorded in the United Kingdom (Jepson & Deaville 2009), France (Dabin et al. 2008, Viricel et al. 2008) and Ireland (Murphy 2004). A range of causes has been proposed (reviewed in Geraci & Lounsbury 2005), including local topography, presence of shifting sandbanks, exceptionally low tides (Murray & Murphy 2003). More recently, proximity to international naval exercises has been established as the most probable cause of a UK common-dolphin MSE (Jepson & Deaville 2009, Jepson et al. in press).

Infectious diseases

All species coevolve with micro- and macroparasites, and common dolphins are no exception (Gibson et al. 1998). A range of macroparasites was recorded in common dolphins in waters around the United Kingdom (Gibson et al. 1998) and NW Spain (Abollo et al. 1998a), and *Anisakis simplex*-associated gastric ulcers were observed in common dolphins in NW Spain (Abollo et al. 1998b), a condition that has been observed in stranded common dolphins from other countries (e.g., Ireland). Non-specific reactive hepatitis and chronic parasitic cholangitis with lymphoid proliferation have been described in common dolphins stranded in the Canary Islands (Jaber et al. 2003). Pulmonary angiomatosis was observed in 71% (25/35) of common dolphins stranded in the Canary Islands, which was strongly associated with pulmonary parasitic infestation (Diaz-Delgado et al. 2012).

Microparasites include CMV, which caused the epizootic event in Black Sea common dolphins (Birkun et al. 1999). A novel *Helicobacter* sp. was isolated and characterized from gastric ulcers in both common and Atlantic white-sided dolphins (*Lagenorhynchus acutus*) in the NW Atlantic, suggesting that *Helicobacter* species may play a role in the etiopathogenesis of gastritis and gastric ulcers in cetaceans (Harper et al. 2000). Subsequently, *Helicobacter cetorum* infection has been characterized from gastric ulcers in UK-stranded dolphins, including the common dolphin (Davison et al. unpublished data). Meningoencephalitis associated with *Brucella ceti* infection was also recently reported in a UK-stranded common dolphin (Davison et al. unpublished data), which is consistent with similar lesions more commonly described in striped dolphins in the United Kingdom (Davison et al. 2009). A range of macro- and microparasitic (bacterial and fungal) infections have sometimes been the cause of stranding and death in UK-stranded common dolphins (Jepson 2005, Deaville & Jepson 2011). A study assessing the presence of epidermal virus lesions, morphologically similar to the pox virus, on common dolphins off the Irish coast revealed a low prevalence, with evidence of sexual variation, as twice as many males than females were infected (Murphy 1999).

Non-infectious diseases

A range of traumatic injuries and other causes of death have been diagnosed in common dolphins, including by-catch (Kuiken et al. 1994), boat collision (Deaville & Jepson 2011) and fatal attack from bottlenose dolphins (interspecies aggression) (Murphy et al. 2005b, Barnett et al. 2009). Tumours are rarely found in common dolphins. The first reported case of a meningioma (tumour) in any cetacean was a microcystic meningioma in a common dolphin stranded on the French Atlantic coast, which was diagnosed by immunohistochemical and ultrastructural analysis (Miclard et al. 2006). A single cavernous hemangioma was found in the lung of one common dolphin in the Canary Islands (Diaz-Delgado et al. 2012). Outside the NE Atlantic, a spontaneous case of renal heterotopia involving the lung was reported in a female adult common dolphin found stranded alive on the northern Adriatic Sea coast of Italy (Di Guardo et al. 2005).

Between 1992 and 2009, a small number of UK-stranded cetaceans, including five common dolphins, were diagnosed with acute and chronic forms of gas embolism (Jepson et al. 2003, Jepson 2005, Deaville & Jepson 2011). The cause of gas embolism is not known but may have a similar mechanism to decompression sickness in humans and experimental animals and be related to excessive supersaturation of tissues with nitrogen on ascent (Jepson et al. 2003, Jepson 2005, Deaville & Jepson 2011). More recently, *in vivo* gas formation has been detected by ultrasound in live-stranded dolphins, including common dolphins stranded in the United States, with off-gassing of supersaturated blood and tissues again considered the most probable origin of the bubbles (Dennison et al. 2011). The close proximity of naval exercises to a mass stranding of beaked whales with acute gas and fat embolic lesions on necropsy indicates a potential role for high-intensity anthropogenic activities (like naval exercises) in the pathogenesis of gas embolism, possibly due to abnormal dive profiles (Jepson et al. 2003, Férnandez et al. 2005, Jepson et al. 2005b). Another suspected case of fatal gas embolism was a mature female common dolphin that became stranded on the western coast of Ireland in 2004 (Murphy & Rogan 2004).

Reproductive failure and abnormalities of the reproductive tract

Disturbance at any point of reproduction in marine mammals can lead to failure, evident as abortion, stillbirth, premature birth, or illness or death of the newborn (Geraci & Lounsbury 2009). Reproductive failure can be caused by genetic defects, congenital disorders, nutritional or environmental stress, systemic infection, high levels of certain anthropogenic contaminants/endocrine disruptors and marine biotoxins (Geraci et al. 1999, Reeves et al. 2001, Geraci & Lounsbury 2009). In the NE Atlantic common dolphin population, there has been no reported increase in occurrence of preterm births, stillbirths or early newborn mortality since the year 2000. Within a control group of 'healthy' by-caught common dolphins that subsequently became stranded on the UK coastline, 8.9% of mature females showed evidence of recent miscarriage during their second trimester (Murphy et al. 2012b). There was also an association between high tissue contaminant levels and the incidence of miscarriage (see 'Pollutants' section).

A six-year-old immature female common dolphin found stranded on the SW coast of the United Kingdom was diagnosed as a true hermaphrodite. The individual in question had one ovotestis containing both ovarian follicles and testicular tubular elements and a contralateral ovary (Murphy et al. 2011). This is the first reported case of an ovotestis in a cetacean species, and it is not known if this disorder of genital development is due to abnormalities of genetic or chromosomal origin or inappropriate hormone exposure. A number of other reproductive tract abnormalities have been identified in female *Delphinus delphis* stranded on the UK coastline, which are currently under investigation. These include vaginal calculi and numerous ovarian abnormalities, such as tumours and an ovarian cyst (Murphy et al. 2012b). Other conditions, such as genital warts (Gottschling et al. 2011) and infection of testicular issue with *Brucella ceti* (Jepson & Deaville 2009), have been reported in males.

The first cases of twins in common dolphins were recorded in the NE Atlantic population. A dead floating female common dolphin, found off the Galician coastline in June 1998, was pregnant with a female fetus measuring 72 cm in body length (with a curved caudal area) located in the left uterine horn and a slightly deformed male fetus measuring 46 cm in body length located in the uterus (González et al. 1999). A 12-year-old by-caught pregnant female common dolphin, in excellent nutritional condition and health status, died in December 1994 off the UK coast. Both fetuses were male, measuring 27 and 31 cm, and were located in the left uterine horn (Murphy et al. 2012b).

Threats and pressures

Overview of past and present threats

A large number of pressures and threats have the potential to have an impact on common dolphins in the NE Atlantic, the most significant being adverse fisheries interactions. Others include climate change, pollutants, noise pollution and habitat disturbance. Additional possible threats are boat collisions and whale and dolphin watching. Although data are currently being collated on contemporary threats and pressures, there is no information on whether the population has been overexploited in the past or if habitat degradation or loss may have reduced carrying capacity.

Direct fisheries for small cetaceans have operated in the NE Atlantic, though most of the information available on catch rates is only anecdotal. The Portuguese small-cetacean fishery officially operated until 1981, when all cetaceans were given legal protection in Portuguese waters; nonetheless, Silva & Sequeira (2003) noted that common dolphins stranded along the Portuguese coasts until 1998 showed signs of having been hunted. In the 1970s and earlier, bow-riding dolphins were deliberately harpooned for food, and a rough extrapolation of activities of the French fleet fishing in the Celtic and Irish Seas at that time, based on an estimated two dolphins captured per month per vessel, suggests a total of 3000 dolphins (not identified to species) were killed (ICES WGMME 2005). In addition, a retired fisherman from the French tuna lining fleet suggested that possibly 2000 dolphins (not just *Delphinus delphis*) were killed every year, a practice that may also have been undertaken by the Spanish tuna fleet (Antoine 1990, ICES WGMME 2005).

In the 1970s, set nets were not mechanized and were relatively short, and pelagic pair trawling had not yet been developed; thus, it is assumed that incidental capture of dolphins would have been less than following the introduction of new commercial fishing practices (ICES WGMME 2005). However, there may have been indirect effects of fishing activities on the local common dolphin population during the twentieth century through overfishing. Reduced availability of prey caused by overfishing and habitat degradation has been proposed as the main reason for the rapid decline in abundance of common dolphins in the Mediterranean Sea since the 1960s (Bearzi et al. 2003, Cañadas & Hammond 2008). Other putative factors that may also have contributed to the decline include incidental mortality in fishing gear and direct catches, contamination by xenobiotic chemicals resulting in immunosuppression and reproductive impairment, and environmental changes, such as increased water temperatures affecting ecosystem dynamics (Bearzi et al. 2003).

Fisheries interactions

Fisheries affect the NE Atlantic common dolphin population in two different ways: through 'operational effects' and 'biological effects'. Operational effects occur when individuals come into physical contact with fishing gear, which may result in serious injury or death (Northridge 2009). If the marine mammal dies and is subsequently discarded, the process is termed *by-catch* (Read 2008). Operational effects pose a serious threat to many populations of marine mammals due to their slow life histories and limited potential rates of increase (Reilly & Barlow 1986, Read 2008). Fisheries also considerably alter the trophic structure, species assemblages and pathways of energy flow (Pauly et al. 1998, Jackson et al. 2001, Myers & Worm 2003), resulting in ecological changes that may have adverse consequences for cetaceans (Read 2008). These interactions result in biological effects.

Operational effects

Even though fisheries observer programmes offer the potential for spatially explicit, effort-corrected and gear-specific estimates of capture and mortality levels (e.g., Tregenza et al. 1997, Morizur et al. 1999, Lewison et al. 2004, Leeney et al. 2008), there is still a lack of understanding regarding why dolphins are caught in nets. This is hampering scientists' ability to develop effective solutions, and overall, a better understanding of the behavioural interactions of marine mammals with fishing gear is required (ICES Sea Study Group for Bycatch of Protected Species [SGBYC] 2008). Furthermore, the sporadic nature of by-catch makes accurate assessment of its nature and the extent, as well as the development of solutions, problematic (Rihan 2010). Changes in fishing tactics may provide some solutions, such as the use of acoustic deterrent devices (i.e., pingers), enhancing acoustic detectability of nets, or simply avoiding setting nets close to cetaceans or in areas where cetaceans are known to be in high density. Although research has been undertaken on a wide variety of these approaches, to date no definitive mitigation measure has come to the forefront, with each having its own significant concerns. For example, pingers have several issues, including high cost, low resilience and relatively short battery life, potential habituation and displacement effects on marine mammals, and there are still questions about the practicalities of their use in commercial fisheries (Rihan 2010, ICES Working Group on Bycatch of Protected Species [WGBYC] 2012). Nevertheless, in recent years, various pinger types have been found to be effective in mitigating by-catch in different gear types. For example, the acoustic deterrent device DDD02-F has been successfully trialled to reduce common dolphin by-catch in the UK component of the midwater pair trawl fishery for sea bass (*Dicentrarchus labrax*) in the western English Channel, and the effectiveness of a variant in UK gill and tangle net fisheries is being assessed (Northridge & Kingston 2010, Kingston & Northridge 2011).

A large number of fisheries operate in the NE Atlantic, using various different gear types, ranging from towed, static to encircling nets, targeting a wide range of species, including prey species of the common dolphin. Where common dolphins and fisheries target the same fish resource, interactions are inevitable, and these may have negative effects on both the fishery, from an economic point of view, and the common dolphin population, through incidental mortality and resource competition (Brophy et al. 2009). In Australian waters, common dolphins have been observed taking fish directly from the cod end and foraging on discarded fish at the surface (Svane 2005). The latter was also reported by Couperus et al. (1997), who noted that common dolphins were either actively feeding in the vicinity of the pelagic trawl nets targeting horse mackerel (scad) off the SW coast of Ireland or may have been scavenging on discards from this fishery as fresh horse mackerel remains were found in the stomachs of dolphins caught in the trawl. Conversely, common dolphins have been observed, via an underwater video camera system, inside trawls targeting sea bass in the English Channel during winter. It was suggested that individuals may be actively feeding on small, non-target fish inside the net (Northridge et al. 2004). In addition, dolphins left and entered the trawl net at will, as there were sightings and resightings of one or more animals for over an hour (Northridge et al. 2004). Tregenza et al. (1997) suggested that common dolphin may be attracted to gill nets during hauling and shooting, especially when the headline floats strike the steel hoop used to spread the net at the stern of the boat as this produces a loud rhythmical 'tonal clatter'. Although results from a recent UK study further suggested that interactions with gear during shooting or hauling may have more of an effect on by-catch than gear characteristics, by-catch rates of common dolphins in bottom-set nets may also be driven by a temporal and spatial overlap of animals and fishing gear, rather than specific characteristics (e.g., soak time, mesh size) of that gear (Mackay 2011). In contrast, Nielsen et al. (2012) found that harbour porpoises avoided a gill net and appeared to be able to detect it at a distance of more than 80 m. Thus, the incidental capture of porpoises may be due to attention shifts or to auditory masking reducing their ability to detect the gill nets by echolocation (Nielsen et al. 2012).

The EU Council Regulation (EC) No. 812/2004 (European Union 2004) states that independent observations of fishing activities are essential to provide reliable estimates of the incidental catch of cetaceans. It is therefore necessary for

> monitoring schemes with independent on-board observers to be set up and for the designation of the fisheries where such monitoring should be given priority. In order to provide representative data on the fisheries concerned, the Member States should design and implement appropriate monitoring programmes for vessels flying their flag engaged in these fisheries. For small-sized fishing vessels less than 15 m overall length, which sometimes are unable to allow an additional person permanently on board as an observer, data on incidental catches of cetaceans should be collected through scientific studies or pilot projects. Common monitoring and reporting tasks also need to be set. (European Union 2004, p. 15)

For vessels less than 15 m, no specified level of precision or coverage or any other guidance on the level of monitoring is given. This has resulted in pilot projects being poorly implemented by some Member States (ICES WGBYC 2011). The regulation also states that the use of acoustic deterrent devices should be required in areas and fisheries with known or foreseeable high levels of by-catch of small cetaceans. However, this only applies to vessels 12 m or over in length.

Currently, monitoring of bottom-set nets in the Celtic shelf and English Channel (ICES subarea VIId–j) is not required under EU Regulation (EC) 812/2004 as pinger deployment is mandated by the regulation in this area. Therefore, since the introduction of pinger deployment in 2005, limited monitoring has been undertaken by some Member States. This has prevented not only an evaluation of the current rate of by-catch in these fisheries but also an evaluation of the effectiveness of pingers as a mitigation practice. By-catch is not a function of vessel length, and since the majority of European set gill net vessels are less than 15 m in length, controversy has arisen over the implementation of certain requirements for some fishers and not others (ICES WGBYC 2012). However, it should be noted that monitoring of by-catch on vessels under 15 m and measures to mitigate by-catch, if necessary, are mandated by the Habitats Directive.

Table 3 outlines the available incidental capture rates for common dolphins in various NE Atlantic fisheries over a 20-year period (1990 to 2009). Data presented are only for those fisheries where observations were actually recorded, and CVs or percentages of observed fishing effort are not presented. EU Regulation (EC) 812/2004 came into force in 2004, and despite notable improvements in reporting and observer coverage, it is still not fully meeting its objective (ICES WGBYC 2012). For all marine mammals, insufficient sampling in the right fisheries or areas has prevented sound management decisions to be made with respect to cetacean by-catch (ICES WGBYC 2012).

As can be seen in Table 3, however, some fisheries/gear types have higher by-catch rates than others. It should be noted that there are other fisheries for which there are no reliable estimates of by-catch. Even the available estimates should be used with caution as they only provide an incomplete assessment due to low and uneven sampling coverage, with some EU Member States still not fulfilling their monitoring objectives (ICES WGBYC 2011, 2012). Common dolphins are caught incidentally in pelagic trawls, drift nets (surface gill nets), static gear and seine nets, with the highest annual by-catch (2317 dolphins) reported in 2009. Ten years earlier, high rates of by-catch were reported (2101 dolphins in 1999) for the combined Irish, UK and French components of the albacore tuna drift net fishery; this fishery is explored in further detail through the conclusion of this section. Common dolphins have also occurred as by-catch in a number of other fisheries not listed in Table 3, such as Portuguese gill, beach seine and trawl nets (Silva & Sequeira 2003), and Spanish trawls (including 'very high vertical opening' bottom pair trawls for hake, *Merluccius merluccius*; ICES WGMME 2005), gill nets, long lines, and seine nets (López et al. 2003). Interviews with fishers from the Galician fleet between 1998 and 1999 suggested an annual by-catch of 200 cetaceans in inshore waters and around 1500 off shore, with the majority of these animals probably being common dolphins (López et al. 2003).

Monitoring of UK and Irish bottom-set gill net fleets operating in the Celtic Sea targeting hake (pollack, *Pollachius pollachius*, and other gadoids were also caught) between 1992 and 1994 indicated a by-catch rate of 1.4 dolphins per 1000 km of net and a total annual by-catch of 234 (95% CI = 78–702) *Delphinus delphis* (Tregenza et al. 1997). A slightly higher by-catch rate was reported for the UK hake gill net fleet during the period 1999–2000, with most common dolphins caught between October and March (ICES WGMME 2005). In more recent years, high (extrapolated) numbers of common dolphins were caught in hake fishing nets throughout the NE Atlantic. A by-catch estimate of 115 common dolphins per annum was estimated for UK hake set nets in 2008, though higher by-catch estimates were calculated for UK monkfish (230 dolphins) and pollack fisheries (214 dolphins) (Northridge & Kingston 2009, Sea Mammal Research Unit [SMRU] 2009). In the Cornish tangle and gill net fisheries, estimated by-catch rates in 2005–2008 were 1.15 per 100 hauls and 0.36 per 100 hauls, respectively (Northridge & Kingston 2009). The annual by-catch of common dolphins in Irish gill net fisheries for hake and cod (*Gadus morhua*) in the Celtic Sea between 2006 and 2007 was approximately double what it had been in 1992–1994 (Tregenza et al. 1997, Cosgrove & Browne 2007). In addition, all common dolphins recorded in the earlier period were caught in late autumn and winter (Tregenza et al. 1997), a period that was not sampled in the later study, which focused on the maximum effective spacing for acoustic deterrent devices (Cosgrove & Browne 2007). Preliminary data suggest that 773 common dolphins were caught by Spanish hake gill nets in 2009 (ICES WGBYC 2011), and more recently, a by-catch rate of 0.055 common dolphins killed per "fishing trip/haul" was determined for Portuguese polyvalent boats using gill or trammel nets targeting hake and sea bream (ICES WGBYC 2012). By-catch estimates for the whole Portuguese fleet were difficult to ascertain as it is a multigear fishery.

Portuguese purse-seine nets fishing primarily for sardines caught 47 (non-extrapolated number) common dolphins in 2010 (ICES WGBYC 2012), and an incidental capture rate of 0.50 dead common dolphins per haul was determined for the French sardine purse-seine net fishery during the same year (Morizur et al. 2011).

Since 2001, monitoring programmes have primarily focused on pelagic trawl fisheries, through projects such as PETRACET (PElagic TRAwl and CETaceans) and EU-funded NECESSITY. The PETRACET project monitored annual fishing effort among the main French, Irish, UK, Danish, and Dutch pelagic trawl fisheries in the Celtic Sea and Bay of Biscay between December 2003 and May 2005. Interestingly, no by-catch of common dolphins was observed in mackerel, horse mackerel (scad) or anchovy fisheries (Northridge et al. 2006). The reasons for this are unknown, though these fisheries did differ in temporal and spatial distribution compared to other pelagic trawl fisheries where by-catch was observed (i.e., albacore tuna and sea bass). During the early to mid-1990s, however, common dolphins were reported as by-catch in Dutch horse mackerel pelagic trawl nets fishing off the SW coast of Ireland and French hake pelagic trawl nets in the inner Bay of Biscay (Couperus 1997, Tregenza & Collet 1998, Morizur et al. 1999). Overall, by-catch events observed during the PETRACET project were clumped in both space and time, with 75 dolphins recorded in 13 tows of the French component of the sea bass fishery, of which 8 were recorded in a relatively small area off Brittany (Northridge et al. 2006). Total by-catch for pelagic trawl fisheries (excluding the UK sea bass fishery), where any cetacean mortality was observed, was estimated to be around 620 common dolphins per annum (Northridge et al. 2006). Numbers of common dolphins caught by the European sea bass pair trawl fishery have fluctuated since monitoring began (see Table 3). While the number of by-catch events has declined in the UK fishery, primarily due to effective mitigation and lower fishing effort, by-catch estimates in the French sea bass fishery increased again in 2008 and 2009. Fernández-Contreras et al. (2010) reported common dolphins as by-catch in Spanish pair trawls between March 2001 and December 2003. This fishery was primarily targeting blue whiting (*Micromesistius poutassou*), with mackerel, hake and horse mackerel as secondary target species.

Table 3 Annual estimates of total by-catch of common dolphin *Delphinus delphis* in ICES areas VI, VII and VIII (1990–2009)

Fishery	1990	1991	1992	1993	1994	1995	1996	1997	1998	1999	2000	2001	2002	2003	2004	2005	2006	2007	2008	2009
Drift nets																				
Irish, UK & French tuna[a]	243	390	608	1347	1580	666	546	947	1706	2101	1589	0	0	0	0	0	0	0	0	0
Pelagic trawls																				
French and Irish tuna[b]														133[1]						
French tuna (ICES VI, VII, VIII)[c–g]					95[2]												60	13	120	900
French sea bass (ICES areas VII and VIII)[b–d,f,h]					25[2]									489[3]				290[4]	300	
French sea bass (ICES areas VII)[e–g]																				40
French sea bass (ICES areas VIII)[e–g]																				300–400
French pelagic trawl (ICES area VIII) (various species)[g]																				13
French midwater otter trawl (ICES areas IV, VII, VIII) (bass, horse mackerel, mackerel, herring and sardine)[f]																	57			
UK sea bass (ICES IIV.1 and ICES IIV.2)[5,i–l]												190	38	115	439	139	84	50–100[6]	7	4[8]
Dutch horse mackerel[c]					101[2]															
French hake pelagic trawls[c]					203[2]															
Spanish blue whiting[m]												394	394							

Other fisheries

Irish & UK bottom-set gill nets (Celtic Sea)[n]				234																
UK set-net and tangle fisheries (ICES area VII)[j,k,l,o]																253	554	114	594	237
French set nets (Bay of Biscay)[f,h]																			100	
Spanish hake set nets (ICES VII and VIII)[g,h]																		23	773	
Total minimum annual estimate	243	390	608	1581	2004	666	546	947	1706	2101	1589	584	432	737	439	392	755	492	1137	23

[a] Rogan & Mackey (2007); [b]Northridge et al. (2006); [c]Tregenza & Collet (1998); [d]Berthou et al. (2008); [e]Demaneche et al. (2010); [f]ICES SGBYC (2010); [g]ICES WGBYC (2011); [h]ICES (2010); [i]SMRU (2008); [j]Northridge & Kingston (2009); [k]SMRU (2009); [l]Northridge & Kingston (2010); [m]Fernández-Contreras et al. (2010); [n]Tregenza et al. (1997); [o]Northridge et al. (2007).

[1] Data from France were from 2003, and data from Ireland were from 2004.

[2] By-catch data obtained by the EU BIOECO project (see Morizur et al. 1999 for further information) and extrapolated by Tregenza & Collet (1998), although these values are only a rough estimate of actual by-catch due to poor sampling during the project as a result of low observer coverage in France.

[3] French bass fleet effort for the 2003–2004 winter season (October 2003–September 2004), including some striped and Risso's dolphins.

[4] Revised estimate.

[5] Not annual data but fishing season, starting from 2000–2001 winter season.

[6] Pinger trial commenced, which continued until the 2008–2009 fishing season.

[7] Fishing effort low, and no observations carried out.

[8] All (46) hauls in this fishery were observed.

Hauling time, fishing depth (all dolphins were captured during tows made in water shallower than 300 m) and season were identified as the key factors possibly influencing by-catch events.

There were high rates of by-catch in the albacore tuna drift net fishery that operated during the 1990s in the NE Atlantic (Goujon et al. 1993b, Goujon 1996, Harwood et al. 1999). Using landings of albacore tuna as an indicator of effort, a by-catch of 11,723 (CI = 7670–15,776) common dolphins was estimated for the period 1990 to 2000 (Rogan & Mackey 2007). In 1991, the Council of the European Union decided to limit the length of surface gill nets to 2.5 km and in 1997 declared its intention to ban the use of drift nets in the tuna fishery. The ban was implemented in 2002 (European Union 1998, Rogan & Mackey 2007). This resulted in BIM (Bord Iascaigh Mhara, Irish Sea Fisheries Board) and the Irish Marine Institute undertaking tests on experimental trawls to develop alternative fishing tactics. In 1999, over 160 days, 313 hauls were observed. No cetacean by-catch was observed in 90% of hauls, though 125 common dolphins (observed, not an extrapolated estimate) were caught in just four pair trawls (BIM 2000). As noted, this highly clustered pattern of by-catch is not unusual for pelagic trawls. The incidental capture of cetaceans declined between 2002 and 2004 (16 in 2002, 1 in 2003 and 2 in 2004), which may have resulted from the implemen-tation of a number of avoidance techniques by the Irish fleet. These included (1) cessation of fishing when cetaceans were active in the area; (2) extinguishing stern lights while towing at night; and (3) lowering the trawl headline to several metres below the surface. These practices were simple to adopt and did not adversely affect fishing for albacore tuna (BIM 2004). The use of acoustic deter-rent devices during 2002 and 2003 may have further reduced cetacean by-catch (BIM 2004). Since 2005, no cetacean by-catch was observed in the Irish albacore tuna pelagic trawl fleet (148 days at sea observed, and 60 of these were by independent observers; R. Cosgrove personal communica-tion, July 2012). Among other things, this was attributed to the use of more powerful sonar, which precluded the need to deploy fishing gear until tuna were reliably detected (ICES WGBYC 2012).

The area of operation of the albacore tuna fishery changed since the introduction of pelagic trawls. The European tuna drift net fishery usually started fishing for tuna in May, north of the Azores, following the migration of juvenile albacore tuna, moving first northwards in June and July and then westwards, to end in September/October along the continental slope off the SW coast of Ireland (Goujon et al. 1993a). In comparison, the pelagic trawl fishery operates in the inner Bay of Biscay along the 1000-m depth contour, up to shallower continental slope waters off the Irish SW coast (BIM 2005). The main reason for the change in fishing location is the larger concentrations of tuna found close to the continental shelf, making it easier to fish with pelagic trawls in this loca-tion. Interestingly, the majority of common dolphins incidentally captured by the Irish albacore tuna pelagic trawl fleet between 1998 and 2003 (though primarily in 1998 and 1999) were caught off the SW coast of Ireland (BIM 2005), as had been the case with the Irish drift net fishery (Rogan & Mackey 2007).

Overall, reduction in by-catch in the Irish component of the pelagic trawl fishery was achieved by carefully targeting tuna instead of towing indiscriminately (ICES WGBYC 2012). In contrast, high numbers of common dolphins were caught in French pelagic trawl nets for tuna in 2009 (see Table 3). This was possibly due to difficulties in finding tuna during that year, which may have resulted in some skippers modifying their fishing operations (ICES WGBYC 2012). By-catch again was clustered, with 94% reported in just two trips, involving two pairs of vessels (Morizur et al. 2010). Cetacean by-catch was lower during the following year (Hassani 2012).

Fisheries selectivity of age-sex maturity classes

It is important to identify what age-sex class of individuals is incidentally captured by each fishery in the NE Atlantic. High mortality of mature (especially pregnant) females, calves and individuals approaching maturity will have a more detrimental effect on the common dolphin population than a high mortality rate of mature males. Analysis of by-caught animals in the predominantly winter European sea bass pelagic trawl fishery revealed a predisposition to capturing juvenile and young adult common dolphins. Of aged common dolphins captured by the French fleet, 85% were less than 11 years of age, and 90% of aged dolphins caught by the UK fleet were less than 13 years, with a reported peak in the age-frequency distribution at 8 and 9 years (Murphy et al. 2007b). These results imply a lack of learned behaviour of juveniles and young mature individuals around nets, whereas mature individuals may have developed suitable behavioural strategies for feeding within trawl nets (Murphy et al. 2007b). Alternatively, some older individuals may not partake in this type of foraging behaviour. A bias towards male common dolphins was observed in nets of Spanish pair trawls targeting blue whiting, mackerel and other species in Galician waters (2001 and 2002), with an average age of 13.4 ± 4.4 (± standard deviation, SD) years for male *Delphinus delphis* and 11.5 ± 4.8 years for females (Fernández-Contreras et al. 2010). Two mass capture events comprising only males (7 and 15 dolphins), with an average age of 7.4 ± 3.2 years, were observed in July 2001 (Fernández-Contreras et al. 2010). This further suggests age and sex segregation of the population during summer (Fernández-Contreras et al. 2010).

Low numbers of calves (<1 year old; 3% of the whole aged by-catch sample) and yearlings (6%) were incidentally captured by both the UK and French sea bass pelagic trawl fleets (Murphy et al. 2007b), and no calves were reported in Spanish pair trawls operating off Galicia (Fernández-Contreras et al. 2010). The low by-catch of calves and weaned juveniles may be due to a lack of association of these individuals with trawl nets (i.e., weaned juveniles not actively feeding within the trawl net) or from the cod end. However, the opposite was found for the (summer) Irish albacore tuna drift net fleet, as common dolphins 2 years old or younger or 165 cm long or less comprised 51.2% of the whole by-catch sample obtained between 1996 and 1999, indicating a strong propensity for calves and yearlings to be captured in drift nets (Murphy & Rogan 2006). A large proportion of calves were also reported in the by-catch of French albacore tuna drift nets, which operated in an area extending from 44°N to 51.5°N and from the Bay of Biscay region, 6°W to 21°W (Goujon et al. 1994). It was suggested that a lack of learned behaviour around nets and lower echolocation capabilities in calves were possible causes for their higher capture rate in the tuna drift net fishery compared to other age classes (Murphy & Rogan 2006). In addition, the high mortality rates of calves in drift nets may also have occurred due to a combination of the length of the net (up to 2.5 km), the lack of discriminating behaviour of this gear type (depending on the habitat usage by age-sex maturity groups), and the timing (during the calving season of the common dolphin) and location of the tuna drift net fishery (Murphy et al. 2007b). Sexually mature individuals of both sexes, including pregnant and recently pregnant females, were also incidentally captured by the Irish tuna drift net fleet, with 43% of the 91 aged common dolphins older than 10 years (Murphy & Rogan 2006). Thus, this fishery, which is now banned, was incidentally capturing the most important age-sex maturity groups in the population.

Biological effects

There are two main forms of interaction resulting in biological effects: exploitative competition by fisheries (removing cetaceans' prey) and interference competition, which involves the disruption of cetacean feeding activities as a result of disturbance (Plaganyi & Butterworth 2005). The long-term impacts of fisheries on the NE Atlantic ecosystem are immense, leading to changes in fish communities due to the loss of larger predators and corresponding ecological function (ICES 2008). A number of fish stocks in the NE Atlantic have been overfished, including both pelagic and demersal stocks (Sparholt et al. 2007), and this resulted in a succession of fisheries for species at lower trophic levels, a process known as 'fishing down the food web'. Common dolphins consume an energy-rich diet, and a decline in suitable prey may cause reduced condition and a decline in reproductive output, with extreme cases leading to starvation and death (Certain et al. 2011). Between 1991 and 2010, only 4% (21/537) of common dolphins necropsied by the UK cetacean stranding investigation programme died from starvation (Deaville & Jepson 2011).

In the last few decades, there has been awareness that fisheries management should consider the broader impact of fisheries on the ecosystem as a whole and the impact of the ecosystem, and other users of the ecosystem, on fisheries (Food and Agriculture Organization of the United Nations [FAO] 2008). The overall goal is the sustainable use of the whole system, and achieving this goal requires the implementation of an ecosystem approach to fisheries (EAF). This is defined by the FAO (2003) as follows:

> An ecosystem approach to fisheries strives to balance diverse societal objectives, by taking account of the knowledge and uncertainties of biotic, abiotic and human components of ecosystems and their interactions and applying an integrated approach to fisheries within ecologically meaningful boundaries. (p. 7)

As part of the EAF, ecosystem models representing a wide range of technological and ecological processes affecting species in the ecosystem (including multispecies and whole-ecosystem models) are used to investigate how the system may change under different future scenarios, including different management options (FAO 2008). Lassalle et al. (2012) undertook an ecosystem approach to assess the impact of fisheries on marine top predators in the Bay of Biscay. Although bottlenose dolphins appeared to be sensitive to resource depletion, common dolphins (and harbour porpoises) were most impacted by their incidental capture in fishing gears (Lassalle et al. 2012). However, results further suggested that the Bay of Biscay was not far from overexploitation at the current fishing rate. The Pianka index value for resource overlap with fisheries was high for common dolphins inhabiting neritic waters of the Bay.

Climate change

Common dolphins are wide ranging and have shown a capacity for range expansion. However, the significance of the effects of climate change on the NE Atlantic population is unknown. In general, climate change is regarded as a key threat to all biodiversity and to the structure and function of ecosystems that may already be subject to significant anthropogenic stress (Graham & Harrod 2009). The NE Atlantic has a temperate-to-subarctic climate, and around the United Kingdom and Ireland, common dolphins have been sighted during summer (calving period) in SSTs ranging between 8.1°C and 18.5°C (mean = 14.9°C; SD = 1.6°C, period 1983–1998) (MacLeod et al. 2008). However, this species has been reported from a wide area around the Strait of Gibraltar to Norway (Øien & Hartvedt 2009) and along the mid-Atlantic ridge, with a mean SST of 15.2°C (CI = 12.3–18.1°C; mean depth 3008.8 m, CI = 2145–3872 m; period June 2004; Doksæter et al. 2008). These are considerably warmer temperatures than those reported during the winter in this region. For example, in the western English Channel, where an increased density of individuals has been reported in winter, SST can fall to between 7°C and 10°C (Gislason & Gorsky 2010, Hughes et al. 2011).

The distribution of the common dolphin in the North Sea fluctuated during the twentieth century. Slight distributional shifts into this sea were observed between the 1920s and 1960s and also since the 1990s (see 'Distribution and abundance' section). The water temperatures in the North Sea fluctuated during the last century, with a period of low water temperatures between 1950 and 1979 (Lambert et al. 2011). Following this, an abrupt ecosystem shift, or regime shift, occurred in both pelagic and benthic ecosystems of the North Sea (Reid & Edwards 2001, Beaugrand et al. 2008). Increased water temperature was proposed as the primary factor influencing the distribution and increased occurrence of common dolphins off the NW coast of Scotland (period 1992–2003; MacLeod et al. 2005). Seawater temperature in that area has risen 0.2–0.4°C per decade since 1981 (Fisheries Research Services 2003, MacLeod et al. 2005). In addition, the recent summertime incursion of common dolphins into the outer Moray Firth and NE North Sea has been anecdotally attributed to increasing regional sea temperatures (Robinson et al. 2010).

An increase in the observed winter abundance of *Delphinus delphis* in the western English Channel between 1996 and 2006 (MacLeod et al. 2009) coincided with an upturn in reported strandings of the species along the SW coast of the United Kingdom (Jepson 2005, Deaville & Jepson 2011). During this period, there was a 1°C rise in the mean annual SST in the western English Channel (1990–2000), which exceeded any other SST change in the area over the last 100 years (Hawkins et al. 2003). By the 2080s, the temperature of these waters is expected to rise by up to 3°C, which may lead to the loss of some economically important cold-adapted species (Graham & Harrod 2009). In recent years, however, there has been a decline in the numbers of common dolphins becoming stranded along the SW coast of England and the Atlantic coast of France, possibly reflecting other variables, in particular fishing effort (see 'Stranding patterns' section).

Although it has been suggested that temperature is a key limiting factor in the northern limit of common dolphins in western European waters, and individuals may shift their distribution to stay within their thermal niche (Lambert et al. 2011), changes in temperature also affect prey species of the common dolphin, influencing physiological and ecological processes in a number of direct, indirect and complex ways (Graham & Harrod 2009). Thus, common dolphins may shift distribution to remain within their ecological niche. The decline in reported strandings off the SW coast of England between the 1930s and 1970s (after an earlier peak in strandings during the 1920s and 30s; see 'Distribution and abundance') followed fluctuations in pelagic assemblages (zooplankton and larval fish) in the English Channel during the 1920s and 1930s. Changes were attributed to an increased SST and a reduced Atlantic flow into the Channel—later known as the 'Russell cycle'—which resulted in a decreased biomass of all higher trophic levels (Southward et al. 2005 and references therein). A decline in the abundance of cold-water fishes in the English Channel was observed at that time (Southward 1963, Evans & Scanlan 1989, Southward et al. 2005), with a northwards shift in their distribution, and it is believed that common dolphins followed (Fraser 1934, Evans & Scanlan 1989, Murphy et al. 2006). In the western English Channel, there has been an alternation in abundance of herring (*Clupea harengus*), a cold-water species, to pilchard (*Sardina pilchardus*), a warm-water species, in response to environmental conditions since at least the fourteenth century. However, the change in fish composition from 1926 to 1936 is seen as a climatically mediated shift exacerbated by intense overfishing, leading to recruitment failure of herring (Southward et al. 2005 and references therein). From the late 1960s onwards, many of the conditions prevailing in the early 1920s returned, along with an increase in common dolphin strandings along the SW coast of England (Evans & Scanlan 1989) and the southern and western coasts of Ireland (Murphy 2004). Since the 1980s, however, conditions in the English Channel have changed again, with warm-water species, such as pilchard, increasing in abundance (Hawkins et al. 2003, Southward et al. 2005 and references therein).

Murphy (2004) linked patterns of common dolphin strandings on the Irish coast (1900–2003) with changing oceanographic conditions due to the North Atlantic Oscillation (NAO). The decline in Irish strandings took place during a negative NAO index phase between the mid-1930s and the mid-1970s (see Figure 9). Following this, there was a sharp reversal to a highly positive NAO index phase (Hurrell et al. 2003), with an associated increase in common dolphin strandings along the southern and western coasts of Ireland (Murphy 2004). Changes in the NAO have had wide-scale effects on the North Atlantic ecosystem, influencing SST and winds—both linked to variation in the production of zooplankton—as well as fluctuations in several important fish stocks across the North Atlantic (Planque & Taylor 1998, O'Brien et al. 2000, Hurrell et al. 2003). Furthermore, the weather conditions associated with the positive phase of the NAO, including an increase in winter storm activity, stronger westerly winds and greater wave heights (Bacon & Carter 1993, Hurrell 1995, Stenseth et al. 2002, Hurrell & Dickson 2004), could also increase the number of strandings of common dolphins directly, not only by driving carcasses ashore but also by contributing to the death of diseased or injured individuals. Recent studies have suggested that variability in the NAO index is due to anthropogenic climate change, and this was an explanation for the intensification (strongly positive) of the NAO up to 1997 (Woollings et al. 2010). This period of intensification was abruptly

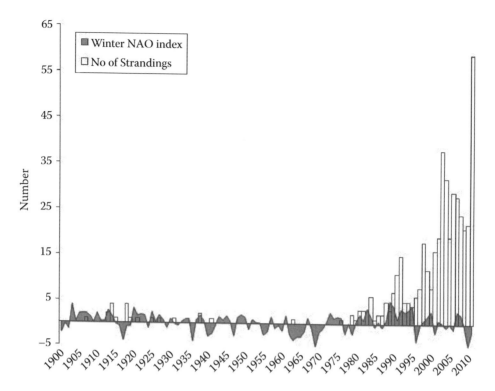

Figure 9 Winter (December–March) North Atlantic Oscillation index and number of common dolphin strandings along the coast of Ireland by year. Winter NAO index based on the difference of normalized sea-level pressure (SLP) between Lisbon, Portugal, and Stykkisholmur/Reykjavik, Iceland, from 1900 to 2010. Data on the winter North Atlantic Oscillation index were obtained from the website of the National Center for Atmospheric Research, USA (Climate and Global Dynamics Division [CGD] 2012). (Adapted from Murphy 2004; updated with strandings data from Berrow et al. 2007, Brophy et al. 2006, O'Connell & Berrow 2012, Philpott et al. 2007 and Philpott & Rogan 2007.)

reversed, and the NAO was weak and variable between 2000 and 2009 due to shifts in atmospheric pressure patterns and then strongly negative in the winters of 2010 and 2011 (Hughes et al. 2012). However, during the 2000s, although the NAO was weak and variable, stranding rates along the Irish coast continued to remain high (>19 per year) (Figure 9). In 2011, strandings of common dolphins were unusually high, with 59 records, the highest in the preceding decade of recorded effort (O'Connell & Berrow 2012). It is not known if this represents a new phenomenon or a temporary peak. Current stranding patterns along the Irish coast may be heavily influenced by anthropogenic activities (see 'Stranding patterns' section), which requires further investigation. There may also be other climatic changes in the ecosystem (e.g., Reid & Beaugrand 2012).

Pollutants

Common dolphins are susceptible to the effects of anthropogenic pollutants, such as persistent organic pollutants (POPs), for instance, PCBs, dichlorodiphenylethanes (e.g., dichlorodiphenyltrichloroethane, DDT, a widely used pesticide), hexabromocyclododecane (HBCD), and various heavy metals, such as cadmium and mercury. Pollutants enter the body almost exclusively through the diet, and toxins such as POPs are lipophilic compounds that accumulate in the lipid-rich blubber layers of marine mammals. Apart from some heavy metals, pollutants both biomagnify (higher levels higher up the food chain) and bioaccumulate (increased concentration with age). A large number of organochlorine compounds (OCs), such as PCBs and DDT, are hormone- or endocrine-disrupting chemicals. Endocrine functions can be altered by these toxins through interference with the synthesis, secretion, transport, binding, action, or elimination of the endogenous natural hormones responsible for homeostasis, reproduction, development, and behaviour (US Environmental Protection Agency [EPA] 1997). As animals can be exposed to a complex mixture of compounds, there may be further significant impacts through additive and synergetic effects.

The production of PCBs and DDT has been limited or completely banned since the 1970s in most developed countries, though DDT is still used in some developing countries for controlling vectors of parasitic diseases (Toft et al. 2004). In addition, OCs, including PCBs, are still being released into the environment through disposal, volatilization of previously released material, and creation of PCBs and dioxins during combustion (Breivik et al. 2002, Katami et al. 2002, Toft et al. 2004).

Persistent organic pollutants

Blubber of common dolphins from the NW coast of Spain, sampled in 1984 and 1996, and of common dolphins entangled in fishing nets in the SW Mediterranean Sea, sampled during 1992–1994, were analysed for organochlorine pollutants (Borrell et al. 2001), which are persistent organic pollutants. Organochlorine levels in both areas were at the mid-to-low end of the range of concentrations detected in common dolphin populations elsewhere in the world and in other delphinids from the same region and were therefore considered unlikely to have played a significant role in the decline of the common dolphin in recent decades in the western Mediterranean Sea.

Retinoids are chemical compounds, related to vitamin A, essential for normal vision, growth, reproduction, immune function, and cellular division and differentiation in mammals. In common dolphins incidentally caught off the NW coast of Spain, age and blubber lipid content were strong determinants of blubber retinoid concentrations in males (Tornero et al. 2006). Retinoid levels were positively correlated with organochlorines in males and negatively in females. As organochlorine pollution levels were only moderate and unlikely to be above proposed threshold levels for mammalian toxicity, the cause-effect relationships between organochlorines and retinoids could not be established (Tornero et al. 2006).

A number of other POPs have been identified in cetaceans, such as butyltins (tributyl tin, TBT; dibutyl tin, DBT; and monobutyl tin, MBT); polyaromatic hydrocarbons (PAHs); and perfluorinated organochemicals. Few studies have measured these in common dolphins in the NE Atlantic region. In the United Kingdom, one of the largest datasets on toxicology in any marine mammal species has been generated on harbour porpoises stranded and incidentally caught between 1990 and 2011 (Jepson 2005, Deaville & Jepson 2011, Law et al. 2012a). The long-term trends in harbour porpoises show stable (and often high) levels of PCBs but declining levels of organochlorine pesticides (such as DDT and dieldrin) (Law et al. 2012a), declining PBDEs (penta-mix brominated diphenyl ether congeners; after an initial increase in the late 1990s) (Law et al. 2010), and only trace levels of butyltins (including TBT) (Law et al. 2012b). Similar trends are likely to be found in common dolphins around UK and northern European waters.

Impacts on reproduction During pregnancy in cetaceans, lipid-soluble contaminants, such as OCs, may be transferred from the mother to the fetus. However, the majority (~80% of OCs) of the pollutant burden accumulated by females (primarily prior to sexual maturity) is believed to be transferred to their firstborn calf during the first 7 weeks of lactation (Cockcroft et al. 1989). In light of this, resting mature females (non-pregnant, non-lactating) with high blubber pollutant burdens and showing signs of recent gravidity may have aborted, or their offspring may have died soon after birth (Murphy et al. 2012a).

An EU-funded Fifth Framework study known as BIOCET investigated the potential impacts of POPs on reproduction in female *Delphinus delphis*. This project pooled samples and data from individuals found stranded in many countries in the NE Atlantic (Ireland, Scotland, France and

Galicia in NW Spain), collected primarily between 2001 and 2003 (Pierce et al. 2008). Factors such as geographic variation in POP burdens in blubber tissue and relationships between POP burdens and age and fatty acid profiles and reproduction were taken into account within the analysis. The most important variable explaining POP profiles in common dolphin blubber was individual feeding history (Pierce et al. 2008). A substantial proportion of individuals in the BIOCET sample had pollutant levels above the threshold of 17 mg kg^{-1} (PCB lipid weight: mass of PCB per unit mass of lipid; Kannan et al. 2000) reported to have adverse health effects, based on experimental studies of both immunological and reproductive effects in seals, otters, and mink. This threshold was frequently exceeded in common dolphins (40%), especially common dolphins inhabiting waters off the French coast (50%). In addition, pregnant females had lower blubber levels of PCBs and PBDEs than other mature females. This relationship could be interpreted as evidence that high POP concentrations inhibit pregnancy; however, Pierce et al. (2008) pointed out that infertility due to other causes may allow high levels of POPs to bioaccumulate. As data on the health status for all the individuals was unavailable, it is not known if the low estimated pregnancy rate in the BIOCET sample (25%; Learmonth et al. 2004a) was due to disease, loss of nutritional status or high contaminant burdens causing an adverse effect on reproductive output, that is, instigating abortions or infertility.

Subsequent studies by Murphy et al. (2010, 2012a) compared samples and data from the BIOCET project with a control group of 'healthy' common dolphins caught in fishing nets that subsequently became stranded on the SW coast of the United Kingdom between 1992 and 2004. The ovarian corpora (ovulation) scar number significantly increased with PCB burdens in sexually mature *D. delphis* in the BIOCET sample. The majority of individuals with contaminant burdens above the threshold level for adverse health effects were resting mature females (83%) with high numbers of ovarian scars (Murphy et al. 2010). This suggests that (1) due to high contaminant burdens, females may be unable to reproduce and thus continue ovulating; or (2) some females are not reproducing for some other reason, either physical or social, and therefore accumulate higher levels of contaminants (Murphy et al. 2010). Within the BIOCET sample, 92% of mature females with contaminant burdens above the threshold level and corresponding high corpora counts (≥15 scars) were obtained from a mass live stranding event in Pleubian, France, in February 2002. As noted, genetic analysis of this group gave no evidence of a matriarchal system, and a lack of genetic relatedness among mature individuals existed (Viricel et al. 2008). Therefore, the existence of non-reproductive females (based on high contaminant loads and high numbers of ovulations) within this social group is noteworthy (Murphy et al. 2010).

The control group of healthy common dolphins also had high PCB burdens, above the threshold level, but these were not inhibiting ovulation, conception or implantation, though the impact on the fetal survival rate required further investigation (Murphy et al. 2010). Studies of the effects of PCBs on reproduction in mink have shown that although ovulation, conception and implantation proceed as normal, fetuses die during gestation or shortly after birth. This results from changes in maternal vasculature in the placenta, leading to degeneration of the trophoblast and fetal vessels and subsequent fetal growth retardation or death (Bäcklin et al. 1997, 1998, Murphy et al. 2012b). A similar situation may be occurring in female common dolphins in the NE Atlantic because within the control group, 8.9% of mature females showed evidence of recent abortion during their second trimester (Murphy et al. 2012b). The association between high contaminant burdens and the incidence of abortion raises serious concerns about the population-level effects of POPs and is currently under further investigation.

Heavy metals

Cetaceans appear to be protected from the effects of many heavy metals due to the presence of metallothioneins as they play a key role in essential metal homeostasis (e.g., Das et al. 2006). Heavy metals accumulate primarily in the liver, kidney and bone. Tissues of fetus-mother pairs of common dolphins stranded along the French coasts (Bay of Biscay and English Channel) were analysed for their cadmium, copper, mercury, selenium and zinc contents (Lahaye et al. 2007). In the kidneys, fetal cadmium levels were extremely low. Strong relationships between copper and zinc suggested the involvement of metallothioneins since early fetal life. There was only limited maternal transfer of mercury during pregnancy. Hepatic mercury levels in fetuses increased with body length and were also proportionate to maternal hepatic, renal and muscular mercury levels. There was evidence of selenium-based mercury detoxification in both adults and fetuses. Metal levels in muscle, liver, fat tissue and skin were studied in 15 *Delphinus delphis* that became stranded along the Atlantic coastline of Portugal (Carvalho et al. 2002). The concentrations of mercury, tin, chromium, zinc, nickel, cobalt, cadmium, manganese, iron and copper were determined in the liver, kidney and muscle of 24 common dolphins stranded on the Portuguese coast between 1995 and 1998 (Zhou et al. 2001). The concentrations of iron, zinc and mercury were relatively high, particularly in liver, whereas chromium, nickel and cadmium were present at much lower levels or even were undetected. Total mercury concentration increased with body length, and concentrations in the kidney, muscle and particularly liver were higher in females than in males. Total zinc and copper concentrations in muscle decreased with dolphin length (Zhou et al. 2001).

Oil spills

There has been only one detailed study of the effects of oil spills on common dolphins. In December 1999, the tanker ERIKA broke up and sank off the coast of Brittany and continued to leak heavy fuel oil for a few months, impacting the local pelagic and coastal ecosystems. No effects of the ERIKA oil spill were observed on the local common dolphin population (Ridoux et al. 2004). Interestingly, vanadium concentrations in common dolphins were chronically high, both before and after the oil spill (Ridoux et al. 2004).

Noise pollution

Concern regarding the impact of anthropogenically derived sound on marine mammals has been rising in recent decades. The range of sources of anthropogenic noise in the marine environment is wide and varied. Some activities (e.g., shipping and other motorized vessels, use of explosives, drilling, dredging and construction) produce noise indirectly. Other sources, such as active sonars operating at a variety of frequencies, airguns and boomers used in seismic surveys, pingers and acoustic harassment devices, are sources of deliberately introduced sound in the marine environment.

The impact of this noise varies from nil (or attraction, e.g., bow riding) to severe depending on the type, frequency and duration of the noise as well as the relation to the species of concern. Noise can be tolerated, with normal activity patterns maintained and no evidence of an overt response (Würsig & Richardson 2009). Cetaceans may sometimes tolerate noise to remain in a preferred location (e.g., feeding ground) even when the noise is strong enough to cause reactions when the same species is engaged in other activities. In their review of responses of cetaceans to anthropogenic noise, Nowacek et al. (2007) divided responses to noise into three main categories:

- Behavioural responses, such as deviation from normal activity, including changes in swimming speed and breathing/diving activity and avoidance of an area (Richardson et al. 1995). Some effects can be subtle, whilst others are more obvious.
- Acoustic responses: Changes in the type or timing of vocalizations in response to the noise source.
- Physiological responses: Exposure to loud sounds can include temporary or permanent reductions in hearing sensitivity (auditory threshold shifts) (Schlundt et al. 2000).

In addition, physiological effects and symptoms associated with decompression sickness (e.g., embolism and tissue separation), including central nervous system defects (e.g., disorientation, visual and auditory dysfunction) have been noted (Crum & Mao 1996, Houser et al. 2001). Chronic exposure may also cause stress reactions.

Another issue that has been identified with respect to noise pollution is masking; the noise obscures other sounds of interest to the individual (e.g., feeding clicks). Continuous noise at a similar frequency as the sound of interest is of greatest concern, particularly if the two sounds are received from similar directions (Würsig & Richardson 2009).

Military activity

The impact of military activity and, in particular, use of low- and midfrequency active sonar of high intensity has become a major issue in recent years. A number of MSEs in the last few decades, usually involving beaked whale species, have been temporally and spatially coincident with military activities using such systems (Simmonds & Lopez-Jurado 1991, Todd et al. 1996, Frantzis 1998, Fernández et al. 2005, Rommel et al. 2006, Weilgart 2007a, Mooney et al. 2009, Tyack 2009, Zirbel et al. 2011).

Responses by odontocetes to the use of military sonar include modifications to vocalizations, with the response differing among species. For example, sperm whales (*Physetermacro cephalus*) and beaked whales become silent (Watkins et al. 1993, Cressey 2008), whilst long-finned pilot whales (*Globicephala melas*) increase whistling (Rendell & Gordon 1999), and humpback whales (*Megaptera novaeangliae*) increase the duration of their songs (Miller et al. 2000, Fristrup et al. 2003). Such changes could potentially have an impact on breeding, feeding, and social cohesion, depending on which calls are affected (Weilgart 2007b).

There have been relatively few experimental studies of the effects of military sonars on odontocetes. Mooney et al. (2009) demonstrated both behavioural and physiological effects of midfrequency sonar on bottlenose dolphins. The behavioural reactions were mild, and temporary hearing loss occurred only after prolonged exposures, whereas beaked whales showed a disruption to foraging behaviour (stopped echolocating during deep foraging dives) and avoidance at exposures well below those used by regulators to define disturbance (Tyack et al. 2011).

In 2008, a mass stranding of 26 common dolphins occurred within days of an international military exercise in UK waters. The naval exercise could have caused this large group to come unusually close to the shore, but based on available evidence, the midfrequency sonar in use during the exercise was considered unlikely to have directly triggered the stranding event (Jepson et al. 2009, in press). Subsequent to this mass stranding, the UK Ministry of Defence developed a real-time alert procedure for naval training operations. This enables local information on unusual cetacean sightings, such as the presence of a cetacean group closer to shore than usual, to be incorporated into the training schedule and for operations to be relocated if necessary. This was successfully implemented in April 2009 in relation to the presence of short-beaked common dolphin in the Falmouth Bay area. Over 20 dolphins were seen 15 minutes after Royal Navy sonar trials started. The Royal Navy immediately modified the exercise until the group of dolphins had returned to open water several hours later. Such continual improvement of mitigation strategies by the military themselves is probably the best way to limit future impacts.

Where strandings have been associated with military sonar usage, the major symptoms observed are generally similar to decompression sickness. Exposure to intense, low-frequency sound is thought to stimulate bubble growth within biological tissue, particularly when the tissue is supersaturated with dissolved gas (Crum & Mao 1996). Diving behaviour (depth of dive, depth at which lung collapse occurs and descent/ascent rates) has a significant effect on tissue gas concentration and will therefore influence the susceptibility of a particular species to acoustic exposure (Houser et al. 2001). In addition, the surface interval between multiple dives is related to gas clearance and initial gas tension on subsequent dives (Houser et al. 2001). The species most at risk are those that regularly dive to depths greater than 70 m (the depth at which lung collapse occurs), those with slow descent/ascent rates (slower rates allow greater amount of gas supersaturation) and short surface intervals (higher initial gas tension on subsequent dives). Bubble growth can cause tissue damage and vascular blockage, which may underlie cetacean stranding events associated with acoustic exposure (Houser et al. 2001) (see 'Health status and cause of death' section). Beaked whales in particular are susceptible to such effects (Fernández et al. 2005, Ketten 2005).

Between 1992 and 2004, however, incidences of acute and chronic gas embolic lesions were also identified in five short-beaked common dolphins, four Risso's dolphins (*Grampus griseus*) and two harbour porpoises in UK waters (Jepson et al. 2005b, Jepson 2005). A decompression-related mechanism involving embolism of intestinal gas or *de novo* gas bubble (emboli) development derived from tissues supersaturated with nitrogen during rapid surfacing was suspected (Jepson et al. 2005b, Jepson 2005). Since these cases, no additional observations of gas emboli were made from animals stranding in UK waters until 2009, when a single case was observed in a Risso's dolphin (Deaville & Jepson 2011).

Seismic surveys

Oil and gas exploration and production generate a variety of noise, including initial geophysical surveys (using seismic methodologies), rig construction and drilling, and finally structure removal. Of greatest concern is the noise associated with the seismic surveys, which use airguns to generate low-frequency sound. The airguns function by venting high-pressure air into the water column, creating an air-filled cavity that expands and contracts violently, creating sound with each oscillation (Richardson et al. 1995). The sound pulses created depend on the size, number and spacing of the airguns in the array and the air pressure utilized.

Small odontocetes demonstrate the strongest avoidance of seismic survey activity of any cetacean species, with significant increases in fast-swimming activity and declines in sighting rates during periods when airguns are firing (Stone & Tasker 2006, Weir 2008, Gray & Van Waerebeek 2011). Some evidence of temporary threshold shifts has also been noted (Finneran et al. 2002). For common dolphins specifically, avoidance reactions to airgun emissions have been noted in the immediate vicinity, although the species is generally able to tolerate the pulses at 1-km distance from the array (Goold 1996, Goold & Fish 1998).

Operators of seismic activities in many countries are required to work in accordance with specific guidelines. For example, in the United Kingdom, operators are required to follow the Joint Nature Conservation Committee (JNCC) (2010a) guidelines, which include conducting marine mammal observations prior to and during seismic activity and utilizing procedures such as 'soft start' (gradually increasing the number of active airguns to allow animals nearby to move away) to reduce and avoid direct harm to animals. Over the years, most recently in 2010, these guidelines have been reviewed and revised in the light of scientific evidence, technical developments and operational understanding.

Aggregate extraction and dredging

The main concern with aggregate extraction is noise generation during survey work. Non-intrusive studies utilize shallow seismic surveys with 'boomers', which are considerably quieter than the deep seismic surveys undertaken by the oil and gas industry. Currently, consideration is being given to the possible impact of aggregate extraction works on cetaceans, with a view to guidelines being developed for UK waters. However, by comparison to other anthropogenic sound in the marine environment, aggregate extraction is not considered to be a major threat.

Renewable energy

Marine renewable energy generation is a rapidly evolving industry, with some developments amongst the largest offshore engineering projects ever undertaken. The marine renewables industry encompasses three major sectors: offshore wind, tidal stream and wave energy. The ICES WGMME assessed the effects of construction and operation of wind farms (2010), tidal devices (2011) and wave energy converters (2012) on marine mammals, work that was synthesized by Murphy et al. (2012b). This section summarizes the main conclusions of the ICES WGMME.

The extraction of energy has many parallels among all three renewable energy sectors, with developments involving placement of substantial structures into the marine environment, requiring large investment and specialized equipment to place and service them. However, there are also fundamental differences when considering the potential interactions with large marine vertebrates, namely, the requirement for submerged moving structures by the 'wet renewables' sectors (tidal stream and wave energy).

The majority of offshore wind turbines have monopile foundations, though other foundations, such as tripod, jacket, and gravity, have been used, depending on the seabed type. Monopiles are steel tubes ranging from 2 to 6 m in diameter, driven into the seabed by some thousand strokes of strong hydraulic hammers, produced at a rate of 30–60 pulses per minute (ICES WGMME 2010). Assessment of the effects of the offshore wind industry on marine mammals has focused on near-shore species, such as the harbour porpoise, bottlenose dolphin and seals. To date, piledriving constitutes the single most important type of impact. Studies have identified a decrease in acoustic activity of harbour porpoises and bottlenose dolphins, up to 20–25 km in one study, following a single pile-driving event (Brandt et al. 2009, Diederichs et al. 2009, Tougaard et al. 2009, Thompson et al. 2010). However, the nature of the behavioural reaction is unknown, as are the consequences of such activities on the long-term survival of individuals (ICES WGMME 2010). During the operational phase, studies have reported either a full recovery of acoustic activity to predisturbance levels (Tougaard et al. 2006) or a significant increase in acoustic activity above baseline levels (Scheidat et al. 2011). In the latter study, increased food availability inside the wind farm (reef effect) or the absence of vessels in an otherwise heavily trafficked part of the North Sea (sheltering effect) were provided as potential reasons for the apparent preference for the wind farm area (Scheidat et al. 2011).

As wet renewable devices are at a relatively early stage of development compared to the offshore wind sector, knowledge of the potential interactions with marine mammals is limited (Murphy et al. 2012a). There are many different concepts (device types) being simultaneously developed within the wet renewables sectors; these devices are extremely diverse in size, shape, method of fixing and many other characteristics (ICES WGMME 2011, 2012). The various designs are at a range of stages of development from conceptual or scale models to a small number of full-scale test rigs deployed at sea (ICES WGMME 2011, 2012). The site requirements for tidal and wave energy extraction are also much more specific than those for offshore wind, as well as being fundamentally different in nature for animals living in these areas compared to other marine areas (Murphy et al. 2012b).

To date, the majority of offshore renewable energy developments within the NE Atlantic have progressed in shallow waters outside the main distributional range of the common dolphin (i.e., Baltic and southern North Seas; Convention for the Protection of the Marine Environment of the North-East Atlantic [OSPAR] Commission 2012), although common dolphins are known occasionally to enter shallow waters. The situation will change in the next few years, however, and the common dolphin population will come into direct contact with the construction and operation of these devices due to the expansion of the industry, both geographically as it moves further off shore and in terms of the number of devices. Many activities of the offshore renewables industry, because of their noise emissions, have been identified as potentially having an effect on marine mammals. These activities may cause hearing damage, disturbance by eliciting behavioural responses, and habitat exclusion (Murphy et al. 2012b). Outlined by Murphy et al. (2012b), these activities include the following:

Site Survey: Noise from seismic surveys, side-scan sonar and survey vessels may cause disturbance or hearing damage.

Construction: Piledriving may cause hearing damage at close range or disturbance and habitat exclusion. Noise from drilling, dredging, and increased vessel traffic may cause behavioural change and disturbance. Possible mortality from vessels involved in renewable device installations, especially those using ducted propellers to manoeuvre accurately at small spatial scales. Levels of turbidity (resuspension of sediments) or pollution may increase.

Operation: There are possible collision risks with wet renewable devices. Disturbance and masking of biologically significant signals may result from turbine noise and increased vessel traffic associated with maintenance. Possible disturbance due to habitat alteration, reduced fishing effort (may be positive or negative) and other ecological effects, such as introduction of hard substrate into other environments, may occur.

Decommissioning: Possible disturbance and behavioural change from increased vessel traffic is of concern. Disturbance or hearing damage may result from bottom profiling (seismic surveys and side-scan sonar) or operation of cutting machinery. The use of explosives may cause acute hearing damage or mortality. Levels of pollution may increase.

As for all marine mammals, it is important to assess the cumulative impacts of renewable energy technology on the common dolphin at a population level. This will necessitate a pan-European assessment as common dolphins are highly mobile and are likely to spend only a small proportion of their time within the effective range of a device or even within an array of these devices (Murphy et al. 2012b). It will also become increasingly important to consider the effects of a number of large marine renewable energy sites, with various device types, numbers and configurations of arrays, being constructed relatively close together in space and time. In addition, deployment of marine renewable energy devices is but one of many concurrent activities that might take place within a given marine area; thus, common dolphins in these areas may also be affected by other local anthropogenic activities as well as the large-scale impacts of climate change (Murphy et al. 2012b).

Other impacts

Collisions with vessels and shipping noise

Potentially, all cetaceans can be subject to collisions with vessels, with anecdotal reports occurring for most species (e.g., Kraus 1990, Perry et al. 1999, Knowlton & Kraus 2001, Nowacek et al. 2004). European waters contain some of the busiest shipping routes in the world, such as those in the English Channel and North Sea (Evans et al. 2010). Since the early 1990s, the number of shipping movements, size of vessels and their average speeds have all increased in the region (OSPAR 2010). For European waters, the main areas of collision risk appear to be in parts of the Celtic Sea, Bay of Biscay, and off NW Spain (Evans et al. 2010). Despite this, relatively few cetacean deaths are recorded as a result of ship, small vessel or propeller strikes. Specifically for short-beaked common dolphins, between 2005 and 2010 (inclusive), only 3 of the 129 necropsies undertaken by the UK cetacean strandings investigation programme were diagnosed as ship, small vessel or propeller strike (Deaville & Jepson 2011). An additional four were identified as physical trauma of unknown cause.

Commercial shipping generates intense low-frequency (long-wavelength) noise that can propagate over dozens to hundreds of kilometres (Würsig & Greene 2002). Many factors influence the intensity and frequency of sound produced by vessels, which can lead to different potential effects on cetaceans. Engine type and mounting, exhaust configuration, type of hull construction, power and frequency of sonar units, operation of the vessels (e.g., abrupt changes in speed or gears) and propeller cavitation all affect the noise created. In addition, submarine topography and physical oceanographic factors influence sound propagation and therefore the distance over which the sound can have an impact on cetaceans. The effects of these sounds on cetaceans have not been studied comprehensively, but they may affect communication and other activities associated with sound production and perception (e.g., Castellote et al. 2012). However, small odontocetes are most sensitive to sounds above about 10 kHz. High-frequency hearing is good, with upper limits of sensitive hearing ranging from about 65 kHz to well above 100 kHz (Richardson et al. 1995), reflecting the use of high-frequency sound pulses for echolocation and moderately high-frequency calls for communication.

The increase over recent years in the recreational use of the sea and activities such as dolphin watching create the potential for an increase in threats through direct physical contact (collisions and propeller damage) and by the sounds introduced into the marine environment.

Whale watching and ecotourism

Few sectors of tourism have experienced the levels of growth in such a short time interval that have been observed in whale- and dolphin-watching operations. Increasingly, concerns have been raised about the impact of such activities on cetaceans (Janik & Thompson 1996, Ananthaswamy 2004, Steckenreuter et al. 2012), but owing to the rapid growth of this sector, the management responses to the potential impacts of cetacean watching have usually been reactive rather than proactive (Lusseau & Higham 2004). Assessing the impacts of tourism on cetaceans is challenging, with few studies attempting to examine more than one aspect of the problem. In addition, few studies have assessed the long-term impact of whale and dolphin watching on cetaceans.

A variety of responses has been observed in marine mammals reacting to tourists, with most studies focusing on changes in behaviour. Common behavioural reactions include schooling animals swimming closer together (e.g., Blane & Jaakson 1994, Nowacek et al. 2001, Steckenreuter et al. 2012); increased swimming speeds (e.g., Blane & Jaakson 1994, Williams et al. 2002a); changes in movement patterns (e.g., Nowacek et al. 2001, Jelinski et al. 2002, Williams et al. 2002b); and changes in resting, feeding, diving and respiratory behaviour (e.g., Baker et al. 1988, Janik & Thompson 1996, Moore & Clarke 2002, Lusseau 2003, Constantine et al. 2004, Dans et al. 2008, Steckenreuter et al. 2012). In addition, changes in habitat use and avoidance of previously preferred areas have been associated with increases in boat traffic (e.g., Glockner-Ferrari & Ferrari 1990, Corkeron 1995, Duffus 1996, Steckenreuter et al. 2012).

A study assessing the impact of tourism on common dolphins in New Zealand revealed that foraging and resting bouts were significantly disrupted (duration and overall time spent in these two states) by boat interactions; the disruptions were to a level that raises concern about the sustainability of this impact (Stockin et al. 2008). Foraging dolphins were more likely to stop foraging and took longer to resume foraging when disturbed by a tour boat compared with periods when a tour boat did not approach.

The presence, density and distance of boats affect cetacean behaviour (Lusseau & Higham 2004). The speed of vessels and their rate of directional change are thought to be critical determinants of the impact of encounters with cetaceans (Nowacek et al. 2001, Williams et al. 2002a,b, Lusseau 2003, Steckenreuter et al. 2012). Unpredictable and erratic vessel movements lead to typical antipredator behaviours in many cetaceans (Lusseau & Higham 2004).

The underwater noise associated with whale-watching operations has been poorly investigated in comparison to noise generated by shipping. Generally, depending on the species concerned, the noise associated with whale-watching vessels is considered likely to mask communication calls and, where large numbers of vessels are continuously present, may cause permanent impairment of hearing (Au & Green 2000, Erbe 2002, Jelinski et al. 2002). There is a growing realization that behavioural changes in marine mammals do occur as a result of engagement with tourist vessels, but the biological significance of such changes has yet to be elucidated (Lusseau & Higham 2004). Behavioural changes may be long term, with impacts on reproductive success and population growth (Janik & Thompson 1996, Notarbartolo-Di-Sciara et al. 2003). If sustainable development of marine ecotourism is to be taken seriously, then rigorous scientific research is required, and the results of such work need to receive action.

With the rapid growth of this industry globally, many nations now have regulations in place that restrict the number of vessels in close proximity to cetaceans and specify minimum approach distances (e.g., Orams 2000, Valentine et al. 2004, Steckenreuter et al. 2012).

Legislation

In the NE Atlantic, common dolphins are covered by a wide variety of legislation, including national, European and international statutes and conventions, all with aims to protect, conserve, manage and study. Although some legislation aims not only to halt deterioration of the 'status' of this species but also to achieve a significant and measurable improvement in it, the lack of information on trends in abundance, incidental capture rates, and other parameters prevents a thorough assessment of status for the common dolphin in this region.

International conventions

United Nations Convention on the Law of the Sea

The United Nations Convention on the Law of the Sea (UNCLOS) lays down a comprehensive regime of law and order in the world's oceans and seas, establishing rules governing all uses of the oceans and their resources (United Nations 2001). It enshrines the notion that all problems of ocean space are closely interrelated and need to be addressed as a whole. The convention governs all aspects of ocean space, including delimitation, environmental regulation, conservation of marine resources, marine scientific research, economic and commercial activities, transfer of technology and the settlement of disputes relating to ocean matters.

Included within the convention (United Nations 2001) are general provisions relating to marine conservation. Specifically, the convention states that contracting parties "shall cooperate with a view to the conservation of marine mammals and in the case of cetaceans shall in particular work through the appropriate international organizations for their conservation, management and study" (p. 48), and that signatories must take measures "necessary to protect and preserve rare or fragile ecosystems as well as the habitat of depleted, threatened or endangered species and other forms of marine life" (p. 101).

Convention on Biological Diversity

The Convention on Biological Diversity (CBD) is one of the three global conventions agreed at the Rio summit in 1992. Its goal is to promote biodiversity, balancing conservation with sustainable use and the sharing of economic benefits that are derived from biodiversity. CBD requires countries to prepare a national biodiversity strategy (or equivalent instrument) and to ensure that this strategy is incorporated into the planning and activities of all those sectors whose activities can have an impact (positive and negative) on biodiversity.

The vision of the CBD Strategic Plan for Biodiversity 2011–2020 is "by 2050, biodiversity is valued, conserved, restored and wisely used, maintaining ecosystem services, sustaining a healthy planet and delivering benefits essential for all people" (CBD 2010 Annex paragraph II). As part of these requirements, the European Commission developed and, in 2011, adopted the EU biodiversity strategy (European Commission 2011), a target of which is "to halt the deterioration in the status of all species and habitats covered by EU nature legislation and achieve a significant and measurable improvement in their status so that, by 2020, compared to current assessments, 100% more habitat assessments and 50% more species assessments under the Habitats Directive show an improved conservation status". As such, it includes the common dolphin.

At the national level, the UK government, for example, launched the UK Biodiversity Action Plan (UK BAP) in 1994. At that time, four plans covering cetaceans were implemented, with a grouped plan for small dolphins, which included the common dolphin. A review of BAP targets was undertaken in 2004, and the Cetacean BAP Steering Group suggested that the United Kingdom should move towards a single Cetacean BAP as many of the targets were generic across all cetacean species, and very few were pertinent to a single species or group of species. This was, however, not implemented, and in 2007 a BAP species and habitat review was undertaken. Under this review, 20 cetacean species were identified, including the common dolphin (JNCC 2010b), for which plans were required. During 2008, priority actions were developed for these species that reflected international obligations. For common dolphins, these are the following:

1. Undertake research on cetaceans using UK waters to identify areas of particular importance for breeding, feeding or migration;
2. Undertake any necessary research and fully implement mitigation measures to reduce by-catch as much as possible;
3. Develop and implement a UK Cetacean Surveillance Strategy;
4. Maintain the UK stranding scheme, which provides an indication of the extent of anthropogenic mortality, and implement appropriate remedial action when necessary; and
5. Undertake research into population structure.

Convention on International Trade in Endangered Species of Wild Fauna and Flora

The Convention on International Trade in Endangered Species of Wild Flora and Fauna (CITES), also less commonly known as the Washington Convention, aims to regulate international trade in species that are endangered or may become endangered if their exploitation is not controlled (CITES Secretariat 2012). CITES is implemented within Europe through two EC regulations (338/97 and 865/06 as amended). Species covered under CITES are listed in three appendices, with common dolphins listed in Appendix 2. This means that trade in the species is permitted as long as the authorities have ascertained that it will not be detrimental to the survival of the species; that the specimen was not obtained in contravention of the laws of that state for the protection of fauna and flora; and that any living specimen will be so prepared and shipped that it minimizes the risk of injury, damage to health or cruel treatment.

The Bonn Convention and the Agreement on the Conservation of Small Cetaceans of the Baltic, North-East Atlantic, Irish and North Seas

The Convention on Migratory Species (CMS), or Bonn Convention, sets out general provisions for the protection and conservation of certain migratory marine mammals (CMS Secretariat 2012). Common dolphins in the North and Baltic Seas are listed in Appendix II; those in the wider North Atlantic are not listed. Appendix II includes species that have an unfavourable conservation status and that require international agreements for their conservation and management, as well as those that have a conservation status that would significantly benefit from the international cooperation that could be achieved by an international agreement. Parties that are 'range states' (countries with waters in the geographical range of the species concerned) of migratory species listed in Appendix II are expected to conclude agreements to benefit the species and should give priority to those species in an unfavourable conservation status. One such agreement is the Agreement on the Conservation of Small Cetaceans in the Baltic, North-East Atlantic, Irish and North Seas (ASCOBANS; ASCOBANS 2012).

ASCOBANS includes a concise Conservation and Management Plan (CMP) that outlines the conservation and management measures to be implemented by signatories. This states that research "shall be conducted in order to (a) assess the status and seasonal movements of the populations and stocks concerned, (b) locate areas of special importance to their survival, and (c) identify present and potential threats to the different species." Besides these requirements to monitor abundance and distribution of small cetacean species, the CMP also states that "each party shall endeavour to establish efficient systems for reporting and retrieving bycatches and stranding specimens and to carry out … full autopsies in order to collect tissues for further studies and reveal possible causes of death and to document food composition". In addition, the CMP also states that "information shall be provided to the general public in order to ensure support for the aims of the agreement in general and to facilitate the reporting of sightings and strandings in particular; and to fishermen in order to facilitate and promote the reporting of bycatches and the delivery of dead specimens to the extent required for research under the agreement" (p. 8).

Besides the CMP, a number of resolutions have been developed by parties to ASCOBANS, the most relevant of which are the following:

- Resolution 7 of the Fourth Meeting of the Parties in 2003, 'Cetacean Populations in the ASCOBANS Area', requires parties to "support further work to elucidate temporal and spatial aspects of distribution of small cetaceans in the ASCOBANS area" (p. 3).
- Resolution 5 of the Fifth Meeting of the Parties in 2006, 'Incidental Take of Small Cetaceans', recommends that "total anthropogenic removal is reduced by the Parties to below the threshold of 'unacceptable interactions' with the precautionary objective to reduce bycatch to less than 1% of the best available abundance estimate and the general aim to minimise bycatch (i.e., to ultimately reduce to zero)" (p. 1).

To date, ASCOBANS has created and developed conservation plans for harbour porpoises in the Baltic Sea, North Sea, and the Western Baltic, Belt and Kattegat Seas. As yet, no conservation plan has been developed for common dolphins in the NE Atlantic under the auspices of ASCOBANS or any other intergovernmental or non-governmental organization. A conservation plan has been developed for common dolphins in the Mediterranean Sea by ACCOBAMS (the Agreement on the Conservation of Cetaceans of the Black Sea, Mediterranean Sea and Contiguous Atlantic Area).

Convention for the Protection of the Marine Environment of the North-East Atlantic

The Convention for the Protection of the Marine Environment of the North-East Atlantic replaced both the Oslo and Paris Conventions, with the intention of providing a comprehensive and simplified approach to addressing issues associated with maritime pollution; it also provides for conservation and protection of habitats and species. Article 2(1)(a) states the following:

> The Contracting Parties shall, in accordance with the provisions of the Convention, take all possible steps to prevent and eliminate pollution and shall take the necessary measures to protect the maritime area against the adverse effects of human activities so as to safeguard human health and to conserve marine ecosystems and, when practicable, restore marine areas which have been adversely affected. (p. 8)

The OSPAR Convention is the mechanism by which 15 governments of the coastal states of NW Europe, together with the European Commission, cooperate to protect the marine environment of the NE Atlantic (OSPAR 2012). Although common dolphins are not listed by OSPAR as a threatened and declining species, the convention clearly states that "definitions of 'biological diversity', 'ecosystem', and 'habitat' are those contained in the Convention of Biological Diversity of 5 June 1992". Therefore, the OSPAR Convention covers all habitats and species of the NE Atlantic maritime area.

The OSPAR Quality Status Report (QSR) states that to

support an ecosystem approach, OSPAR must extend its focus beyond protecting individual species and habitats or specific sites. Given the array of different actors managing the pressures that impact upon biodiversity and ecosystems, OSPAR should prioritise the development of an effective scheme for monitoring and assessing wider biodiversity status and ecosystem function. This must be linked with the concept of "Good Environmental Status" under the EU Marine Strategy Framework Directive. (OSPAR Commission 2010, p. 141)

Such an approach should benefit common dolphins.

The Bern Convention

The Convention on the Conservation of European Wildlife and Natural Habitats (or the Bern Convention) is a binding international legal instrument in the field of nature conservation that covers most of the natural heritage of the European continent and extends to some states of Africa (European Union 2007). Common dolphins in the North Atlantic are listed in Appendix 2 'strictly Protected Fauna Species', for which the following activities are prohibited:

1. All forms of deliberate capture and keeping and deliberate killing;
2. The deliberate damage to or destruction of breeding or resting sites;
3. The deliberate disturbance of wild fauna, particularly during the period of breeding, rearing and hibernation, insofar as disturbance would be significant in relation to the objectives of this Convention;
4. The deliberate destruction or taking of eggs from the wild or keeping these eggs even if empty;
5. The possession of and internal trade in these animals, alive or dead, including stuffed animals and any readily recognisable part or derivative thereof, where this would contribute to the effectiveness of the provisions of this article.

There is also a requirement for contracting parties to coordinate "efforts for the protection of the migratory species specified in Appendices II and III whose range extends into their territories".

For Member States of the European Union, the provisions of the Bern Convention are largely taken up in the 1992 Directive on the Conservation of Natural Habitats and of Wild Fauna and Flora (92/43/EEC), otherwise known as the 'Habitats Directive' (see 'European Legislation').

International Convention for the Regulation of Whaling

The IWC was set up under the International Convention for the Regulation of Whaling, which was signed in Washington, D.C., in December 1946 (IWC 2012). The purpose of the convention is to "provide for the proper conservation of whale stocks and thus make possible the orderly development of the whaling industry" (p. 1).

The commission has since its inception regulated the catches of the large whale species. Different views, however, are held concerning the legal competence of the IWC to regulate direct and incidental catches of small cetaceans, including the common dolphin. Despite the different views on the question of legal competence, the IWC does recognize the need for further international cooperation to conserve and rebuild depleted stocks of small cetaceans. Each year, the IWC Scientific Committee, through its Sub-Committee on Small Cetaceans, identifies priority species/regions for consideration by a review. Topics considered include distribution, stock structure, abundance, seasonal movements, life history, ecology, and directed and incidental takes. In 2009, the Sub-Committee on Small Cetaceans undertook a worldwide review of the common dolphin (IWC 2009).

European legislation

Directive on the Conservation of Natural Habitats and of Wild Fauna and Flora

The Directive on the Conservation of Natural Habitats and of Wild Fauna and Flora (92/43/EEC), commonly known as the Habitats Directive, is one of the most important pieces of European legislation aimed at the conservation of wildlife in the European Union. Common dolphins are listed in Annex IV, 'Animal and Plant Species of Community Interest in Need of Strict Protection'.

Article 11 requires Member States to monitor the conservation status of the habitats and species listed in the annexes; Article 17 requires a report on this work to be sent to the European Commission every 6 years. In the directive, conservation status is defined as "the sum of the influences acting on the species that may affect the long-term distribution and abundance of its populations" (Article I(i), p. 5). Conservation status can be considered favourable if

- population dynamics data indicate that the species is maintaining itself on a long-term basis as a viable component of its natural habitats,
- the natural range of the species is neither being reduced nor is likely to be reduced in the foreseeable future, and
- there is, and will probably continue to be, a sufficiently large habitat to maintain its populations on a long-term basis.

Assessment of favourable conservation status (FCS) therefore requires consideration of range, population, main pressures and threats, habitat and future prospects of the species, including any identifiable trends. These assessments should be undertaken using a standard methodology, designed to facilitate aggregation and comparisons between Member States and biogeographical regions (see the section 'Conservation status' for further information on the FCS assessments of the common dolphin).

Under Article 12:

Member States shall take the requisite measures to establish a system of strict protection for the animal species listed in Annex IV(a) in their natural range, prohibiting: (a) all forms of deliberate capture or killing of specimens of these species in the wild; (b) deliberate disturbance of these species, particularly during the period of breeding, rearing, hibernation and migration; and (d) deterioration or destruction of breeding sites or resting places. (pp. 9–10)

Under Article 12(4) of the Habitats Directive:

Member States should establish a system to monitor the incidental capture and killing of the animal species listed in Annex IV(a), and in the light of the information gathered, Member States shall take further research or conservation measures as required to ensure that incidental capture and killing does not have a significant negative impact on the species concerned. (p. 10)

These apply to common dolphins, with the most significant anthropogenic impact being by-catch.

EC Council Regulation 812/2004 (the 'Fisheries Regulation')

EC Council Regulation 812/2004 (the 'Fisheries Regulation') lays down measures concerning incidental catches of cetaceans in fisheries and amends Regulation (EC) No. 88/98. The measures pertinent to common dolphins in the North Atlantic are the coordinated monitoring of cetacean by-catch through compulsory onboard observers for given fisheries and the mandatory use of acoustic deterrent devices ('pingers') in certain fisheries.

EC Regulation 812/2004 requires that sampling should be such that a by-catch estimate with a CV of less than 0.3 can be achieved. However, for the CV to be determinable, the mean by-catch rate must be non-zero, so one or more by-catch events must be observed (Northridge & Thomas 2003). In the absence of any observed by-catch, and assuming continued monitoring is needed, the United Kingdom uses the 'pilot study' levels of 10% and 5% for the various fishery segments as the most appropriate approach to setting monitoring requirement levels. The United Kingdom is recognized as having one of the best by-catch observer schemes in Europe (European Commission 2009) and secured dispensation from the monitoring requirements in certain fisheries that had been demonstrated to have no by-catch over a 5-year period (e.g., pelagic trawls in the Celtic Sea targeting a variety of species, including mackerel, herring, blue whiting, horse mackerel, sardine, sprat and bass) to focus on those fisheries not covered by the legislation but known or suspected as having high levels of by-catch (Northridge & Kingston 2010) (see 'Fisheries interactions' section).

Marine Strategy Framework Directive and Good Environmental Status

The Marine Strategy Framework Directive (MSFD, Directive 2008/56/EC) requires Member States of the European Union to develop marine strategies that apply

> an ecosystem-based approach to the management of human activities while enabling a sustainable use of marine goods and services, priority should be given to achieving or maintaining good environmental status in the Community's marine environment, to continuing its protection and preservation, and to preventing subsequent deterioration. (European Union 2008, p. 20).

To determine good environmental status (GES), 11 qualitative descriptors have been selected, as outlined in Annex I. These cover biological diversity, non-indigenous species, population of commercial fish/shellfish, elements of marine food webs, eutrophication, seafloor integrity, alteration of hydrographical conditions, contaminants, contaminants in fish and seafood for human consumption, marine litter, and introduction of energy, including underwater noise.

In November 2011, OSPAR brought together its contracting parties to discuss proposed indicators and targets of GES for Descriptor 1: Biodiversity. The majority of relevant Member States proposed cetacean indicators and targets associated with abundance and distribution and by-catch. These are currently being further developed through the auspices of ICG-COBAM's (Intersessional Correspondence Group–Coordination of Biodiversity Assessment and Monitoring) expert group on marine mammals and reptiles with support from the ICES Working Group on Marine Mammal Ecology.

Management of the North-east Atlantic population

Management unit

One common dolphin population exists within the NE Atlantic. As a consequence of a lack of sampling of 'offshore' common dolphins for genetic analysis, the actual distributional range of the NE Atlantic population is undetermined. Thus, the ASCOBANS/HELCOM Small Cetacean Population Structure Workshop (Murphy et al. 2009a) and ICES WGMME (2009) recommended that the MU/area for the *Delphinus delphis* population in the NE Atlantic be confined to the continental shelf and slope waters and the oceanic waters of the Bay of Biscay. Taking into account recent abundance estimates, this area was extended to include the surveyed blocks of SCANS-II and CODA (see Figure 4) for the purposes of estimating by-catch limits for this population (Winship et al. 2009). This region also encompasses some of the main locations of commercial fishery operations in the NE Atlantic. As noted in the discussion in the ecological stock section, separate 'neritic' and 'oceanic' stocks were proposed for the Bay of Biscay (Lahaye et al. 2005). As the sample size of the putative oceanic stock in this study was only 10 individuals, a two-stock approach to management was not proposed by the ASCOBANS/HELCOM workshop (Murphy et al. 2009a) or ICES

WGMME (2009, 2012). A subsequent working paper presented to the IWC Sub-Committee on Small Cetaceans further delineated stock structure within the NE Atlantic using various ecological tracers and proposed that the continental shelf constituted an ecological boundary with some degree of permeability (Caurant et al. 2011). The IWC noted that some links and subunits were questionable due to very small sample sizes, sex bias in sampling, and temporal differences among the samples not being taken into account (IWC 2009). If further analysis supports the designation of oceanic and neritic stocks in the Bay of Biscay, delineation of stock boundaries will prove problematic, primarily as the main concentration of common dolphins in this region during summer (mating period) is along the continental shelf edge and adjacent oceanic waters (SCANS-II 2008, CODA 2009). In addition, the increased abundance of common dolphins on the continental shelf (e.g., western English Channel) during winter suggests onshore movements of individuals (ICES WGMME 2005), thus increasing the available (SCANS-II summertime) abundance estimate for these waters at that time.

Population status

Genetic analysis of stranded and by-caught common dolphins from continental shelf and adjacent oceanic waters of the Bay of Biscay waters revealed a high haplotype diversity of the control region sequences, suggesting a large effective population size within the NE Atlantic (Natoli et al. 2006, Viricel 2006). Combining the SCANS-II abundance estimate for shelf waters (56,221 dolphins, CV = 0.23) and CODA estimate for offshore waters (116,709 dolphins, CV = 0.34) produces an abundance estimate of 172,920, considered to be representative of the abundance of common dolphins for the summer of 2006 (after Winship et al. 2009). Common dolphins are one of the most abundant cetacean species occurring in continental shelf waters of the NE Atlantic, second only to the harbour porpoise with a population abundance of 385,617 (CV = 0. 20) individuals in European continental shelf waters (SCANS-II 2008). Some historical abundance estimates exist for common dolphins (see section on population abundance), and 273,159 (CV = 0.26) individuals were estimated for the NASS-west survey block for the year 1995 (Cañadas et al. 2009), a region located outside the MU area.

Worldwide population abundance estimates for the common dolphin vary widely, with 2,963,000 (CV = 0.24) short-beaked common dolphins inhabiting the ETP (encompassing the northern, central, and southern stocks) (Gerrodette & Forcada 2002); 411,211 (CV = 0.21) individuals in waters off California, Oregon and Washington (Carretta et al. 2011); and 120,743 (CV = 0.23) dolphins in the NW Atlantic. The last estimate listed was based on two surveys undertaken in 2004 that had the most complete coverage of the species' habitat (Waring et al. 2011). As no previous abundance estimate exists for the NE Atlantic MU area and the current net productivity rate based on trends in abundance is undetermined, it is not known if the population is currently depleted compared to historical levels. To date, evaluations of the current status of the species have been based on *ad hoc* assessment methods, and as with other cetacean species, in the future it is essential that long-term (decadal) management will encompass, inter alia, appropriate long-term research planning, with the incorporation of a good monitoring programme (Donovan 2005). As part of the improvements in the approach to assessing species status within the NE Atlantic, a collaborative project, the Joint Cetacean Protocol (JCP) has been developed, which should deliver information on the distribution, abundance and population trends of cetacean species. The JCP brings together effort-related cetacean sightings data from a variety of sources, including large-scale international surveys, surveys based on platforms of opportunity, as well as more localized non-governmental data and industry data. These data, collected between 1979 and 2010, represent the largest NE Atlantic cetacean sightings resource ever collated and have been standardized to a common format, checked and corrected (see Paxton & Thomas 2010, Paxton et al. 2011).

Survival can be estimated on the basis of age and reproductive state determined from teeth sections and gonads of stranded and by-caught cetaceans (Murphy et al. 2007a, Mannocci et al. 2012). These parameters can then be used as inputs in demographic models to conduct population projections and risk analyses, as well as potentially providing indicators of population status in their own right (ICES WGMME 2012). However, use of age-at-death distributions for producing life tables, survivorship curves and determining population growth rates has its limitations and biases. It is important when constructing life tables that the sample/population represents a stable age distribution, that is, where the age structure of the population is constant with time, is not growing (or if growing, it is assumed that the rate of growth is constant and known) or subject to any density-dependent processes (Caughly 1966, Barlow & Boveng 1991).

An assessment of available age-at-death data from the NE Atlantic common dolphin population revealed a large sampling bias due to a high immature (and young adult) mortality from anthropogenic activities (Murphy et al. 2007a). Samples obtained directly from fisheries and from stranding projects exhibited this sampling bias—even the sample of stranded individuals for which cause of death was not attributed to incidental capture. However, a time series of such mortality data could provide an indicator of changing population status, with increasing mortality potentially indicating an undesirable trend (ICES WGMME 2012). R_{MAX} is the potential rate of population increase under optimal environmental conditions (i.e., the maximum net productivity rate) and is determined by the intrinsic life-history characteristics of the species (Murphy 2009a). For cetaceans, the default value of maximum net productivity rate, when no specific estimate is available, is 4% per annum. This is based on theoretical modelling showing that cetacean populations may not grow at rates much greater than 4% given the constraints of their reproductive life history (Barlow & Boveng 1991, Waring et al. 2011). Estimates of R_{MAX} for NE Atlantic common dolphins range from 4% to 4.5% per year (Murphy et al. 2007a, Mannocci et al. 2012).

The NE Atlantic common dolphin population exhibits evidence of age-sex segregation (especially outside the mating period), reproductive seasonality, a low pregnancy rate of 26%, extended calving interval of 3–4 years, and a low potential lifetime reproductive output of about 4 calves. Monitoring trends in life-history parameters is an important requirement of both the Habitats Directive and ASCOBANS as this can also be used to assess conservation status (ICES WGMME 2010). Understanding the causes of change is essential for the design and implementation of conservation and management measures, and purely monitoring trends in abundance will not provide this information. With respect to the findings based solely on mortality data, Murphy et al. (2009b) found no evidence of density dependence in reproductive parameters. The low annual pregnancy rate reported throughout the 16-year sampling period may suggest either that the level of anthropogenic mortality did not cause a substantial population decline or that available prey declined at approximately the same rate as the dolphin population. Even if the low pregnancy rate observed in the NE Atlantic is, in fact, close to the natural rate for a common dolphin population in a temperate region, it cannot be ruled out that other factors, such as environmental and anthropogenic activities, including the introduction and release of physical and chemical pollutants, may be contributing to the population's low reproductive output (Murphy 2009b). The impact of pollutants on fetal and newborn survival rates in *Delphinus delphis* is currently being investigated, and preliminary analysis suggests an association between the incidence of abortion and high pollutant burdens (Murphy et al. 2012b).

Summary of main pressures and threats

A number of environmental and anthropogenic factors are considered to affect cetacean population growth rates, including anthropogenic mortality, food availability, disease, pollutants and climate change (Murphy 2009a). The effects of disturbance on cetaceans at the population level are not well understood.

By-catch and entanglement in fishing gear (including discarded gear) are considered main current threats to the NE Atlantic common dolphin population, and even now these effects cannot be quantified due to a lack of data on incidental capture rates in some fisheries and limited sampling in other fisheries. If pollutants have an adverse effect on individual reproductive capabilities, the population would be more vulnerable to other pressures than is normally assumed, especially other anthropogenic activities, such as incidental capture, and would not necessarily recover from these adverse interactions in a predictable way (Murphy 2009b). Within the NE Atlantic, the degree of human disturbance to cetaceans (offshore construction activities, boat traffic), including the level of underwater noise (e.g., seismic surveys, active sonar, piledriving), has been increasing in recent years and will continue to increase due to expansion of the marine renewables industry. Although short, intense noise can cause injury or death to marine organisms, long-term exposure to less-intense sounds can have sublethal effects, including effects related to stress (Tasker et al. 2010). The long-term cumulative effects of increased noise and disturbance on individual cetaceans are currently unknown.

Common dolphins primarily consume energy-rich prey and show both interannual and seasonal dietary variations depending on prey resource availability. The effects of climate change will alter prey species distribution and abundance, which will have a direct effect on the local common dolphin population. In the NE Atlantic, there is evidence of both seasonal movements and long-term distributional patterns in common dolphins, possibly reflecting changes in resource availability. Therefore, it is expected that in the future common dolphins will adapt to effects of climate change; some studies have already shown changes in contemporary distribution and occurrence related to environmental factors (e.g., MacLeod et al. 2005). In the eastern North Pacific, common dolphins have crossed stock boundaries during periods of significant environmental change. A decrease in the abundance of the northern *Delphinus delphis* stock in the ETP and an associated increase off Southern California, starting in the late 1970s, suggest that a large-scale shift in the distribution of common dolphins may have occurred in this region (Anganuzzi & Buckland 1994, Danil & Chivers 2006). There is evidence that *D. delphis* in the southern stock of the ETP moved into the higher-quality habitat of the 'Costa Rica dome' (an area of upwelling, with a shallow thermocline, associated with high biological productivity), which was within the distributional range of the central stock, during the strong El Niño of 1972–1973 (Danil & Chivers 2006). Even if (gradual) changes in environmental conditions and the distribution or density of preferred prey species are not limiting factors, the health status of the NE Atlantic population and ongoing anthropogenic perturbations all affect the population's ability to both adapt to and recover from significant environmental and resource variability (e.g., during periods of abrupt changes in the ecosystem caused by climate, overfishing, or a combination of these) (Beaugrand et al. 2008).

The contemporary NE Atlantic population is exposed to a variety of novel stressors (including chemical, physical and biological). The effect of any given stressor will be conditional on multiple factors, including sex, age and reproductive condition, as well as other stressors currently affecting the individual. The timing of such effects is also important, as the closer an individual is to allostatic overload when subjected to an additional stressor, the more likely it will have an adverse effect (National Research Council [NRC] 2005). Assessing the impacts of these stressors is complex as stressors rarely act independently. Generally, identification of causal relationships between a stressor and a change in survival and reproductive success in individuals has been confounded by the interacting effects of multiple stressors.

Management framework

Human impacts can be classified into two groups that result in either instantaneous or near-instantaneous death (e.g., incidental capture, ship strikes) or those that affect the overall 'fitness' of a population (e.g., pollution, overfishing of prey species and habitat loss) (Donovan 2005). Conservation management is the process of regulating human activities to minimize their negative impacts on natural environments while allowing sustainable exploitation and development. To maintain species in a favourable conservation status, a comprehensive management framework approach must be developed that clearly outlines management objectives and implements research and monitoring programmes to obtain scientific information necessary to inform management. Two types of management objectives might be set: targets (to achieve certain conditions) and limits (to avoid certain conditions). Objectives can be established both for the state of the population and for impacts on that population (Tasker 2006). The establishment of management objectives and the relative weight given to those objectives (the trade-offs) ultimately require political rather than scientific decisions, though it should be remembered that scientists clearly have an obligation to explain the implications of any decisions by, in part, providing politicians with a range of specific options and their likely consequences (Donovan 2005). Monitoring is a fundamental part of management, and in 2010 the ICES WGMME recommended the adoption of an adaptive monitoring and surveillance approach for cetaceans in the ICES area, under which objectives, monitoring and outcomes are regularly reviewed and updated by a steering group composed of representatives from all relevant stakeholders (ICES WGMME 2010). This requires a coordinated international approach for developing a single assessment for each cetacean species at an appropriate biological scale.

If the European Union is to achieve the goals of its Habitats Directive (i.e., maintaining or restoring FCS for listed species), it will need a comprehensive conservation plan for those individual cetacean species for which conservation issues arise due to anthropogenic causes. Such plans have been produced by ASCOBANS for the harbour porpoise in the North Sea and by ACCOBAMS for common dolphins in the Mediterranean Sea and are based on the available information on status, trends and threats. They identify research needs and set conservation targets to respond to the key threats through threat reduction measures, improved regulations or other mitigation strategies (IWC 2008, Reijnders et al. 2008). If no measure exists for specific threats, it has been recommended that a programme should be established involving stakeholders, with all aspects of mitigation measures considered, including science, practicalities, the legal framework, education and awareness (Reijnders et al. 2008). To date, only measures to mitigate the effects of mortality in fishing gear have been proposed for the NE Atlantic common dolphin population.

Management objectives

Management objectives based solely on detecting trends in abundance are inadequate, since a change in population size does not necessarily signify a change in the optimum sustainable level of that population owing to the fact that the carrying capacity may have changed due to natural causes (Gerrodette & DeMaster 1990). To manage and monitor a population appropriately, both the population condition index and the abundance index need to be assessed to detect demographic changes at an early stage (Gerrodette & DeMaster 1990). The optimum sustainable population (OSP), described by the US Marine Mammal Protection Act, is the number of animals that will result in the maximum productivity of the population or the species, keeping in mind the carrying capacity of the habitat and the health of the ecosystem of which they form a constituent element (Wade 1998). The US National Marine Fisheries Service defined OSP as a population level between carrying capacity K and the population size at maximum net productivity (Gehringer 1976), that is, the maximum net productivity level (MNPL) forms the lower boundary of the OSP range (Gerrodette & DeMaster 1990).

The management goal of the US Marine Mammal Protection Act is to prevent populations from 'depletion' and maintain populations above MNPL, estimated to be between 50% and 85% of carrying capacity (and is more likely to be in the lower portion of that range) (Taylor & DeMaster 1993) or between 50 and 70% of a historic population size thought to represent carrying capacity (Gerrodette & DeMaster 1990, Wade 1998). To estimate the historical population size, at a time prior to the directed (direct fisheries or incidental captures) or indirect (habitat deterioration or harvest or competition for similar prey) impacts by humans, information on vital rates, numbers of animals killed by humans and a current population abundance estimate are required (Gerrodette & DeMaster 1990).

In European waters, the aim of ASCOBANS is to "restore and/or maintain biological or management stocks of small cetaceans at the level they would reach when there is the lowest possible anthropogenic influence" (ASCOBANS 2000, p. 94), and the contracting parties have agreed that a suitable short-term practical objective is to "restore and/or maintain stocks/populations to 80% or more of carrying capacity" (ASCOBANS 2000, p. 94; 2006). In contrast, the Habitats Directive does not set explicit objectives.

Estimating by-catch limits   To enforce management objectives to maintain a species at a certain fraction of carrying capacity and to reduce the negative effects of incidental capture, it is necessary to set by-catch limits for specific species, fisheries and areas. A number of approaches can be used to estimate by-catch limits for the defined MU. Based on the ASCOBANS conservation objective (restoring or maintaining the population at 80% or more of K), a harbour porpoise by-catch limit reference point of 1.7% of population size per year was derived, assuming the maximum annual rate of increase was 4% (IWC 2000). ASCOBANS uses this value for representing 'unacceptable interactions', and 1% of the population size is used as an 'intermediate precautionary objective' (ASCOBANS 2000, 2006). Although this reference point was originally produced for harbour porpoises, it has subsequently been applied to numerous other cetacean species, including the common dolphin. The limit of 1.7% was derived using a simple deterministic population dynamics model, assuming a single stock with more-or-less independent dynamics. When this is not the case, the limit is liable to be inappropriate (ICES WGMME 2012).

In 2010, the ICES WGMME recommended that we should "move away from implicit and automated conservation targets and towards the explicit definition and justification of target population sizes and management objectives" (p. 146). One aim of the SCANS-II (2008) and CODA (2009) projects was to develop a robust framework using all available information to generate safe by-catch limits for harbour porpoises and common dolphins (Winship et al. 2009). Two candidate management procedures were developed: adaptations of the US potential biological removal (PBR) method, and the IWC's catch limit algorithm (CLA) (part of the revised management procedure) approach. The PBR procedure takes a single, current estimate of absolute population size as input, while the CLA, a more complicated approach, takes time series of estimates of absolute population size and estimates of absolute by-catch as inputs, and thus should be more conservative. Both procedures, however, explicitly incorporate uncertainty in the estimates of population size (SCANS-II 2008, CODA 2009), unlike the deterministic 1.7% by-catch limit reference point. Three different tunings were developed, ranging from conservative to worst-case scenarios. Before this management procedure can be implemented for a particular species in a particular region, several steps need to be taken. These involve agreement by policy makers on the exact conservation/management objective(s), including what fraction of carrying capacity to maintain or restore populations to, and over what period of time; generation by scientists of by-catch limits for a specified period; and establishment of a feedback mechanism for informing the next phase of implementation of the procedure (SCANS-II 2008).

To use the PBR approach correctly, it has been stipulated that population size estimates should not be older than 8 years (Wade & Angliss 1997) as a population that declines at 10% per year from carrying capacity would be reduced to less than 50% of its original abundance after 8 years (Wade & Angliss 1997). In the NE Atlantic, an abundance estimate for continental shelf waters was determined for July 2005 (SCANS-II) and adjacent offshore waters in July 2007 (CODA). Although large-scale decadal surveys were planned for the NE Atlantic, at present (2013), funding for subsequent repeat large-scale survey(s) of these regions has not been obtained.

Conservation status

To assess conservation status, it is essential to understand population structure, including distribution and abundance, key drivers of population dynamics, key resources and the effects of stressors, especially those caused by anthropogenic interactions.

The International Union for Conservation of Nature (IUCN), founded in 1948, is the world's oldest and largest global environmental organization. IUCN's mission is "to influence, encourage and assist societies throughout the world to conserve the integrity and diversity of nature and to ensure that any use of natural resources is equitable and ecologically sustainable" (http://www.iucn.org/about/).

Probably the best known of the IUCN's publications is the Red List of threatened species (IUCN 2012). This has become an increasingly powerful tool in conservation management and decision making worldwide (Rodrigues et al. 2006, Currey et al. 2009, Butchart & Bird 2010). The Red List provides information on population size and trends, geographic range and habitat needs of species. The list also outlines the threats and pressures to which the species is exposed and whether it is considered sacred or whether it is protected by international law.

Short-beaked common dolphins are classified as 'least concern' throughout most of their range "despite ongoing threats to local populations. The species is widespread and very abundant (with a total population in excess of four million), and none of these threats are believed to be resulting in a major global species decline" (Hammond et al. 2008). The common dolphin is classified as 'endangered' in the IUCN Regional Red List for the Mediterranean Sea (Hammond et al. 2008).

Conservation status assessments are also required by Article 17 of the Habitats Directive for European waters (see 'Legislation' section). In 2007, the first FCS reports were submitted by Member States (Figure 10). Status in the combined Marine Atlantic biogeographic region assessment for common dolphins was 'unknown' due to the lack of data on current trends in the population and future prospects. Status in the Mediterranean Sea was considered to be 'unfavourable bad' and in the Macronesia biogeographic region 'unknown (but not favourable)'. The next round of FCS reports is due in 2013.

For the NE Atlantic common dolphin population, historical population size is not known, and due to a lack of data cannot be calculated. Furthermore, a time series of population size estimates for the MU area (see 'Population abundance' section) is not available. There is no information on whether the population has been subject to excessive anthropogenic mortality in the past or if habitat degradation or loss may have reduced carrying capacity. All sightings data used to estimate abundance have been obtained during the summer, whereas pelagic trawl fisheries (and a large number of other fisheries) operate predominantly during the winter. Apart from high rates of incidental captures reported by the tuna drift net fishery during the 1990s, there is a lack of adequate information on by-catch rates in a large number of other fisheries. Even though individual countries are now collecting incidental capture rate data under legislative requirements, there are still large data gaps due to uneven and insufficient sampling of fisheries (see 'Fisheries interactions' section). Consequently, we are unable to estimate reliably an annual population incidental mortality rate. The minimum estimated annual incidental mortality rate for the NE Atlantic MU in 2009 was above the 'intermediate precautionary objective' of ASCOBANS (see Table 3). Large data gaps also exist for other stressors.

Even taking all these unknowns and uncertainties onboard, indicators and targets can be developed to maintain or recover the species to GES in support of the MSFD. These indicators would monitor the impacts of pressures, and ensure that the structure and functions of the ecosystems are safeguarded and are not adversely affected, through coordinated marine monitoring programmes with integrated periodic assessments.

Indicators in support of conservation status assessments

OSPAR's ICG-COBAM proposed a number of potential MSFD indicators for cetaceans, including (1) distribution range and distributional pattern within range, (2) population abundance (detections

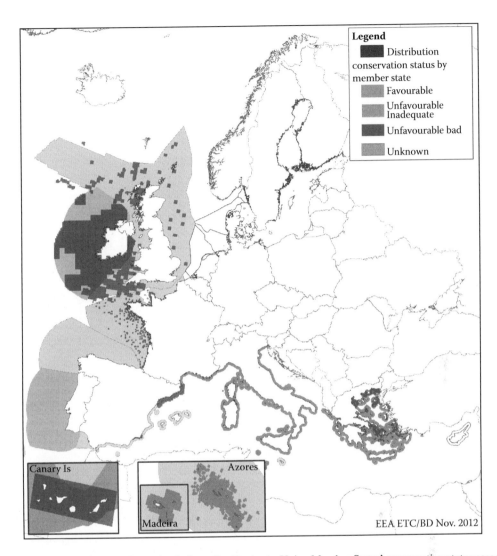

Figure 10 (See also colour figure in the insert)   European Union Member States' conservation status assessments for common dolphin in 2007. (Figure produced by Brian Mac Sharry, EEA-European Topic Centre on Biological Diversity [ETC/BD]. With permission.)

of trends in abundance), and (3) mortality rate (anthropogenic mortality in fishing gear) (ICES WGMME 2012). These proposed indicators are largely based on current monitoring requirements for other European legislation. For common dolphins and other cetaceans, unless there is knowledge and continued assessment of population diversity, structure and biological parameters, as well as an understanding of the pressure-state relationships, the usefulness of indicators such as 'population abundance' is questionable, as understanding the root cause of a significant decline in population abundance is necessary for successfully managing that population and achieving GES. Therefore, indicators focusing on pressures and changes in population condition should also be explored. However, it should be noted that data on population abundance are necessary for evaluating other indicators, such as by-catch.

Further development of a by-catch management procedure (based on those developed by SCANS-II and CODA) and creation of a potential MSFD by-catch indicator for common dolphins will be undertaken jointly by the ICES WGMME and WGBYC. For further assessment of population condition, indicators that use mortality data (e.g., obtained from strandings and fisheries by-catch programmes) can be developed. Two such population condition indicators proposed here are (1) blubber PCB toxicity threshold concentration of 13 mg kg^{-1} lipid weight (summed ICES-7 chlorinated biphenyl [CB] congener—numbers 28, 52, 101, 118, 138, 153, and 180) and (2) assessing changes in demographic characteristics. Other indicators, not reviewed here, can be obtained from strandings data, such as number of common dolphins whose death was caused by anthropogenic activity.

Blubber PCB toxicity threshold indicator

Detailed research on UK-stranded cetaceans conducted under the UK Cetacean Strandings Investigation Programme has shown strong links between elevated blubber PCB levels and mortality from infectious disease (Jepson & Deaville 1999, Jepson et al. 2005a, Hall et al. 2006) consistent with fatal PCB-induced immunosuppression. In one case-control study of UK-stranded harbour porpoises, the risk of infectious disease mortality increased by 2% for every 1% increase in the summed concentration of 25 CB congeners (Hall et al. 2006). A doubling of risk occurred at approximately 45 mg kg^{-1} (blubber) lipid. In a second case-control study of UK-stranded harbour porpoises, mean summed 25 CB congeners in the 'healthy' control group (death due to physical trauma) was 13.6 mg kg^{-1}, compared with 27.6 mg kg^{-1} for the animals that died of infectious diseases (Jepson et al. 2005a).

The levels of PCBs in tissues are easily and accurately measured, provided blubber samples from dead stranded animals or biopsies from live animals are available and appropriate sampling and analytical methodologies are in place. Previous studies by Jepson and colleagues (Jepson et al. 2005a, Hall et al. 2006) estimated a threshold of toxicity (including immunosuppression and reproductive impairment) for blubber PCB concentrations in harbour porpoises of 20 mg kg^{-1} weight (for summed 25 CB congeners). This equates to a blubber PCB toxicity threshold concentration of 13 mg kg^{-1} (for summed ICES-7 CB congeners) based on standard regressions between summed 25 CBs and summed ICES-7 CBs. This threshold (13 mg kg^{-1} concentration for summed ICES-7 congeners) could be used for other marine mammal species, including the common dolphin, to assess populations that may show risks of toxic effects at individual and population levels. For target setting, it is recommended that the biological effects from contaminants are kept within safe limits so that there are no significant impacts on, or risks to, marine mammals. The cause-and-effect relationships need to be established and monitored, as well as the impacts of accumulated (independent and interactive) effects. To undertake these tasks, knowledge of information on population growth rates, population structure, life-history parameters and density-dependent changes in these parameters is required. As female cetaceans transfer the majority of their PCB burden to their first calf during pregnancy and lactation (see 'Ecological stocks' section), data on age and reproductive status, and whether a female was previously gravid (i.e., pregnant), have to be assessed to provide context to the estimated contaminant burden. For males, age is the most important criterion.

Assessing changes in demographic characteristics indicator

Temporal variations in reproductive parameters can occur due to alterations in the availability of prey resources and population density. Cetacean populations are regulated through density-dependent changes in reproduction and survival, and it has been proposed that food resources are the main causative agent in the expression of density dependence, resulting in an increase in population growth rates (and reproductive output) at low densities (e.g., following large-scale incidental mortality in fishing gear) and a decrease in growth rates (and reproductive output) at high densities (see Murphy et al. 2009b and references therein). However, anthropogenic toxins and disease can alter reproductive rates by decreasing fertility and causing abortions, premature parturition and neonatal mortality (Murphy 2009b).

As part of the various European cetacean stranding programmes, cause of death, health status, nutritional condition and the status of the reproductive tract of individuals are investigated. Teeth, ovaries and testes are collected for subsequent analyses to assess maturity status and age. Strandings following encounters with fishing gear provide the most reliable samples for estimating biological parameters in the NE Atlantic common dolphin population, such as population pregnancy rates, proportion of mature individuals, proportion of females simultaneously pregnant and lactating, average age attained at sexual maturity, nutritional condition, and variations in reproductive parameters with age (see 'Population parameters' section). Dolphins stranded for other reasons, such as old age, disease and parasitic infestation, are likely to be unrepresentative with respect to these parameters (see 'Stranding patterns' section). However, there is evidence that pelagic fishing gear may be somewhat selective with respect to different categories of common dolphin, perhaps due to differences in behaviour, sensory and locomotory capabilities or geographical distribution (see 'Fisheries interactions' section), so the possibility of biases in stranding samples needs to be considered in relation to the parameters of interest when interpreting data from by-catch and fishing-induced strandings.

The conservation target for demographic indicators is no statistically significant deviation from long-term variation. It is important to consider the sample sizes required to detect deviations reliably. For example, pregnancy rate can be readily estimated from stranding samples, but power analysis suggests that changes in the pregnancy rate of the NE Atlantic common dolphin population would need to be extremely large to be detected statistically (Murphy et al. 2009b). With statistical power of 80% and an initial pregnancy rate of 0.26, a sample size of 150 mature females would be required to detect an absolute decrease of 13% or greater (pregnancy rate at 0.13 or below) between two time periods, whereas a sample size of more than 100 mature females would allow detection of a decrease of 15% or greater (Murphy et al. 2009b). With a sample size of only 50 mature females, however, the minimum detectable decrease would be 20%. In contrast, if there was an increase in the pregnancy rate, a sample size of 150 mature females would be needed to detect an increase of 15% or greater with statistical power of 80% (Murphy et al. 2009b). In addition, adequate age and reproductive data from males and females (at least 50 individuals of each sex) are vital for estimating the average age attained at sexual maturity.

Obtaining such large sample sizes of sexually immature and mature individuals is difficult and requires that cetacean stranding and by-catch observer programmes in NW Europe continue to sample all available and suitable carcasses and should be coordinated to standardize procedures. One compromise would be to alter the criterion used for statistically detecting differences. Many managers remain unaware that the standard criterion used for statistical 'significance', the probability of making a 'type I error', $\alpha = 0.05$ (i.e., incorrectly rejecting the null hypothesis in 5% of tests in the long run), is not an objective scientific value but a policy choice based on the most commonly used level of statistical significance (Taylor & Gerrodette 1993).

In addition to these sampling requirements, to interpret reproductive data correctly, population abundance estimates, trends in abundance and data on parameters that affect the dynamics of the population, such as annual mortality rates in fisheries, temporal variations in prey abundance, and levels of anthropogenic toxins, are required.

Recommended research

Numerous different pressures and threats, both anthropogenic and environmental, have affected and will affect common dolphins in the NE Atlantic. Many studies undertaken to date, however, have sampling biases and other limitations, and large uncertainties still exist, primarily due to the difficulties in understanding the true nature and complexity of adaptive responses to stressors, such as noise, on vital functions and rates. In addition, the numerous potential ways in which such multiple and diverse stressors can interact remain poorly understood. Over the next few years, large-scale expansion of marine infrastructure is foreseen within the NE Atlantic, and when passing through multiple areas with marine infrastructure, dolphins will be exposed to a variety of stressors, varying widely in their nature and impact (ICES WGMME 2010, 2011, 2012). Research is required to support these developments, such that they do not have negative conservation impacts for common dolphins or other cetaceans (as outlined in Murphy et al. 2012b).

Research in areas such as population diversity, structure, abundance and range, seasonal and long-term movements, health status, pollutants, life history, feeding ecology, and mortality rates, including exploring novel causes of death and effective by-catch mitigation practices, should be continued, as should the ongoing evaluation of ecological stocks using tracers that integrate over tens of years (see 'Ecological stocks' section). Novel ways of assessing population range and health status through remote biopsying could be explored (Fossi & Marsili 2011). Finally, with the development of an ecosystem approach to fisheries management, integration of data on common dolphins into ecosystem models not only will allow further elucidation of ecosystem dynamics but also will enable investigation of the effects of climate change, as well as ecosystem and regime shifts, on the local common dolphin population.

Recommended conservation actions

It is crucial that current legal requirements and obligations are fully met and existing management measures implemented and enforced. This could be achieved and aided through the development of an international conservation plan for NE Atlantic common dolphins, which would enable EU Member States to focus on conservation priorities in their waters. Such a plan could be developed through the auspices of ASCOBANS. This plan would continue the identification and evaluation of present and potential threats and reduce potential impacts through the development of threat reduction measures. This conservation plan would accompany other initiatives, such as the ICES surveillance and monitoring framework for marine mammals, and development of MSFD indicators based on population size, distribution, mortality, and population condition.

Acknowledgements

This review was supported by a Marie Curie International Outgoing Fellowship within the Seventh European Community Framework Programme. This review originated from work and helpful discussions undertaken by the ICES Working Group on Marine Mammal Ecology. Robin Law participated in the development of the blubber PCB toxicity threshold indicator. We wish to thank Yvon Morizur for providing information on incidental capture estimates, Simon Berrow and the Irish Whale and Dolphin Group for providing data for Figure 9, and Defra and Devolved Administrations in the United Kingdom for UK strandings and other data. Thanks to Rene Swift for producing Figure 4, Phil Hammond for permission to use Figure 6, and Brian Mac Sharry for producing Figure 10. Sightings data for Figure 1 were provided by Debbie Palka (Northeast Fisheries Science Center, NOAA); William A. McLellan (University of North Carolina at Wilmington); Robert Kenney (University of Rhode Island); Falk Huettmann (Canadian Wildlife Service); Thorvaldur Gunnlaugsson (Marine Research Institute, Iceland); Ana Cañadas (North Atlantic Marine Mammal Commission); Geneviève Desportes (Museum of Natural History, Faroe Islands); Tim Dunn (Joint Nature Conservation Committee); Peter Evans (Sea Watch Foundation); Phil Hammond (Sea Mammal Research Unit, St. Andrews University); Michel Goujon (Institut Français de Recherche pour l'Exploitation de la Mer); Oliver Ó Cadhla (University College Cork); Nøttestad Leif, Henrik Skov, Erik Olsen and Gordon Waring (Mid-Atlantic Ridge Ecosystem Project, MAR-ECO, survey data); and Lisa Steiner (Whale Watch Azores). Luca Mirimin, Ronan Cosgrove, Phil Hammond and John Brophy reviewed sections of previous drafts. We wish to thank Philip Smith for detailed comments and suggestions that greatly improved this review.

References

Abollo, E., Lopez, A., Gestal, C., Benavente, P. & Pascual, S. 1998a. Macroparasites in cetaceans stranded on the northwestern Spanish Atlantic coast. *Diseases of Aquatic Organisms* **32**, 227–231.

Abollo, E., Lopez, A., Gestal, C., Benavente, P. & Pascual, S. 1998b. Long-term recording of gastric ulcers in cetaceans stranded on the Galician (NW Spain) coast. *Diseases of Aquatic Organisms* **32**, 71–73.

Agreement on the Conservation of Small Cetaceans in the Baltic, North East Atlantic, Irish and North Seas (ASCOBANS). 2000. Resolution no. 3: incidental take of small cetaceans. 3rd Session of the Meeting of Parties to ASCOBANS, Bristol, England, 26–28 July 2000. Bonn, Germany: ASCOBANS. Online. http://www.service-board.de/ascobans_neu/files/2000-6.pdf (accessed 4 June 2012).

Agreement on the Conservation of Small Cetaceans in the Baltic, North East Atlantic, Irish and North Seas (ASCOBANS). 2006. Resolution no. 5: incidental take of small cetaceans. 5th Session of the Meeting of Parties to ASCOBANS, Netherlands, 18–20 September and 12 December 2006. Bonn, Germany: ASCOBANS. Online. http://www.service-board.de/ascobans_neu/files/mop5-final-5.pdf (accessed 4 June 2012).

Agreement on the Conservation of Small Cetaceans in the Baltic, North East Atlantic, Irish and North Seas (ASCOBANS). 2012. Agreement on the Conservation of Small Cetaceans in the Baltic, North East Atlantic, Irish and North Seas. Bonn, Germany: ASCOBANS. Online. http://www.ascobans.org/ (accessed 16 July 2012).

Aguilar, A., Borrell, A. & Pastor, T. 1999. Biological factors affecting variability of persistent pollutant levels in cetaceans. *Journal of Cetacean Research and Management* **Special Issue 1**, 83–116.

Amaha, A. 1994. *Geographic variation of the common dolphin,* Delphinus delphis *(Odontoceti: Delphinidae).* PhD thesis, Tokyo University of Fisheries, Japan.

Amaral, A.R., Beheregaray, L.B., Bilgmann, K., Boutov, D., Freitas, L., Robertson, K.M., Sequeira, M., Stockin, K.A., Coelho, M.M. & Möller, L.M. 2012a. Seascape genetics of a globally distributed, highly mobile marine mammal: the short-beaked common dolphin (*Delphinus*). *PLoS one* **7**, e31482, doi:10.1371/journal.pone.0031482.

Amaral, A., Beheregaray, L., Bilgmann, K., Freitas, L., Robertson, K.M., Sequeira, M., Stockin, K.A., Coelho, M.M. & Moller, L.M. 2012b. Influences of past-climate changes on historical population structure and demography of a cosmopolitan marine predator, the common dolphin (genus *Delphinus*). *Molecular Ecology* **21**, 4854–4871.

Amaral, A.R., Beheregaray, L.B., Sequeira, M., Robertson, K.M., Coelho, M.M. & Möller, L.M. 2009. Worldwide phylogeography of the genus *Delphinus* revisited. IWC Scientific Committee Document, SC/61/SM11. Cambridge, UK: International Whaling Commission.

Amaral, A., Sequeira, M., Martínez-Cedeira, J. & Coelho, M. 2007. New insights on population genetic structure of *Delphinus delphis* from the northeast Atlantic and phylogenetic relationships within the genus inferred from two mitochondrial markers. *Marine Biology* **151**, 1967–1976.

Ananthaswamy, A. 2004. Massive growth of ecotourism worries biologists. *New Scientist* **4 March**. Online. http://www.newscientist.com/article/dn4733-massive-growth-of-ecotourism-worries-biologists.html (accessed 4 June 2012).

Anganuzzi, A. & Buckland, S. 1994. Relative abundance of dolphins associated with tuna in the eastern Pacific Ocean: analysis of 1992 data. *Reports of the International Whaling Commission* **44**, 361–366.

Antoine, L. 1990. Des dauphins, des thons et des pêcheurs; le filet dérivant en Atlantique-Nord. *Equinoxe* **33**, 11–14.

Au, W.W.L. & Green, M. 2000. Acoustic interaction of humpback whales and whale-watching boats. *Marine Environmental Research* **49**, 469–481.

Bäcklin, B.-M., Madej, A. & Forsberg, M. 1997. Histology of ovaries and uteri and levels of plasma progesterone, oestradiol-17β and oestrone sulphate during the implantation period in mated and gonadotrophin-releasing hormone-treated mink (*Mustela vison*) exposed to polychlorinated biphenyls. *Journal of Applied Toxicology* **17**, 297–306.

Bäcklin, B.-M., Persson, E., Jones, C.J.P. & Dantzer, V. 1998. Polychlorinated biphenyl (PCB) exposure produces placental vascular and trophoblastic lesions in the mink (*Mustela vison*): a light and electron microscope study. *Acta Pathologica, Microbiologica et Immunologica Scandinavica* **106**, 785–799.

Bacon, S. & Carter, D.J.T. 1993. A connection between mean wave height and atmospheric pressure gradient in the North Atlantic. *International Journal of Climatology* **13**, 423–436.

Baines, M.E. & Evans, P.G.H. 2012. Atlas of the marine mammals of Wales. Marine Monitoring Report No. 68. Bangor, UK: Countryside Council for Wales, 2nd edition.

Baker, C.S., Perry, A. & Vequist, G. 1988. Humpback whales of Glacier Bay, Alaska. *Whalewatcher* **21**, 13–17.

Bakker, J. & Smeenk, C. 1987. Time-series analysis of *Tursiops truncatus*, *Delphinus delphis*, and *Lagenorhynchus albirostris* strandings on the Dutch coast. *ECS Newsletter* **1**, 14–19.

Barlow, J. & Boveng, P. 1991. Modeling age-specific mortality for marine mammal populations. *Marine Mammal Science* **7**, 50–65.

Barnett, J., Davison, N., Deaville, R., Monies, R., Loveridge, J., Tregenza, N. & Jepson, P.D. 2009. Postmortem evidence of interactions of bottlenose dolphins (*Tursiops truncatus*) with other dolphin species in south-west England. *Veterinary Record* **165**, 441–444.

Bearzi, G., Reeves, R.R., Notarbartolo di Sciara, G., Politi, E., Cañadas, A., Frantzis, A. & Mussi, B. 2003. Ecology, status and conservation of short-beaked common dolphins (*Delphinus delphis*) in the Mediterranean Sea. *Mammal Review* **33**, 224–252.

Beaugrand, G., Edwards, M., Brander, K., Luczak, C. & Ibanez, F. 2008. Causes and projections of abrupt climate-driven ecosystem shifts in the North Atlantic. *Ecology Letters* **11**, 1157–1168.

Beerli, P. 2012. *Migrate-n: estimation of population sizes and gene flow using the coalescent.* Tallahassee, Florida: Department of Scientific Computing, Florida State University. Online. http://popgen.sc.fsu.edu/Migrate/ (accessed 14 April 2013).

Bell, C.M., Kemper, C.M. & Conran, J.G. 2002. Common dolphins *Delphinus delphis* in Southern Australia: a morphometric study. *Australian Mammalogy* **24**, 1–10.

Berman-Kowalewski, M. & Newsome, S.D. 2009. You are what you eat but what are you? Utilizing stable isotopes to validate morphometric determination of common dolphin (*Delphinus*) species. In *Abstracts of the 18th Biennial Conference on the Biology of Marine Mammals, 12–16 October 2009, Québec, Canada.* San Francisco: Society of Marine Mammalogy, 32 only.

Berrow, S., Atkinson, J., O'Connell, M. & Wall, D. 2007. Additional unpublished stranding records from the Irish Whale and Dolphin Group. *The Irish Naturalists' Journal* **28**, 517–519.

Berrow, S.D. & Rogan, E. 1997. Review of cetaceans stranded on the Irish coast, 1901–95. *Mammal Review* **27**, 51–76.

Berthou, P., Morizur, Y. & Laurans, M. 2008. Observation en mer de captures de cétacés. Ref. 08-2337. Paris: DPMA, Direction des Pêches Maritimes et de l'Aquaculture.

Bilgmann, K., Möller, L.M., Harcourt, R.G., Gales, R. & Beheregaray, L.B. 2008. Common dolphins subject to fisheries impacts in Southern Australia are genetically differentiated: implications for conservation. *Animal Conservation* **11**, 518–528.

Birkun, A., Jr., Kuiken, T., Krivokhizhin, S., Haines, D.M., Osterhaus, A.D.M.E., van de Bildt, M.W.G., Joiris, C.R. & Siebert, U. 1999. Epizootic of morbilliviral disease in common dolphins (*Delphinus delphis ponticus*) from the Black Sea. *Veterinary Record* **144**, 85–92.

Blane, J.M. & Jaakson, R. 1994. The impact of ecotourism boats on the St. Lawrence beluga whales. *Environmental Conservation* **21**, 267–269.

Bord Iascaigh Mhara (BIM). 2000. Diversification trials with alternative tuna fishing techniques including the use of remote sensing technology. Final report to the Commission of the European Communities Directorate General for Fisheries. EU contract No. 98/010. Dun Laoghaire, Co. Dublin: Bord Iascaigh Mhara (Irish Sea Fisheries Board).

Bord Iascaigh Mhara (BIM). 2004. Report on the development of prototype cetacean deterrent systems for the albacore tuna pair pelagic fishery 2002–2003. NDP supporting measures, programme 2002–2006, project no. 01.SM.T1.15 & 03.SM.T1.04, Report No. 2.04, issued April 2004. Dun Laoghaire, Co. Dublin: Bord Iascaigh Mhara (Irish Sea Fisheries Board), 15 pp.

Bord Iascaigh Mhara (BIM). 2005. Effect of wind and depth on cetacean bycatch from the Irish albacore tuna pair pelagic fishery 1998–2003. Unpublished report. Dun Laoghaire, Co. Dublin: Bord Iascaigh Mhara (Irish Sea Fisheries Board), 4 pp.

Borrell, A., Cantos, G., Pastor, T. & Aguilar, A. 2001. Organochlorine compounds in common dolphins (*Delphinus delphis*) from the Atlantic and Mediterranean waters of Spain. *Environmental Pollution* **114**, 265–274.

Brandt, M.J., Diederichs, A. & Nehls, G. 2009. Harbour porpoise responses to pile driving at the Horns Rev II offshore wind farm in the Danish North Sea. Final report to DONG Energy. Husum, Germany: BioConsult.

Breivik, K., Sweetman, A., Pacyna, J.M. & Jones, K.C. 2002. Towards a global historical emission inventory for selected PCB congeners—a mass balance approach. 1. Global production and consumption. *Science of the Total Environment* **290**, 181–198.

Brereton, T., Williams, A. & Martin, C. 2005. Ecology and status of the common dolphin *Delphinus delphis* in the English Channel and Bay of Biscay 1995–2002. In *Proceedings of the Workshop on Common Dolphins: Current Research, Threats and Issues, Special Issue April 2005, Kolmarden, Sweden 1st April, 2004*, K. Stockin et al. (eds). Cambridge, UK: European Cetacean Society, 15–22.

Brito, C., Vieire, N., Sa, E. & Carvalho, I. 2009. Cetaceans' occurrence off the west central Portugal coast: a compilation of data from whaling, observations of opportunity and boat-based surveys. *Journal of Marine Animals and Their Ecology* **2**, 10–13.

Brophy, J. 2003. *Diet of the common dolphin* Delphinus delphis. MSc thesis, University College Cork, Ireland.

Brophy, J., Murphy, S. & Rogan, E. 2006. Records from the Irish Whale and Dolphin Group for 2003. *Irish Naturalists' Journal* **28**, 214–219.

Brophy, J., Murphy, S. & Rogan, E. 2009. The diet and feeding ecology of the common dolphin (*Delphinus delphis*) in the northeast Atlantic. IWC Scientific Committee Document SC/61/SM14. Cambridge, UK: International Whaling Commission.

Burgess, E.A. 2006. *Foraging ecology of common dolphins (*Delphinus sp.*) in the Hauraki Gulf, New Zealand*. MSc thesis, Massey University, Auckland, New Zealand.

Butchart, S.H.M. & Bird, J.P. 2010. Data deficient birds on the IUCN Red List: what don't we know and why does it matter? *Biological Conservation* **143**, 239–247.

Cañadas, A., Donovan, G.P., Desportes, G. & Borchers, D.L. 2009. A short review of the distribution of short beaked common dolphins (*Delphinus delphis*) in the central and eastern North Atlantic with an abundance estimate for part of this area. *North Atlantic Sightings Surveys. NAMMCO Scientific Publications* **7**, 201–220.

Cañadas, A. & Hammond, P.S. 2008. Abundance and habitat preferences of the short-beaked common dolphin *Delphinus delphis* in the southwestern Mediterranean: implications for conservation. *Endangered Species Research* **4**, 309–331.

Carrera, P. & Porteiro, C. 2003. Stock dynamic of the Iberian sardine (*Sardina pilchardus*, W.) and its implication on the fishery off Galicia (NW Spain). *Scientia Marina* **67**, 245–258.

Carretta, J.V., Forney, K.A., Oleson, E., Martien, K., Muto, M.M., Lowry, M.S., Barlow, J., Baker, J., Hanson, B., Lynch, D., Carswell, L., Brownell, R.L., Jr., Robbins, J., Mattila, D.K., Ralls, K. & Hill, M.C. 2011. Short-beaked common dolphin (*Delphinus delphis delphis*): California/Oregon/Washington Stock. In *U.S. Pacific Marine Mammal Stock Assessments: 2011*. NOAA-Technical Memorandum-NMFS-SWFSC-488. La Jolla, California: Southwest Fisheries Science Center, National Marine Fisheries Service, 111–117.

Carvalho, M.L., Pereira, R.A. & Brito, J. 2002. Heavy metals in soft tissues of *Tursiops truncatus* and *Delphinus delphis* from west Atlantic Ocean by X-ray spectrometry. *Science of the Total Environment* **292**, 247–254.

Castège, I., Soulier, L., Hémery, G., Mouchès, C., Lalanne, Y., Dewez, A., Pautrizel, F., d'Elbée, J. & D'Amico, F. 2012. Exploring cetacean stranding pattern in light of variation in at-sea encounter rate and fishing activity: lessons from time surveys in the south Bay of Biscay (East-Atlantic; France). *Journal of Marine Systems* **109–110**(supplement), S284–S292, doi:10.1016/j.jmarsys.2012.04.007.

Castellote, M., Clark, C.W. & Lammers, M.O. 2012. Acoustic and behavioural changes by fin whales (*Balaenoptera physalus*) in response to shipping and airgun noise. *Biological Conservation* **147**, 115–122.

Caughly, G. 1966. Mortality patterns in mammals. *Ecology* **47**, 906–918.

Caurant, F., Aubail, A., Lahaye, V., Van Canneyt, O., Rogan, E., Lopez, A., Addink, M., Churlaud, C., Robert, M. & Bustamante, P. 2006. Lead contamination of small cetaceans in European waters—the use of stable isotopes for identifying the sources of lead exposure. *Marine Environmental Research* **62**, 131–148.

Caurant, F., Chouvelon, T., Lahaye, V., Mendez-Fernandez, P., Rogan, E., Spitz, J. & Ridoux, V. 2009. The use of ecological tracers for discriminating dolphin population structure: the case of the short-beaked common dolphin *Delphinus delphis* in European Atlantic waters. IWC Scientific Committee Document SC/61/SM34. Cambridge, UK: International Whaling Commission.

Caurant, F., Chouvelon, T., Lahaye, V., Mendez-Fernandez, P., Rogan, E., Spitz, J. & Ridoux, V. 2011. The use of ecological tracers for discriminating dolphin population structure: the case of the short-beaked common dolphin *Delphinus delphis* in European Atlantic waters. AC18/Doc.5-02(P). 18th ASCOBANS Advisory Committee Meeting. Bonn, Germany: ASCOBANS. Online. http://www.service-board.de/ascobans_neu/files/ac18/AC18_5-02_EcologicalTracersCommonDolphins.pdf (accessed 4 June 2012).

Certain, G., Masse, J., Van Canneyt, O., Petitgas, P., Doremus, G., Santos, M. & Ridoux, V. 2011. Investigating the coupling between small pelagic fish and marine top predators using data collected from ecosystem-based surveys. *Marine Ecology Progress Series* **422**, 23–39.

Certain, G., Ridoux, V., van Canneyt, O. & Bretagnolle, V. 2008. Delphinid spatial distribution and abundance estimates over the shelf of the Bay of Biscay. *ICES Journal of Marine Science* **65**, 656–666.

Cetacean Offshore Distribution and Abundance in the European Atlantic (CODA). 2009. St. Andrews, UK: Sea Mammal Research Institute. Online. http://biology.st-andrews.ac.uk/coda/documents/CODA_Final_Report_11-2-09.pdf (accessed 16 July 2012).

Cherel, Y., Hobson, K.A., Bailleul, F. & Groscolas, R. 2005. Nutrition, physiology, and stable isotopes: new information from fasting and molting penguins. *Ecology* **86**, 2881–2888.

Chouvelon, T., Spitz, J., Caurant, F., Mèndez-Fernandez, P., Chappuis, A., Laugier, F., Le Goff, E. & Bustamante, P. 2012. Revisiting the use of $\delta^{15}N$ in meso-scale studies of marine food webs by considering spatio-temporal variations in stable isotopic signatures—the case of an open ecosystem: the Bay of Biscay (North-East Atlantic). *Progress in Oceanography* **101**, 92–105.

Climate and Global Dynamics Division (CGD). 2012. Climate indices. Boulder, Colorado: NCAR Earth System Laboratory, National Center for Atmospheric Research. Online. http://www.cgd.ucar.edu/cas/catalog/climind/ (accessed 16 July 2012).

Clua, E. & Grosvalet, F. 2001. Mixed-species feeding aggregation of dolphins, large tunas and seabirds in the Azores. *Aquatic Living Resources* **14**, 11–18.

Cockcroft, V.G., De Kock, A.C., Lord, D.A. & Ross, G.J.B. 1989. Organochlorines in bottlenose dolphins *Tursiops truncatus* from the east coast of South Africa. *South African Journal of Marine Science* **8**, 207–217.

Collet, A. 1981. *Biologie du dauphin commun* Delphinus delphis *L. en Atlantique Nord-Est.* PhD thesis, L'Université de Poitiers, France.

Constantine, R., Brunton, D.H. & Dennis, T. 2004. Dolphin-watching tour boats change bottlenose dolphin (*Tursiops truncatus*) behaviour. *Biological Conservation* **117**, 299–307.

Convention on Biological Diversity (CBD). 2010. COP 10 decision X/2. Strategic Plan for Biodiversity 2011–2020. Montreal: Convention on Biological Diversity. Online. http://www.cbd.int/decision/cop/?id=12268 (accessed 16 July 2012).

Convention on International Trade of Endangered Species of Wild Fauna and Flora (CITES) Secretariat. 2012. Convention on International Trade in Endangered Species of Wild Flora and Fauna. Signed at Washington, DC, on 3 March 1973. Amended at Bonn on 22 June 1979. Geneva, Switzerland: CITES Secretariat. Online. http://www.cites.org/eng/disc/text.php (accessed 16 July 2012).

Convention on Migratory Species (CMS) Secretariat. 2012. Convention on Migratory Species. Bonn, Germany: UNEP/CMS Secretariat. Online. http://www.cms.int/documents/convtxt/cms_convtxt.htm (accessed 16 July 2012).

Convention for the Protection of the Marine Environment of the North-East Atlantic (OSPAR). 2012. Convention for the Protection of the Marine Environment of the North-East Atlantic. London: OSPAR Commission. Online. http://www.ospar.org/html_documents/ospar/html/ospar_convention_e_updated_text_2007.pdf (accessed 16 July 2012).

Convention for the Protection of the Marine Environment of the North-East Atlantic (OSPAR) Commission. 2010. Quality status report 2010. London: OSPAR Commission. Online. http://qsr2010.ospar.org/en/index.html (accessed 4 December 2012).

Convention for the Protection of the Marine Environment of the North-East Atlantic (OSPAR) Commission. 2012. OSPAR database on off-shore wind farms. Data 2011/2012 (updated in 2012). Biodiversity Series. Publication no. 573/2012. London: OSPAR Commission.

Corkeron, P.J. 1995. Humpback whales (*Megaptera novaeangliae*) in Hervey Bay, Queensland: behaviour and responses to whale-watching vessels. *Canadian Journal of Zoology* **73**, 1290–1299.

Cosgrove, R. & Browne, D. 2007. Cetacean by-catch rates in Irish gillnet fisheries in the Celtic Sea. Marine Technical Report, June 2007. Dun Laoghaire, Co. Dublin: Bord Iascaigh Mhara.

Couperus, A.S. 1997. Interactions between Dutch midwater trawl and Atlantic white-sided dolphins (*Lagenorhynchus acutus*) southwest of Ireland. *Journal of Northwest Atlantic Fishery Science* **22**, 209–218.

Cressey, D. 2008. Sonar does affect whales, military report confirms. *Nature* 1 August, doi:10.1038/news.2008.997.

Crum, L.A. & Mao, Y. 1996. Acoustically enhanced bubble growth at low frequencies and its implications for human diver and marine mammal safety. *Journal of the Acoustical Society of America* **99**, 2898–2907.

Currey, R.J.C., Dawson, S.M. & Slooten, E. 2009. An approach for regional threat assessment under IUCN Red List criteria that is robust to uncertainty: the Fiordland bottlenose dolphins are critically endangered. *Biological Conservation* **142**, 1570–1579.

Dabin, W., Cossais, F., Pierce, G. & Ridoux, V. 2008. Do ovarian scars persist with age in all cetaceans: new insight from the short-beaked common dolphin (*Delphinus delphis* Linnaeus, 1758). *Marine Biology* **156**, 127–139.

Danil, K. & Chivers, S.J. 2006. Habitat-based spatial and temporal variability in life history characteristics of female common dolphins *Delphinus delphis* in the eastern tropical Pacific. *Marine Ecology Progress Series* **318**, 277–286.

Danil, K. & Chivers, S.J. 2007. Growth and reproduction of female short-beaked common dolphins, *Delphinus delphis*, in the eastern tropical Pacific. *Canadian Journal of Zoology* **85**, 108–121.

Dans, S., Crespo, E., Pedraza, S., Degrati, M. & Garaffo, G. 2008. Dusky dolphin and tourist interaction: effect on diurnal feeding behavior. *Marine Ecology Progress Series* **369**, 287–296.

Das, K., Beans, C., Holsbeek, L., Mauger, G., Berrow, S.D., Rogan, E. & Bouquegneau, J.M. 2003. Marine mammals from northeast Atlantic: relationship between their trophic status as determined by $\delta^{13}C$ and $\delta^{15}N$ measurements and their trace metal concentrations. *Marine Environmental Research* **56**, 349–365.

Das, K., Groof, A.D., Jauniaux, T. & Bouquegneau, J.M. 2006. Zn, Cu, Cd and Hg binding to metallothioneins in harbour porpoises *Phocoena phocoena* from the southern North Sea. *BMC Ecology* **6**, 1–7.

Das, K., Siebert, U., Fontaine, M., Jauniaux, T., Holsbeek, L. & Bouquegneau, J.M. 2004. Ecological and pathological factors related to trace metal concentrations in harbour porpoises *Phocoena phocoena* from the North Sea and adjacent areas. *Marine Ecology Progress Series* **281**, 283–295.

Davison, N.J., Cranwell, M.P., Perrett, L.L., Dawson, C.E., Deaville, R., Stubberfield, E.J., Jarvis, D.S. & Jepson, P.D. 2009. Meningoencephalitis associated with *Brucella* species in a live-stranded striped dolphin (*Stenella coeruleoalba*) in south-west England. *Veterinary Record* **165**, 86–89.

Deaville, R. & Jepson, P.D. (compilers) 2011. UK Cetacean Strandings Investigation Programme. Final report for the period 1st January 2005–31st December 2010. (Covering contract numbers CR0346 and CR0364.) Bristol, UK: Department for Environment, Food and Rural Affairs.

Demaneche, S., Gaudou, O. & Miossec, D. 2010. Les captures accidentelles de cétacés dans les pêches françaises en 2009: contribution au rapport national sur la mise en oeuvre du règlement européen (CE) No. 812/2004–(année 2009). Brest, France: Ifremer Centre de Brest, Sciences et Technologies Halieutiques.

Dennison, S., Moore, M.J., Fahlman, A., Moore, K., Sharp, S., Harry, C.T., Hoppe, J., Niemeyer, M., Lentell, B. & Wells, R.S. 2011. Bubbles in live-stranded dolphins. *Proceedings of the Royal Society B: Biological Sciences* **279**, 1041–1050.

De Pierrepont, J.F., Dubois, B., Desmonts, S., Santos, M.B. & Robin, J.P. 2005. Stomach contents of English Channel cetaceans stranded on the coast of Normandy. *Journal of the Marine Biological Association of the United Kingdom* **85**, 1539–1546.

Diaz-Delgado, J., Arbelo, M., Sacchini, S., Quesada-Canales, O., Andrada, M., Rivero, M. & Fernandez, A. 2012. Pulmonary angiomatosis and hemangioma in common dolphins (*Delphinus delphis*) stranded in Canary Islands. *Journal of Veterinary Medical Science* **74**, 1063–1066.

Diederichs, A., Brandt, M.J. & Nehls, G. 2009. Auswirkungen des baus des umspannwerks am offshore-testfeld "alpha ventus" auf Schweinswale. Report to Stiftung Offshore-Windenergie (German Offshore Wind Energy Foundation). Husum, Germany: BioConsult.

Di Guardo, G., Marruchella, G., Affronte, M., Zappulli, V. & Benazzi, C. 2005. Heterotopic kidney tissue in the lung of a free-living common dolphin (*Delphinus delphis*). *Veterinary Pathology Online* **42**, 213–214.

Doksæter, L., Olsen, E., Nøttestad, L. & Fernö, A. 2008. Distribution and feeding ecology of dolphins along the Mid-Atlantic Ridge between Iceland and the Azores. *Deep-Sea Research Part II: Topical Studies in Oceanography* **55**, 243–253.

Donovan, G.P. 2005. Cetaceans: can we manage to conserve them? The role of long-term monitoring. In *Long-Term Monitoring: Why, What, Where, When and How?* J. Solbe (ed.). Sherkin Island, Ireland: Sherkin Island Marine Station, 161–175.

Dransfeld, L., Dwane, O., McCarney, C., Kelly, C.J., Danilowicz, B.S. & Fives, J.M. 2004. Larval distribution of commercial fish species in waters around Ireland. *Irish Fisheries Investigations* **13**. Galway, Ireland: Marine Institute.

Duffus, D.A. 1996. The recreational use of grey whales in southern Clayoquot Sound, Canada. *Applied Geography* **16**, 179–190.

Erbe, C. 2002. Underwater noise of whale-watching boats and potential effects on killer whales (*Orcinus orca*), based on an acoustic impact model. *Marine Mammal Science* **18**, 394–418.

European Commission. 2009. Communication from the Commission to the European Parliament and the Council. Cetacean incidental catches in fisheries: report on the implementation of certain provisions of Council Regulation (EC) No. 812/2004 and on a scientific assessment of the effects of using in particular gillnets, trammel nets and entangling nets on cetaceans in the Baltic Sea as requested through Council Regulation (EC) No. 2187/2005. 16.7.2009 COM(2009) 368 final. Brussels: European Commission. Online. http://eur-lex.europa.eu/LexUriServ/LexUriServ.do?uri=COM:2009:0368:FIN:EN:PDF (accessed 11 December 2012).

European Commission. 2011. Communication from the Commission to the European Parliament, the Council, the Economic and Social Committee, and the Committee of the Regions. Our life insurance, our natural capital: an EU biodiversity strategy to 2020. Brussels: European Commission. Online. http://ec.europa.eu/environment/nature/biodiversity/comm2006/pdf/2020/1_EN_ACT_part1_v7%5b1%5d.pdf (accessed 16 July 2012).

European Union. 1998. Council Regulation (EC) no. 1239/98 of 8 June 1998. Amending Regulation (EC) no. 894/97 laying down certain technical measures for the conservation of fishery measures. *Official Journal of the European Communities, Series L* **171**, 17.6.1998: 1–4. Online. http://eur-lex.europa.eu/LexUriServ/LexUriServ.do?uri=OJ:L:1998:171:0001:0004:EN:PDF (accessed 16 July 2012).

European Union. 2004. Corrigendum to Council Regulation (EC) No. 812/2004 of 26 April 2004 laying down measures concerning incidental catches of cetaceans in fisheries and amending regulation (EC) No. 88/98. *Official Journal of the European Union, Series L* **150**, 30.4.2004: 12-31. Online. http://eur-lex.europa.eu/LexUriServ/LexUriServ.do?uri=OJ:L:2004:150:0012:0031:EN:PDF (accessed 16 July 2012).

European Union. 2007. Summary of treaty. Convention on the Conservation of European Wildlife and Natural Habitats (No. 104, Council of Euope). Brussels: European Union. Online. http://ec.europa.eu/world/agreements/prepareCreateTreatiesWorkspace/treatiesGeneralData.do?step=0&redirect=true&treatyId=497 (accessed 16 July 2012).

European Union. 2008. Directive 2008/56/EC of the European Parliament and of the Council of 17 June 2008 establishing a framework for community action in the field of marine environmental policy (Marine Strategy Framework Directive). *Official Journal of the European Union, Series L* **164** 25.6.2008: 19–40. Online. http://eur-lex.europa.eu/LexUriServ/LexUriServ.do?uri=OJ:L:2008:164:0019:0040:EN:PDF (accessed 4 December 2012).

Evans, P.G.H., Anderwald, P., Ansmann, I., Bush, N. & Baines, M. 2007. Abundance of common dolphins in the Celtic Deep/St. George's Channel 2004–06. Unpublished report to Countryside Council for Wales. Oxford, UK: Sea Watch Foundation.

Evans, P.G.H., Baines, M.E. & Anderwald, P. 2010. Risk assessment of potential conflicts between shipping and cetaceans in the ASCOBANS Region. Bonn, Germany: ASCOBANS. Online. http://www.ascobans.org/pdf/ac18/AC18_6–04_rev1_ProjectReport_ShipStrikes.pdf (accessed 4 June 2012).

Evans, P.G.H. & Scanlan, G.M. 1989. Historical review of cetaceans in British and Irish waters. U.K. Cetacean Group, c/o Zoology Department, University of Oxford, South Parks Road, Oxford, OX1 3PS, England.

Evans, P.G.H. & Teilmann, J. 2009. Report of ASCOBANS/HELCOM Small Cetacean Population Structure Workshop. Bonn, Germany: ASCOBANS.

Evans, W.E. 1975. *Distribution, differentiation of populations, and other aspects of the natural history of* Delphinus delphis Linnaeus *in the northeastern Pacific*. PhD thesis, University of California, Los Angeles.

Evans, W.E. 1982. Distribution and differentiation of stocks of *Delphinus delphis* Linnaeus in the northeastern Pacific. In *Mammals in the Seas*, FAO Fisheries Series No. 5, Volume 4. Rome: Food and Agriculture Organisation of the United Nations, 45–66.

Fernández, A., Edwards, J.F., Rodriguez, F., de los Monteros, A.E., Herraez, P., Castro, P., Jaber, J.R., Martin, V. & Arbelo, M. 2005. 'Gas and fat embolic syndrome' involving a mass stranding of beaked whales (Family Ziphiidae) exposed to anthropogenic sonar signals. *Veterinary Pathology* **42**, 446–457.

Fernández-Contreras, M.M., Cardona, L., Lockyer, C.H. & Aguilar, A. 2010. Incidental bycatch of short-beaked common dolphins (*Delphinus delphis*) by pairtrawlers off northwestern Spain. *ICES Journal of Marine Science* **67**, 1732–1738.

Finneran, J.J., Schlundt, C.E., Dear, R., Carder, D.A. & Ridgway, S.H. 2002. Temporary shift in masked hearing thresholds in odontocetes after exposure to single underwater impulses from a seismic watergun. *Journal of the Acoustical Society of America* **111**, 2929–2940.

Fisheries Research Services. 2003. Scottish ocean climate status report 2000–2001. Report 05/03. Aberdeen, UK: Fisheries Research Services.

Fontaine, M.C., Baird, S.J.E., Piry, S., Ray, N., Tolley, K.A., Duke, S., Birkun, A., Jr., Ferreira, M., Jauniaux, T., Llavona, A., Ozturk, B., Ozturk, A.A., Ridoux, V., Rogan, E., Sequeira, M., Siebert, U., Vikingsson, G.A., Bouquegneau, J.-M. & Michaux, J.R. 2007. Rise of oceanographic barriers in continuous populations of a cetacean: the genetic structure of harbour porpoises in Old World waters. *BMC Biology* **5**, 30, doi:10.1186/1741-7007-5-30.

Food and Agriculture Organization of the United Nations (FAO). 2003. Fisheries management. 2. The ecosystem approach to fisheries. FAO Technical Guidelines for Responsible Fisheries. No. 4, Suppl. 2. Rome: FAO.

Food and Agriculture Organization of the United Nations (FAO). 2008. Fisheries management. The ecosystem approach to fisheries. Best practices in ecosystem modelling for informing an ecosystem approach to fisheries. FAO Technical Guidelines for Responsible Fisheries. No. 4, Suppl. 2, Add. 1. Rome: FAO.

Fossi, M.C. & Marsili, L. 2011. Multi-trial ecotoxicological diagnostic tool in cetacean skin biopsies. In *Skin Biopsy—Perspectives*, U. Khopkar (ed.). Rijeka, Croatia: InTech, 317–336.

Frantzis, A. 1998. Does acoustic testing strand whales? *Nature* **392**, 29.

Fraser, F.C. 1934. Report on Cetacea stranded on the British coasts from 1927 to 1932. No. 11. London: British Museum (Natural History).

Fraser, F.C. 1937. Common dolphins in the North Sea. *The Scottish Naturalist* **1937**, 103–105.

Fraser, F.C. 1946. Report on Cetacea stranded on the British coasts from 1933 to 1937. No. 12. London: British Museum (Natural History).

Fraser, F. C. 1974. Report on Cetacea stranded on the British coasts from 1948 to 1966. No. 14. London: British Museum.

Fristrup, K.M., Hatch, L.T. & Clark, C.W. 2003. Variation in humpback whale (*Megaptera novaeangliae*) song length in relation to low-frequency sound broadcasts. *Journal of the Acoustical Society of America* **113**, 3411–3424.

Gehringer, J.W. 1976. Part 216 – Regulations governing the taking and importing of marine mammals. Marine Mammal Protection Act of 1972; Definition of "Optimum Sustainable Population". *Federal Reserve* **41**(246) 21 December 1976, 55536 only.

Genov, T., Bearzi, G., Bonizzoni, S. & Tempesta, M. 2012. Long-distance movement of a lone short-beaked common dolphin *Delphinus delphis* in the central Mediterranean Sea. *Marine Biodiversity Records* **5**, 30, doi:10.1186/1741-7007-5-30.

Geraci, J.R., Harwood, J. & Lounsbury, V.J. 1999. Marine mammal die-offs: causes, investigations, and issues. In *Conservation and Management of Marine Mammals*, J.R. Twiss, Jr. & R.R. Reeves (eds). Washington, DC: Smithsonian Institution Press, 367–395.

Geraci, J.R. & Lounsbury, V.R. 2005. *Marine Mammals Ashore: a Field Guide for Strandings*. Baltimore: National Aquarium in Baltimore, 2nd edition.

Geraci, J.R. & Lounsbury, V.J. 2009. Health. In *Encyclopedia of Marine Mammals*. F.P. William et al. (eds). London: Academic Press, 2nd edition, 546–553.

Gerrodette, T. & DeMaster, D.P. 1990. Quantitative determination of optimum sustainable population level. *Marine Mammal Science* **6**, 1–16.

Gerrodette, T. & Forcada, J. 2002. Estimates of abundance of striped and common dolphins, and pilot, sperm and Bryde's whales in the eastern tropical Pacific Ocean. Administrative Report LJ-02-20. La Jolla, California: Southwest Fisheries Science Center.

Gibson, D.I., Harris, E.A., Bray, R.A., Jepson, P.D., Kuiken, T., Baker, J.R. & Simpson, V.R. 1998. A survey of the helminth parasites of cetaceans stranded on the coast of England and Wales during the period 1990–1994. *Journal of Zoology* **244**, 563–574.

Gislason, A. & Gorsky, G. 2010. Proceedings of the Joint ICES/CIESM Workshop to Compare Zooplankton Ecology and Methodologies between the Mediterranean and the North Atlantic (WKZEM). ICES Cooperative Research Report No. 300. Copenhagen, Denmark: International Council for the Exploration of the Sea.

Glanville, E., Bartlett, P. & Berrow, S.D. 2003. Common dolphin *Delphinus delphis* (L.). *Irish Naturalist Journal* **27**, 241–242.

Glockner-Ferrari, D.A. & Ferrari, M.J. 1990. Reproduction in the humpback whale (*Megaptera novaeangliae*) in Hawaiian waters, 1975–1988: the life history, reproductive rates and behaviour of known individuals identified through surface and underwater photography. *Report to the International Whaling Commission (Special Issue)* **12**, 161–169.

González, A.F., López, A. & Benavente, P. 1999. A multiple gestation in a *Delphinus delphis* stranded on the north-western Spanish coast. *Journal of the Marine Biological Association of the United Kingdom* **79**, 1147–1148.

Goold, J.C. 1996. Acoustic assessment of populations of common dolphin (*Delphinus delphis*) in conjunction with seismic surveying. *Journal of the Marine Biological Association of the United Kingdom* **76**, 811–820.

Goold, J.C. 1998. Acoustic assessment of populations of common dolphin off the west Wales coast, with perspectives from satellite infrared imagery. *Journal of the Marine Biological Association of the United Kingdom* **78**, 1353–1364.

Goold, J.C. & Fish, P.J. 1998. Broadband spectra of seismic survey air-gun emissions, with reference to dolphin auditory thresholds. *Journal of the Acoustical Society of America* **103**, 2177–2184.

Gosselin, M. 2001. *Aspects of the biology of common dolphins (*Delphinus delphis*) subject to incidental capture in fishing gears in the Celtic Sea and Channel.* MSc thesis, Heriot-Watt University, Edinburgh, UK.

Gottschling, M., Bravo, I.G., Schulz, E., Bracho, M.A., Deaville, R., Jepson, P.D., Bressem, M.-F.V., Stockfleth, E. & Nindl, I. 2011. Modular organizations of novel cetacean papillomaviruses. *Molecular Phylogenetics and Evolution* **59**, 34–42.

Goujon, M. 1996. *Captures accidentelles du filet maillant derivant et dynamique des populations de dauphins au large du Golfe de Gascogne.* PhD thesis, École Nationale Superieure Agronomique de Rennes, France.

Goujon, M., Antoine, L. & Collet, A. 1993a. Incidental catches of cetaceans by the French albacore tuna driftnet fishery: preliminary results. ICES Conference and Meeting (CM) Document 1993/N:13. Copenhagen, Denmark: International Council for the Exploration of the Sea.

Goujon, M., Antoine, L., Collet, A. & Fifas, S. 1993b. Approche de l'mpact de la ecologique de la pecherie thoniere au filet maillant derivant en Atlantique nord-est. Rapport interne de la Direction des Resources Vivantes de l'Ifremer. Plouzane, France: Ifremer.

Goujon, M., Antoine, L., Collet, A. & Fifas, S. 1994. A study of the ecological impact of the French tuna driftnet fishery in the North-east Atlantic. In *European Research on Cetaceans—8. Proceedings of the 8th Annual Conference of the European Cetacean Society, Montpellier, France, 2–5 March 1994,* P.G.H. Evans (ed.). Cambridge, UK: European Cetacean Society, 47–48.

Graham, B.S., Koch, P.L., Newsome, S.D., McMahon, K.W. & Aurioles, D. 2010. Using isoscapes to trace the movements and foraging behavior of top predators in oceanic ecosystems. In *Isoscapes: Understanding Movement, Pattern, and Process on Earth through Isotope Mapping,* J.B. West et al. (eds). Dordrecht, the Netherlands: Springer Science, 299–318.

Graham, C.T. & Harrod, C. 2009. Implications of climate change for the fishes of the British Isles. *Journal of Fish Biology* **74**, 1143–1205.

Gray, H. & Van Waerebeek, K. 2011. Postural instability and akinesia in a pantropical spotted dolphin, *Stenella attenuata,* in proximity to operating airguns of a geophysical seismic vessel. *Journal for Nature Conservation* **19**, 363–367.

Habran, S., Cathy, D., Daniel, E.C., Dorian, S.H., Gilles, L., Jean-Marie, B. & Krishna, D. 2010. Assessment of gestation, lactation and fasting on stable isotope ratios in northern elephant seals (*Mirounga angustirostris*). *Marine Mammal Science* **26**, 880–895.

Hall, A.J., Hugunin, K., Deaville, R., Law, R.J., Allchin, C.R. & Jepson, P.D. 2006. The risk of infection from polychlorinated biphenyl exposure in the harbor porpoise (*Phocoena phocoena*): a case-control approach. *Environmental Health Perspectives* **114**, 704–711.

Hammond, P.S., Bearzi, G., Bjørge, A., Forney, K., Karczmarski, L., Kasuya, T., Perrin, W.F., Scott, M.D., Wang, J.Y., Wells, R.S. & Wilson, B. 2008. *Delphinus delphis.* In *IUCN 2012. IUCN Red List of Threatened Species,* Version 2012.2. Cambridge, UK: IUCN Global Species Programme Red List Unit. Online. http://www.iucnredlist.org/ (accessed 21 November 2012).

Hammond, P.S., Berggren, P., Benke, H., Borchers, D.L., Collet, A., Heide-Jørgensen, M.P., Heimlich, S., Hiby, A.R., Leopold, M.F. & Øien, N. 2002. Abundance of harbour porpoise and other cetaceans in the North Sea and adjacent waters. *Journal of Applied Ecology* **39**, 361–376.

Hammond, P.S., Macleod, K., Berggren, P., Borchers, D.L., Burt, M.L., Cañadas, A., Desportes, G., Donovan, G.P., Gilles, A., Gillespie, D., Gordon, J., Hiby, L., Kuklik, I., Leaper, R., Lehnert, K., Leopold, M., Lovell, P., Øien, N., Paxton, C.G.M., Ridoux, V., Rogan, E., Samarra, F., Scheidat, M., Sequeira, M., Siebert, U., Skov, H., Swift, R., Tasker, M.L., Teilmann, J., Van Canneyt, O. & Vázquez, J.A. in press. Cetacean abundance and distribution in European Atlantic shelf waters to inform conservation and management, *Biological Conservation*.

Harmer, S.F. 1927. Report on Cetacea stranded on the British coasts from 1913 to 1926. No. 10. London: British Museum (Natural History).

Harper, C.M.G., Dangler, C.A., Xu, S., Feng, Y., Shen, Z., Sheppard, B., Stamper, A., Dewhirst, F.E., Paster, B.J. & Fox, J.G. 2000. Isolation and characterization of a *Helicobacter* sp. from the gastric mucosa of dolphins, *Lagenorhynchus acutus* and *Delphinus delphis. Applied and Environmental Microbiology* **66**, 4751–4757.

Harwood, J., Andersen, L.W., Berggren, P., Carlstrom, J., Kinze, C.C., McGlade, J., Metuzals, K., Larsen, F., Lockyer, C.H., Northridge, S., Rogan, E., Walton, M. & Vinther, M. 1999. Assessment and reduction of the by-catch of small cetaceans (BY-CARE). Final report to the European Commission under contract FAIR-CT05-0523. Brussels, Belgium: European Commission Directorate-General.

Hassani, S. 2012. Annual national report 2011 France. Document 2-04. 19th ASCOBANS Advisory Committee Meeting, Galway, Ireland, 20–22 March 2012. Bonn, Germany: ASCOBANS.

Hassani, S., Antoine, L. & Ridoux, V. 1997. Diets of albacore, *Thunnus alalunga*, and dolphins, *Delphinus delphis* and *Stenella coerulaeoalba*, caught in the Northeast Atlantic albacore drift-net fishery: a progress report. *Journal of Northwest Atlantic Fishery Science* **22**, 119–123.

Hawkins, S.J., Southward, A.J. & Genner, M.J. 2003. Detection of environmental change in a marine ecosystem-evidence from the western English Channel. *Science of the Total Environment* **310**, 245–256.

Hershkovitz, P. 1966. Catalog of living whales. *Bulletin—United States National Museum* **246**, 1–259.

Heyning, J.E. & Perrin, W.F. 1994. Evidence for two species of common dolphins (genus *Delphinus*) from the eastern North Pacific. *Contributions in Science (Los Angeles County Museum)* **442**, 1–35.

Houser, D.S., Howard, R. & Ridgway, S.H. 2001. Can diving-induced tissue nitrogen supersaturation increase the chance of acoustically driven bubble growth in marine mammals? *Journal of Theoretical Biology* **213**, 183–195.

Hughes, S.L., Holliday, N.P. & Beszczynska-Möller, A. 2011. ICES report on ocean climate 2010 prepared by the Working Group on Oceanic Hydrography. ICES Cooperative Research Report No. 309. Copenhagen, Denmark: International Council for the Exploration of the Sea.

Hughes, S.L., Holliday, N.P., Gaillard, F. & the ICES Working Group on Oceanic Hydrography. 2012. Variability in the ICES/NAFO region between 1950 and 2009: observations from the ICES Report on Ocean Climate. *ICES Journal of Marine Science* **69**, 706–719.

Hurrell, J.W. 1995. Decadal trends in the North Atlantic Oscillation regional temperatures and precipitation. *Science* **269**, 676–679.

Hurrell, J.W. & Dickson, R.R. 2004. Climate variability over the North Atlantic. In *Ecological Effects of Climatic Variations in the North Atlantic Ocean*, N.C. Stenseth et al. (eds). Oxford: Oxford University Press, 15–31.

Hurrell, J.W., Kushnir, Y., Visbeck, M. & Ottersen, G. 2003. An overview of the North Atlantic Oscillation. In *The North Atlantic Oscillation: Climate Significance and Environmental Impact, Geophysical Monograph Series 134*, J.W. Hurrell et al. (eds). Washington, DC: American Geophysical Union, 1–35.

International Council for the Exploration of the Sea (ICES). 2008. Assessment of the impact of fisheries on the marine environment of the OSPAR maritime area. ICES Advice 2008, Book 1, 1.5.5.9. Copenhagen, Denmark: International Council for the Exploration of the Sea.

International Council for the Exploration of the Sea (ICES). 2010. Special request advice May 2010. New information regarding small cetaceans, marine mammals, seabirds, and sensitive habitats and impact of fisheries. ICES Advice 2010, Book 1, 1.5.1.2. Copenhagen, Denmark: International Council for the Exploration of the Sea.

International Council for the Exploration of the Sea Study Group for Bycatch of Protected Species (ICES SGBYC). 2008. Report of the Study Group for Bycatch of Protected Species (SGBYC), 29–31 January 2008. Copenhagen, Denmark: International Council for the Exploration of the Sea.

International Council for the Exploration of the Sea Study Group for Bycatch of Protected Species (ICES SGBYC). 2010. Report of the Study Group for Bycatch of Protected Species (SGBYC), 1–4 February 2010. Copenhagen, Denmark: International Council for the Exploration of the Sea.

International Council for the Exploration of the Sea Working Group on Bycatch of Protected Species (ICES WGBYC). 2011. Report of the Working Group on Bycatch of Protected Species (WGBYC), 1–4 February 2011. Copenhagen, Denmark: International Council for the Exploration of the Sea.

International Council for the Exploration of the Sea Working Group on Bycatch of Protected Species (ICES WGBYC). 2012. Report of the Working Group on Bycatch of Protected Species (WGBYC), 7–10 February 2012. Copenhagen, Denmark: International Council for the Exploration of the Sea.

International Council for the Exploration of the Sea Working Group on Marine Mammal Ecology (ICES WGMME). 2005. Report of the Working Group on Marine Mammal Ecology, (WGMME), 9–12 May 2005, Savolinna, Finland. Copenhagen, Denmark: International Council for the Exploration of the Sea.

International Council for the Exploration of the Sea Working Group on Marine Mammal Ecology (ICES WGMME). 2009. Report of the Working Group on Marine Mammal Ecology (WGMME), 2–6 February 2009, Vigo, Spain. Copenhagen, Denmark: International Council for the Exploration of the Sea.

International Council for the Exploration of the Sea Working Group on Marine Mammal Ecology (ICES WGMME). 2010. Report of the Working Group on Marine Mammal Ecology (WGMME), 12–15 April 2010, Horta, the Azores. Copenhagen, Denmark: International Council for the Exploration of the Sea.

International Council for the Exploration of the Sea Working Group on Marine Mammal Ecology (ICES WGMME). 2011. Report of the Working Group on Marine Mammal Ecology (WGMME), 21–24 February 2011, Berlin, Germany. Copenhagen, Denmark: International Council for the Exploration of the Sea.

International Council for the Exploration of the Sea Working Group on Marine Mammal Ecology (ICES WGMME). 2012. Report of the Working Group on Marine Mammal Ecology (WGMME), 5–8 March 2012. Copenhagen, Denmark: International Council for the Exploration of the Sea.

International Union for the Conservation of Nature (IUCN). 2012. The IUCN Red List of Threatened Species. Version 2012.2. Cambridge, UK: IUCN Global Species Programme Red List Unit. Online. http://www.iucnredlist.org (accessed 3 November 2012).

International Whaling Commission (IWC). 2000. Report of the Scientific Committee. Annex O. Report of the IWC-ASCOBANS Working Group on Harbour Porpoises. *Journal of Cetacean Research and Management* **2**(Supplement), 297–305.

International Whaling Commission (IWC). 2008. Submitted by the Government of Australia. Conservation management plans for improved cetacean management. 60th Annual Meeting of the International Whaling Commission, June 2008. International Whaling Commission, SC/60/15. Cambridge, UK: International Whaling Commission.

International Whaling Commission (IWC). 2009. Annex L. Report of the Sub-Committee on Small Cetaceans 2009. Report of the International Whaling Commission, IWC/61/Rep 1. Annex L. Cambridge, UK: International Whaling Commission.

International Whaling Commission (IWC). 2012. International Whaling Commission. Cambridge, UK: International Whaling Commission. Online. http://iwcoffice.org/ (accessed 16 July 2012).

Jaber, J.R., Pérez, J., Arbelo, M., Herráez, P., Espinosa de los Monteros, A., Rodríguez, F., Fernández, T. & Fernández, A. 2003. Immunophenotypic characterization of hepatic inflammatory cell infiltrates in common dolphins (*Delphinus delphis*). *Journal of Comparative Pathology* **129**, 226–230.

Jackson, J.B.C., Kirby, M.X., Berger, W.H., Bjorndal, K.A., Botsford, L.W., Bourque, B.J., Bradbury, R.H., Cooke, R., Erlandson, J., Estes, J.A., Hughes, T.P., Kidwell, S., Lange, C.B., Lenihan, H.S., Pandolfi, J.M., Peterson, C.H., Steneck, R.S., Tegner, M.J. & Warner, R.R. 2001. Historical overfishing and the recent collapse of coastal ecosystems. *Science* **293**, 629–637.

Jákupsstovu, S.H.í. 2002. The pelagic fish stocks, pilot whales and squid in Faroese waters—migration pattern, availability to fisheries and possible links to oceanographic events. ICES Conference and Meeting (CM) Document 2002/N:07. Copenhagen, Denmark: International Council for the Exploration of the Sea.

Janik, V.M. & Thompson, P.M. 1996. Changes in surfacing patterns of bottlenose dolphins in response to boat traffic. *Marine Mammal Science* **12**, 597–602.

Jefferson, T.A. & Van Waerebeek, K. 2002. The taxonomic status of the nominal dolphin species, *Delphinus tropicalis* van Bree, 1971. *Marine Mammal Science* **18**, 787–818.

Jelinski, D.E., Krueger, C.C. & Duffus, D.A. 2002. Geostatistical analyses of interactions between killer whales (*Orcinus orca*) and recreational whale-watching boats. *Applied Geography* **22**, 393–411.

Jepson, P.D., Arbelo, M., Deaville, R., Patterson, I.A.P., Castro, P., Baker, J.R., Degollada, E., Ross, H.M., Herraez, P., Pocknell, A.M., Rodriguez, F., Howie, F.E., Espinosa, A., Reid, R.J., Jaber, J.R., Martin, V., Cunningham, A.A. & Fernandez, A. 2003. Gas-bubble lesions in stranded cetaceans. *Nature* **425**, 575–576.

Jepson, P.D., Bennett, P.M., Allchin, C.R., Law, R.J., Kuiken, T., Baker, J.R., Rogan, E. & Kirkwood, J.K. 1999. Investigating potential associations between chronic exposure to polychlorinated biphenyls and infectious disease mortality in harbour porpoises from England and Wales. *Science of the Total Environment* **243–244**, 339–348.

Jepson, P.D., Bennett, P.M., Deaville, R., Allchin, C.R., Baker, J.R. & Law, R.J. 2005a. Relationships between polychlorinated biphenyls and health status in harbour porpoises (*Phocoena phocoena*) stranded in the United Kingdom *Environmental Toxicology and Chemistry* **24**, 238–248.

Jepson, P.D. & Deaville, R. (compilers), Acevedo-Whitehouse, K., Barnett, J., Brownell, R.L., Colloff, A., Clare, F.C., Davison, N., Law, R.J., Loveridge, J., Macgregor, S.K., Morris, S., Penrose, R., Perkins, M., Pinn, E., Simpson, V., Tasker, M., Tregenza, N., Cunningham, A.A. & Fernández, A. 2009. Investigation of the common dolphin mass stranding event in Cornwall, 9th June 2008. Report to the Department for Environment, Food and Rural Affairs (under a variation to Contract CR0364). Online. http://randd.defra.gov.uk/Document.aspx?Document=WC0601_8031_TRP.pdf (accessed 6 June 2012).

Jepson, P.D., Deaville, R., Acevedo-Whitehouse, K., Barnett, J., Brownlow, A., Brownell, R.L., Jr Clare, F.C., Davison, N., Law, R.J., Loveridge, J., Macgregor, S.K., Morris, S., Murphy, S., Penrose, R., Perkins, M.W., Pinn, E., Seibel, H., Siebert, U., Sierra, E., Simpson, V., Tasker, M.L., Tregenza, N., Cunningham, A.A. & Fernández, A. in press. What caused the UK's largest common dolphin (*Delphinus delphis*) mass stranding event? *PLoS one*.

Jepson, P.D., Deaville, R., Patterson, I.A.P., Pocknell, A.M., Ross, H.M., Baker, J.R., Howie, F.E., Reid, R.J., Colloff, A. & Cunningham, A.A. 2005b. Acute and chronic gas bubble lesions in cetaceans stranded in the United Kingdom. *Veterinary Pathology* **42**, 291–305.

Jepson, P.D. (ed.) 2005. Sabin, R.C., Spurrier, C.J.H., Chimonides, P.D.J., Jepson, P.D., Deaville, R., Perkins, M., Cunningham, A.A., Reid, R.J., Patterson, I.A.P., Foster, G., Barley, J., Penrose, R. & Law, R.J., contributing authors. Trends in cetacean strandings around the U.K. coastline and cetacean and marine turtle post-mortem investigations, 2000 to 2004 inclusive (Contract CRO 238). Cetaceans Strandings Investigation and Co-ordination in the U.K. Report to Defra for the period 1st January 2000–31st December 2004. Bristol, UK: Department for Environment, Food and Rural Affairs.

Joint Nature Conservation Committee (JNCC). 2010a. JNCC guidelines for minimising the risk of disturbance and injury to marine mammals from seismic surveys. Peterborough, UK: Joint Nature Conservation Committee. Online. http://jncc.defra.gov.uk/pdf/JNCC_Guidelines_Seismic%20Guidelines_August%20 2010.pdf (accessed 16 July 2012).

Joint Nature Conservation Committee (JNCC). 2010b. UK priority species data collation Delphinus delphis version 2. UK priority species pages—Version 2. Peterborough, UK: Joint Nature Conservation Committee. Online. http://jncc.defra.gov.uk/_speciespages/258.pdf (accessed 16 July 2012).

Jordan, F.F. 2012. *Skull morphometry of the common dolphin,* Delphinus *sp., from New Zealand waters.* MSc thesis, Massey University, Auckland, New Zealand.

Kannan, K., Blankenship, A.L., Jones, P.D. & Giesy, J.P. 2000. Toxicity reference values for the toxic effects of polychlorinated biphenyls to aquatic mammals. *Human and Ecological Risk Assessment* **6**, 181–201.

Katami, T., Yasuhara, A., Okuda, T. & Shibamoto, T. 2002. Formation of PCDDs, PCDFs, and coplanar PCBs from polyvinyl chloride during combustion in an incinerator. *Environmental Science and Technology* **36**, 1320–1324.

Kennedy, S. 1998. Morbillivirus infections in aquatic mammals. *Journal of Comparative Pathology* **119**, 201–225.

Ketten, D.R. (ed.) 2005. Beaked whale necropsy findings for strandings in the Bahamas, Puerto Rico, and Madeira, 1999–2002. WHOI technical reports. WHOI-2005–09. Woods Hole, Massachusetts: Woods Hole Oceanographic Institute.

Kingston, A. & Northridge, S. 2011. Extension trial of an acoustic deterrent system to minimise dolphin and porpoise bycatch in gill and tangle net fisheries. Report to the Cornish Fisheries Producers Organisation Ltd. St. Andrews, UK: Sea Mammal Research Unit.

Kingston, S.E. & Rosel, P.E. 2004. Genetic differentiation among recently diverged delphinid taxa determined using AFLP markers. *Journal of Heredity* **95**, 1–10.

Kinze, C.C. 1995. Danish whale records 1575–1991 (Mammalia, Cetacea). *Steenstrupia* **21**, 155–196.

Kiszka, J., Hassani, S. & Pezeril, S. 2004. Distribution and status of small cetaceans along the French Channel coasts: using opportunistic records for a preliminary assessment. *Lutra* **47**, 33–45.

Kiszka, J., Macleod, K., Van Canneyt, O., Walker, D. & Ridoux, V. 2007. Distribution, encounter rates, and habitat characteristics of toothed cetaceans in the Bay of Biscay and adjacent waters from platform-of-opportunity data. *ICES Journal of Marine Science* **64**, 1033–1043.

Knowlton, A.R. & Kraus, S.D. 2001. Mortality and serious injury of northern right whales (*Eubalaena glacialis*) in the western North Atlantic Ocean. *Journal of Cetacean Research and Management (special issue)* **2**, 193–208.

Kraus, S.D. 1990. Rates and potential causes of mortality in North Atlantic right whales (*Eubalaena glacialis*). *Marine Mammal Science* **6**, 278–291.

Kuiken, T., Simpson, V.R., Allchin, C.R., Bennet, P.M., Codd, G.A., Harris, E.A., Howes, G.J., Kennedy, S., Kirkwood, J.K., Law, R.J., Merret, N.R. & Philips, S. 1994. Mass mortality of common dolphins (*Delphinus delphis*) in south west England due to incidental capture in fishing gear. *Veterinary Record* **134**, 81–89.

Kunito, T., Watanabe, I., Yasunaga, G., Fujise, Y. & Tanabe, S. 2002. Using trace elements in skin to discriminate the populations of minke whales in southern hemisphere. *Marine Environmental Research* **53**, 175–197.

Lahaye, V., Bustamante, P., Dabin, W., Churlaud, C. & Caurant, F. 2007. Trace element levels in foetus-mother pairs of short-beaked common dolphins (*Delphinus delphis*) stranded along the French coasts. *Environment International* **33**, 1021–1028.

Lahaye, V., Bustamante, P., Spitz, J., Dabin, W., Das, K., Pierce, G.J. & Caurant, F. 2005. Long-term dietary segregation of common dolphins *Delphinus delphis* in the Bay of Biscay, determined using cadmium as an ecological tracer. *Marine Ecology Progress Series* **305**, 275–285.

Lambert, E., MacLeod, C.D., Hall, K., Brereton, T., Dunn, T.E., Wall, D., Jepson, P.D., Deaville, R. & Pierce, G.J. 2011. Quantifying likely cetacean range shifts in response to global climatic change: implications for conservation strategies in a changing world. *Endangered Species Research* **15**, 205–222.

Lassalle, G., Gascuel, D., Le Loc'h, F., Lobry, J., Pierce, G.J., Ridoux, V., Santos, M.B., Spitz, J. & Niquil, N. 2012. An ecosystem approach for the assessment of fisheries impacts on marine top predators: the Bay of Biscay case study. *ICES Journal of Marine Science* **69**, 925–938.

Law, R.J., Barry, J., Barber, J.L., Bersuder, P., Deaville, R., Reid, R.J., Brownlow, A., Penrose, R., Barnett, J., Loveridge, J., Smith, B. & Jepson, P.D. 2012a. Contaminants in cetaceans from UK waters: status as assessed within the Cetacean Strandings Investigation Programme from 1990 to 2008. *Marine Pollution Bulletin* **64**, 1485–1494.

Law, R.J., Barry, J., Bersuder, P., Barber, J.L., Deaville, R., Reid, R.J. & Jepson, P.D. 2010. Levels and trends of brominated diphenyl ethers in blubber of harbor porpoises (*Phocoena phocoena*) from the U.K., 1992–2008. *Environmental Science & Technology* **44**, 4447–4451.

Law, R.J., Bolam, T., James, D., Barry, J., Deaville, R., Reid, R.J., Penrose, R. & Jepson, P.D. 2012b. Butyltin compounds in liver of harbour porpoises (*Phocoena phocoena*) from the UK prior to and following the ban on the use of tributyltin in antifouling paints (1992–2005 & 2009). *Marine Pollution Bulletin* **64**, 2576–2580.

Lawson, J., Gosselin, J.-F., Desportes, G., Acquarone, M., Heide-Jørgensen, M.P., Mikkelsen, B., Pike, D., Víkingsson, G., Zabavnikov, V. & Øien, N. 2009. A note on the distribution of short-beaked common dolphins, *Delphinus delphis*, observed during the 2007 T-NASS (Trans North Atlantic Sightings Survey). IWC Scientific Committee Document SC/61/SM35. Cambridge, UK: International Whaling Commission.

Lea, M.-A., Nichols, P.D. & Wilson, G. 2002. Fatty acid composition of lipid-rich myctophids and mackerel icefish (*Champsocephalus gunnari*)—Southern Ocean food-web implications. *Polar Biology* **25**, 843–854.

Learmonth, J.A., Murphy, S., Dabin, W., Addink, M., Lopez, A., Rogan, E., Ridoux, V., Guerra, A. & Pierce, G.J. 2004a. Measurement of reproductive output in small cetaceans from the Northeast Atlantic. BIOCET workpackage 5–final report. Project Reference: EVK3-2000-00027. Brussels, Belgium: European Commission.

Learmonth, J.A., Santos, M.B., Pierce, G.J., Moffat, C.F., Rogan, E., Murphy, S., Ridoux, V., Meynier, L., Lahaye, V., Pusineri, C. & Spitz, J. 2004b. Dietary studies on small cetaceans in the NE Atlantic using stomach contents and fatty acid analyses. BIOCET workpackage 6—final report. Project reference: EVK3-2000-00027. Aberdeen, UK: University of Aberdeen.

LeDuc, R.G., Perrin, W.F. & Dizon, A.E. 1999. Phylogenetic relationships among the delphinid cetaceans based on full cytochrome B sequences. *Marine Mammal Science* **15**, 619–648.

Leeney, R., Amies, R., Broderick, A., Witt, M., Loveridge, J., Doyle, J. & Godley, B. 2008. Spatio-temporal analysis of cetacean strandings and bycatch in a UK fisheries hotspot. *Biodiversity and Conservation* **17**, 2323–2338.

Lewis, T. 1991. The development of a postanal hump in male common dolphins, *Delphinus delphis*. In *Abstracts of the 9th Biennial Conference on the Biology of Marine Mammals, Chicago, USA*. San Francisco: Society of Marine Mammalogy, 42 only.

Lewison, R.L., Crowder, L.B., Read, A.J. & Freeman, S.A. 2004. Understanding impacts of fisheries bycatch on marine megafauna. *Trends in Ecology & Evolution* **19**, 598–604.

López, A., Pierce, G.J., Santos, M.B., Gracia, J. & Guerra, A. 2003. Fishery by-catches of marine mammals in Galician waters: results from on-board observations and an interview survey of fishermen. *Biological Conservation* **111**, 25–40.

López, A., Pierce, G.J., Valeiras, X., Santos, M.B. & Guerra, A. 2004. Distribution patterns of small cetaceans in Galician waters. *Journal of the Marine Biological Association of the United Kingdom* **84**, 283–294.

López, A., Santos, M.B., Pierce, G.J., González, A.F., Valeiras, X. & Guerra, A. 2002. Trends in strandings and by-catch of marine mammals in north-west Spain during the 1990s. *Journal of the Marine Biological Association of the United Kingdom* **82**, 513–521.

Lusseau, D. 2003. Male and female bottlenose dolphins (*Tursiops* spp.) have different strategies to avoid interactions with tour boats in Doubtful Sound, New Zealand. *Marine Ecology Progress Series* **257**, 267–274.

Lusseau, D. & Higham, J.E.S. 2004. Managing the impacts of dolphin-based tourism through the definition of critical habitats: the case of bottlenose dolphins (*Tursiops* spp.) in Doubtful Sound, New Zealand. *Tourism Management* **25**, 657–667.

Mackay, A.I. 2011. *An investigation of factors related to the bycatch of small cetaceans in fishing gear*. PhD thesis, University of St. Andrews, UK.

MacLeod, C.D. 2010. The relationship between body mass and relative investment in testes mass in cetaceans: implications for inferring interspecific variations in the extent of sperm competition. *Marine Mammal Science* **26**, 370–380.

MacLeod, C.D., Bannon, S.M., Pierce, G.J., Schweder, C., Learmonth, J.A., Herman, J.S. & Reid, R.J. 2005. Climate change and the cetacean community of north-west Scotland. *Biological Conservation* **124**, 477–483.

MacLeod, C.D., Brereton, T. & Martin, C. 2009. Changes in the occurrence of common dolphins, striped dolphins and harbour porpoises in the English Channel and Bay of Biscay. *Journal of the Marine Biological Association of the United Kingdom* **89**, 1059–1065.

MacLeod, C.D., Weir, C.R., Santos, M.B. & Dunn, T.E. 2008. Temperature-based summer habitat partitioning between white-beaked and common dolphins around the United Kingdom and Republic of Ireland. *Journal of the Marine Biological Association of the United Kingdom* **88**, 1193–1198.

Mannocci, L., Dabin, W., Augeraud-Véron, E., Dupuy, J.-F., Barbraud, C. & Ridoux, V. 2012. Assessing the impact of bycatch on dolphin populations: the case of the common dolphin in the eastern North Atlantic. *PLoS one* **7**, e32615.

Marcos, E., Salazar, J.M. & De Stephanis, R. 2010. Cetacean diversity and distribution in the coast of Gipuzkoa and adjacent waters, southeastern Bay of Biscay. *Munibe (Ciencias Naturales-Natur Zientziak)* **58**, 221–231.

Marcos-Ipiña, E., Salazar-Sierra, J.M. & De Stephanis, R. 2005. Cetacean population in coast of the Basque country: diversity and distribution spring-summer 2003–2004. In *European Research on Cetaceans—19. Proceedings of the 19th Annual Conference of the European Cetacean Society, La Rochelle, France, 2–7 April 2005*. P.G.H. Evans & V. Ridoux (eds). Cambridge, UK: European Cetacean Society.

McBrearty, D.A., Message, M.A. & King, G.A. 1986. Observations on small cetaceans in the north-east Atlantic Ocean and the Mediterranean Sea: 1978–1982. In *Research on Dolphins*, M.M. Bryden & R. Harrison (eds). Oxford, UK: Clarendon Press, 225–249.

Mèndez-Fernandez, P., Bustamante, P., Bode, A., Chouvelon, T., Ferreira, M., López, A., Pierce, G.J., Santos, M.B., Spitz, J., Vingada, J.V. & Caurant, F. 2012. Foraging ecology of five toothed whale species in the Northwest Iberian Peninsula, inferred using carbon and nitrogen isotope ratios. *Journal of Experimental Marine Biology and Ecology* **413**, 150–158.

Mendolia, C. 1989. *Reproductive biology of common dolphins* (Delphinus delphis *Linnaeus) off the south east coast of southern Africa*. MSc thesis, University of Port Elizabeth, South Africa.

Meynier, L. 2004. *Food and feeding ecology of the common dolphin,* Delphinus delphis, *in the Bay of Biscay: intraspecific dietary variation and food transfer modelling*. MSc thesis, University of Aberdeen, UK.

Meynier, L., Pusineri, C., Spitz, J., Santos, M.B., Pierce, G.J. & Ridoux, V. 2008. Intraspecific dietary variation in the short-beaked common dolphin *Delphinus delphis* in the Bay of Biscay: importance of fat fish. *Marine Ecology Progress Series* **354**, 277–287.

Miclard, J., Mokhtari, K., Jouvion, G., Wyrzykowski, B., Van Canneyt, O., Wyers, M. & Colle, M.A. 2006. Microcystic meningioma in a dolphin (*Delphinus delphis*): immunohistochemical and ultrastructural study. *Journal of Comparative Pathology* **135**, 254–258.

Miller, P.J.O., Biassoni, N., Samuels, A. & Tyack, P.L. 2000. Whale songs lengthen in response to sonar. *Nature* **405**, 903.

Mirimin, L., Viricel, A., Amaral, A.R., Murphy, S., Ridoux, V. & Rogan, E. 2009a. Population genetic structure of common dolphins in the north-east Atlantic using microsatellite loci and mtDNA control region markers. IWC Scientific Committee Document SC/61/SM27. Cambridge, UK: International Whaling Commission.

Mirimin, L., Westgate, A.J., Rogan, E., Rosel, P., Read, A.J., Coughlan, J. & Cross, T. 2009b. Population structure of short-beaked common dolphins (*Delphinus delphis*) in the North Atlantic Ocean as revealed by mitochondrial and nuclear genetic markers. *Marine Biology* **156**, 821–834.

Mirimin, L., Westgate, A., Stockin, K., Murphy, S., Northridge, S., Cross, T. & Rogan, E. 2012. Genetic analyses of groups of short-beaked common dolphins by-caught in fisheries in the North Atlantic and Southwest Pacific oceans. In *26th Annual Conference of the European Cetacean Society, Galway, Ireland, 26 March 2012–28 March 2012*, B. McGovern et al. (eds). Cambridge, UK: European Cetacean Society, 86 only.

Möller, L., Valdez, F., Allen, S., Bilgmann, K., Corrigan, S. & Beheregaray, L. 2011. Fine-scale genetic structure in short-beaked common dolphins (*Delphinus delphis*) along the East Australian Current. *Marine Biology* **158**, 113–126.

Mooney, T.A., Nachtigall, P.E. & Vlachos, S. 2009. Sonar-induced temporary hearing loss in dolphins. *Biology Letters* **5**, 565–567.

Moore, S.E. & Clarke, J.T. 2002. Potential impact of offshore human activities on gray whales. *Journal of Cetacean Research and Management* **4**, 19–25.

Morizur, Y., Berrow, S.D., Tregenza, N.J.C., Couperus, A.S. & Pouvreau, S. 1999. Incidental catches of marine-mammals in pelagic trawl fisheries of the northeast Atlantic. *Fisheries Research* **41**, 297–307.

Morizur, Y., Demaneche, S., Gaudou, O. & Miossec, D. 2010. Les captures accidentelles de cétacés dans les pêches françaises en 2009: contribution au rapport national sur la mise en oeuvre du règlement européen (CE) No. 812/2004–(année 2009). Brest, France: Ifremer Centre de Brest Sciences et Technologies Halieutiques.

Morizur, Y., Demaneche, S., Fauconnet, L., Gaudou, O. & Badts, V. 2011. Les captures accidentelles de cétacés dans les pêches professionnelles françaises en 2010: contribution au rapport national sur la mise en oeuvre du règlement européen (CE) No. 812/2004–(année 2010). Rapport contractuel Ifremer/DPMA. Convention socle No. 10/1218641/NF. Brest, France: Ifremer.

Moura, A.E. 2010. *Investigating the relative influence of genetic drift and natural selection in shaping patterns of population structure in Delphinids* (Delphinus delphis; Tursiops *spp.*). PhD thesis, University of Durham, UK.

Moura, A.E., Sillero, N. & Rodrigues, A. 2012. Common dolphin (*Delphinus delphis*) habitat preferences using data from two platforms of opportunity. *Acta Oecologica* **38**, 24–32.

Murphy, S. 1999. *Morphometric studies and skin lesions on the common dolphin* Delphinus delphis. BSc dissertation, University College Cork, Ireland.

Murphy, S. 2004. *The biology and ecology of the common dolphin* Delphinus delphis *in the North-east Atlantic*. PhD thesis, University College Cork, Ireland.

Murphy, S. 2006. Sexual dimorphism in cranial measurements of *Delphinus delphis* in the eastern North Atlantic. *Marine Mammal Science* **22**, 1–4.

Murphy, S. 2009a. Environmental and anthropogenic factors linked to influencing or controlling cetacean population growth rates. Task 3 Deliverable 'Cetacean stock assessment in relation to exploration and production industry sound'. JIP Cetacean Stock Assessment. St. Andrews, UK: Sea Mammal Research Unit.

Murphy, S. 2009b. Effects of contaminants on reproduction in small cetaceans. Final report of Phase I to ASCOBANS. St. Andrews, UK: Sea Mammal Research Unit.

Murphy, S., Collet, A. & Rogan, E. 2005a. Mating strategy in the male common dolphin *Delphinus delphis*: what gonadal analysis tells us. *Journal of Mammalogy* **86**, 1247–1258.

Murphy, S., Dabin, W., Ridoux, V., Morizur, Y., Larsen, F. & Rogan, E. 2007a. Estimation of R_{max} for the common dolphin in the Northeast Atlantic. Report to the European Commission, NECESSITY Contract 501605 Periodic Activity Report No. 2–Annex 8.4. Ijmuiden, The Netherlands: IMARES Wageningen UR, Institute for Marine Resources & Ecosystem Studies.

Murphy, S., Deaville, R., Monies, R.J., Davison, N. & Jepson, P.D. 2011. True hermaphroditism: first evidence of an ovotestis in a cetacean species. *Journal of Comparative Pathology* **144**, 195–199.

Murphy, S., Evans, P.G.H. & Collet, A. 2008. Common dolphin *Delphinus delphis*. In *Mammals of the British Isles: Handbook*. S. Harris & D.W. Yalden (eds). Southampton, UK: Mammal Society, 4th edition, 719–724.

Murphy, S., Herman, J.S., Pierce, G.J., Rogan, E. & Kitchener, A.C. 2006. Taxonomic status and geographical cranial variation of common dolphins (*Delphinus*) in the eastern North Atlantic. *Marine Mammal Science* **22**, 573–599.

Murphy, S., Jepson, P.D. & Deaville, R. 2012a. Effects of contaminants on reproduction in small cetaceans, phase II. Final report of Phase II to ASCOBANS. St. Andrews, UK: Sea Mammal Research Unit.

Murphy, S., Mirimin, L., Englund, A. & Mackey, M. 2005b. Evidence of a violent interaction between *Delphinus delphis* L. and *Tursiops truncatus* (Montagu). *Irish Naturalists' Journal* **28**, 42–43.

Murphy, S., Natoli, A., Amaral, A.R., Mirimin, L., Viricel, A., Caurant, F., Hoelzel, R. & Evans, P.G.H. 2009a. Short-beaked common dolphin *Delphinus delphis*. In *Report of ASCOBANS/HELCOM Small Cetacean Population Structure Workshop, 8–10 October 2007, U.N. Campus, Bonn, Germany*. Bonn, Germany: ASCOBANS, 111–130.

Murphy, S., Northridge, S., Dabin, W., Van Canneyt, O., Ridoux, V., Rogan, E., Philpott, E., Jepson, P., Deaville, R., Reid, B. & Morizur, Y. 2007b. Biological parameters of common dolphin population resulting from stranded or bycaught animals in the Northeast Atlantic. Report to the European Commission, NECESSITY Contract 501605 Periodic Activity Report No. 2–Annex 6.2. Ijmuiden, The Netherlands: IMARES Wageningen UR, Institute for Marine Resources & Ecosystem Studies.

Murphy, S., Pierce, G.J., Law, R.J., Bersuder, P., Jepson, P.D., Learmonth, J.A., Addink, M., Dabin, W., Santos, M.B., Deaville, R., Zegers, B.N., Mets, A., Rogan, E., Ridoux, V., Reid, R.J., Smeenk, C., Jauniaux, T., López, A., Farré, J.M.A., González, A.F., Guerra, A., García-Hartmann, M., Lockyer, C. & Boon, J.P. 2010. Assessing the effect of persistent organic pollutants on reproductive activity in common dolphins and harbour porpoises. *Journal of Northwest Atlantic Fishery Science* **42**, 153–173.

Murphy, S. & Rogan, E. 2004. Marine Mammal Stranding Project 2003–2004. Final report to National Parks and Wildlife (Ireland). Cork, Ireland: University College Cork.

Murphy, S. & Rogan, E. 2006. External morphology of the short-beaked common dolphin, *Delphinus delphis*: growth, allometric relationships and sexual dimorphism. *Acta Zoologica* **87**, 315–329.

Murphy, S. (ed.) Tougaard, J., Wilson, B., Benjamins, S., Haelters, J., Lucke, K., Werner, S., Brensing, K., Thompson, D., Hastie, G., Geelhoed, S., Braeger, S., Lees, G., Davies, I., Graw, K.-U. & Pinn, E. 2012b. Assessment of the marine renewables industry in relation to marine mammals: synthesis of work undertaken by the ICES Working Group on Marine Mammal Ecology (WGMME). Report to the International Whaling Commission, IWC/64/SC MRED1. Cambridge, UK: International Whaling Commission.

Murphy, S., Winship, A., Dabin, W., Jepson, P.D., Deaville, R., Reid, R.J., Spurrier, C., Rogan, E., López, A., González, A.F., Read, F.L., Addink, M., Silva, M., Ridoux, V., Learmonth, J.A., Pierce, G.J. & Northridge, S.P. 2009b. Importance of biological parameters in assessing the status of *Delphinus delphis*. *Marine Ecology Progress Series* **388**, 273–291.

Murray, T. & Murphy, S. 2003. Common dolphin *Delphinus delphis* L. strandings on the Mullet Peninsula. *Irish Naturalists' Journal* **27**, 240–241.

Myers, R.A. & Worm, B. 2003. Rapid worldwide depletion of predatory fish communities. *Nature* **423**, 280–283.

National Research Council (NRC). 2005. *Marine Mammal Populations and Ocean Noise: Determining When Noise Causes Biologically Significant Effects*. Washington, DC: National Academies Press.

Natoli, A., Cañadas, A., Peddemors, V.M., Aguilar, A., Vaquero, C., Fernandez-Piqueras, P. & Hoelzel, A.R. 2006. Phylogeography and alpha taxonomy of the common dolphin (*Delphinus* sp.). *Journal of Evolutionary Biology* **19**, 943–954.

Natoli, A., Cañadas, A., Vaquero, C., Politi, E., Fernandez-Navarro, P. & Hoelzel, A.J. 2008. Conservation genetics of the short-beaked common dolphin (*Delphinus delphis*) in the Mediterranean Sea and in the eastern North Atlantic Ocean. *Conservation Genetics* **9**, 1479–1487.

Natural History Museum. 1995. Studies on the biology of Cetacea. A report for the Welsh Office. Ref. No. WEP/100/154/6. London: Natural History Museum.

Neumann, D.R. 2001. The activity budget of free-ranging common dolphins (*Delphinus delphis*) in the northwestern Bay of Plenty, New Zealand. *Aquatic Mammals* **27**, 121–136.

Neumann, D.R. & Orams, M.B. 2003. Feeding behaviour of short-beaked common dolphins, *Delphinus delphis*, in New Zealand. *Aquatic Mammals* **29**, 137–149.

Neumann, D.R., Russell, K., Orams, M.B., Baker, C.S. & Duignan, P. 2002. Identifying sexually mature, male short-beaked common dolphins (*Delphinus delphis*) at sea, based on the presence of a postanal hump. *Aquatic Mammals* **28**, 181–187.

Nielsen, T.P., Wahlberg, M., Heikkilä, S., Jensen, M., Sabinsky, P. & Dabelsteen, T. 2012. Swimming patterns of wild harbour porpoises *Phocoena phocoena* show detection and avoidance of gillnets at very long ranges. *Marine Ecology Progress Series* **453**, 241–248.

Nordstrom, C.A., Wilson, L.J., Sara, I.J. & Tollit, D.J. 2008. Evaluating quantitative fatty acid signature analysis (QFASA) using harbour seals *Phoca vitulina richardsi* in captive feeding studies. *Marine Ecology Progress Series* **360**, 245–263.

Noren, D.P. & Mocklin, J.A. 2012. Review of cetacean biopsy techniques: factors contributing to successful sample collection and physiological and behavioral impacts. *Marine Mammal Science* **28**, 154–199.

Northridge, S. 2009. Fishing industry, effects of. In *Encyclopedia of Marine Mammals*. F.P. William et al. (eds). London: Academic Press, 2nd edition, 443–447.

Northridge, S. & Kingston, A. 2009. Common dolphin bycatch in UK fisheries. IWC Scientific Committee Document SC/61/SM37. Cambridge, UK: International Whaling Commission.

Northridge, S. & Kingston, A. 2010. Annual report on the implementation of Council Regulation (EC) No. 812/2004–2009. Report to the European Commission on the Implementation of Regulation 812/204 by the United Kingdom for the calendar year 2009. St. Andrews, UK: Sea Mammal Research Unit.

Northridge, S., Kingston, A., Thomas, L. & Mackay, A. 2007. Second annual report on the U.K. cetacean bycatch monitoring scheme. Contract report to Defra on the work conducted 2005–2006, June 2007. St. Andrews, UK: Sea Mammal Research Unit.

Northridge, S., Mackay, A., Sanderson, D., Woodcock, R. & Kingston, A. 2004. A review of dolphin and porpoise bycatch issues in the Southwest of England. An occasional report to the Department for Environment Food and Rural Affairs. St. Andrews: Sea Mammal Research Unit.

Northridge, S., Morizur, Y., Souami, Y. & Van Canneyt, O. 2006. Final PETRACET report to the European Commission. Project EC/FISH/2003/09, 1735R07D. June 2006. Lymington, New Hampshire, UK: MacAlister Elliott and Partners Ltd.

Northridge, S. & Thomas, L. 2003. Monitoring levels required in European fisheries to assess cetacean bycatch, with particular reference to U.K. fisheries. Final report to DEFRA (EWD). St. Andrews, UK: Sea Mammal Research Unit.

Notarbartolo-Di-Sciara, G., Zanardelli, M., Jahoda, M., Panigada, S. & Airoldi, S. 2003. The fin whale *Balaenoptera physalus* (L. 1758) in the Mediterranean Sea. *Mammal Review* **33**, 105–150.

Nowacek, D.P., Johnson, M.P. & Tyack, P.L. 2004. North Atlantic right whales (*Eubalaena glacialis*) ignore ships but respond to alerting stimuli. *Proceeding of the Royal Society of London B* **271**, 227–231.

Nowacek, D.P., Thorne, L.H., Johnston, D.W. & Tyack, P.L. 2007. Responses of cetaceans to anthropogenic noise. *Mammal Review* **37**, 81–115.

Nowacek, S.M., Wells, R.S. & Solow, A.R. 2001. Short-term effects of boat traffic on bottlenose dolphins, *Tursiops truncatus*, in Sarasota Bay, Florida. *Marine Mammal Science* **17**, 673–688.

O'Brien, C.M., Fox, C.J., Planque, B. & Casey, J. 2000. Climate variability and North Sea cod. *Nature* **404**, 142.

Ó Cadhla, O., Mackey, M., Aguilar de Soto, N., Rogan, E. & Connolly, N. 2003. Cetaceans and seabirds of Ireland's Atlantic Margin. Volume II—Cetacean distribution and abundance. Report on research conducted under the 1997 Irish Petroleum Infrastructure Programme (PIP): Rockall Studies Group (RSG) projects 98/6, 99/38 and 00/13. Cork, Ireland: Coastal and Marine Research Group, University College, Cork.

O'Connell, M. & Berrow, S. 2012. Report of the Cetacean Strandings Scheme January–December 2011. Irish Whale and Dolphin Group. Report to the Irish National Parks and Wildlife Service. Online. http://www.npws.ie/marine/marinereports/Cetacean%20Strandings%20Scheme%20report_2011.pdf (accessed 22 November 2012).

Øien, N. & Hartvedt, S. 2009. Common dolphins *Delphinus delphis* in Norwegian waters. IWC Scientific Committee Document SC/61/SM9. Cambridge, UK: International Whaling Commission.

Orams, M.B. 2000. Tourists getting close to whales, is it what whale-watching is all about? *Tourism Management* **21**, 561–569.

Pascoe, P.L. 1986. Size data and stomach contents of common dolphins, *Delphinus delphis*, near Plymouth. *Journal of the Marine Biological Association of the United Kingdom* **66**, 319–322.

Pauly, D., Christensen, V., Dalsgaard, J., Frose, R. & Torres, J.F. 1998. Fishing down the marine food webs. *Science* **279**, 860–863.

Paxton, C.G.M., Mackenzie, M., Burt, M.L., Rexstad, E. & Thomas, L. 2011. Phase II data analysis of Joint Cetacean Protocol data resource. Report to Joint Nature Conservation Committee contract number C11-0207-0421. St. Andrews, UK: Centre for Research into Ecological and Environmental Modelling, University of St. Andrews. Online. http://jncc.defra.gov.uk/pdf/JCP_Phase_II_report.pdf (accessed 14 March 2012).

Paxton, C.G.M. & Thomas, L. 2010. Phase I data analysis of Joint Cetacean Protocol data. Report to Joint Nature Conservation Committee on JNCC contract No. C09-0207-0216. St. Andrews, UK: Centre for Research into Ecological and Environmental Modelling, University of St. Andrews. Online. http://www.creem.st-and.ac.uk/len/papers/PaxtonJNCC2010.pdf (accessed 11 December 2012).

Peltier, H., Dabin, W., Daniel, P., Van Canneyt, O., Dorémus, G., Huon, M. & Ridoux, V. 2012. The significance of stranding data as indicators of cetacean populations at sea: modelling the drift of cetacean carcasses. *Ecological Indicators* **18**, 278–290.

Perrin, W.F. 2009. Common dolphins: *Delphinus delphis* and *D. capensis*. In *Encyclopedia of Marine Mammals*. F.P. William et al. (eds). London: Academic Press, 2nd edition, 255–259.

Perry, S.L., DeMaster, D.P. & Silber, G.K. 1999. The great whales: history and the status of six species listed as endangered under the U.S. Endangered Species Act of 1973. *Marine Fisheries Review* **61**, 1–74.

Peschak, T.P. 2005. *Currents of Contrast: Life in Southern Africa's Two Oceans*. Cape Town, South Africa: Struik.

Philpott, E. & Rogan, E. 2007. Records from the Irish Whale and Dolphin Group for 2005. *The Irish Naturalists' Journal* **28**, 414–418.

Philpott, E., Wall, D. & Rogan, E. 2007. Records from the Irish Whale and Dolphin Group for 2004. *The Irish Naturalists' Journal* **28**, 379–385.

Pierce, G.J., Caldas, M., Cedeira, J., Santos, M.B., Llavona, Á., Covelo, P., Martinez, G., Torres, J., Sacau, M. & López, A. 2010. Trends in cetacean sightings along the Galician coast, north-west Spain, 2003–2007, and inferences about cetacean habitat preferences. *Journal of the Marine Biological Association of the United Kingdom* **90**, 1547–1560.

Pierce, G.J., Santos, M.B., Murphy, S., Learmonth, J.A., Zuur, A.F., Rogan, E., Bustamante, P., Caurant, F., Lahaye, V., Ridoux, V., Zegers, B.N., Mets, A., Addink, M., Smeenk, C., Jauniaux, T., Law, R.J., Dabin, W., López, A., Alonso Farré, J.M., González, A.F., Guerra, A., García-Hartmann, M., Reid, R.J., Moffat, C.F., Lockyer, C. & Boon, J.P. 2008. Bioaccumulation of persistent organic pollutants in female common dolphins (*Delphinus delphis*) and harbour porpoises (*Phocoena phocoena*) from western European seas: geographical trends, causal factors and effects on reproduction and mortality. *Environmental Pollution* **153**, 401–415.

Pikesley, S.K., Witt, M.J., Hardy, T., Loveridge, J., Loveridge, J., Williams, R. & Godley, B.J. 2011. Cetacean sightings and strandings: evidence for spatial and temporal trends. *Journal of the Marine Biological Association of the United Kingdom* **92**, 1809–1820.

Pinela, A.M., Borrell, A. & Aguilar, A. 2011. Common dolphin morphotypes: niche segregation or taxonomy? *Journal of Zoology* **284**, 239–247.

Plaganyi, E.E. & Butterworth, D.S. 2005. Indirect fishery interactions. In *Marine Mammal Research. Conservation Beyond Crisis*, J.E. Reynolds III et al. (eds). Baltimore: Johns Hopkins University Press, 19–45.

Planque, B. & Taylor, A.H. 1998. Long term changes in zooplankton and the climate of the North Atlantic. *ICES Journal of Marine Science* **55**, 644–654.

Pusineri, C., Magnin, V., Meynier, L., Spitz, J., Hassani, S. & Ridoux, V. 2007. Food and feeding ecology of the common dolphin (*Delphinus delphis*) in the oceanic northeast Atlantic and comparison with its neritic areas. *Marine Mammal Science* **23**, 30–47.

Quérouil, S., Freitas, L., Cascão, I., Alves, F., Dinis, A., Almeida, J., Prieto, R., Borràs, S., Matos, J., Mendonça, D. & Santos, R. 2010. Molecular insight into the population structure of common and spotted dolphins inhabiting the pelagic waters of the Northeast Atlantic. *Marine Biology* **157**, 2567–2580.

Read, A.J. 2008. The looming crisis: interactions between marine mammals and fisheries. *Journal of Mammalogy* **89**, 541–548.

Read, A.J., Halpin, P.N., Crowder, L.B., Best, B.D. & Fujioka, E. (eds) 2007. Ocean biogeographic information system—spatial ecological analysis of megavertebrate populations. Beaufort, North Carolina: Duke University Marine Lab. Online. http://seamap.env.duke.edu/ (accessed 13 November 2007).

Read, F.L., Santos, B., González, A.F., Martínez-Cedeira, J., López, A. & Pierce, G.J. 2009. Common dolphin (*Delphinus delphis*) in Galicia, NW Spain: distribution, abundance, life history and conservation. IWC Scientific Committee Document SC/61/SM5. Cambridge, UK: International Whaling Commission.

Reeves, R.R., Rolland, R. & Clapham, P.J. 2001. Causes of reproductive failure in North Atlantic right whales: new avenues of research. Reference document 01-16. Falmouth, Massachusetts: Northeast Fisheries Science Centre.

Reid, J.B., Evans, P.G.H. & Northridge, S.P. 2003. *Atlas of Cetacean Distribution in North-West European Waters*. Peterborough, UK: Joint Nature Conservation Committee.

Reid, P.C. & Beaugrand, G. 2012. Global synchrony of an accelerating rise in sea surface temperature. *Journal of the Marine Biological Association of the United Kingdom* **92**, 1435–1450.

Reid, P.C. & Edwards, M. 2001. Long-term changes in the pelagos, benthos and fisheries of the North Sea. *Marine Biodiversity* **31**, 107–115.

Reijnders, P.J.H., Donovan, G.P., Bjørge, A., Kock, K.-H. & Tasker, M.L. 2008. Draft ASCOBANS conservation plan for harbour porpoises (*Phocoena phocoena* L.) in the North Sea. 15th Meeting of the ASCOBANS Advisory Committee, U.N. Campus, Bonn, Germany, 31 March–3 April 2008, Advisory Committee Document 14. Bonn, Germany: ASCOBANS.

Reilly, S.B. & Barlow, J. 1986. Rates of increase in dolphin population size. *Fishery Bulletin* **84**, 527–533.

Rendell, L.E. & Gordon, J.C.D. 1999. Vocal response of long-finned pilot whales (*Globicephala melas*) to military sonar in the Ligurian Sea. *Marine Mammal Science* **15**, 198–204.

Richardson, W.J., Greene, C.R., Jr., Malme, C.I. & Thomson, D.H. 1995. *Marine Mammals and Noise*. London: Academic Press.

Ridoux, V., Lafontaine, L., Bustamante, P., Caurant, F., Dabin, W., Delcroix, C., Hassani, S., Meynier, L., Pereira da Silva, V., Simonin, S., Robert, M., Spitz, J. & Van Canneyt, O. 2004. The impact of the "Erika" oil spill on pelagic and coastal marine mammals: combining demographic, ecological, trace metals and biomarker evidences. *Aquatic Living Resources* **17**, 379–387.

Rihan, D. 2010. Measures to reduce interactions of marine megafauna with fishing operations. In *Behavior of Marine Fishes: Capture Processes and Conservation Challenges*, P. He (ed.). Singapore: Wiley-Blackwell, 315–342.

Robinson, K.P., Eisfeld, S.M., Costa, M. & Simmonds, M.P. 2010. Short-beaked common dolphin (*Delphinus delphis*) occurrence in the Moray Firth, north-east Scotland. *Marine Biodiversity Records* **3**, 1–4.

Rodrigues, A.S.L., Pilgrim, J.D., Lamoreux, J.F., Hoffmann, M. & Brooks, T.M. 2006. The value of the IUCN Red List for conservation. *Trends in Ecology and Evolution* **21**, 71–76.

Rogan, E. & Mackey, M. 2007. Megafauna bycatch in drift-nets for albacore tuna (*Thunnus alalunga*) in the NE Atlantic. *Fisheries Research* **86**, 6–14.

Rohr, J.J., Fish, F.E. & Gilpatrick, J.W. 2002. Maximum swim speeds of captive and free-ranging delphinids; critical analysis of extraordinary performance. *Marine Mammal Science* **18**, 1–19.

Rommel, S.A., Costidis, A.M., Fernandez, A., Jepson, P.D., Pabst, D.A., McLellan, W.A., Houser, D.S., Cranford, T.W., van Helden, A.L., Allen, D.M. & Barros, N.B. 2006. Elements of beaked whale anatomy and diving physiology and some hypothetical causes of sonar-related stranding. *Journal of Cetacean Research and Management* **7**, 189–209.

Rosel, P.E., Dizon, A.E. & Heyning, J.E. 1994. Genetic analysis of sympatric morphotypes of common dolphins (genus *Delphinus*). *Marine Biology* **119**, 159–167.

Sabin, R., Jepson, P.D., Reid, P.D.J., Chimindies, R. & Deaville, R. 2002. Trends in cetacean strandings around the UK coastline and marine mammals post-mortem investigations for the year 2002 (contract CRO 238). Report to Defra No. ECM 516F00/03. London: Natural History Museum.

Saito, H. & Murata, M. 1998. Origin of the monoene fats in the lipid of midwater fishes: relationship between the lipids of myctophids and those of their prey. *Marine Ecology Progress Series* **168**, 21–23.

Santos, M.B., Pierce, G.J., Lopez, A., Martinez, M.T., Fernandez, M.T., Ieno, E., Mente, E., Porteiro, P., Carrera, P. & Meixide, M. 2004. Variability in the diet of common dolphins (*Delphinus delphis*) in Galician waters 1991–2003 and relationships with prey abundance. ICES Conference and Meeting (CM) Document 2004/Q:09. Copenhagen, Denmark: International Council for the Exploration of the Sea.

Scheidat, M., Tougaard, J., Brasseur, S., Carstensen, J., van Polanen Petel, T., Teilmann, J. & Reijnders, P. 2011. Harbour porpoises (*Phocoena phocoena*) and wind farms: a case study in the Dutch North Sea. *Environmental Research Letters* **6**, 025102, doi:10.1088/1748-9326/6/2/025102.

Schlundt, C.E., Finneran, J.J., Carder, D.A. & Ridgway, S.H. 2000. Temporary shift in masked hearing thresholds of bottlenose dolphins, *Tursiops truncatus*, and white whales, *Delphinapterus leucas*, after exposure to intense tones. *Journal of the Acoustical Society of America* **107**, 3496–3508.

Scott, M., Chivers, S., Olson, R., Fiedler, P. & Holland, K. 2012. Pelagic predator associations: tuna and dolphins in the eastern tropical Pacific Ocean. *Marine Ecology Progress Series* **458**, 283–302.

Sea Mammal Research Unit (SMRU). 2008. Annual report of the United Kingdom to the European Commission on the implementation of Council Regulation 812/2004 on cetacean bycatch. Results of fishery observations collection during 2007. St. Andrews, UK: Sea Mammal Research Unit.

Sea Mammal Research Unit (SMRU). 2009. Annual report of the United Kingdom to the European Commission on the implementation of Council Regulation 812/2004 on cetacean bycatch. Results of fishery observations collection during 2008. St. Andrews, UK: Sea Mammal Research Unit.

Sheldrick, M.C. 1976. Trends in the strandings of Cetacea on the British coasts 1913–72. *Mammal Review* **6**, 15–23.

Silva, M.A. 1999. Diet of common dolphins, *Delphinus delphis*, off the Portuguese continental coast. *Journal of the Marine Biological Association of the United Kingdom* **79**, 531–540.

Silva, M.A., Prieto, R., Magalhaes, R., Cabecinhas, A., Cruz, A., Goncalves, J.M. & Santos, R.S. 2003. Occurrence and distribution of cetaceans in the waters around the Azores (Portugal), Summer and Autumn 1999–2000. *Aquatic Mammals* **29.1**, 77–83.

Silva, M.A. & Sequeira, M. 2003. Patterns in the mortality of common dolphins (*Delphinus delphis*) on the Portuguese coast, using strandings records 1975–1998. *Aquatic Mammals* **29.1**, 88–98.

Simmonds, M.P. & Lopez-Jurado, L.F. 1991. Whales and the military. *Nature* **351**, 448.

Small Cetaceans in the European Atlantic and North Sea (SCANS-II). 2008. Final report to the European Commission under contract LIFE04NAT/GB/000245. St. Andrews, UK: Sea Mammal Research Unit.

Southall, B.L., Braun, R., Gulland, F.M.D., Heard, A.D., Baird, R.W., Wilkin, S.M. & Rowles, T.K. 2006. Hawaiian melon-headed whale (*Peponocephala electra*) mass stranding event of 3–4 July 2004. NOAA Technical Memorandum NMFS-ORP-31. Silver Spring, Maryland: National Oceanic and Atmospheric Administration National Marine Fisheries Service, Office of Protected Resources.

Southward, A.J. 1963. The distribution of some plankton animals in the English Channel and Western Approaches. III. Theories about long term biological changes, including fish. *Journal of the Marine Biological Association of the United Kingdom* **43**, 1–29.

Southward, A.J., Langmead, O., Hardman-Mountford, N.J., Aiken, J., Boalch, G.T., Dando, P.R., Genner, M.J., Joint, I., Kendall, M.A. & Halliday, N.C. 2005. Long-term oceanographic and ecological research in the western English Channel. *Advances in Marine Biology* **47**, 1–105.

Sparholt, H., Bertelsen, M. & Lassen, H. 2007. A meta-analysis of the status of ICES fish stocks during the past half century. *ICES Journal of Marine Science* **64**, 707–713.

Spitz, J., Mourocq, E., Leauté, J.P., Quéro, J.C. & Ridoux, V. 2010. Prey selection by the common dolphin: fulfilling high energy requirements with high quality food. *Journal of Experimental Marine Biology and Ecology* **390**, 73–77.

Spitz, J., Richard, E., Meynier, L., Pusineri, C. & Ridoux, V. 2006. Dietary plasticity of the oceanic striped dolphin, *Stenella coeruleoalba*, in the neritic waters of the Bay of Biscay. *Journal of Sea Research* **55**, 309–320.

Spyrakos, E., Santos-Diniz, T.C., Martinez-Iglesias, G., Torres-Palenzuela, J.M. & Pierce, G.J. 2011. Spatiotemporal patterns of marine mammal distribution in coastal waters of Galicia, NW Spain. *Hydrobiologia* **670**, 87–109.

Steckenreuter, A., Möller, L. & Harcourt, R. 2012. How does Australia's largest dolphin-watching industry affect the behaviour of a small and resident population of Indo-Pacific bottlenose dolphins? *Journal of Environmental Management* **97**, 14–21.

Stenseth, N.C., Mysterud, A., Ottersen, G., Hurrell, J.W., Chan, H.M. & Lima, M. 2002. Ecological effects of climate fluctuations. *Science* **297**, 1292–1296.

Stockin, K. & Orams, M.B. 2009. The status of common dolphins (*Delphinus delphis*) within New Zealand waters. IWC Scientific Committee Document SC/61/SM20. Cambridge, UK: International Whaling Commission.

Stockin, K.A., Lusseau, D., Binedell, V., Wiseman, N. & Orams, M.B. 2008. Tourism affects the behavioural budget of the common dolphin *Delphinus* sp. in the Hauraki Gulf, New Zealand. *Marine Ecology Progress Series* **355**, 287–295.

Stone, C.J. & Tasker, M.L. 2006. The effects of seismic airguns on cetaceans in U.K. waters. *Journal of Cetacean Research and Management* **8**, 255–263.

Sundarama, B., Pojea, A.C., Veitb, R.R. & Nganguia, H. 2006. Acoustical dead zones and the spatial aggregation of whale strandings. *Journal of Theoretical Biology* **238**, 764–770.

Svane, I. 2005. Occurrence of dolphins and seabirds and their consumption of by-catch during prawn trawling in Spencer Gulf, South Australia. *Fisheries Research* **76**, 317–327.

Tasker, M.L. 2006. Marine management: can objectives be set for marine top predators? In *Top Predators in Marine Ecosystems*, I.L. Boyd et al. (eds). Cambridge, UK: Cambridge University Press, 361–369.

Tasker, M.L., Amundin, M., Andre, M., Hawkins, A., Lang, W., Merck, T., Scholik-Schlomer, A., Teilmann, J., Thomsen, F., Werner, S. & Zakharia, M. 2010. Marine Strategy Framework Directive Task Group 11 Report: underwater noise and other forms of energy. European Commission Joint Research Centre/ICES Report EUR 24341 EN-2010. Online. http://www.ices.dk/projects/MSFD/TG11final.pdf (accessed 11 December 2012).

Taylor, B.L. & DeMaster, D.P. 1993. Implications of non-linear density dependence. *Marine Mammal Science* **9**, 360–371.

Taylor, B.L. & Gerrodette, T. 1993. The uses of statistical power in conservation biology: the vaquita and the northern spotted owl. *Conservation Biology* **7**, 489–500.

Thompson, P.M., Lusseau, D., Barton, T., Simmons, D., Rusin, J. & Bailey, H. 2010. Assessing the responses of coastal cetaceans to the construction of offshore wind turbines. *Marine Pollution Bulletin* **60**, 1200–1208.

Todd, S., Stevick, P., Lien, J., Marques, F. & Ketten, D. 1996. Behavioural effects of exposure to underwater explosions in humpback whales (*Megaptera novaeanlgiae*). *Canadian Journal of Zoology* **74**, 1661–1672.

Toft, G., Hagmar, L., Giwercman, A. & Bonde, J.P. 2004. Epidemiological evidence on reproductive effects of persistent organochlorines in humans. *Reproductive Toxicology* **19**, 5–26.

Tornero, V., Borrell, A., Aguilar, A., Forcada, J. & Lockyer, C. 2006. Organochlorine contaminant and retinoid levels in blubber of common dolphins (*Delphinus delphis*) off northwestern Spain. *Environmental Pollution* **140**, 312–321.

Tougaard, J., Carstensen, J., Teilmann, J., Skov, H. & Rasmussen, P. 2009. Pile driving zone of responsiveness extends beyond 20 km for harbor porpoises (*Phocoena phocoena* (L.)). *The Journal of the Acoustical Society of America* **126**, 11–14.

Tougaard, J., Carstensen, J., Wisz, M.S., Teilmann, J., Bech, N.I. & Skov, H. 2006. Harbour porpoises on Horns Reef in relation to construction and operation of Horns Rev Offshore Wind Farm. Technical report to Elsam Engineering A/S. Roskilde, Denmark: National Environmental Research Institute.

Tregenza, N.J.C., Berrow, S.D., Hammond, P.S. & Leaper, R. 1997. Common dolphin, *Delphinus delphis* L., bycatch in bottom set gillnets in the Celtic Sea. SC/48/SM48. *Reports of the International Whaling Commission* **47**, 835–839.

Tregenza, N.J.C. & Collet, A. 1998. Common dolphin *Delphinus delphis* bycatch in pelagic trawl and other fisheries in the northeast Atlantic. *Reports of the International Whaling Commission* **48**, 453–459.

Tyack, P. 2009. Acoustic playback experiments to study behavioral responses of free-ranging marine animals to anthropogenic sound. *Marine Ecology Progress Series* **395**, 187–200.

Tyack, P.L., Zimmer, W.M.X., Moretti, D., Southall, B.L., Claridge, D.E., Durban, J.W., Clark, C.W., D'Amico, A., DiMarzio, N., Jarvis, S., McCarthy, E., Morrissey, R., Ward, J. & Boyd, I.L. 2011. Beaked whales respond to simulated and actual navy sonar. *PLoS one* **6**, e17009.

United Nations. 2001. Convention on the Law of the Sea of 10 December 1982. New York: Division for Ocean Affairs and the Law of the Sea, Office of Legal Affairs, United Nations. Online. http://www.un.org/Depts/los/convention_agreements/texts/unclos/unclos_e.pdf (accessed 16 July 2012).

U.S. Environmental Protection Agency (EPA). 1997. Special report on environmental endocrine disruption: an effects assessment and analysis. 630/R-96/012. Washington, DC: U.S. Environmental Protection Agency, Risk Assessment Forum.

Valentine, P.S., Birtles, A., Curnock, M., Arnold, P. & Dunstan, A. 2004. Getting closer to whales—passenger expectations and experiences, and the management of swim with dwarf minke whale interactions in the Great Barrier Reef. *Tourism Management* **25**, 647–655.

Van Canneyt, O., Dabin, W., Demaret, F., Doremus, G., Dussus, C. & Gonzalez, L. 2012. Les échouages de mammifères marins sur le littoral français en 2011. Rapport CRMM pour le Ministère de l'Ecologie, du Développement Durable des Transports et du Logement, Direction de l'eau et de la biodiversité, Programme Observatoire du Patrimoine Naturel. La Rochelle, France: Centre de Recherche sur les Mammifères Marins (Centre for Research on Marine Mammals), Université de La Rochelle.

Van Canneyt, O., Dabin, W., Demaret, F., Doremus, G. & Gonzalez, L. 2011. Les échouages de mammifères marins sur le littoral français en 2010. Rapport CRMM pour le Ministère de l'Ecologie, du Développement Durable des Transports et du Logement, Direction de l'eau et de la biodiversité, Programme Observatoire du Patrimoine Naturel. La Rochelle, France: Centre de Recherche sur les Mammifères Marins (Centre for Research on Marine Mammals), Université de La Rochelle.

Van Waerebeek, K. 1997. Long-beaked and short-beaked common dolphins sympatric off central-west Africa. IWC Scientific Committee Document SC/49/SM46. Cambridge, UK: International Whaling Commission.

Vieira, N., Carvalho, I. & Brito, C. 2009. Occurrence and relative abundance of common dolphins in three sites of the Portuguese shore. 61st Annual Meeting of the International Whaling Commission, Funchal, Madeira. IWC Scientific Committee Document SC/61/SM16. Cambridge, UK: International Whaling Commission.

Viricel, A. 2006. *Spatial and social structure of the common dolphin* Delphinus delphis *in the Northeast Atlantic inferred from genetic data.* MSc thesis, Graduate School of the College of Charleston, South Carolina.

Viricel, A., Strand, A.E., Rosel, P.E., Ridoux, V. & Garcia, P. 2008. Insights on common dolphin (*Delphinus delphis*) social organization from genetic analysis of a mass-stranded pod. *Behavioral Ecology and Sociobiology* **63**, 173–185.

Wade, P.R. 1998. Calculating limits to the allowable human-caused mortality of cetaceans and pinnipeds. *Marine Mammal Science* **14**, 1–37.

Wade, P.R. & Angliss, R.P. 1997. Guidelines for assessing marine mammal stocks: report of the GAMMS Workshop April 3–5, 1996, Seattle, Washington. NOAA Technical Memorandum NMFS-OPR-12. Silver Spring, Maryland: National Oceanic and Atmospheric Administration National Marine Fisheries Service, Office of Protected Resources. Online. http://www.nmfs.noaa.gov/pr/pdfs/sars/gamms_report.pdf (accessed 19 April 2013).

Wall, D., Channon, N., Williams, D., Enlander, I., Ryan, C. & Wilson, C. 2011. Seasonal distribution of harbour porpoise, common dolphin and minke whale in the Irish Sea as determined from long-term monitoring using commercial ferries. In *Abstract Book, 25th Conference of the European Cetacean Society. Long Term Datasets on Marine Mammals: Learning from the Past to Manage the Future. 21–23 March 2011. Cadiz, Spain.* P. Gauffier & P. Verborgh (eds). Cambridge, UK: European Cetacean Society, 231 only.

Waples, R.S. & Gaggiotti, O.E. 2006. What is a population? An empirical evaluation of some genetic methods for identifying the number of gene pools and their degree of connectivity. *Molecular Ecology* **15**, 1519–1439.

Waring, G.T., Josephson, E., Maze-Foley, K. & Rosel, P.E. (eds) 2012. Short-beaked common dolphin (*Delphinus delphis delphis*): Western North Atlantic Stock. In *U.S. Atlantic and Gulf of Mexico Marine Mammal Stock Assessments—2011.* NOAA Technical Memorandum NMFS-NE-221. Woods Hole,

Massachusetts: National Oceanic and Atmospheric Administration National Marine Fisheries Service, Northeast Fisheries Science Center, 91–98. Online. http://www.nefsc.noaa.gov/publications/tm/tm221/ (accessed 19 April 2013).

Watkins, W.A., Daher, M.A., Fristrup, K.M., Howald, T.J. & Notarbartolo Di Sciara, G. 1993. Sperm whales tagged with transponders and tracked underwater by sonar. *Marine Mammal Science* **9**, 55–67.

Weilgart, L.S. 2007a. The impacts of anthropogenic ocean noise on cetaceans and implications for management. *Canadian Journal of Zoology* **85**, 1091–1116.

Weilgart, L.S. 2007b. A brief review of known effects of noise on marine mammals. *International Journal of Comparative Psychology* **20**, 159–168.

Weir, C.R. 2008. Overt responses of humpback whales (*Megaptera novaeangliae*), sperm whales (*Physeter macrocephalus*), and Atlantic spotted dolphins (*Stenella frontalis*) to seismic exploration off Angola. *Aquatic Mammals* **34**, 71–83.

Westgate, A.J. 2007. Geographic variation in cranial morphology of short-beaked common dolphins (*Delphinus delphis*) from the North Atlantic. *Journal of Mammalogy* **88**, 678–688.

Williams, R., Bain, D.E., Ford, J.K.B. & Trites, A.W. 2002b. Behavioural responses of male killer whales to a 'leapfrogging' vessel. *Journal of Cetacean Research and Management* **4**, 305–310.

Williams, R., Trites, A.W. & Bain, D.E. 2002a. Behavioural responses of killer whales (*Orcinus orca*) to whale-watching boats: opportunistic observations and experimental approaches. *Journal Zoology* **256**, 255–270.

Winship, A.J., Murphy, S., Deaville, R., Jepson, P.D., Rogan, E. & Hammond, P.S. 2009. Preliminary assessment and bycatch limits for northeast Atlantic common dolphins. IWC Scientific Committee Document SC/61/SM19. Cambridge, UK: International Whaling Commission.

Woollings, T., Hannachi, A., Hoskins, B. & Turner, A. 2010. A regime view of the North Atlantic Oscillation and its response to anthropogenic forcing. *Journal of Climate* **23**, 1291–1307.

Würsig, B. & Greene, C.R. 2002. Underwater sounds near a fuel receiving facility in western Hong Kong: relevance to dolphins. *Marine Environmental Research* **54**, 129–145.

Würsig, B. & Richardson, W.J. 2009. Noise, effects of. In *Encyclopedia of Marine Mammals*. F.P. William et al. (eds). London: Academic Press, 2nd edition, 765–773.

Young, D.D. & Cockcroft, V.G. 1994. Diet of common dolphin (*Delphinus delphis*) off the south east coast of southern Africa: opportunism or specialization? *Journal of Zoology* **234**, 41–53.

Zagzebski, K.A., Gulland, F.M.D., Haulena, M., Lander, M.E., Greig, D.J., Gage, L.J., Hanson, M.B., Yochem, P.K. & Stewart, B.S. 2006. Twenty-five years of rehabilitation of odontocetes stranded in central and northern California, 1977 to 2002. *Aquatic Mammals* **32**, 334–345.

Zhou, J.L., Salvador, S.M., Liu, Y.P. & Sequeira, M. 2001. Heavy metals in the tissues of common dolphins (*Delphinus delphis*) stranded on the Portuguese coast. *Science of the Total Environment* **273**, 61–76.

Zirbel, K., Balint, P. & Parsons, E.C.M. 2011. Public awareness and attitudes towards naval sonar mitigation for cetacean conservation: a preliminary case study in Fairfax County, Virginia (the DC Metro area). *Marine Pollution Bulletin* **63**, 49–55.

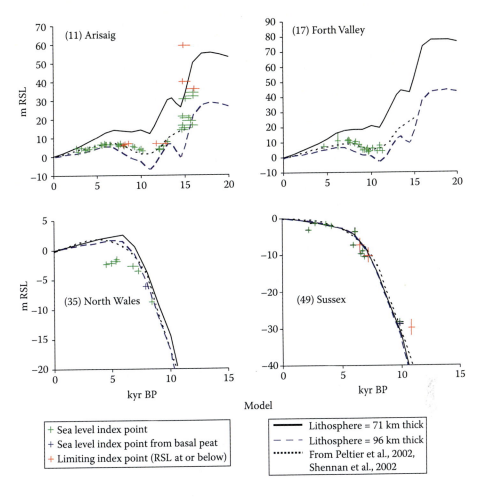

Colour Figure 7 (Scourse)

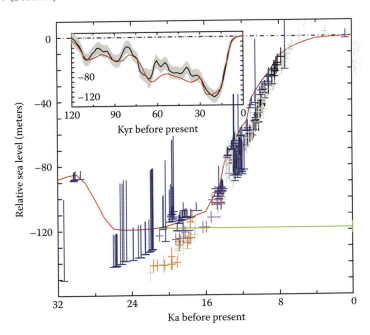

Colour Figure 11 (Scourse)

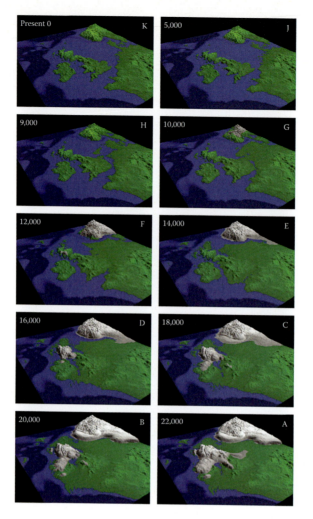

Colour Figure 12 (Scourse)

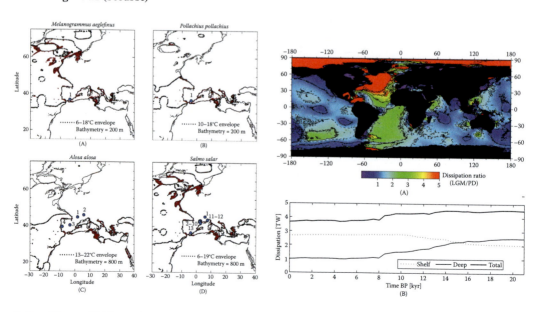

Colour Figure 18 (Scourse)

Colour Figure 19 (Scourse)

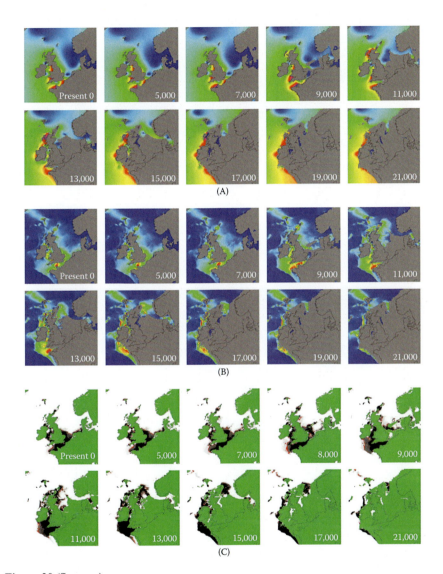

(A)

(B)

(C)

Colour Figure 20 (Scourse)

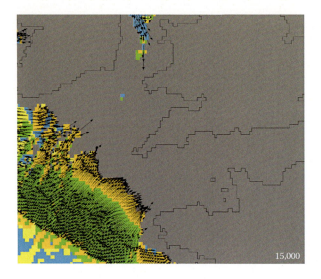

Colour Figure 21 (Scourse)

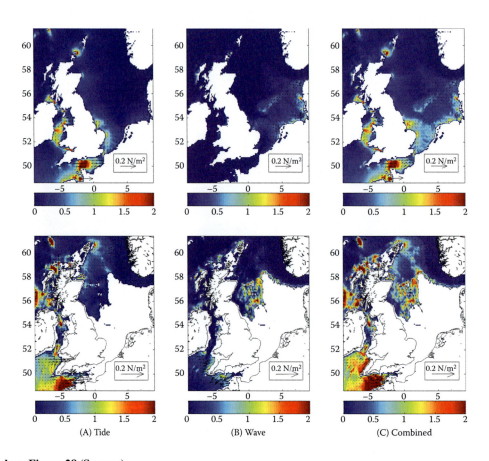

(A) Tide

(B) Wave

(C) Combined

Colour Figure 28 (Scourse)

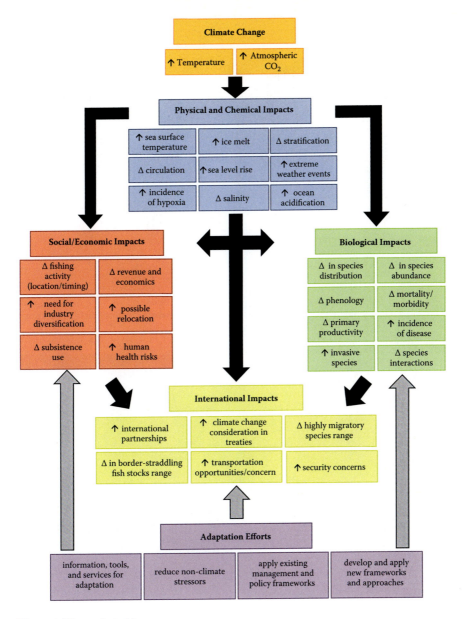

Colour Figure 1 (Howard et al.)

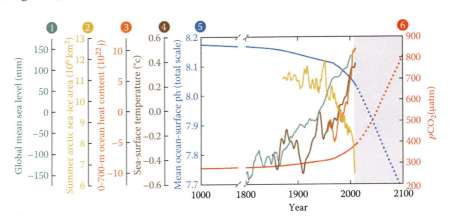

Colour Figure 2 (Howard et al.)

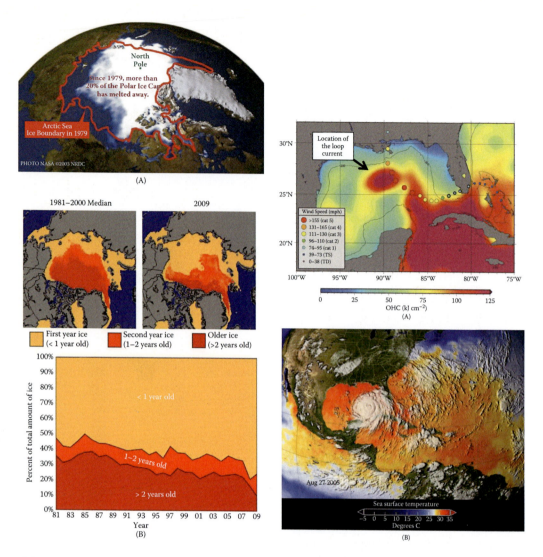

Colour Figure 4 (Howard et al.)

Colour Figure 5 (Howard et al.)

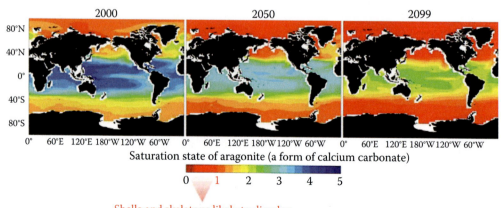

Saturation state of aragonite (a form of calcium carbonate)

Shells and skeletons likely to dissolve

Colour Figure 6 (Howard et al.)

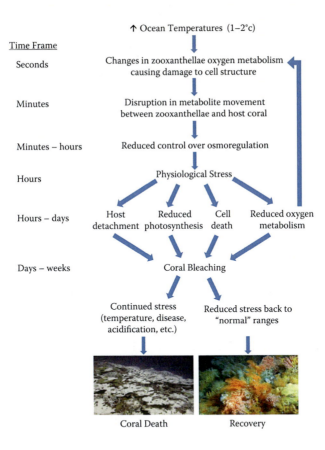

Colour Figure 7 (Howard et al.)

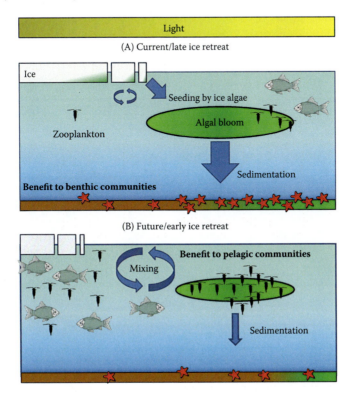

Colour Figure 8 (Howard et al.)

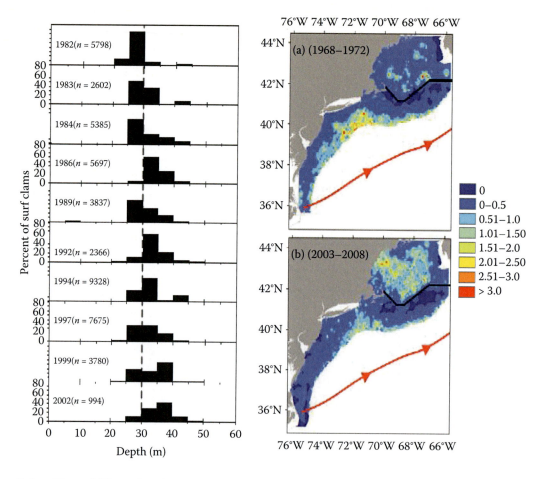

Colour Figure 9 (Howard et al.)

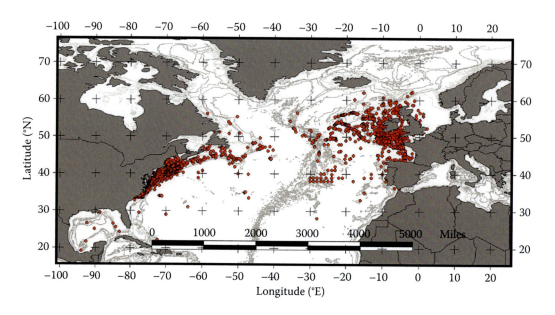

Colour Figure 1 (Murphy, Pinn & Jepson)

Short-Beaked Common Dolphin *(Delphinus delphis)*

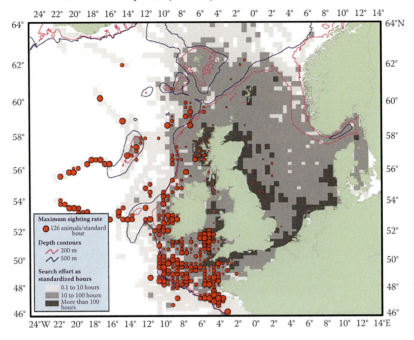

Colour Figure 2 (Murphy, Pinn & Jepson)

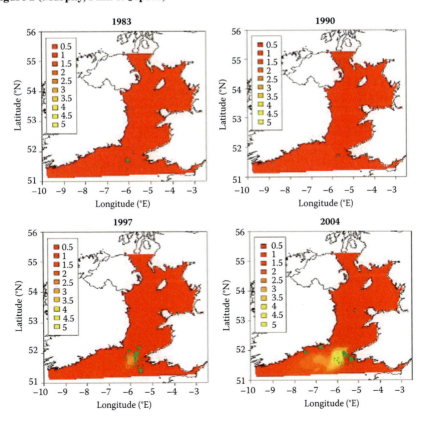

Colour Figure 3 (Murphy, Pinn & Jepson)

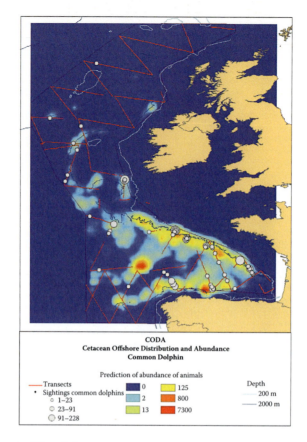

Colour Figure 6 (Murphy, Pinn & Jepson)

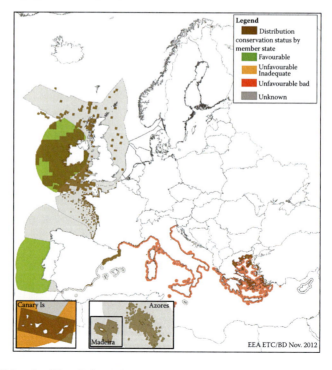

Colour Figure 10 (Murphy, Pinn & Jepson)

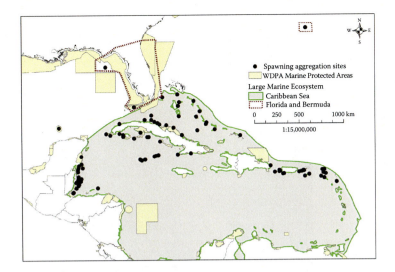

Colour Figure 1 (Kobara, Heyman, Pittman & Nemeth)

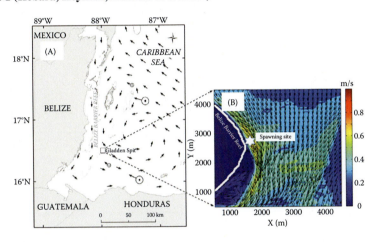

Colour Figure 5 (Kobara, Heyman, Pittman & Nemeth)

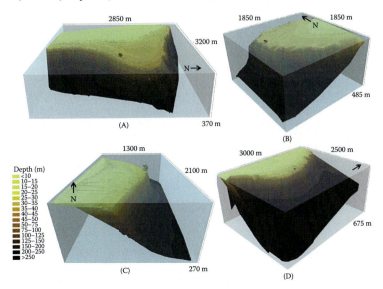

Colour Figure 7 (Kobara, Heyman, Pittman & Nemeth)

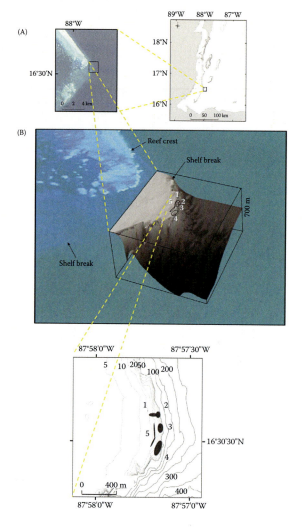

Colour Figure 8 (Kobara, Heyman, Pittman & Nemeth)

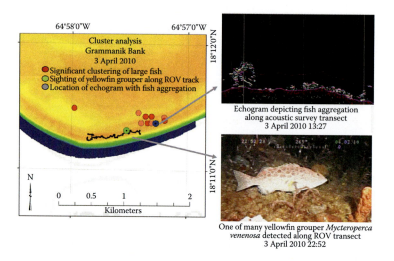

Echogram depicting fish aggregation
along acoustic survey transect
3 April 2010 13:27

One of many yellowfin grouper *Mycteroperca
venenosa* detected along ROV transect
3 April 2010 22:52

Colour Figure 9 (Kobara, Heyman, Pittman & Nemeth)

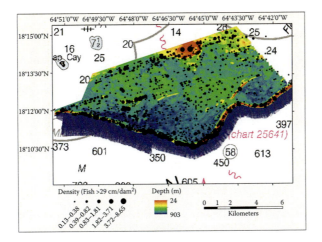

Colour Figure 10 (Kobara, Heyman, Pittman & Nemeth)

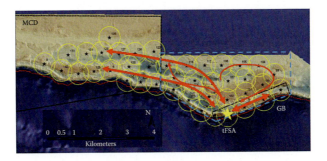

Colour Figure 11 (Kobara, Heyman, Pittman & Nemeth)

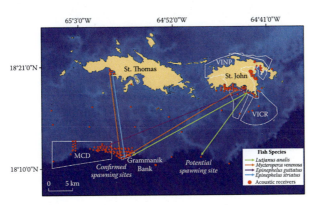

Colour Figure 12 (Kobara, Heyman, Pittman & Nemeth)

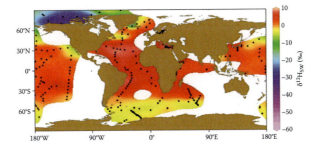

Colour Figure 1 (McMahon, Hamady & Thorrold)

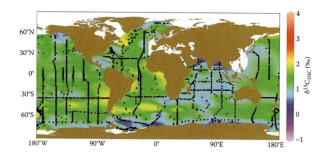

Colour Figure 2 (McMahon, Hamady & Thorrold)

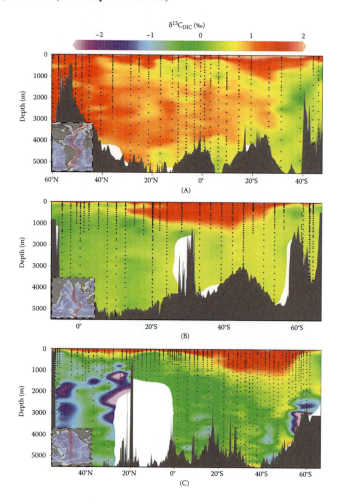

Colour Figure 3 (McMahon, Hamady & Thorrold)

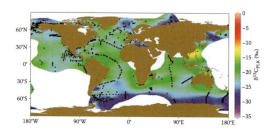

Colour Figure 4 (McMahon, Hamady & Thorrold)

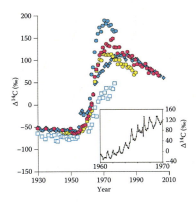

Colour Figure 5 (McMahon, Hamady & Thorrold)

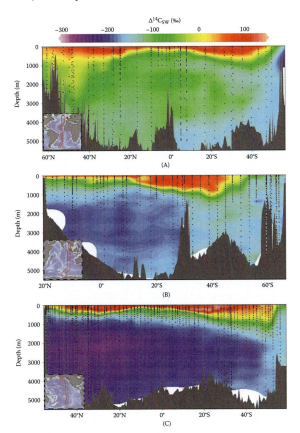

Colour Figure 6 (McMahon, Hamady & Thorrold)

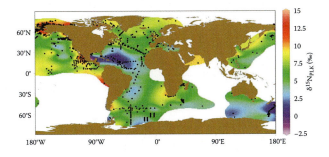

Colour Figure 7 (McMahon, Hamady & Thorrold)

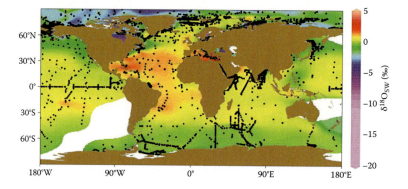

Colour Figure 8 (McMahon, Hamady & Thorrold)

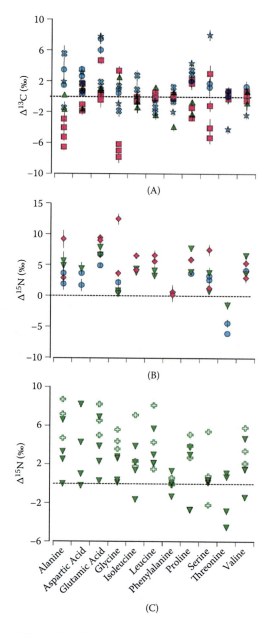

Colour Figure 9 (McMahon, Hamady & Thorrold)

Oceanography and Marine Biology: An Annual Review, 2013, **51**, 281-326
© Roger N. Hughes, David Hughes, and I. Philip Smith, Editors
Taylor & Francis

BIOGEOGRAPHY OF TRANSIENT REEF-FISH SPAWNING AGGREGATIONS IN THE CARIBBEAN: A SYNTHESIS FOR FUTURE RESEARCH AND MANAGEMENT

SHINICHI KOBARA[1], WILLIAM D. HEYMAN[2],
SIMON J. PITTMAN[3,5] & RICHARD S. NEMETH[4]

[1]*Department of Oceanography, Texas A&M University, College Station, TX 77843-3146, USA*
E-mail: shinichi@tamu.edu (corresponding author)
[2]*Department of Geography, Texas A&M University, College Station, TX 77843-3147, USA*
[3]*Biogeography Branch, Center for Coastal Monitoring and Assessment,*
National Oceanic and Atmospheric Administration,
1305 East West Highway, Silver Spring, MD 20910, USA
[4]*Center for Marine and Environmental Studies, University of the Virgin Islands,*
2 John Brewer's Bay, St. Thomas, VI 00802, USA
[5]*The Marine Institute, Marine Building, Plymouth University,*
Drake Circus, Plymouth, PL4 8AA, United Kingdom

Transient fish spawning aggregations (tFSAs) are critical life-history phenomena where fish migrate to specific locations at predictable times of year to reproduce en masse. In the Wider Caribbean region, 37 species of fish from 10 families are now known to form tFSAs. Although tFSAs likely occur at times and in places that maximize reproductive success, little is known about the complex suite of interacting environmental patterns and ecological processes that dictate the timing and locations of tFSAs. This review synthesizes the latest advances in the study of tFSAs in the Wider Caribbean to (1) illustrate the current state of knowledge; (2) highlight gaps in our understanding of the geography and ecology of aggregation sites; and (3) suggest future research needs and conservation strategies. We have compiled multidisciplinary data on 108 tFSAs across 14 states and territories in the Wider Caribbean and reviewed the full range of approaches and technologies applied to study tFSAs. Existing research and associated hypotheses are grouped and examined by data type. We propose a multitier research framework that provides an incremental approach to information gathering at individual sites and suites of sites. We advocate applying the framework to facilitate consistent and coordinated data collection and monitoring across a Wider Caribbean network of tFSAs.

Introduction

Spatial and temporal information on the reproductive ecology of marine species is vital to the development of effective strategies for marine resource management and biodiversity conservation (Vincent & Sadovy 1998, Pittman & McAlpine 2003). An estimated 164 species of fish globally have evolved a reproductive strategy that involves the mass aggregation (hundreds to thousands) of sexually mature males and females at specific geographical locations to spawn (Claydon 2004). Many fishes that are associated with coral reefs can travel relatively long distances over days and weeks to an aggregation site during a very specific portion of the day, month or year and then return to their routine home range area after spawning (Nemeth 2009). This life-history strategy creates a temporary or transient, yet critical, reproductive area in time and space (Johannes 1978, Domeier & Colin 1997) referred to here as a transient fish spawning aggregation (tFSA). Many of these species are key carnivores in tropical coral reef ecosystems and form a commercially and socially important component of reef fisheries (Hamilton et al. 2012, Sadovy de Mitcheson et al. 2012).

Species that form spawning aggregations can be grouped into two broad categories: (1) those that form transient aggregations (tFSA) and are typically large-bodied, slow-maturing, commercially valuable species, including several members of the snapper (Lutjanidae) and grouper (Serranidae) families, which can migrate tens to hundreds of kilometres to visit aggregation sites (Colin 1992, Bolden 2000); and (2) those that form resident spawning aggregations within or close to (<2 km) their routine home range. Fish families known to form resident aggregations include ecologically important species like some parrotfish (Scaridae), surgeonfish (Acanthuridae), goatfish (Mullidae), and wrasses (Labridae) among others (Sadovy de Mitcheson et al. 2008, Nemeth 2009). Resident spawners also tend to spawn more frequently than transient spawners, sometimes on a monthly, weekly or even daily basis (e.g., bluehead wrasse, *Thalassoma bifasciatum*; yellowtail parrotfish, *Sparisoma rubripinne*) (Randall & Randall 1963, Colin & Clavijo 1988, Warner 1995). The distinction between transient and resident spawning life-history strategies is based on not only interspecies similarities and differences in the frequency and longevity of aggregations and the relative distances travelled to the aggregation (Domeier & Colin 1997, Domeier 2012), but also how much reproductive effort is expended during a single aggregation event. Resident aggregating species invest only about 1–8% of their reproductive effort in a single spawning aggregation (Nemeth 2009). By contrast, transient aggregating species, which aggregate only a few times a year, invest from 33% to 100% of their reproductive effort per spawning episode (Domeier & Colin 1997, Nemeth 2009).

The tFSA sites are most often discovered first by fishers (Johannes 1978, Sadovy de Mitcheson et al., 2008). When fishers encounter these highly concentrated (in space and time) groups of large fish, they typically obtain high catch per unit of fishing effort. Most of the fish brought aboard fishing vessels have fully developed gonads and are often visibly reproductive, with eggs or sperm flowing from their vents. During these periods, local fish markets are flooded with large adult fish and roe.

Direct scientific evidence of a spawning site is gained through observations of spawning and the presence of hydrated oocytes within the gonads of females collected from the site (Heyman et al. 2004). Since tFSA species typically spawn in low ambient light at around sunset, scientists also consider indirect evidence of tFSA occurrence, including conspicuous and characteristic courtship behaviours, such as changes in colour (Archer et al. 2012, Heppell et al., 2012), nudging, shaking, or pair chasing; vocalization (Schärer et al. 2012a,b); biometric features, such as swollen abdomens; and an increase in the gonadosomatic index (Claro & Lindeman 2003, Heyman et al. 2004a,b); as well as elevated abundance, whereby the number of individuals at the site is at least three times greater than expected under non-spawning conditions (Domeier & Colin 1997). The indicative abundance has recently been redefined as a 4-fold increase at the site compared with other times of year (Domeier 2012).

The number of aggregating species and their co-occurrence in space and time vary among sites, but the majority of aggregation sites seem to have multispecies spawning aggregations. For instance, as many as 10 fish species are known to spawn at one particular spawning aggregation in Puerto Rico (Ojeda-Serrano et al. 2007), and as many as 17 fish species have been reported to spawn at different times of the year at a single site in Belize (Heyman & Kjerfve 2008). This suggests that some generality may exist across species in the physical suitability of a site for spawning. An understanding of exactly which physical characteristics make one area more suitable than another is just beginning to emerge, and the topic is reviewed here.

Aggregating behaviour, however, makes fish highly vulnerable to extraction by fisheries (Sadovy & Domeier 2005). The impact of fishing mortality on a local population is more acute for late-maturing species, and this is of particular concern to fisheries management when insufficient information exists about the species' population biology and tFSA site characteristics.

Unregulated fishing at tFSA sites is thought to have led to local extirpations and even the collapse of some fisheries (e.g., Nassau grouper, *Epinephelus striatus*) (Sadovy et al. 1994, Sadovy & Domeier 2005, Sadovy de Mitcheson et al. 2012). Nassau grouper was once the most valuable grouper species in the Caribbean fishery, but it is estimated that approximately 35% of all known Nassau grouper spawning aggregations have been extirpated, and population abundance and fishery landings of the species remain low (Claro et al. 2001), resulting in an 'endangered species' designation (International Union for Conservation of Nature and Natural Resources [IUCN] 2011). In Belize, catches from Nassau grouper spawning aggregation sites have declined by as much as 81% (Sala et al. 2001). Marked declines have also been reported for other aggregating species. For example, in Belize catch per unit effort from a mutton snapper (*Lutjanus analis*) spawning aggregation declined by nearly 60% (Graham et al. 2008). In Puerto Rico, declines in fish catch were reported by fishers at approximately 25% of all known spawning aggregation sites (Ojeda-Serrano et al. 2007). It appears that, in many areas, fishing spawning aggregations is an unsustainable practice that can, if unregulated, lead to the eventual loss of the resources (Sadovy & Domeier 2005). Conversely, enhanced protection of spawning aggregations can be a viable strategy for restoring the trophic structure of coral reef ecosystems and rebuilding fisheries and associated livelihoods. Nemeth (2005) reported a 60% increase in red hind (*Epinephelus guttatus*) density and biomass at a spawning site closed to fishing and increased catch rates outside the protected area just a few years after fishing was prohibited.

In the Caribbean region, one of the critical science gaps impeding the development of effective ecosystem-based fisheries management (EBFM) strategies is the lack of information on the specific geographical locations of spawning aggregations (Sale et al. 2005, Appeldoorn 2008, Crowder & Norse 2008) and on the environmental factors that may make some sites optimal for spawning. The identification and characterization of tFSAs can provide useful information to guide the strategic placement of marine reserves, seasonal closures for species or groups, redistribution of fishing effort, parameterization of connectivity models and community-based management initiatives.

The tFSAs occur at places and times that exhibit a complex suite of interacting environmental patterns and processes that are thought to maximize reproductive success. Yet, little is known about the key physical and biological factors and the interlinked ecological mechanisms that promote survival of spawning fish populations. The most actively studied explanatory factors are the geomorphological features of a site, seafloor habitat types and oceanography (Shcherbina et al. 2008, Kobara & Heyman 2010); yet, the interactions among these physical patterns and processes and the importance of coupling between benthic and pelagic realms remains largely unknown. In the Caribbean region and elsewhere, tFSAs typically occur at distinctive bathymetric features (slopes, promontories, channels, outer reef edges) in close proximity to deeper water (Johannes 1978, Heyman et al. 2007, Kobara & Heyman 2008, 2010, Wright & Heyman 2008). For example, geomorphological studies of Nassau grouper spawning aggregations in the Cayman Islands found that all five known sites were located within 1 km of the tips of reef promontories in water depths of 25 to 45 m and less than 50 m from the horizontal 'shelf edge' identified from bathymetric data (Kobara & Heyman 2008). We use the term 'shelf edge' (also 'shelf break') to refer to the seaward edge of a bank or reef, where there is a steep or vertical downward slope in the seabed.

Although existing studies suggest that geomorphological characteristics of the seabed may determine site suitability more than any other single variable (Harris & Baker 2011), studies are also emerging that suggest that other ecological patterns and processes, such as hydrodynamics, seawater temperature and proximity to suitable benthic habitats for settlement, could also be important

(Nemeth 2009). For example, several recent studies indicate that the structure and movement of the water mass, including the current speed and the prevailing direction of flow, are important in the effective dispersal of eggs and larvae from a spawning site (Heyman & Kjerfve 2008, Shcherbina et al. 2008), but may also provide a mechanism for retaining eggs and larvae near natal habitats (Nemeth et al. 2008, Cherubin et al. 2011, Ezer et al. 2011). Thus, conditions that result in optimal oceanographic connectivity between spawning sites and nursery areas could form an important evolutionary driver in the site selection process and consequently determine the regional distribution of the species, as has been postulated for other species of fish (Symonds & Rogers 1995).

Several theories have been proposed to explain the location and timing of spawning, including the hypothesis that spawning coincides with conditions that enhance the entrainment of larvae by ocean currents, which in turn increase their chances of finding food in patchy environments, avoiding predators, and finding suitable habitat for settlement (Johannes 1978, Lobel 1978, Barlow 1981). Others have proposed that tFSA sites might be correlated with optimal environmental parameters that confer some genetic or developmental advantages or that sites may be selected primarily to increase the likelihood that males and females will find one another when ready to reproduce (Zaiser & Moyer 1981, Shapiro et al. 1988).

Owing to the geographical remoteness of many tFSAs, their ephemeral nature and the fact that studies are generally disparate and uncoordinated, many gaps exist in our understanding of why fish select certain areas for spawning aggregations and exactly how these areas function to optimize reproductive success. Furthermore, a better understanding of the physical and biological characteristics of tFSAs will provide the information required to develop predictive models that can locate key sites, some of which may currently be unmanaged or even undiscovered. This dearth of information will limit the effectiveness of place-based management strategies such as marine spatial planning and could result in suboptimal marine protected area (MPA) performance and network design, specifically where productive fisheries and protection of essential fish habitat are goals (Friedlander et al. 2003, Sale et al. 2005, Crowder & Norse 2008). For example, lack of geographic information on grouper spawning aggregations in Florida resulted in the establishment of an MPA with a specific goal of protecting black grouper (*Mycteroperca bonaci*), yet unintentionally excluded an important black grouper spawning aggregation later discovered less than 100 m outside the MPA boundary (Eklund et al. 2000). Further compounding the error, the MPA was bounded by the 18-m depth contour, whereas the black grouper aggregation was found at depths of 18–28 m. Greater understanding of the physical characteristics of sites that support spawning aggregations will ensure that they can be easily and accurately identified and assessed during the MPA design phase. For population monitoring, tFSAs provide the best opportunity to collect data cost effectively on the status of adult fish populations that at other times are widely dispersed across tens to hundreds of square kilometres of ocean.

The Society for the Conservation of Reef Fish Aggregations (SCRFA) maintains a global database on FSAs based on data collected through fisher interviews and literature searches to document all known aggregation sites (Sadovy de Mitcheson et al. 2008). SCRFA compiled useful information on the biological characteristics of fish species, such as seasonality, photoperiod and lunar cycles of spawning, and spawning behaviours, but due to limited physical data for most sites, has limited information on site characteristics, such as bathymetry, geomorphology, oceanography and benthic communities. To date, no research has been conducted to synthesize the environmental characteristics of tFSAs across the Wider Caribbean region, including Florida, the Bahamas and Bermuda.

The overall goals of this review are to (1) synthesize the key biological and environmental characteristics of all currently known tFSA sites in the Wider Caribbean region; (2) provide a hierarchically ordered list of the data collection techniques commonly used to study individual tFSAs and suggest a minimum level of information needed for management action; (3) suggest future research directions for improving our understanding of tFSAs and their connectivity at the regional level; and (4) provide recommendations for a coherent network of MPAs designed strategically to monitor and protect multispecies tFSA sites across the Caribbean. A mutually replenishing network of well-managed and monitored tFSAs will support recovery of depleted predatory reef fish populations and thus contribute to restoring ecological integrity, sustainable coastal fisheries and community livelihoods at the regional scale.

Methodology for compiling existing knowledge of Caribbean tFSA sites

Fish species that form tFSAs in the Wider Caribbean

This review focuses on characterizing transient spawning aggregations in the Caribbean Large Marine Ecosystem (CLME) management unit, with some information from peripheral locations, such as Florida and Bermuda (Figure 1). We refer to this combined region as the Wider Caribbean, consistent with the area definition from the Regional Seas Programme of the United Nations Environment Programme. The Wider Caribbean region includes all island states and territories in the Caribbean Sea and the Caribbean coasts of Mexico, Central America and South America, as well as waters of the Atlantic Ocean adjacent to these states and territories (28 island and continental countries). Owing to similarities in fish communities associated with coral reefs of the Caribbean Sea and the presence of active spawning aggregations, we include Bermuda in our use of the term *Wider Caribbean*.

The SCRFA database served as the starting point for the analysis (Sadovy de Mitcheson et al. 2008, SCRFA 2011) and included 45 tFSAs for the Wider Caribbean region. No information was

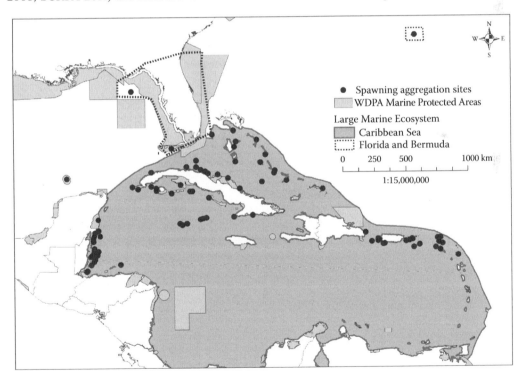

Figure 1 (See also colour figure in the insert)   Historically known spawning aggregation sites of grouper and snapper with/without direct/indirect evidence since 1884. Filled circles represent the general area of FSA sites, not the exact location. The darker shaded area is the Caribbean Large Marine Ecosystem region. In addition, the study area includes Florida and Bermuda regions, shown by thick dotted lines. The lighter shading indicates marine protected areas listed in the World Database on Protected Areas (WDPA, www.wdpa.org).

available for the south-eastern and southern Caribbean, although species distribution records from FishBase (Froese & Pauly 2010) indicated that species known to form aggregations do inhabit these areas, suggesting a major geographic gap in the regional information on tFSAs. For some records, the type of spawning aggregation (resident vs. transient) was unspecified. For example, Nassau grouper (*Epinephelus striatus*) is well known as a transient aggregation spawner but was occasionally categorized as 'unspecified'. When these entries were corrected, 143 of 151 records were classified as transient spawning aggregations. Most records in the SCRFA database were for Nassau grouper (32 of 143, or 22%), mutton snapper (*Lutjanus analis*), cubera snapper (*L. cyanopterus*), gray snapper (*L. griseus*), red hind (*Epinephelus guttatus*), black grouper (*Mycteroperca bonaci*) and yellowfin grouper (*M. venenosa*). It is important to note, however, that the SCRFA database is biased towards the commercially important species, which make up the large majority of the database records, in large part because of the heavy reliance on fisher interviews (Sadovy de Mitcheson et al. 2008).

To address the data gap in the biogeography of tFSAs, we invited specialists to a regional workshop, "Characterization and Prediction of Transient Reef Fish Spawning Aggregations in the Gulf and Caribbean Region", held in Cumaná, Venezuela, as part of the 62nd Annual Gulf and Caribbean Fisheries Institute meeting on 5 November 2009. The workshop, sponsored by the National Science Foundation's Virgin Islands Experimental Program to Stimulate Competitive Research (VI-EPSCoR), included over 50 participants from 18 countries around the Caribbean. Participants compiled data on the location of multispecies tFSAs in their respective regions, the species-specific timing of spawning and conservation status, and reported on the types of information collected through scientific investigations and local knowledge. The results revealed an additional 13 tFSAs from six countries and territories that were not included in the SCRFA database at the time of review. To broaden our information gathering on the specific characteristics of individual tFSAs, we also conducted an online survey designed to collect metadata for each tFSA. Data requested included lists of fish species that aggregate to spawn, the timing and specific location of spawning and a wide range of ancillary information generated from fishery-dependent surveys at ports and landing sites, underwater visual censuses, visual and acoustic tag and recapture studies, bathymetric and habitat mapping exercises, *in situ* and remotely sensed oceanographic studies (chemical, biological and physical) and modelling.

In addition to the SCRFA database, this review included several additional records documented in the grey literature. For example, we included Nassau grouper tFSA sites that were documented in Honduras (Fine 1990, 1992), Dominican Republic (Sadovy 1997, Sadovy et al. 2008) and Antigua and Barbuda (Munro & Blok 2003). In addition, eight sites on Alacranes Reef, off northern Yucatan Peninsula, Mexico, were identified by local fishers' traditional ecological knowledge (Aguilar-Perera et al. 2008). Ojeda-Serrano et al. (2007) identified 94 potential spawning sites in Puerto Rico from interviews with fishers, although these sites have not been verified. We include five sites that Ojeda-Serrano et al. (2007) described as known spawning aggregations. Aguilar-Perera et al. (2009) provided 28 potential Nassau grouper tFSA sites at the Quintana Roo coasts and Banco de Chinchorro, Mexico, based on fisher interviews, of which four sites have been scientifically verified as active spawning sites and are included here.

In total, 37 species of fish from 10 families were documented to aggregate and spawn in tFSAs in the Wider Caribbean (Table 1). The major fish families included groupers (Serranidae), snappers (Lutjanidae) and jacks (Carangidae).

Geographical distribution of documented tFSAs

A total of 108 geographically discrete spawning aggregations were identified from 14 states and territories throughout the Wider Caribbean (Figure 1, Table 2). Twenty-eight sites were identified in the western Atlantic Ocean, including the Bahamas and Bermuda; 47 in the northern Caribbean;

Table 1 Species that form transient spawning aggregations in the Wider Caribbean, by country or geographic area

Family	Species name	Common name	Countries
Serranidae	*Epinephelus striatus*	Nassau grouper	BM, BZ, CI, MX, STT, (BH), (CU), (DR), (HnD), (PR), (TCI)
	E. guttatus	Red hind	BM, PR, STT, STX, MX, NA (AB), (AG), (BZ), (CU)
	E. adscensionis	Rock hind	(BVI), (PR)
	E. itajara	Goliath grouper	(FL), (MX)
	E. morio	Red grouper	(FL), (CU), (MX)
	Mycteroperca bonaci	Black grouper	BM, BZ, BH, (CI), (CU), (FL), (PR), (MX)
	M. venenosa	Yellowfin grouper	BZ, PR, STT, (BH), (CI), (CU), (TCI), (MX), (FL)
	M. tigris	Tiger grouper	CI, PR, STT, TCI, (BH), (BZ), (MX), (CU)
	M. interstitialis	Yellowmouth grouper	(PR)
	M. phenax	Scamp	(FL)
	M. microlepis	Gag	(FL)
Lutjanidae	*Lutjanus analis*	Mutton snapper	BH, BZ, CU, STX, (FL), (TCI)
	L. jocu	Dog snapper	BZ, STT, (CI), (CU), (FL), (PR), (MX)
	L. synagris	Lane snapper	CU, STT, (FL)
	L. cyanopterus	Cubera snapper	BZ, STT, (CU), (FL)
	L. griseus	Gray snapper	STT, (CU), (FL), (MX)
	L. campechanus	Red snapper	(FL)
	L. apodus	Schoolmaster	(FL), (NA), (STT), (MX)
	Ocyurus chrysurus	Yellowtail snapper	BZ, (FL)
Carangidae	*Caranx ruber*	Bar jack	CI, BZ
	C. bartholomaei	Yellow jack	BZ
	C. lugubris	Black jack	CI
	C. latus	Horse-eye jack	CI, BZ, (PR)
	C. hippos	Crevalle jack	BZ
	Seriola dumerili	Greater amberjack	BZ
	Trachinotus falcatus	Permit	BZ
	Decapterus macarellus	Mackerel scad	CI
Ephippidae	*Chaetodipterus faber*	Atlantic spadefish	(BZ)
Scombridae	*Scomberomorus cavalla*	King mackerel	(BZ)
Labridae	*Lachnolaimus maximus*	Hogfish	BZ
Haemulidae	*Haemulon album*	White margate	BZ
Balistidae	*Canthidermis sufflamen*	Ocean triggerfish	BZ
	Xanthichthys ringens	Sargassum triggerfish	BZ
	Balistes vetula	Queen triggerfish	(NA), (STX), (PR)
Sparidae	*Calamus bajonado*	Jolthead porgy	BZ
Ostraciidae	*Lactophrys trigonus*	Buffalo trunkfish	BZ
	Rhinesomus triqueter	Smooth trunkfish	BZ, (PR)

Notes: Those countries for which there is direct evidence of spawning aggregation (i.e., observations of gamete release or the presence of hydrated eggs in mature females) are listed without parentheses. Abbreviated country names are given in brackets where only indirect evidence of aggregations exists for a species, for example, a 3-fold increase in abundance at the site over other times of year (Domeier & Colin 1997), courtship behaviours and colouration changes, abnormally high catch per unit effort with more than 70% mature individuals containing ripe gonads.

Antigua-Barbuda (AB), Anguilla (AG), the Bahamas (BH), Bermuda (BM), Belize (BZ), British Virgin Islands (BVI), the Cayman Islands (CI), Cuba (CU), Dominican Republic (DR), Florida (FL), Honduras (HnD), Mexico (MX), Netherlands Antilles (NA), Puerto Rico (PR), St. Thomas, VI (STT), St. Croix, VI (STX), and Turks and Caicos Islands (TCI).

Table 2 Documented transient fish spawning aggregations (tFSAs) for reef fish at 108 sites in the Wider Caribbean based on both direct and indirect evidence

Country	Total no.	Spawning sites/general area name		Reference
Bermuda	3	Challenger and Argus Banks (3 sites): western, eastern hind ground, black grouper spawning site		Sadovy 1997, Sadovy de Mitcheson et al. 2008, Luckhurst 2010
Gulf of Mexico	3	Southern Florida Keys (Riley's Hump, Madison Swanson, Steamboat Lumps) Eastern Gulf of Mexico (north-western coast of Florida)		Coleman et al. 1996, 2011, Koenig et al. 1996, 2000, Sadovy 1997, Lindeman et al. 2000
The Bahamas	23	Andros Island (5) Long Island (3) Exuma Berry Island (4) New Providence	Ragged Island Cay Sal Cat Cay/Bimini (2) Eleuthera (4) Acklins	Smith 1972, Colin 1992, Sadovy 1997, C. McKinney personal communication, November 2009
Turks and Caicos Islands	2	Phillips Reef	Northwest Point	Sadovy 1997, Sadovy de Mitcheson et al. 2008
Antigua-Barbuda	2	Green Island	Knolls between two islands	Munro & Blok 2003
Netherlands Antilles	4	Saba Bank St. Maarten	St. Eustatius St. Barthélemy Channel	Munro & Blok 2003, Kadison et al. 2009a
Anguilla	3	Seal Island	Scrub Island (2)	Munro & Blok 2003
US Virgin Islands	4	St. Thomas: Hind Bank St. Croix: Lang Bank	Grammanik Bank Southwest area	Sadovy 1997, Nemeth 2005, R.S. Nemeth et al. 2007, Sadovy de Mitcheson et al. 2008 Beets & Friedlander 1999
Puerto Rico	7	El Seco, Vieques Island		Sadovy et al. 1994, White et al. 2002, Matos-Caraballo et al. 2006, Ojeda-Serrano et al. 2007
		Multiple sites around main island (5): La Parguera, El Hoyo, Bajo de Sico, Abrir la Sierra, Tourmaline		Colin et al. 1987, Colin & Clavijo 1988, Shapiro et al. 1993, Sadovy 1997, Sadovy de Mitcheson et al. 2008
		Mona Island		Aguilar-Perera et al. 2006, Schärer et al. 2010
Dominican Republic	1	Punta Rusia		Sadovy 1997, Sadovy de Mitcheson et al. 2008
Cuba	21	Bajo Mandinga Cabo Cruz Cayo Bretón Banco de Jagua Cay Guano Cayo Diego Pérez Cayo Avalos Punta Francés Cayos Los Indios Cayo San Felipe Cabo Corrientes	Cabo San Antonio Corona de San Carlos Punta Hicacos-Cayo Mono Cayo Megano de Nicolao Boca de Sagua Cayo Lanzanillo Cayo Fragoso Cayo Calmán Grande Cayo Paredón Cayo Sabinal	Sadovy 1997, Claro et al. 2001, 2009, Claro & Lindeman 2003, Sadovy de Mitcheson et al. 2008

Table 2 (continued)   Documented transient fish spawning aggregations (tFSAs) for reef fish at 108 sites in the Wider Caribbean based on both direct and indirect evidence

Country	Total no.	Spawning sites/general area name		Reference
Cayman Islands	6	Grand Cayman	North-eastern end and Sand Caye	Colin et al. 1987, Tucker et al. 1993, Sadovy 1997, Whaylen et al. 2004, Sadovy de Mitcheson et al. 2008
		Twelve Mile Bank		
		Little Cayman	Eastern and western end	
		Cayman Brac East		
Mexico	13	Majahual		Aguilar-Perera & Aguilar-Davila 1996, Sosa-Cordero & Cárdenas-Vidal 1996, Sadovy 1997, Aguilar-Perera 2006, Sadovy de Mitcheson et al. 2008
		Nichiehabin, San Juan-Chenchomac, Blanquizal-Sta Julia		Aguilar-Perera et al. 2009
		Alacranes Reef (8)		Aguilar-Perera et al. 2008
		Banco Chinchorro		Aguilar-Perera & Aguilar-Davila 1996
Belize	14	Rocky Point	Sandbore	Carter & Perrine 1994, Sadovy 1997, Paz & Grimshaw 2001, Sala et al. 2001, Heyman et al. 2005, Graham & Castellanos 2005, Heyman & Kjerfve 2008, Sadovy de Mitcheson et al. 2008, Kobara & Heyman 2010
		Dog Flea Caye	Halfmoon Caye	
		Mauger Caye	South Point	
		Soldier Caye	Northern Glovers	
		Caye Bokel	Long Caye	
		Caye Glory	Gladden Spit	
		Nicholas Caye	Rise and Fall Bank	
Honduras	2	Guanaja	Cayos Cochinos	Fine 1990, 1992, Sadovy 1997, Sadovy de Mitcheson et al. 2008

29 in Central America and an additional 3 tFSAs sites in the Gulf of Mexico. Of these, 55 sites were Nassau grouper tFSAs, and at least 32 sites were shared with other species. No evidence of tFSAs was found for the southern Caribbean.

Distributions of tFSA sites vary geographically among species, but it is not clear if the distribution of tFSAs provides ecologically meaningful information, or if it simply represents the patchy availability of information about tFSA locations. For instance, some species records in our database shared the same sites, although the timing of spawning varied among species. Detailed observations of multispecies tFSAs were available from Belize (Heyman & Requena 2002, Heyman & Kjerfve 2008, Kobara & Heyman 2010), the Cayman Islands (Whaylen et al. 2004, 2006, Kobara & Heyman 2008), Cuba (Claro & Lindeman 2003), Puerto Rico (M.I. Nemeth et al. 2007, Schärer et al. 2010), US Virgin Islands (Nemeth 2005, Kadison et al. 2006, R.S. Nemeth et al. 2007, Nemeth 2009) and Florida (Coleman et al. 1996, 2011, Koenig et al. 1996, 2000).

The 108 documented sites are almost certainly a considerable underestimation of the true number of tFSAs in the Caribbean region and also supposedly represent only a small proportion of those known within the traditional knowledge of Caribbean fishing communities. Evidence for a geographical knowledge gap is highlighted in Figures 2–4, which reveals a large disparity between the spatial distributions of aggregating species and the spatial distribution of known tFSAs. For example, Nassau grouper, red hind and mutton snapper sightings have occurred across a much broader

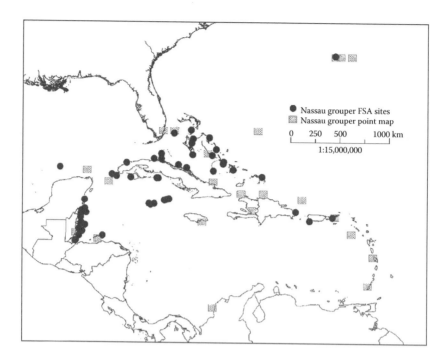

Figure 2 Historically known spawning aggregation sites of Nassau grouper (*Epinephelus striatus*) with/without direct/indirect evidence (circles). Rectangles represent a computer-generated distribution point map for each species from FishBase (Froese & Pauly 2010).

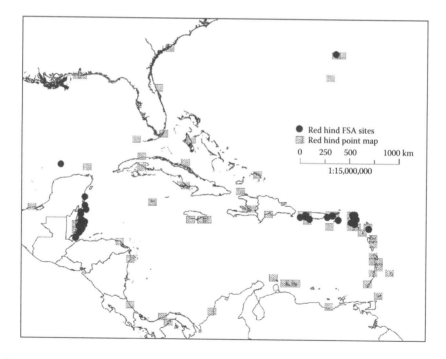

Figure 3 Historically known spawning aggreation sites of red hind (*E. guttatus*) with/without direct/indirect evidence (circles). Rectangles represent a computer-generated distribution point map for each species from FishBase (Froese & Pauly 2010).

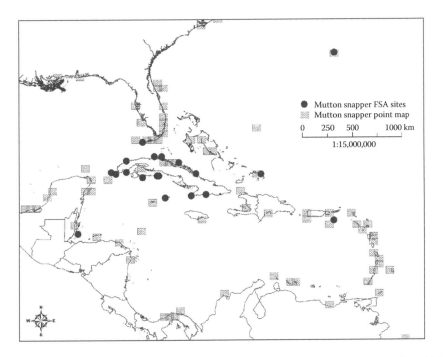

Figure 4 Historically known spawning aggregation sites of mutton snapper (*Lutjanus analis*) with/without direct/indirect evidence (circles). Rectangles represent a computer-generated distribution point map for each species from FishBase (Froese & Pauly 2010).

spatial extent than would be feasible to sustain based only on known spawning aggregations. More specifically, documented red hind tFSAs are only reported for the eastern Caribbean, Bermuda and Mexico, while red hind are distributed throughout the Wider Caribbean. This knowledge gap is particularly evident in the southern Caribbean islands, where few aggregations are known, yet populations of spawning fish clearly exist and in some cases form an important component of the fisheries landings (e.g., red hind and mutton snapper in Grenada (Jeffrey 2000)). Islands most underrepresented in the synthesis, but likely to support tFSAs, include Jamaica, Hispaniola, St. Vincent and the Grenadines, Netherlands Antilles, Barbados and Trinidad and Tobago. In addition, little has been reported on tFSAs off the Caribbean coast of South America or southern Central America (Panama, Costa Rica, Nicaragua, Honduras). A synthesis of local knowledge in fishing communities and tFSA-focussed research and surveys of bathymetric features are needed in these locations. Local fisher knowledge can also be used to determine if optimal sites exist that were once tFSAs with potential for recovery (e.g., Gleason et al. 2011).

Biophysical characteristics of tFSA sites

Timing of tFSAs in the Wider Caribbean

Since the 1970s, a number of papers have summarized the available information on the environmental and biological factors influencing the timing of reef FSAs, primarily from the Caribbean and the tropical western Pacific (Johannes 1978, Robertson 1991, Claydon 2004, Nemeth 2009, Colin 2012). Far fewer studies have examined the timing and location of tFSAs in the Gulf of Mexico (Heyman & Wright 2011). Synthesis of existing data on transient aggregating fish species suggests that the timing of migration, formation of aggregations and spawning are regulated by cues operating on at least three nested temporal scales: intraannual, lunar and diel (Nemeth 2009). Hierarchical cues for the scheduling of spawning-related activities have also been documented in other marine organisms. For example, spawning in some tropical corals is influenced by sea temperature (time of year), lunar or tidal cycles (time of month) and diel light cycles (time of day) (Mendes & Woodley 2002). Some evidence from long-term monitoring studies also suggests a fourth temporal scale, that of interannual periodicity; a portion of the adult population may only aggregate to spawn every other year (Nemeth 2005). Little is known about mechanisms and occurrence of biennial spawning, but this strategy could be an adaptive response to local environmental conditions or related to optimizing longer-term cost and benefits related to reproductive energetics. Studies of Atlantic salmon (*Salmo salar*) found that some fish spawn annually and some biennially, but that the proportion of biennial spawners increased with fish size, whereby the smallest repeat spawners (<60 cm) spawned annually and the largest (>90 cm) biennially (Jonsson et al. 1991). If such population dynamics characterize species that form tFSAs, then understanding population age structure could shed light on differential vulnerability to fishing and oscillations in fisheries catch and aid in forecasting recovery rates after protection from fishing.

The multiscale environmental and biological cues synchronizing behaviours leading to spawning may include photoperiod, water temperature, current speed and direction, lunar phase, tidal phase, ambient lighting (particularly times of sunset and sunrise), fish density, presence of mates displaying breeding colouration, intensity of courtship behaviour and production of specific pheromones and sounds (Colin 1992, Lobel 1992, Mann et al. 2009). Many of these cues are interrelated, thus confounding our ability to determine the relative importance of individual factors. For example, photoperiod and water temperature are both influenced by the amount of solar radiation reaching the earth. The depletion of gravid females and changes in these environmental and biological factors are also important for determining reproductive activity at daily, monthly and annual timescales (Thresher 1984, Domeier & Colin 1997, Heyman et al. 2005, R.S. Nemeth et al. 2007, Starr et al. 2007). These cues also operate across different spatial scales. Changing photoperiod and water temperature influence organisms at the broadest spatial scales (global to regional). Changes in water temperature, current speed, tidal cycle and ambient lighting operate at intermediate spatial scales (regional to local), and fish density and intraspecific sexual behaviours operate at the finest spatial scales (site-specific).

Determining the relative importance of these many complex interrelated spatial and temporal factors is an important challenge for applied research. Nevertheless, some generalities emerge for certain species and functional groups. At the broadest spatial scales, water temperature has a greater influence on timing of reproduction than photoperiod. For example, Nassau grouper and red hind in the Caribbean form spawning aggregations from December to March when water temperature cools to less than about 27°C (Colin et al. 1987, Carter 1989, Colin 1992, Carter et al. 1994, Sadovy & Eklund 1999, R.S. Nemeth et al. 2007). In contrast, spawning aggregations of Nassau grouper, red hind and black grouper farther north, in Bermuda, take place between May and August when water temperatures increase beyond 25°C (Smith 1971, Luckhurst 1998, 2010). In both regions, the ideal temperature range for spawning of these three species is between 25°C and 26.5°C (Tucker et al. 1993).

The importance of temperature to spawning may be related to the physiological limitations of vitellogenesis and egg development at higher water temperatures (Lam 1983). Watanabe et al. (1995b) examined the effects of temperature on eggs and yolk sac larvae of the Nassau grouper under controlled hatchery conditions. Development and survival of newly hatched larvae to first feeding was inversely related to temperatures of 26°C, 28°C and 30°C. A temperature of 26°C was deemed optimal for incubating Nassau grouper eggs and larvae, although even lower temperatures may provide additional benefits to survival (Watanabe et al. 1995a). Ellis et al. (1997) compared the feed utilization and growth of hatchery-reared, postsettlement-stage Nassau grouper juveniles at temperatures of 22°C, 25°C, 28°C and 31°C under controlled laboratory conditions. Final weights and growth rates were higher at 28°C and 31°C than at 22°C or 25°C; thus, a temperature range of 28–31°C was recommended for culture of early juveniles (Ellis et al. 1997). This experimental evidence suggests that Nassau grouper (and possibly red hind) may synchronize spawning to seasonal time periods when water temperature is not only optimal for egg development and larval survival but also sufficiently warm to maximize growth of postsettlement larvae. This may explain why Nassau grouper and red hind in Bermuda do not spawn in the autumn when temperatures for reproduction are optimal but conditions for juvenile growth and survival are poor due to cold water in the winter. Such thermal synchronization will likely have implications for the onset of migrations and the timing of spawning in a changing global climate, with as-yet-unknown consequences (Sims et al. 2005).

At regional and local spatial scales, the lunar cycle and its influence on tides is a widely recognized temporal cue for spawning. Although most aggregating fish species spawn during a distinct period of the lunar cycle (full moon, new moon, etc.), the tidal influence in the Caribbean may be relatively minor because of low tidal amplitude (40–55 cm daily) (Kjerfve 1981, Heyman et al. 2005). Indeed, tidal timing, as a trigger for formation of aggregations and spawning, is far more important in the Indo-western Pacific (Pet et al. 2005), where tidal amplitudes may range 1–3 m. However, many of the species that synchronize spawning activity to tidal patterns are resident aggregators that spawn during the day on outgoing tides within reef channels (Sancho et al. 2000b). Although no clear tidal cue has yet been detected, most transient aggregations in the Wider Caribbean, as well as in the Indo-Pacific, are formed in association with a time of day, usually with spawning at sunset or at night (Domeier & Colin 1997), and with a specific time of month around full moon or other phases of the lunar cycle (Claydon 2004, Nemeth 2009). In addition, field measurements and oceanographic models at Gladden Spit in Belize showed that spring tides can excite significant high-frequency flow variations near the reef, indicating that the preference for spawning in the days immediately following full moon may not be coincidental (Ezer et al. 2011). The strategies of spawning on an outgoing tide and when light levels are low are both advantageous to egg survival and dispersal. Low light enhances survival because egg predators are typically visual predators (Hamner et al. 2007).

Oceanographic characteristics of tFSAs in the Wider Caribbean

Interacting site characteristics such as physical geomorphology, hydrodynamics and the composition of benthic biological communities, together with geographical factors, such as connectivity to suitable settlement sites, appear to combine to make some locations function as tFSA sites. The exact physical conditions remain unknown for most sites and may vary geographically, but it is likely that a subset of variables exists that explains much of why a spawning site persists at a single location over generations. Clearly, a diverse multidisciplinary approach is required to unravel the complexities of tFSA structure and function. The majority of research on site characteristics has focused on geographical location across the reef, seafloor geomorphology and oceanography.

The local geomorphology can have a very strong influence on oceanographic features that fish can use to their advantage when releasing eggs (Kingsford et al. 1991). For example, some hypotheses suggest that the location and timing of spawning aggregations coincide with preferable currents that increase larval dispersal to suitable places (Barlow 1981), enhancing connectivity within a patchy coral reef environment (Sale 2004). Ocean currents flowing towards and around reef promontories often result in the formation of gyres or eddies (Lobel & Lobel 2008). In contrast to the off-reef transport hypothesis, the presence of gyres during spawning seasons has generated alternative hypotheses that spawning is synchronized to exploit hydrographic features that entrain and retain developing eggs and larvae and return them to their natal seascapes (Johannes 1978, Lobel 1978, Lobel & Robinson 1988, Karnauskas et al. 2011). Levels of larval retention and the spatial extent of connectivity in fish are thought to be strongly influenced by physical setting (Jones et al. 2009). Local retention of fish larvae has been shown for several small-bodied reef fish species (Swearer et al. 1999, Planes et al. 2009, Hogan et al. 2012), but there have been very few ocean

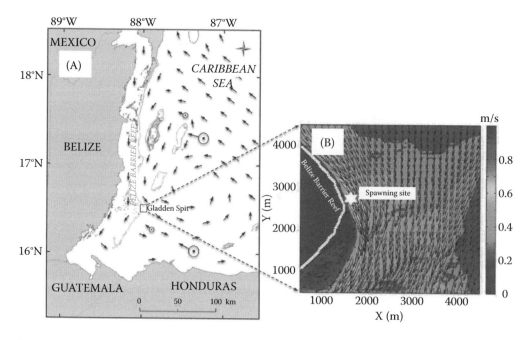

Figure 5 (See also colour figure in the insert) Currents at the Gladden Spit spawning aggregation site on the Belize Barrier Reef. (A) Mean currents (arrows) and winds (arrows with circles) as redrawn from Craig (1966) and Ezer et al. (2005). Note the recirculation eddy south of the reef promontory spawning site. (B) Detailed circulation at the Gladden Spit spawning aggregation site as modelled by Ezer et al. (2011, with permission from Springer).

current studies seeking to identify mechanisms of local retention and dispersal in relation to tFSA sites in the Caribbean. Studies of oceanic patterns (regional to local: >100-km distance) in the Bahamas (Colin 1995) and Belize (Figure 5) (Ezer et al. 2005, Heyman & Kjerfve 2008) suggest that currents around tFSA sites in those areas are oriented along the shelf edge. Ezer et al. (2005) illustrated that when westward-propagating cyclonic eddies approached the Mesoamerican Barrier Reef, the Caribbean Current shifted offshore, the cyclonic circulation in the Gulf of Honduras intensified, and a strong southwards flow resulted along the reefs (Figure 5). Conversely, when anti-cyclonic eddies approached the reef, the Caribbean Current moved shorewards, and the flow was predominantly northwards and westwards across the reefs. Local to small-scale studies (50 m x 50 m grid cells in a model domain of 5 km x 5 km) revealed potential mechanisms for larval transport and retention (Ezer et al. 2011). Ezer et al. (2012) further showed that the particular shape of the promontory amplified coastal currents at the spawning site, through non-linear interactions between flow and topography. Karnauskas et al. (2011) used a hydrodynamic model to explore the effects of topography on currents at the Gladden Spit reef promontory along the Mesoamerican Barrier Reef and predicted local retention of larvae.

In addition, internal waves play an important role in transport and concentration of plankton and other organisms across topographic features, such as shelf edges (Stevick et al. 2008). Ezer et al. (2011) demonstrated the occurrence of internal waves at a multispecies tFSA site in Belize (Figure 6). These high-frequency occurrences may affect larval transport and intermittent upwelling and are worthy of further study. Remote sensing (e.g., synthetic aperture radar [SAR] to map internal wave patterns) and *in situ* oceanographic sensors, such as acoustic current profilers, can be used to determine if the timing of spawning coincides with periods of high-frequency internal waves that may help transport the offspring to inshore settlement habitats. An alternative hypothesis

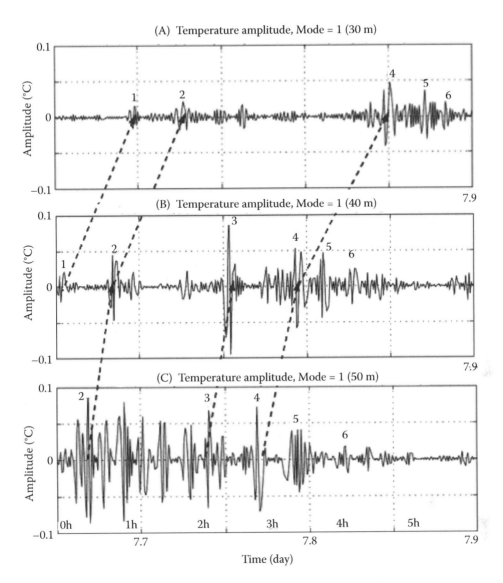

Figure 6 An example of a 6-h period of the first mode (highest frequency) of the Hilbert-Huang transform (HHT). (A)–(C) The observed records at 30, 40 and 50 m depths, respectively. Dashed lines and numbers indicate examples of groups of waves that seem to propagate from deep to shallower water at about the speed of internal waves. (After Ezer et al. 2011 with permission from Springer.)

worthy of testing is that spawning is timed to avoid internal waves because they concentrate offspring at the surface where they are more susceptible to predators.

The interaction between these physical features likely explains why some sites host spawning aggregations by multiple species while other sites have little or no aggregation activity. Characterizations of existing sites will help identify the most relevant environmental predictors to input to models that predict locations of structurally suitable sites, some of which may be unfished and unknown. This then provides a cost-effective tool to address an information gap that will be important in developing comprehensive management strategies and understanding connectivity (Heyman & Kobara 2012). Here, we examine the structural characteristics hierarchically from broad-scale geological patterns and processes that shape continental shelves, to the finer-scale characteristics of benthic communities that provide food, refuge from predation and other resources for spawning fishes.

Seafloor geomorphology of tFSAs in the Wider Caribbean

Seafloor geomorphology has emerged as a useful predictor of tFSA locations. Seafloor geomorphology is typically mapped from bathymetry, and derivative metrics (sometimes referred to as morphometrics) have been applied to quantify various physical characteristics (Prior & Hooper 1999, Pittman et al. 2009). The most useful metrics include water depth and surface geometry of the seafloor (slope, sinuosity, topographic complexity) and the orientation, or aspect, relative to waves and other water movements. A key characteristic of most multispecies tFSAs is their location in close proximity to sharp shelf edges and deeper water, typically where a steep (>20°) downward slope or abrupt change in depth occurs, commonly referred to as a drop-off. These structural edges in the seafloor terrain were formed as sea level rose and stabilized. Transgressive fringing and barrier reefs first grew upwards and then outwards between successive uplift events, forming wide reef crests bounded by relatively steep forereef slopes (Burbank & Anderson 2001). The environmental edge effect around tFSAs can occur at the shelf edge, along channels, along the slopes of fringing or barrier reefs, and at other distinctive bathymetric features such as promontories. A promontory was defined as a distinct turning point, or bend in the shelf edge, where the steep terrain protrudes prominently into deeper waters like a submerged headland (Figures 7 and 8; Kobara & Heyman 2008, 2010).

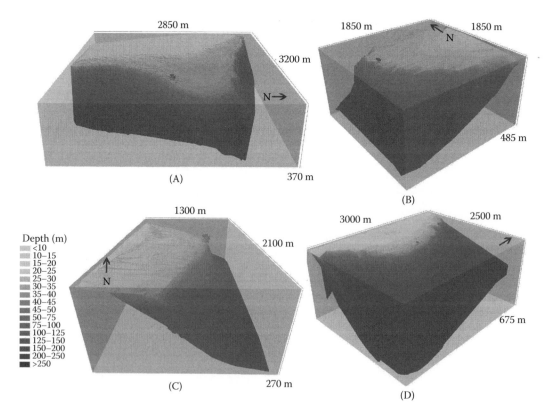

Figure 7 (See also colour figure in the insert) Three-dimensional bathymetry of two representative spawning sites in the Cayman Islands, (A) Grand Cayman East and (B) Little Cayman West, and two in Belize, (C) Emily (or Caye Glory) and (D) Gladden Spit. (Modified from Heyman & Kobara 2012.)

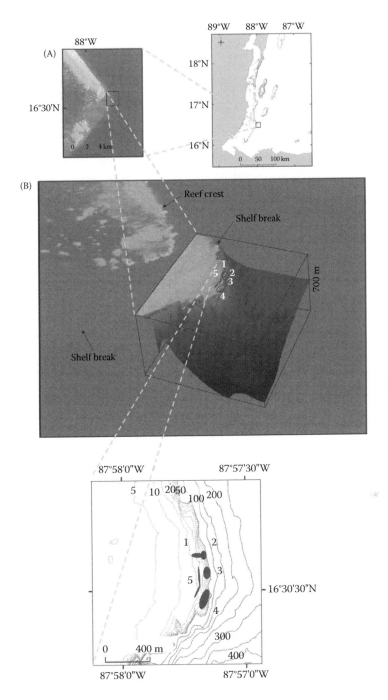

Figure 8 (See also colour figure in the insert) (A) Multispecies spawning aggregations at Gladden Spit in Belize. This oblique aerial view from the south-east was created in ARCGIS 10.1 and ARCScene and combines a Landsat™ satellite image with field-collected bathymetric and species-specific spawning aggregation location data. The primary spawning aggregation sites for various species are shown in relation to the shelf edge and labelled as follows: 1, *Epinephelus striatus*; 2, *Mycteroperca bonaci*, *M. venenosa*, *M. tigris*; 3, *Lutjanus jocu*; 4, *Lutjanus cyanopterus*; 5, *Lachnolaimus maximus*, *Lactophrys trigonus*, *Rhinesomus* (=*Lactophrys*) *triqueter*. (B) Detailed bathymetry of the site (depth contours in metres) showing the location of the same spawning aggregation sites as in Figure 8A.

Table 3 Geomorphological type assigned to transient spawning aggregation sites in the Wider Caribbean based on site descriptions in the literature.

Geographic region	tFSA site	Shelf edge	Reef promontory	Adjacent to drop-off	On reef crest	Near reef channel
Bahamas	Bimini	y				
	Long Island East	y				
	Long Island South			250 m away		
Turks and Caicos Islands		y	y	y		
US Virgin Islands	Grammanik Bank	y	y	y		
	Red Hind Bank	y	n	300 m away		
	Lang Bank	n	y	y		
Puerto Rico	El Hoyo	y	y	y		
	El Seco	y	y	250 m away		
Cuba	Bajo Mandinga	y	y	y		
	Cabo Cruz	y		y		
	Cayo Bretón	y		y		
	Banco de Jagua	y		y		
	Cayo Diego Pérez			y		
	Cayo Avalos	y		y		
	Cayos Los Indios	y				
	Cayo San Felipe	y				
	Cabo Corrientes	y				
	Cabo San Antonio	y				
	Corona de San Carlos	y		y		
	Punta Hicacos-Cayo Mono	y				
	Cayo Megano de Nicolao	y		y		
	Cayo Caimán Grande				y	
	Cayo Paredón				y	y
	Cayo Sabinal					y
Mexico	Mahahual			n		
Honduras	Guanaja	y		y		

Notes: y = yes, n = no, blank = no explicit description. These classifications are subjective, based on descriptions given in the original studies, since there is no consensus on how close a tFSA must be to a geomorphological feature to be considered 'associated' with it.

Our review revealed that most tFSAs were described as occurring close to shelf edges (80.6%) or drop-offs (63.9%) (Table 3). Two tFSAs (in Mexico and the Bahamas), however, were described as not located close to a drop-off (Colin 1992, Aguilar-Perera & Aguilar-Dávila 1996). It remains unclear if these exceptions are linked to species-specific preferences or unique site-specific characteristics, but possible causal mechanisms are being examined (S. Kobara et al. unpublished data). In most cases, however, the definition of 'shelf edge' and the actual distances from the shelf edge or drop-off were unspecified in the data. In addition, there are insufficient quantitative data to define 'proximity' accurately, i.e., the maximum distance that can be considered 'close' to a geomorphological feature. Kobara & Heyman (2008) observed that all Nassau grouper tFSA sites in the Cayman Islands were between 0 and 50 m from the shelf edge where (1) promontories occurred, and (2) water depth was between 25 and 45 m. These promontories also served as tFSA sites for at least two other fish species and in some cases up to 17 fish species aggregated to spawn. Distinct promontories supported tFSAs in Belize, Turks and Caicos Islands, Puerto Rico, Cuba and the Cayman Islands. In Belize, 12 multispecies tFSAs occurred in water deeper than 20 m, less than

100 m horizontally from shelf edges, less than 200 m from reef promontory tips and at a mean distance of 78 m from the 100-m isobath (Figures 7 and 8; Kobara & Heyman 2010). Each fish species appeared to maintain site fidelity to a specific spawning location within a distinct area of the greater promontory feature. At Gladden Spit, for example, geographically distinct spawning locations for 10 species have been observed and mapped within a 6-ha area at the tip of the reef promontory (Figure 8; Heyman & Kjerfve 2008).

The suitability for spawning of broad-scale and relatively easily detected geomorphological features provides a realistic, cost-effective and fisheries-independent method for locating not only existing tFSAs but also those that may have been extirpated or those remaining undiscovered. Using this fishery-independent method, Kobara & Heyman (2010) identified South Point in Belize as a multispecies tFSA site. Much of the geomorphological characterization was achieved using an acoustic depth sounder that recorded bathymetry with 30–50 m spatial resolution (Kobara & Heyman 2008, 2010), but it is also possible to detect promontories and other shelf-edge features through examination of the sinuosity and curvature of the shelf-edge contour or isobath (Kobara & Heyman 2006). Locations with shelf-edge curvature in a certain range (e.g., a reef promontory within a 1-km radius circle) were considered as potential tFSA sites. Some relatively large sites could be identified from smaller-scale bathymetry; however, it is likely that identification of many other biologically important sites would require higher resolution (<30 m) bathymetric data. Investigations of scale dependency in recognition of potential spawning sites from bathymetry alone are under way, and the results will determine the appropriate scale(s) for future bathymetric mapping missions aimed at characterizing tFSAs (J. Blondeau unpublished data, S. Kobara et al. unpublished data).

Not all tFSAs, however, occur at distinctive promontories. For example, red hind tFSA sites off western Puerto Rico, US Virgin Islands and Saba (Netherland Antilles) have been reported over relatively flat, non-promontory reef areas (Nemeth 2005, R.S. Nemeth et al. 2007, Kadison et al. 2009b). Boomhower et al. (2010) evaluated the predicted occurrence of tFSAs at reef promontories in Los Roques, Venezuela, based on a 'reef promontory' search image. Results did not show large tFSAs at predicted aggregation sites and were inconclusive. Closer examination revealed that while the sites were reef promontories, the shelf edges occurred in water less than 15 m deep, and although adjacent to a steep drop-off, the bottom levelled onto a shelf at 30–40 m, rather than dropping directly into deeper water (e.g., >70 m depth). Gag grouper (*Mycteroperca microlepis*) aggregate for spawning along the crests of ridges, adjacent to drop-offs in the eastern Gulf of Mexico (Coleman et al. 2011). These findings suggest that shelf-edge adjacency to deep water may be a critically important factor for the formation of tFSAs. In summary, although there are several geomorphological features that support spawning aggregations for a range of fish species, it is clear that bathymetry and its derivatives alone can be a valuable data type for identifying and studying the biogeography of tFSAs.

Use of a relatively static characteristic such as bathymetry, which is common to many multispecies tFSAs sites, provides an opportunity to conduct detailed comparative investigations to examine generalities in geomorphological structure. Sufficiently detailed bathymetric maps, however, are available from only a few countries in the region (Wright & Heyman 2008), thus limiting development of broad-scale pan-Caribbean predictive models. Relatively low-cost analytical options for mapping broad-scale shelf-edge geomorphology include modelled bathymetry or pseudobathymetry based on algorithms applied to reflected light in optical imagery, such as satellite data and aerial photography (Stumpf et al. 2003). Optical sensors, however, are limited by water clarity, turbidity and water roughness, which influence the light penetration/reflectivity through the water column (Jensen 2007). A better understanding of the relationship between tFSA occurrence and function and specific features of coastal geomorphology can be achieved through further comparative investigation with quantitative data. Establishing a clear definition of 'shelf edge', 'deep water' and other features, together with identifying ecologically meaningful spatial scale(s), are key challenges for future analyses.

Table 4 Detailed descriptions of the benthic habitat found at each transient tFSA site in the Caribbean (from literature cited in Table 2)

Country	tFSA sites	Benthic habitat
Bahamas	Bimini	A thin sand veneer over limestone base rock with abundant soft corals, sponges and occasional colonies of stony coral
	Long Island (eastern side)	Rocky shelf
	Long Island (southern point)	Rubble plain extended seawards to the actual rocky shelf edge, which dropped away vertically to great depth
Turks and Caicos		Shallow fringing reefs dominated by *Montastraea annularis* and *Acropora palmata*
United States: Florida	Western Florida	High-relief, shelf-edge reefs of the south-eastern United States. A series of clustered limestone pinnacles 5–30 m in height, separated by a flat, soft-sediment bottom. The pinnacles were topped by the ivory tree coral, *Oculina varicosa*, which grows in spherical heads 1–2 m in diameter. Two 6-km long pronounced rocky ridges projecting up to 15 m off the seabed.
US Virgin Islands	St. Thomas	High topographic complexity and coral cover, primarily flat-surface colonies of *Montastraea annularis* species complex
	St. Croix	Low-relief spur-and-groove reef, low coral cover
Puerto Rico	El Hoyo	High coral cover and diversity
	El Seco	High topographically complex coral, predominantly *Montastraea annularis* species complex, extended over a relatively level area
Cuba	Bajo Mandinga	High coral cover
	Cabo Cruz	A rocky-sandy bottom until the drop-off at about 20–25 m depth
	Cayo Bretón	Reef slope: high coral cover
	Banco de Jagua	Oceanic bank: rocky bottom, moderate coral cover
	Puntalon de Cay Guano	Reef slope: high coral cover
	Cayo Diego Perez	Reef slope: high coral cover
	Cayo Avados	Reef slope: high coral cover
	Punta Francés	Reef slope: high coral cover
	Cayos Los Indios	Reef slope: high coral cover
	Cayo San Felipe	Reef slope: high coral cover
	Cabo Corrientes	Sandy, rocky, coral heads
	Cabo San Antonio	Reef slope: high coral cover
	Corona San Carios	Reef slope: sandy, rocky, coral heads
	Punta Hicacos-Cayo Mono	Reef slope: sandy, rocky, coral heads
	Cayo Megano de Nicolao	Reef slope: high coral cover
	Boca de Sagua	Reef slope: moderate coral cover
	Cayo Lanzanillo	Reef slope: moderate coral cover
	Cayo Fragoso	Reef slope: moderate coral cover
	Cayo Calmán Grande	Reef slope: high coral cover
	Cayo Paredón	Reef slope: high coral cover
	Cayo Sabinal	Reef slope: high coral cover
Cayman Islands	Little Cayman southeast	Sandy depression
	Little Cayman southwest	Low-relief broad ridges with hard and soft corals and sponges
Mexico	Mahahual	Low-relief, patchy, hard corals interspersed with plexaurids and gorgonians. Hard corals present were mountain coral, *Montastraea annularis,* and leaf coral, *Agaricia* spp., growing between sandy areas.

Table 4 (continued) Detailed descriptions of the benthic habitat found at each transient tFSA site in the Caribbean (from literature cited in Table 2)

Country	tFSA sites	Benthic habitat
Belize	Rocky Point	Hard substratum with sparse coral
	Dog Flea Caye	Low-relief spur-and-groove reef
	Caye Bokel	High-relief spur-and-groove reef
	Caye Glory	Low-relief spur-and-groove reef. The bottom consisted primarily of sand with scattered patches of hard and soft corals
	Sandbore	Low-relief spur-and-groove reef
	Half Moon Caye	High-relief spur-and-groove reef
	North Glover's	Coral ridges together with sandbars made up a spur-and-groove reef
	Gladden spit	Sand floor with low-profile mound
Honduras	Guanaja	Sandy plain began at about 135 ft (41 m)
Saba, NA	Saba Bank	Low-relief spur-and-groove reef with sparse coral cover

Benthic habitat characteristics of tFSAs in the Wider Caribbean

Well-developed deep coral reefs and areas with high topographic complexity characterized by ledges, undercuts and caves have been observed at most Caribbean tFSAs and are thought to offer refuge from predators during prespawning periods and during active mate selection by aggregating species (Carter et al. 1994, Sancho et al. 2000a, Nemeth 2005, Nemeth et al. 2006, Kadison et al. 2009b). Synthesis from site descriptions revealed that scleractinian corals were the dominant biogenic structural organisms at 87.5% of sites; however, sand or rock was also observed (Table 4). In the Turks and Caicos Islands, US Virgin Islands, Puerto Rico and Mexico, reef-building corals in the genus *Montastraea* dominate the coral community at tFSA sites. The *Montastraea* species complex is the most prevalent reef-building coral in the Caribbean, so its presence at tFSA sites may be coincidental. In the US Virgin Islands, extensive *Montastraea*-dominated mesophotic coral reefs occur on the shelf and along the shelf edge at tFSA and non-tFSA sites (Herzlieb et al. 2006, Armstrong et al. 2006, Smith et al. 2010). High coral cover at many sites was not always associated with high reef fish abundance (Nemeth & Quandt 2005). More research is needed to determine the importance of benthic communities at spawning sites (Rivera et al. 2006) and whether benthic components besides coral cover may be necessary to support the formation of tFSAs. For example, although red grouper (*Epinephelus morio*) in the Gulf of Mexico do not form large spawning aggregations like other groupers (Brule et al. 1999), they physically create suitable spawning habitat by excavating pits in soft sediments near the small-group aggregation sites (Coleman et al. 2011).

Data types, availability and research needs for understanding and managing tFSAs

The overall aim of this section is to describe the data needed and methods to acquire those data to support effective information-based decision making and ecosystem-based management of individual sites and a network of sites. In the foregoing review, we identified two major information gaps that hinder the characterization and thus effective management of tFSAs. First is the insufficient fundamental baseline data for known, protected sites. Second is the lack of a synthesis of the available data for comparative purposes and for modelling. Though some specific data exist for many sites, there is high within-region variability in the amount, type and quality of information. This uneven data availability is a product of variability in research and management capacity, access to infrastructure and equipment and individual researchers' and managers' interests and priorities. Prior to this synthesis and review, regional studies of biogeography or comparative management status have been impossible. This section aims to (1) classify the data types and available methods to collect those data, (2) identify the minimum data requirement for management implementation and monitoring and (3) identify future research needs that will improve understanding and management of tFSAs across the Wider Caribbean region.

In this synthesis, a multitier research framework is suggested to classify existing methods and guide future research activities. We classify eight types of information most frequently collected at and around tFSA sites and organize them hierarchically with increasing technical complexity and cost. Examples for each information type are provided (Table 5). Although several sites have been intensively surveyed, much of our information about tFSAs across the Wider Caribbean has come from local fishers' knowledge and port surveys; these formed the base of our framework (Table 5, level 1).

The framework provides a consistent way of measuring and evaluating the information status of individual sites. When populated with information for each site across a network, it can be used to set strategic priorities, evaluate costs and select appropriate partners for the sharing of knowledge, data, expertise and equipment. We propose that for effective and appropriate decision making, and for comprehensive regional management, all known tFSA sites in the Wider Caribbean should eventually be documented using techniques from levels 1 to 4 inclusive. We consider these first four levels to include the minimum necessary information on which managers can confidently base management decisions. We advocate the establishment of a pan-Caribbean research and management network for tFSAs through scientist-community-manager partnerships that can share data up to level 4 in support of site and regional management and multinational research and hypothesis testing aimed at understanding tFSA phenomena and connectivity more broadly.

Research level 1: fisher interviews and port surveys

We recognize the importance of local knowledge and the active participation of local fishing communities in studying and sustainably managing tFSAs. Detailed studies of fishers' knowledge of spawning aggregations were initiated by Johannes (1978). Numerous studies of fishers' knowledge of spawning aggregations have since been completed (Heyman & Hyatt 1996, Hamilton et al. 2005, 2011) and can often offer a baseline from which to begin field observations and measurements (Colin et al. 2003). Additional conversations with fishers can also provide auxiliary information on perceptions of historical trends and environmental change, geographical areas and specific habitat types preferentially targeted and other observations on the health and physical condition of fish.

In this case, however, our focus is primarily on physical and biological data collection methods. Surveys and macroscopic examination of landed catch can provide general biological information and conservation status for individual species at relatively low cost. Common direct measurements and information types include fish species, body length and weight, seasonality of reproduction, as well as landings per unit time and landings per unit effort where effort can be recorded accurately and fishing gear and area are consistent. Additional analyses of extracted body parts such as otoliths can provide information on age and growth rate, and tissue samples can provide information on trophic connectivity and habitat affinities via chemical signatures of accumulated radioisotopes (Elsdon et al. 2008). Proper assessments require knowledge of the size and age structure of fish populations in general. Fish that have been marked or tagged as part of mark-recapture studies can also be recorded during port surveys. Rates of growth, mortality and reproduction can only be calculated if changes in length at age can be monitored over time. Required equipment generally includes a measuring board and weighing scale (Table 5). Otolith age analysis requires more sophisticated and expensive equipment (diamond blade saw, microscopes, software), but otoliths can also be processed at national laboratories at a reasonable cost. Some examples of research questions are the following: Is there a seasonal pattern in landings of certain species? Are they gravid? What size ranges are being harvested and with what gear? Are there changes in species composition, size and age of individual species over time?

Table 5 FSA data collection methods, equipment needs, expected outputs and examples studies

Level	Data collection method	Equipment needs	Data and information generated	Expected output	References, examples and use of these techniques
1	Fisher interviews and port surveys	Access to and trust of fishers, survey instruments or interview forms	Approximate timing, species composition, location, historical exploitation and present status of tFSAs	Traditional ecological knowledge about tFSAs, observations from markets, unverified indications of tFSAs	Johannes 1978, Sadovy et al. 1994, Claro & Lindeman 2003, Hamilton et al. 2005, 2011, Ojeda-Serrano et al. 2007, Rhodes & Tupper 2007, Heyman 2011
2	Fishery-dependent surveys	Access to landing sites, measuring boards, weighing scales, otolith-processing equipment and microscopes; access to historical catch data	Landings data, gear types, catch per unit effort (CPUE); length-weight frequencies, otolith analysis for age and growth, sex ratios, histological data, biometric information (e.g., gonadosomatic indices)	Detailed data on extraction from tFSA sites; valuable data for stock assessment; can provide verification of FSA	Beets & Friedlander 1999, Erisman & Allen 2006, Matos-Caraballo et al. 2006, R.S. Nemeth et al. 2007, Rhodes & Tupper 2007, Rhodes et al. 2011
3	Underwater visual census (UVC)	Scuba gear, appropriate breathing mixtures, closed-circuit rebreathers (option: sound recorders); video cameras, video laser calipers, remotely operated vehicles (ROVs); video drop cameras, GPS (on boat to give approximate location)	Area; habitat use; depth; timing; density; length and abundance; behaviour (spawning, courtship, colouration, sound); temperature; qualitative description and benthic mapping of habitat and reef structures; site fidelity	Fish densities, timing of spawning and courtship behaviours; verification that aggregation is for spawning; valuable data for stock assessment; effects of divers on behaviour	Colin 1992, Sadovy et al. 1994, Samoilys 1997, Whaylen et al. 2004, 2006, Burton et al. 2005, Heyman et al. 2005, 2010, Pet et al. 2005, Erisman & Allen 2006, R.S. Nemeth et al. 2007, Heyman & Kjerfve 2008, Boomhower et al. 2010, Heppell et al. 2012

continued

Table 5 (continued)   FSA data collection methods, equipment needs, expected outputs and examples studies

Level	Data collection method	Equipment needs	Data and information generated	Expected output	References, examples and use of these techniques
4	Mapping bathymetry, benthic habitat, and fish utilization of tFSAs	Remotely sensed data (Landsat, IKONOS, Quickbird, lidar); existing nautical charts and bathymetric maps; GPS; sonar (single beam, dual beam, multibeam, or side scan); UVC data; seabed-mounted or drop video cameras; autonomous underwater vehicles (AUVs)	Bathymetric maps, benthic habitat maps showing substratum and biotic cover, rugosity, and slope	General site information can be valuable for MPA delineation; detailed map of spawning aggregation sites showing bathymetry, habitat types, and use by tFSAs	Stumpf et al. 2003, Armstrong et al. 2006, Taylor et al. 2006, Heyman et al. 2007, Kobara & Heyman 2008, 2010, Shcherbina et al. 2008, Coleman et al. 2011
5	Mark and recapture studies using conventional and acoustic tagging and tracking	Standard identification tags, continuous and coded acoustic transmitters, acoustic receivers (e.g., VR2W), omni- and uni-directional hydrophones	Site utilization and site fidelity of fishes at FSA sites; time and date of arrival, residency time, and departure at tFSA site by species, sex, lunar period	Detailed information on temporal aspects of fish utilization of tFSA site, including site fidelity, movement patterns, migration pathways	Zeller 1998, Bolden 2000, Nemeth 2005, R.S. Nemeth et al. 2007, Starr et al. 2007, Mann et al. 2009, Rhodes et al. 2012
6	Acoustic monitoring of courtship sounds of fishes	Digital sound recorders (both installed *in situ* as a monitoring station and mobile, attached to a video camera); rebreathers; baseline of spawning sounds for key species	Quantitative assessments of species' timing and level of participation in spawning based on acoustic monitoring	Monitoring of tFSA site use by various species	Lobel 1992, Mann & Lobel 1995, Holt 2008, Mann et al. 2009, Rowell et al. 2012, Schärer et al. 2012a,b

Table 5 (continued)   FSA data collection methods, equipment needs, expected outputs and examples studies

Level	Data collection method	Equipment needs	Data and information generated	Expected output	References, examples and use of these techniques
7	Oceanographic and meteorological data collection from *in situ* measurements and remotely sensed data	Remotely sensed data (sea-surface temperature, sea-surface height, ocean colour from MODIS, others), weather stations, underwater temperature loggers, acoustic Doppler current profilers (ADCPs), electromagnetic current meters, conductivity temperature depth (CTD) meters, AUVs, surface drifters, underwater gliders, light meters	Weather patterns, including air temperature, wind speed and direction and rainfall; current speed and direction; hydrostatic pressure, which can offer tide and wave height information; temperature variability and profiles; light intensity; chlorophyll concentrations	Understanding of oceanographic variability and forcing factors at tFSA sites	Lobel 1978, Colin 1992, 1995, Ezer et al. 2005, 2011, Nemeth et al. 2008
8	Modelling oceanographic and biophysical connectivity; predicting the location of tFSA sites based on geomorphology; predicting larval transport from known tFSA sites	Three-dimensional numerical simulation models (e.g., Regional Ocean Model [ROMS], Princeton Ocean Model [POM]); oceanographic and bathymetric data; spawning dates, times and locations; larval behaviour; otolith microchemistry; genetics	Predictions of larval connectivity to nursery habitats; local and regional hypothesis testing about connectivity	Models of larval transport and connectivity; models predicting timing and location of undiscovered tFSAs	Jones et al. 1999, 2005, Paris & Cowen 2004, Ezer et al. 2005, 2011, Paris et al. 2005, Kobara & Heyman 2006, Elsdon et al. 2008, Boomhower et al. 2010, Kobara & Heyman 2010, Cherubin et al. 2011, Karnauskas et al. 2011

Research level 2: fishery-dependent surveys

Level 2 aims to obtain information on the spawning aggregation process. A common type of fishery data collection consists of formal observer programmes; personnel trained in scientific data collection work alongside fishers in their boats. Because not all fish that are caught are brought to market, this technique allows for a more accurate assessment of landings that include undersize, damaged or otherwise unmarketable fish, as well as by-catch that may be used as bait or discarded. Sampling on-board fishing vessels potentially provides data on the species composition and morphometrics by species (e.g., length and weight), age and growth, histology, fecundity, seasonality and lunar periodicity. While level 1 would provide a general idea of seasonal trends and the approximate location of fish aggregations, level 2 will determine more detailed biological information and approximate site fidelity. Species- and age-specific distribution patterns and catch per unit of effort (CPUE) can all be calculated and used to assess the fishery. Matos-Caraballo et al. (2006) provided an excellent example of a fishery-dependent characterization of a tiger grouper (*Mycteroperca tigris*) aggregation, and several others are also listed in Table 5. A spawning-aggregation study in the Gulf of California is another great example of cooperative research with fishers, who willingly put global positioning system (GPS) receivers on their boats and recorded aspects of the fishery, such as capture rates and volumes in space and time, as the aggregation developed and subsided (Erisman et al. 2012).

Fishery-independent surveys utilize either hired commercial fishing vessels or research vessels and use the same gear types as commercial or artisanal fishers. These surveys differ from fishery-dependent surveys in that the sampling is typically random or stratified-random, with strata being fishing areas or habitat types; the types and amounts of bait are standardized, and fishing effort (i.e., trap soak time) is carefully recorded. In addition to the equipment used as discussed previously, line fishing gear or baited fish traps, tags, tagging guns, and a GPS receiver are needed.

Fishery-dependent and -independent surveys can address questions such as the following: What is the distribution of effort and species among habitat types? How has CPUE or species composition changed over time? What is the percentage of by-catch relative to gear type and other factors? Can fishing gear be modified to reduce or eliminate by-catch (e.g., by incorporating escape vents in fish traps)?

Research level 3: underwater visual census and echo-sounding surveys

Level 3 focuses on acquiring detailed *in situ* information on the spawning process to gain a greater understanding of when, where and how spawning aggregations of particular species occur. Underwater visual survey either by snorkelling (Mazeroll & Montgomery 1995) or using scuba gear (Nemeth 2005) is among the most common techniques used to verify the existence and status of tFSAs (e.g., Sala et al. 2001). Because many tFSA sites are located in depths of 30–70 m, they can be surveyed by divers, but in the deeper part of this range, advanced diving techniques using helium-based breathing mixtures and perhaps closed-circuit rebreathers are required to conduct underwater visual surveys safely and effectively. Underwater diver surveys are usually associated with disturbance to normal fish behaviour, yet observations suggest that small groups of experienced divers would not negatively affect some species of aggregating fishes (Heyman et al. 2010). Furthermore, the lack of bubbles from closed-circuit rebreathers is thought to reduce disturbance to fish and allow for a more realistic observation of spawning behaviours (Lobel 2005, R. Nemeth, personal observation).

Survey techniques will vary depending on species and habitat but may range from total counts of fish within an aggregation, stationary point counts or belt transects to estimate fish population density (Colin et al. 2003). Surveys can also utilize diver propulsion vehicles (underwater scooters) to count fish or search for aggregation sites over large areas, but the associated motor noise can influence fish behaviour and disrupt aggregation formation for certain species (Nemeth et al. 2006). The use of diver-operated video and still cameras has aided these efforts dramatically in recent years, and the increasing availability of low-cost drop cameras and remotely operated vehicles (ROVs) has facilitated observations of tFSAs in deep water. Stereo-cameras or laser measuring devices mounted on a video housing can also be used to estimate the size of fish within an aggregation (Rochet et al. 2006, Heppell et al. 2012). Since migration and spawning are often synchronized to seasonal and daily lunar, temperature and light cycles, underwater visual surveys are invaluable for detecting the arrival and departure of spawning fish from a tFSA site in relation to environmental conditions. In addition to observations of spawning behaviours (including courtship and colouration), underwater surveys can illustrate school density and abundance, spawning areas, timing and local habitat usage (Colin et al. 2003, Archer et al. 2012). Comparing data from this type of survey completed in various locations can address questions like the following: How has the abundance, size or behaviour of fish within the aggregation changed over time? How does lunar timing of spawning differ among sites? At what depth does spawning occur, and is this related to specific oceanographic conditions?

As an optional study, for deeper water and rapid surveys across large areas by day and night, echo-sounding technology provides a cost-effective solution for instantaneous mapping of fish size and abundance at tFSAs. An additional benefit from echo-sounding technology is that the vertical position of fish and other fauna, such as plankton, can be mapped and tracked over short distances. In the US Caribbean, a fisheries acoustic study of spawning aggregations was conducted through simultaneous application of a ship-mounted echo sounder (Simrad 120-kHz split-beam transducer) to detect fish swim bladders alongside a multibeam sonar (Reson 7125) to map seafloor bathymetry (Kracker et al. 2011). Researchers were able to collect large volumes of 3-dimensional data from known fish aggregations and to explore potential aggregations at shelf-edge promontories (Figure 9).

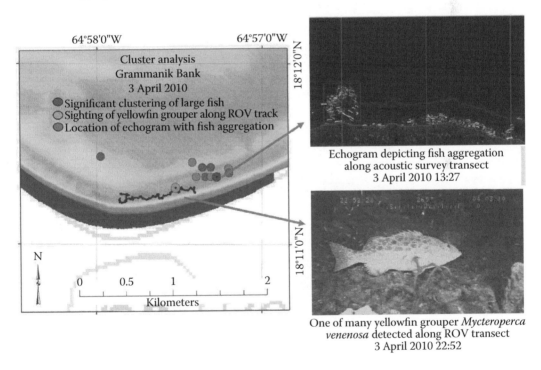

Figure 9 (See also colour figure in the insert) Locations with statistically significant clustering of large fish (>29 cm) from acoustic surveys. Top inset shows fish aggregations on echogram. Bright purple horizontal line denotes the contour of the seafloor, with thin purple line representing 5 m above seafloor. Bottom inset shows video still image providing direct evidence of yellowfin grouper (*Mycteroperca venenosa*) taken simultaneously by remotely operated vehicle (ROV) at locations with significant clustering of fish. (Adapted with permission from Kracker et al. 2011.)

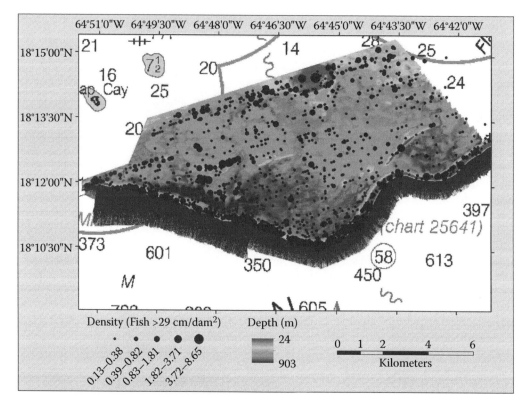

Figure 10 (See also colour figure in the insert) Population density of large-bodied (>30 cm) fish along the shelf edge south of St. John, US Virgin Islands. The target strength-length relationship of Love (1977) was used to estimate fish size. Fish density is expressed as number per square decametre (dam⁻²). (Figure provided by C. Taylor, National Oceanic & Atmospheric Administration.)

These data allow for coupling of water column data with information on the heterogeneity of seafloor structure across an entire spawning site and neighbouring areas (Figure 10). Furthermore, repeat surveys at different times of the day can provide detailed information on the 3-dimensional spatial movements of fish at various stages of the spawning event.

Research level 4: mapping bathymetry, benthic habitat and fish utilization of tFSAs

A comprehensive understanding of the seafloor characteristics associated with spawning aggregation sites is of great interest to environmental managers tasked with identifying priority areas for protecting and rebuilding sustainable fish populations and ecosystem resilience. In conservation prioritization and in ecosystem-based marine spatial planning, the locations of tFSAs are typically considered high-priority information to inform the decision-making process (Crowder & Norse 2008, Geselbracht et al. 2008). Within sites, bathymetric data provide an opportunity to study the influence of geomorphology on aggregating fish distributions, including species-specific distributions and movement patterns, if tracking data are also available. In addition, collecting geomorphological data at tFSA sites will contribute to the data necessary to understand and model the interacting effects of geomorphology and oceanography on biological aggregations (Hyrenbach et al. 2000, Heyman & Wright 2011). The 3-dimensional structure or bathymetry can be mapped with a wide range of sensors, both in the water and from aircraft and satellites. Several low-cost options for mapping the bathymetry of tFSAs range from the creation of pseudobathymetry from high-resolution satellite data (e.g., IKONOS; Stumpf et al. 2003) in optically suitable waters to directly collecting depth soundings using standard sonar from small boats (Heyman et al. 2007). Airborne lidar (light detection and ranging) can provide more detailed bathymetry over large areas to allow production of a highly accurate digital terrain model, but the cost would be prohibitive for many researchers. Lidar bathymetry has been used to map habitat suitability for fish associated with coral reefs, including grouper (Pittman & Brown 2011).

Geomorphology from multiple sites can be compared to evaluate the generality that multispecies tFSA sites occur at reef promontories and near shelf edges, adjacent to deep water. For example, can seafloor characteristics that are similar to the reef promontories described for Belize (Kobara & Heyman 2010) and the Cayman Islands (Kobara & Heyman 2008) serve as reliable proxies for the presence of tFSAs across the Caribbean?

Benthic habitat maps usually produced from interpretations of optical and acoustic mapping technologies provide useful spatial information on the types of benthic community that exist at and near tFSA sites. Benthic habitat maps of moderate-to-deep seabed (depths >35 m) can be rapidly and reliably classified from high-resolution bathymetry using semi-automated classification techniques (Costa & Battista 2008). Mapping benthic habitats also provides a baseline to evaluate future disturbances to coral reef ecosystems. Analysing seafloor habitat and seascape structure surrounding spawning sites will likely provide insights on the navigational routes and the structural features of the seascape that influence connectivity to suitable settlement habitats (Grober-Dunsmore et al. 2009). Few sites, however, have been mapped at sufficient resolution to provide useful ecological information for studying fish distributions at tFSA sites.

Research level 5: mark and recapture studies and ultrasonic tracking

Individual and group movements by fish are intrinsically linked to the function of tFSAs. An understanding of space use by aggregating fish is essential to identifying habitat requirements at tFSAs and for assessing the amount of protection provided by place-based management strategies, such as no-take or seasonal MPAs. At finer temporal and spatial scales, movement information can provide insights into the behavioural mechanisms of individual fish at spawning sites. Little information is available on movement patterns during courtship and spawning or between spawning peaks (Nemeth 2009). At a few sites, acoustic tagging using a fixed acoustic array has provided detailed information on site fidelity, residence time, migration pathways, migration area, habitat use and fine-scale movement patterns during both night and day (Nemeth 2009, Rhodes et al. 2012). These types of study are helping better define spatial components of FSAs, including catchment area, staging area, courtship arena and the spawning aggregation site (Nemeth 2012). Mapping the migratory pathways to and from tFSAs provides valuable information on regional connectivity. Understanding of daily home range movements and seasonal migrations can be used at research level 5 to appropriately design the size, shape and placement of MPAs and to examine the efficacy of established MPAs.

Studies using acoustic transmitters and conventional tags show that movements of aggregating species are much more extensive than previously realized (Zeller 1998, Nemeth 2005, Starr et al. 2007, Rhodes & Tupper 2008). A mutton snapper moved 255 km from a tFSA site in 3 weeks (W.D. Heyman unpublished data); a Nassau grouper moved over 200 km to a tFSA site in the Bahamas (Bolden 2000). Nemeth (2009) produced direct evidence that yellowfin grouper (*Mycteroperca venenosa*) and Nassau grouper (*Epinephelus striatus*) swam distances of between 1 and 5 km during the day before returning to the spawning aggregation site by late afternoon. Some of the same individuals moved over 10 km per day between monthly spawning peaks (Figure 11). These species used deep linear coral reefs as pathways to and from their spawning aggregation site (R.S.

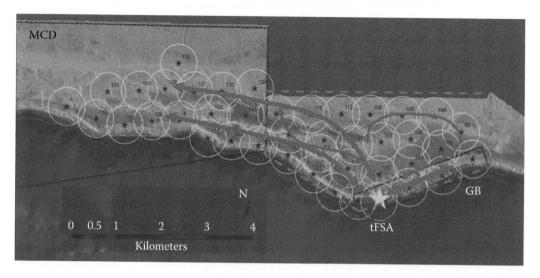

Figure 11 (See also colour figure in the insert) Daily movements and migration pathways of Nassau (*Epinephelus striatus*) and yellowfin (*Mycteroperca venenosa*) grouper. The red arrows in different thicknesses represent hourly (thickest), daily and weekly movements to and from the spawning aggregation site (large star). The Marine Conservation District (MCD; area within black dashed line on left) and the Grammanik Bank (GB; small rectangle on right) are two marine protected areas in the US Virgin Islands. Small stars and numbers indicate the positions of acoustic receivers. The thin red line along reefs is shelf edge. Migration pathways followed deep linear reefs, and the study indicated that an area of approximately 11 km² protected area boundary (blue dashed line) was required to fully protect both species during the spawning season. (After Nemeth 2009.)

Nemeth unpublished data). Tagging studies can also provide insights into the connectivity between MPAs and distant spawning sites. For example, several tagged grouper and snapper using shallow nearshore embayment within the US Virgin Islands National Park at St. John were also detected at acoustic receivers moored at shelf-edge spawning sites (S.J. Pittman & R.S. Nemeth unpublished data; Figure 12).

In addition to extensive horizontal migration, some fish at tFSAs exhibit vertical movements at different stages of the aggregation (pre- and postspawning and time of day). For example, Starr et al. (2007) observed Nassau grouper descending from an average depth of 20 m during the first month of spawning to over 50 m during the remaining 2 months of the spawning season. It is not clear what these fish were doing during these extensive periods of roaming or depth changes, but several hypotheses are emerging. Directed movements between spawning periods may represent foraging activity in particular habitats (i.e., staging areas), prespawning aggregations or regular visits to cleaning stations (Samoilys 1997, Rhodes & Sadovy 2002, Nemeth et al. 2006, Semmens et al. 2006, Coleman et al. 2011). Grouper movements may also be associated with attracting or leading conspecific adults to the spawning aggregation site, a behaviour that has been reported for resident aggregating species (Mazeroll & Montgomery 1998). Finally, dramatic changes in depth, as seen in Nassau grouper, may allow adult fish to enter cooler water, forage for preferred prey, avoid higher parasite or predation rates or release larvae in optimal currents or water strata (Starr et al. 2007). Studies that combine tracking and mapping of movement pathways with detailed benthic habitat maps and bathymetry over the entire staging area will likely offer new insights to our understanding of how fish use tFSAs (Hitt et al. 2011).

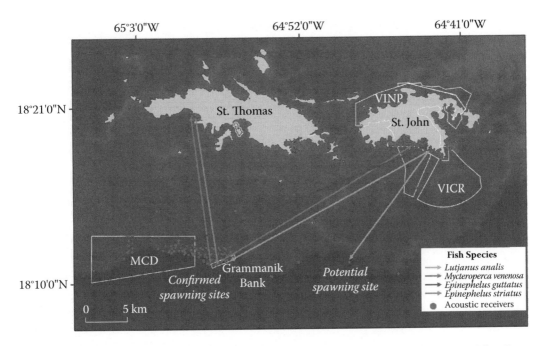

Figure 12 (See also colour figure in the insert) Acoustic array detections of tagged fishes providing direct evidence of connectivity between nearshore coral reef ecosystems and shelf-edge spawning aggregations for three species of grouper (*Mycteroperca venenosa*, *Epinephelus guttatus* and *E. striatus*) and a mutton snapper (*Lutjanus analis*) (Pittman & Legare 2010, CCMA 2011). Straight lines represent direct links between receiver locations rather than actual movement pathways. Marine Conservation District (MCD) and Grammanik Bank are sites of known spawning aggregations. Virgin Islands Coral Reef National Monument (VICR) and Virgin Islands National Park (VINP) are marine protected areas.

Research level 6: hydrophone monitoring of courtship sounds of fishes

In addition to visual and echo-sounding surveys, the ability to detect fish vocalization and chemical cues can be useful in level 6 research. Hydroacoustic surveys of tFSAs have proven valuable in recent years because of species-specific courtship and spawning sounds emitted by some species (Lobel 1992, Mann & Lobel 1995, Locascio & Mann 2008, Mann et al. 2009, Nelson et al. 2011). Seafloor-mounted hydrophones at known or potential spawning sites (Rowell et al. 2012) provide utility as continuous listening stations regardless of sea conditions and visibility, although data storage capacity limits the duration of recordings, and interpretation of marine animal sounds at spawning sites is still in its infancy. In Puerto Rico, seafloor-mounted hydrophones have been used to record fish vocalization at a known spawning site for Nassau grouper (Schärer et al. 2012b), yellowfin grouper (Schärer et al. 2012a) and red hind (Mann et al. 2010, Rowell et al. 2012). Little is known about the sound production of snappers and other fish that aggregate to spawn. Boat-towed hydrophone arrays have been used to locate spawning aggregations and to estimate the abundance of fish across large spatial areas, such as for red drum (*Sciaenops ocellatus*) in the western Gulf of Mexico (Holt 2008), Nassau grouper in the Cayman Islands (Taylor et al. 2006) and red hind in Puerto Rico (Johnston et al. 2006). Autonomous underwater gliders have also been used successfully as mobile remote platforms to carry hydrophones over large distances for recording fish sounds (Wall et al. 2012). The technology offers great potential for exploratory deployments along shelf edges and promontories at known or suspected spawning times. Much can be learned from other more developed fields in animal vocalization and the broader field of acoustical engineering, and sophisticated algorithms can be developed for automated search and recognition of specific acoustic signatures of spawning fish (Mann et al. 2009).

Olfactory cues, such as pheromones, may be involved in attracting fish to spawning aggregations and synchronizing ovulation and milt production. Males could test the female for spawning readiness through an olfactory cue, which in turn elicits a behavioural response in the male (Pankhurst & Fitzgibbon 2006). In Pacific herring, olfactory detection of a pheromonal component of milt stimulates spawning behaviour in both sexes (Carolsfeld et al. 1997). However, little is known about the use of chemical signalling in Caribbean groupers and snappers. There have been no investigations of the possible role of species-specific pheromones in the formation of aggregations and initiation of spawning in Wider Caribbean fishes (Stacey et al. 2003, Ganias 2008).

Research level 7: oceanographic and meteorological measurements

Research on oceanographic patterns and processes at tFSAs is required to understand the multiscale variables that influence the initial dispersal and destination of fertilized eggs. Techniques described in level 7 will collect *in situ* and satellite-derived information on the oceanography of a spawning site and the physical forcing factors that operate at spatial scales broader than the site. Suitable questions to address include testing the hypothesis that the interaction between far-field oceanic forces and localized physical-oceanographic conditions promote connectivity and contribute to making a location suitable for reproduction. Because recent findings indicate that many large, multispecies tFSAs occur at geomorphologically distinctive areas such as reef promontories, there may be specific advantages for spawning in these areas, indicating the wide applicability of research at this level. Ezer et al. (2011) developed a high-resolution 3-dimensional numerical circulation model using current data measured during spawning times, with a suite of *in situ* physical measurements from acoustic current profilers, electromagnetic current profilers, conductivity-temperature-depth instruments and tide gauges, complemented by remotely sensed sea-surface height anomaly and sea-surface temperature. Further studies are under way to calibrate and validate the numerical model under various conditions using Eulerian and Lagrangian field measurements of local currents, winds, tides, waves and temperature/salinity profiles at several locations surrounding the reef promontory.

Tracking the movement of fertilized eggs from spawning sites is difficult. Quantitative plankton tows or drifters can be used to analyse the track, an approach that typically assumes all larval movement is passive advection by currents. The transport fate of spawned material under various conditions can be predicted using particle-tracking algorithms within the numerical simulation model described in the next level (level 8).

To date, marine connectivity studies through movement of eggs and larvae largely rely on numerical simulation models. In situ and remotely sensed data to support these models are scarce. Although ocean current data could be used to evaluate the relative merits of several competing hypotheses, no extended data records or descriptions of local current patterns are yet available to test models at the resolution and scale at which biological activities occur (ca 100–1000 m). One study, however, simultaneously measured current profiles at three distant red hind (*Epinephelus guttatus*) spawning aggregation sites in the north-eastern Caribbean (St. Thomas, St. Croix, Saba Bank). In that study, Nemeth et al. (2008) found that the prevailing currents at each site tended to carry larvae to the shelf during the week of spawning, but not at other times of the lunar cycle.

Research level 8: modelling connectivity and ecosystem processes at FSA sites

Mesoscale and nested numerical ocean circulation models are continually being refined, and resolution and predictive accuracy are being increased. Such models can be used to examine source-sink mechanisms and pathways to settlement sites and to estimate the importance of local retention versus broader dispersal. There are myriad ways to model biological and physical parameters that could be helpful for the management of tFSAs. This review, however, focuses primarily on data collection, analysis and use rather than modelling. Numerical and spatially explicit models can be applied at any stage of research, and although listed here as a final level, we acknowledge that if sufficient data are available, models can play a valuable exploratory role in understanding the environmental context of spawning sites, the potential connectivity across networks of sites and even processes at Caribbean basin-wide scales (Levin & Lubchenco 2008). Spawning at specific locations may increase the likelihood of successful recruitment by carrying fertilized eggs and developing larvae into favourable ocean currents that promote survival and growth while trans-porting settling larvae towards suitable nursery habitats. Although this idea is intuitive, evidence is insufficient to support any generalities on the oceanographic and larval transport processes that make one location more suitable than another for multispecies spawning aggregations.

Numerical ocean simulation models are based on mesoscale ocean currents (ca 10- to 100-km scales), although advances in higher-resolution hydrodynamic modelling, coupled with sophisti-cated biophysical individual-based simulations, provide increasingly reliable predictions of larval pathways (Paris et al. 2005, Werner et al. 2007, Cowen & Sponaugle 2009, Karnauskas et al. 2011). For example, in the Eastern Caribbean, Cherubin et al. (2011) found that virtual particles released in an oceanographic model of a spawning aggregation site were entrained in downwelling currents and returned to their release site within 8–10 days, suggesting a possible retention mechanism for red hind (*Epinephelus guttatus*).

From an ecosystem perspective, an understanding of the influence of large ephemeral spawning aggregations on energy transfer across coral reef environments and at aggregation sites will help to define the role that intact FSAs have on ecosystem integrity, trophic balance and marine ecosystem health. Larger populations of spawning fishes at tFSAs may influence food webs, both upwards to apex predators feeding on spawning fish (e.g., sharks and dolphins) and planktivores feeding on newly released eggs (e.g., whale sharks and manta rays) (Heyman et al. 2001), and downwards on the prey of spawning fish. Finally, larger spawning biomass will increase egg and larval production, which will sustain distant populations via dispersal and local populations via retention. Modelling these trophic interactions and the ecological ramifications of restoring populations of species from the snapper and grouper families represents an interesting avenue for further research. Scenario models that can show managers the ecosystem effects of various interventions might be particu-larly valuable.

Existing status and management of Caribbean tFSAs

Many different strategies have been applied to better manage tFSAs, including fishery closures, species closures, seasonal closures, gear regulations and restrictions, size and weight restrictions, marketing restrictions, landing quotas, encouragement of alternative livelihoods and more (Sadovy de Mitcheson et al. 2008, Claro et al. 2009). Some of these restrictions are implemented through comanagement, voluntary compliance, taboo systems or national and international initiatives, with notable successes emerging from all levels of management. Suitability of a particular strategy or set of strategies will vary geographically depending on history, culture, prosperity, education and the local ecology. Regulatory strategies must be carefully considered as part of a more holistic, ecosystem-based fisheries and marine biodiversity management strategy.

MPAs, including no-take zones, are widely considered to protect habitat from destructive fishing practices; allow the recovery of overexploited fish populations and protected species dur-ing critical life-history stages, such as reproduction; and preserve biological and genetic diversity (Roberts 1995, 2000, Lauck et al. 1998, Agardy 2000, Roberts et al. 2001, McLeod et al. 2005, Grober-Dunsmore & Keller 2008) (Figure 1). There are examples of such closures in Belize, the

Cayman Islands, US Virgin Islands, and the Bahamas; seasonal and area closures have been used in Florida, Bermuda, and across the US Caribbean.

Importantly, some tFSA sites have shown remarkable recovery after their protection (Beets & Friedlander 1999, Whaylen et al. 2004, Burton et al. 2005, Nemeth 2005, Kadison et al. 2006, 2009b). Even sites in which certain species have been extirpated have seen recovery. The best example of this is provided from a well-studied, largely extirpated multispecies tFSA in the Virgin Islands. Kadison et al. (2009b) illustrated that once protected, aggregations of several species of groupers began to recover, and although Nassau grouper had been extirpated from the site long ago, this species had returned to the protected site. A similar example occurred at Caye Glory in Belize, where the Nassau grouper population dwindled to only 21 fish in 2001 (Sala et al. 2001) but has subsequently rebounded into the thousands (Belize Spawning Aggregations Working Committee unpublished data). These examples illustrate that protection of multispecies tFSA sites can provide protection for extant aggregations and recovery of extirpated species. Indeed, protection of defunct or extirpated sites may be an excellent strategy as the habitat is ideal for spawning in a variety of species at various times of year and may attract new reproductive adults. If sufficient numbers of regionally important tFSA sites can be protected, then evidence suggests that the spawning populations will recover rapidly and contribute to rebuilding of local fisheries and have positive impacts on the trophic integrity of Caribbean coral reef ecosystems through an increase in ecologically important carnivores.

Future research to support management

To manage tFSA sites effectively, a minimum amount of information about their status and condition is required. A better ecological understanding of the environmental characteristics of tFSAs will enable scientists, resource managers and fishers to predict the consequences of environmental and management change on spawning behaviour and reproductive success at these ecologically and socioeconomically important locations. Clearly, monitoring population recovery after protection is required and forms a core component of adaptive management. Data gathered from multiple levels of organization and complexity can be synthesized into a compelling case for management and a baseline for future monitoring. We suggest that a minimum amount of information might include bathymetric maps, year-round characterizations of the numbers and species spawning using underwater visual assessments, baseline port surveys or fishery-dependent data, some amount of physical environmental data and supporting video and still imagery (levels 1–4 in Table 5). These data, coupled with data from other locations, can be used to illustrate the importance of any tFSA site and track the recovery rate of a wide range of ecologically and economically important species that aggregate to spawn, as well as those key species that visit aggregations to feed, such as sharks and manta rays.

Although beyond the scope of the present review, understanding life-cycle connectivity is important and may lead to enhanced recovery of aggregative fish stocks by allowing the correct identification and adequate protection of essential fish habitats of each life stage, especially for the areas in which tFSAs once existed. For example, Nassau grouper require a suite of habitat types to support changing requirements for food and refuge through the various stages of the life cycle (ontogenetic habitat shifts). In the Bahamas, Nassau grouper settled exclusively in clumps of macroalgae and not in seagrass or on sand; postsettlement fish (25–35 mm total length, TL) resided within the algae clumps, early juveniles (60–150 mm TL) resided adjacent to the algae, and juveniles (>150 mm TL) colonized natural and artificial patch reefs in areas apparently removed from the postsettlement and early juvenile habitat (Eggleston 1995, Eggleston et al. 1998, Grover et al. 1998, Dahlgren & Eggleston 2001). The relationship between the health of nursery areas and recovery rates of adult populations at tFSA sites remains unclear; however, it is likely that the condition of both of these essential habitats will have an impact on overall population recovery.

Owing to the recent history of extirpation of some once very productive tFSAs and the vulnerability of many tropical species that aggregate to spawn, we concur with Sadovy et al. (2008) and advocate caution in communicating the locations of newly found or unprotected tFSAs. Information in the present review, however, was gained entirely from well-known and documented tFSAs. Predicting the location of tFSAs can create the risk of exposing undisturbed spawning stocks to opportunistic and severe fishing and rapid depletion before evaluation of sustainable management options takes place. Conversely, sites unregulated because they are unknown to environmental managers can also be vulnerable to overexploitation. From the perspective of science-driven, ecosystem-based management, fully detailed maps and descriptions of tFSAs are crucial to understanding the broader ecological patterns and processes operating within a region and between regions.

This review highlighted future directions for research, monitoring and management of Caribbean tFSAs. In the Caribbean region, enforcement capacity may be greater than in many remote areas of the Pacific but is still rather limited. Data sharing can be an essential component for management of regional or national resources, as demonstrated in Belize (Heyman 2011). By sharing information on site ecology and habitat maps, as well as enforcement techniques, legislation and policies, and techniques and tools for research, scientists, fishing communities and environmental managers can learn from others' successes and challenges.

Acknowledgements

R.S. Nemeth was supported in part by the National Science Foundation's Virgin Islands Experimental Program to Stimulate Competitive Research (VI-EPSCoR EPS-0346483) and the Center for Marine and Environmental Studies at the University of the Virgin Islands. S. Kobara was supported in part by VI-EPSCoR and the Department of Geography at Texas A&M University. W.D. Heyman was supported in part by the Department of Geography at Texas A&M University, the Marine Managed Area Science Program of Conservation International, and the Summit and Oak Foundations. S.J. Pittman was supported by National Oceanic and Atmospheric Administration (NOAA) Coral Reef Conservation Program. We would like to thank all of those who contributed unpublished data for this analysis, including A. Aguilar-Perera, F.F. Peralta, B. Reveles, R. Claro, K. Lindeman, A.Q. Espinosa, P. Bush, C. McCoy, B. Johnson, L. Whaylen, S. Heppel, J. Mateo, K. Baldwin, M. Schärer, R. Appeldoorn, M. Nemeth, N. Foster, A. Acosta, T. Kellison, M. Burton, M. Feeley, B. Glazer, A. Gleason, B. Keller, T. Battista, C. Taylor, K. Kingon, C. Jeffery, C. Koenig, F. Coleman, A. Oviedo, C. Cepeda, T. Trott, B. Kojis, L. Carr, E. Kadison, J. Claydon, N. Requena, N. Catzim, I. Majil, J. Posada, E. Ron, M. Prada, N. Zenny and A. Williams. We would like to thank J. Blondeau for early contributions to the manuscript. Finally, we wish to thank the myriad fishers throughout the Wider Caribbean who have contributed crucial anecdotal information and field assistance to scientists and managers in the characterization and conservation of tFSAs. Some of these include D. Samuel, G. Martinez, P. Gladding, D. DeMaria, E. Cuevas, E. Cuevas Jr., K.K. Nunez, A.J. Nunez, J. Cabral, J. Young, B. Young, S. Garbutt, K. Modera and V. Jacobs. This is contribution 81 from the Center for Marine and Environmental Studies at the University of the Virgin Islands.

References

Agardy, T. 2000. Information needs for marine protected areas: scientific and societal. *Bulletin of Marine Science* **66**, 875–888.

Aguilar-Perera, A. 2006. Disappearance of a Nassau grouper spawning aggregation off the southern Mexican Caribbean coast. *Marine Ecology Progress Series* **327**, 289–296.

Aguilar-Perera, A. & Aguilar-Dávila, W. 1996. A spawning aggregation of Nassau grouper *Epinephelus striatus* Pisces: Serranidae in the Mexican Caribbean. *Environmental Biology of Fishes* **45**, 351–361.

Aguilar-Perera, A., González-Salas, C., Tuz-Sulb, A., Villegas-Hernández, H. & López-Gómez, M. 2008. Identifying reef fish spawning aggregations in Alacranes reef, off Northern Yucatan peninsula, using the fishermen traditional ecological knowledge. *Proceedings of the Gulf and Caribbean Fisheries Institute* **60**, 554–558.

Aguilar-Perera, A., González-Salas, C. & Villegas-Hernández, H. 2009. Fishing, management, and conservation of the Nassau grouper, *Epinephelus striatus*, in the Mexican Caribbean. *Proceedings of the Gulf and Caribbean Fisheries Institute* **61**, 313–319.

Aguilar-Perera, A., Schärer, M. & Nemeth, M. 2006. Occurrence of juvenile Nassau grouper, *Epinephelus striatus* (Teleostei: Serranidae), off Mona Island, Puerto Rico: considerations of recruitment potential. *Caribbean Journal of Science* **42**, 264–267.

Appeldoorn, R. 2008. Transforming reef fisheries management: application of an ecosystem-based approach in the USA Caribbean. *Environmental Conservation* **35**(3), 232–241.

Archer, S.K., Heppell, S.A., Semmens, B.X., Pattengill-Semmens, C.V., Bush, P.G., McCoy, C.M. & Johnson, B.C. 2012. Patterns of color phase indicate spawn timing at a Nassau grouper *Epinephelus striatus* spawning aggregation. *Current Zoology* **58**, 73–83.

Armstrong, R.A., Singh, H., Torres, J., Nemeth, R.S., Can, A., Roman, C., Eustice, R., Riggs, L. & Garcia-Moliner, G. 2006. Characterizing the deep insular shelf coral reef habitat of the Hind Bank Marine Conservation District (U.S. Virgin Islands) using the seabed autonomous underwater vehicle. *Continental Shelf Research* **26**, 194–205.

Barlow, G.W. 1981. Patterns of parental investment, dispersal and size among coral-reef fishes. *Environmental Biology of Fishes* **6**, 65–85.

Beets, J. & Friedlander, A. 1999. Evaluation of a conservation strategy: a spawning aggregation closure for red hind, *Epinephelus guttatus*, in the U.S. Virgin Islands. *Environmental Biology of Fishes* **55**, 91–98.

Bolden, S.K. 2000. Long distance movement of a Nassau grouper (*Epinephelus striatus*) to a spawning aggregation in the central Bahamas. *Fisheries Bulletin* **98**, 642–645.

Boomhower, J., Romero, M., Posada, M., Kobara, S. & Heyman, W. 2010. Prediction and verification of possible reef-fish spawning aggregation sites in Los Roques Archipelago National Park, Venezuela. *Journal of Fish Biology* **77**(4), 822–840.

Brule, T., Deniel, C., Colas-Marrufo, T. & Sanches-Crespo, M. 1999. Red grouper reproduction in the southern Gulf of Mexico. *Transactions of the American Fisheries Society* **128**, 385–402.

Burbank, D.W. & Anderson, R.S. 2001. *Tectonic Geomorphology*. Malden, Massachusetts: Blackwell.

Burton, M.L., Brennan, K.J., Muñoz, R.C. & Parker, R.O., Jr. 2005. Preliminary evidence of increased spawning aggregations of mutton snapper (*Lutjanus analis*) at Riley's Hump two years after establishment of the Tortugas South Ecological Reserve. *Fishery Bulletin* **103**, 404–410.

Carolsfeld, J., Scott, A.P. & Sherwood, N.M. 1997. Pheromone-induced spawning of Pacific herring. *Hormones and Behavior* **31**, 269–276.

Carter, J. 1989. Grouper sex in Belize. *Natural History* **October**, 60–69.

Carter, J., Marrow, G.J. & Pryor, V. 1994. Aspects of the ecology and reproduction of Nassau grouper, *Epinephelus striatus*, off the coast of Belize, Central America. *Gulf and Caribbean Fisheries Institute* **43**, 64–111.

Carter, J. & Perrine, D. 1994. A spawning aggregation of dog snapper, *Lutjanus jocu* (Pisces, Lutjanidae) in Belize, Central America. *Bulletin of Marine Science* **55**, 228–234.

Center for Coastal Monitoring and Assessment (CCMA). 2011. Acoustic tracking of fish movements in coral reef ecosystems. Silver Spring, Maryland: The National Centers for Coastal Ocean Science. Online. http://ccma.nos.noaa.gov/ecosystems/coralreef/acoustic_tracking/ (accessed 25 October 2012).

Cherubin, L.M., Nemeth, R.S. & Idrisi, N. 2011. Flow and transport characteristics at an *Epinephelus guttatus* (red hind grouper) spawning aggregation site in St. Thomas (US Virgin Islands). *Ecological Modelling* **222**, 3132–3148.

Claro, R., Baisre, J., Lindeman, K.C. & Garcia-Artaga, J.P. 2001. Cuban fisheries: historical trends and current status. In *Ecology of the Marine Fishes of Cuba*, R. Claro et al. (eds). Washington, DC: Smithsonian Institution Press, 194–219.

Claro, R. & Lindeman, K.C. 2003. Spawning aggregation sites of snapper and grouper species (Lutjanidae and Serranidae) on the insular shelf of Cuba. *Gulf and Caribbean Research* **14**, 91–106.

Claro, R., Sadovy de Mitcheson, Y., Lindeman, K.C. & Garcia-Cagide, A.R. 2009. Historical analysis of Cuban commercial fishing effort and the effects of management interventions on important reef fishes from 1960–2005. *Fisheries Research* **99**, 7–16.

Claydon, J. 2004. Spawning aggregations of coral reef fishes: characteristics, hypotheses, threats and management. *Oceanography and Marine Biology: An Annual Review* **42**, 265–302.

Coleman, F.C., Koenig, C.C. & Collins, L.A. 1996. Reproductive styles of shallow-water groupers (Pisces: Serranidae) in the eastern Gulf of Mexico and the consequences of fishing spawning aggregations. *Environmental Biology of Fishes* **47**, 129–141.

Coleman, F.C., Koenig, C.C. & Scanlon, K.M. 2011. Groupers on the edge: shelf-edge spawning habitat in and around marine reserves of the Northeastern Gulf of Mexico. *The Professional Geographer* **63**(4), 1–19.

Colin, P.L. 1992. Reproduction of the Nassau grouper, *Epinephelus striatus* (Pisces, Serranidae) and its relationship to environmental conditions. *Environmental Biology of Fishes* **34**, 357–377.

Colin, P.L. 1995. Surface currents in Exuma Sound, Bahamas and adjacent areas with reference to potential larval transport. *Bulletin of Marine Sciences* **56**, 48–57.

Colin, P.L. 2012. Timing and location of aggregation and spawning in reef fishes. In *Reef Fish Spawning Aggregations: Biology, Research and Management*, Y. Sadovy de Mitcheson & P.L. Colin (eds). Fish & Fisheries Series **35**. Dordrecht, The Netherlands: Springer, 117–158.

Colin, P.L. & Clavijo, I.E. 1988. Spawning activity of fishes producing pelagic eggs on a shelf edge coral reef, southwestern Puerto Rico. *Bulletin of Marine Science* **43**, 249–279.

Colin, P.L., Sadovy, Y. & Domeier, M.L. 2003. *Manual for the Study and Conservation of Reef Fish Spawning Aggregations*. Society for the Conservation of Reef Fish Aggregations Special Publication No. 1 (Version 1.0). Hong Kong: Society for the Conservation of Reef Fish Aggregations.

Colin, P.L., Shapiro, D.Y. & Weiler, D. 1987. Aspects of the reproduction of two groupers, *Epinephelus guttatus* and *E. striatus* in the West Indies. *Bulletin of Marine Science* **40**, 220–230.

Costa, B. & Battista, T. 2008. Semi-automated classification of acoustic and optical remotely sensed imagery in the U.S. Caribbean. In *Proceedings of 11th International Coral Reef Symposium, Ft. Lauderdale, Florida, 7–11 July 2008*, **1**, 673–677.

Cowen, R.K. & Sponaugle, S. 2009. Larval dispersal and marine population connectivity. *Annual Reviews in Marine Science* **1**, 443–466.

Craig, A.K. 1966. *Geography of Fishing in British Honduras and Adjacent Coastal Waters*. Baton Rouge: Louisiana State University Press.

Crowder, L. & Norse, E. 2008. Essential ecological insights for marine ecosystem-based management and marine spatial planning. *Marine Policy* **32**(5), 772–778.

Dahlgren, C.P. & Eggleston, D.B. 2001. Spatio-temporal variability in abundance, size and microhabitat associations of early juvenile Nassau grouper *Epinephelus striatus* in an off-reef nursery system. *Marine Ecology Progress Series* **217**, 145–156.

Domeier, M.L. 2012. Revisiting spawning aggregations: definitions and challenges. In *Reef Fish Spawning Aggregations: Biology, Research and Management*, Y. Sadovy de Mitcheson & P.L. Colin (eds). Fish & Fisheries Series **35**. Dordrecht, The Netherlands: Springer, 1–20.

Domeier, M.L. & Colin, P.L. 1997. Tropical reef fish spawning aggregations: defined and reviewed. *Bulletin of Marine Science* **60**, 698–726.

Eggleston, D.B. 1995. Recruitment of Nassau grouper *Epinephelus striatus*: post-settlement abundance, microhabitat features, and ontogenetic habitat shifts. *Marine Ecology Progress Series* **124**, 9–22.

Eggleston, D.B., Grover, J.J. & Lipcius, R.N. 1998. Ontogenetic diet shifts in Nassau grouper: trophic linkages and predatory impact. *Bulletin of Marine Science* **63**, 111–126.

Eklund, A.M., McClellan, D.B. & Harper, D.E. 2000. Black grouper aggregations in relation to protected areas within the Florida Keys National Marine Sanctuary. *Bulletin of Marine Science* **66**, 721–728.

Ellis, S.C., Watanabe, W.O. & Ellis, E.P. 1997. Temperature effects on feed utilization and growth of postsettlement stage Nassau grouper. *Transactions of the American Fisheries Society* **126**, 309–315.

Elsdon, T.S., Wells, B.K., Campana, S.E., Gillanders, B.M., Jones, C.M., Limburg, K.E., Secor, D.H., Thorrold, S.R. & Walther, B.D. 2008. Otolith chemistry to describe movements and life-history parameters of fishes: hypotheses, assumptions, limitations, and inferences. *Oceanography and Marine Biology: An Annual Review* **46**, 297–330.

Erisman, B.E., Aburto-Oropeza, O., Gonzalez-Abraham, C., Mascareños-Osorio, I., Moreno-Báez, M. & Hastings, P.A. 2012. Spatio-temporal dynamics of a fish spawning aggregation and its fishery in the Gulf of California. *Scientific Reports* **2**(284), 11 pp. doi:10.1038/srep00284.

Erisman, B.E. & Allen, L.G. 2006. Reproductive behaviour of a temperate serranid fish, *Paralabrax clathratus* (Girard), from Santa Catalina Island, California, U.S.A. *Journal of Fish Biology* **68**, 157–184.

Ezer, T., Heyman, W.D., Houser, C. & Kjerfve, B. 2011. Modeling and observations of high-frequency flow variability and internal waves at a Caribbean reef spawning aggregation site. *Ocean Dynamics* **61**, 581–598.

Ezer, T., Heyman, W.D., Houser, C. & Kjerfve, B. 2012. Extreme flows and unusual water levels near a Caribbean coral reef: was this the case of a "perfect storm"? *Ocean Dynamics* **62**, 1043–1057.

Ezer, T., Thattai, D.V., Kjerfve, B. & Heyman, W.D. 2005. On the variability of the flow along the Meso-American Barrier Reef system: a numerical model study of the influence of the Caribbean current and eddies. *Ocean Dynamics* **55**, 458–475.

Fine, J.C. 1990. Grouper in love: spawning aggregations of Nassau groupers in Honduras. *Sea Frontiers* **January/February**, 42–45.

Fine, J.C. 1992. Greedy for groupers. *Wildlife Conservation* **November/December**, 68–71.

Friedlander, A.M., Brown, E.K., Jokiel, P.L., Smith, W.R. & Rodgers, K.S. 2003. Effects of habitat, wave exposure, and marine protected area status on coral reef fish assemblages in the Hawaiian archipelago. *Coral Reefs* **22**, 291–305.

Froese, R. & Pauly, D. (eds) 2010. FishBase. Online. http://www.fishbase.org/ (accessed January 2011).

Ganias, K. 2008. Ephemeral spawning aggregations in the Mediterranean sardine, *Sardina pilchardus*: a comparison with other multiple-spawning clupeoids. *Marine Biology* **155**, 293–301.

Geselbracht, L., Torres, R., Cumming, G.S., Dorfman, D., Beck, M. & Shaw, D. 2008. Identification of a spatially efficient portfolio of priority conservation sites in marine and estuarine areas of Florida. *Aquatic Conservation: Marine and Freshwater Ecosystems* **19**, 408–420.

Gleason, A.C.R., Kellison, G.T. & Reid, R.P. 2011. Geomorphic characterization of reef fish aggregation sites in the upper Florida Keys, USA, using single-beam acoustics. *The Professional Geographer* **63**, 443–445.

Graham, R.T., Carcamo, R., Rhodes, K.L., Roberts, C.M. & Requena, N. 2008. Historical and contemporary evidence of a mutton snapper (*Lutjanus analis* Cuvier, 1828) spawning aggregation fishery in decline. *Coral Reefs* **27**, 311–319.

Graham, R.T. & Castellanos, D.W. 2005. Courtship and spawning behaviors of carangid species in Belize. *Fishery Bulletin* **103**, 426–432.

Grober-Dunsmore, R. & Keller, B.D. 2008. Caribbean connectivity: implications for marine protected area management. In *Proceedings of a Special Symposium, 9–11 November 2006, 59th Annual Meeting of the Gulf and Caribbean Fisheries Institute, Belize City, Belize*, R. Grober-Dunsmore & B.D. Keller (eds). Marine Sanctuaries Conservation Series ONMS-08-07. Silver Spring, Maryland: U.S. Department of Commerce, National Oceanic and Atmospheric Administration, Office of National Marine Sanctuaries, 4–7.

Grober-Dunsmore, R., Pittman, S.J., Caldow, C., Kendall, M.A. & Fraser, T. 2009. A landscape ecology approach for the study of ecological connectivity across tropical marine seascapes. In *Ecological Connectivity Among Tropical Coastal Ecosystems*, I. Nagelkerken (ed.). Dordrecht, The Netherlands: Springer, 493–530.

Grover, J.J., Eggleston, D.B. & Shenker, J.M. 1998. Transition from pelagic to demersal phase in early-juvenile Nassau grouper, *Epinephelus striatus*: pigmentation, squamation, and ontogeny of diet. *Bulletin of Marine Science* **62**, 97–113.

Hamilton, R.J., Matawai, M., Totuku, T., Kama, W., Lahui, P., Warku, J. & Smith, A.J. 2005. Applying local knowledge and science to the management of grouper aggregation sites in Melanesia. *SPC Live Reef Fish Information Bulletin* **14**, 7–19.

Hamilton, R.J., Potuku, T. & Montambault, J.R. 2011. Community-based conservation results in the recovery of reef fish spawning aggregations in the Coral Triangle. *Biological Conservation* **144**, 1850–1858.

Hamilton, R.J., Sadovy de Mitcheson, Y. & Aguilar-Perera, A. 2012. The role of local ecological knowledge in the conservation and management of reef fish spawning aggregations. In *Reef Fish Spawning Aggregations: Biology, Research and Management*, Y. Sadovy de Mitcheson & P.L. Colin (eds). Fish & Fisheries Series **35**. Dordrecht, The Netherlands: Springer, 331–369.

Hamner, W.M., Colin, P.L. & Hamner, P.P. 2007. Export-import dynamics of zooplankton on a coral reef in Palau. *Marine Ecology Progress Series* **334**, 83–92.

Harris, P.T. & Baker, E.K. (eds) 2011. *Seafloor Geomorphology as Benthic Habitat: Geohab Atlas of Seafloor Geomorphic Features and Benthic Habitats*. Amsterdam: Elsevier, 947 pp.

Heppell, S., Semmens, B.X., Archer, S.K., Pattengill-Semmens, C.V., Bush, P.G., McCoy, C.M., Heppell, S.S. & Johnson, B.C. 2012. Documenting recovery of a spawning aggregation through size frequency analysis from underwater laser calipers measurements. *Biological Conservation* **155**, 119–127.

Herzlieb, S., Kadison, E., Blondeau, J. & Nemeth, R.S. 2006. Comparative assessment of coral reef systems located along the insular platform of St. Thomas, U.S. Virgin Islands and the relative effects of natural and human impacts. In *The International Coral Reef Symposium (ICRS). Proceedings of the 10th International Coral Reef Conference, June 28–July 2 2004, Okinawa, Japan*, Y. Suzuki et al. (eds). Tokyo: Japanese Coral Reef Society, 1144–1151.

Heyman, W.D. 2011. Elements for building a participatory, ecosystem-based marine reserve network. *The Professional Geographer* **63**, 475–488.

Heyman, W.D., Azueta, J., Lara, O.F., Majil, I., Neal, D., Luckhurst, B., Paz, M., Morrison, I., Rhodes, K.L., Kjerfve, B., Wade, B. & Requena, N. 2004. *Spawning Aggregation Monitoring Protocol for the Meso-American Reef and the Wider Caribbean*. Version 2.0. Belize City, Belize: Meso-American Barrier Reef Systems Project.

Heyman, W.D., Carr, L. & Lobel, P.S. 2010. Diver ecotourism and disturbance to reef fish spawning aggregations: it is better to be disturbed than to be dead. *Marine Ecology Progress Series* **419**, 201–210.

Heyman, W.D., Ecochard, J.L.B. & Biasi, F.B. 2007. Low-cost bathymetric mapping for tropical marine conservation: a focus on reef fish spawning aggregation sites. *Marine Geodesy* **30**, 37–50.

Heyman, W.D. & Hyatt, T. 1996. *An Analysis of Commercial and Sport Fishing in the Proposed Port Honduras Marine Reserve*. Belize City: Belize Center for Environmental Studies.

Heyman, W.D. & Kjerfve, B. 2008. Characterization of transient multi-species reef fish spawning aggregations at Gladden Spit, Belize. *Bulletin of Marine Science* **83**, 531–551.

Heyman, W.D., Kjerfve, B., Graham, R.T., Rhodes, K.L. & Garbutt, L. 2005. Spawning aggregations of *Lutjanus cyanopterus* (Cuvier) on the Belize Barrier Reef over a 6 year period. *Journal of Fish Biology* **67**, 83–101.

Heyman, W.D., Kjerfve, B, Johannes, R.E. & Graham, R. 2001. Whale sharks, *Rhincodon typus*, aggregate to feed on fish spawn in Belize. *Marine Ecology Progress Series* **215**, 275–282.

Heyman, W.D. & Kobara, S. 2012. Geomorphology of reef fish spawning aggregations in Belize and the Cayman Islands (Caribbean). In *Seafloor Geomorphology as Benthic Habitat*, P.T. Harris & E.K. Baker (eds). London: Elsevier, 388–396.

Heyman, W.D. & Requena, N. 2002. *Status of Multi-Species Spawning Aggregations in Belize*. Punta Gorda, Belize: Nature Conservancy.

Heyman, W.D. & Wright, D.J. 2011. Marine geomorphology in the design of marine reserve networks. *The Professional Geographer* **63**, 1–14.

Hitt, S., Pittman, S.J. & Nemeth, R.S. 2011. Diel movements of fish are linked to benthic seascape structure in a Caribbean coral reef ecosystem. *Marine Ecology Progress Series* **247**, 275–291.

Hogan, J.D., Thiessen, R.J., Sale, P.F. & Heath, D.D. 2012. Local retention, dispersal and fluctuating connectivity among populations of a coral reef fish. *Oecologia* **168**, 61–71.

Holt, S.A. 2008. Distribution of red drum spawning sites identified by a towed hydrophone array. *Transactions of the American Fisheries Society* **137**, 551–561.

Hyrenbach, K.D., Forney, K.A. & Dayton, P.K. 2000. Marine protected areas and ocean basin management. *Aquatic Conservation: Marine and Freshwater Ecosystems* **10**, 437–458.

International Union for Conservation of Nature and Natural Resources (IUCN). 2011. *IUCN Red List of Threatened Species*. Version 2011.1. London: IUCN Global Species Programme Red List Unit. Online. http://www.iucnredlist.org (accessed 25 October 2011).

Jeffrey, C.F.G. 2000. Annual, coastal and seasonal variation in Grenadian demersal fisheries (1986–1993) and implications for management. *Bulletin of Marine Science* **66**, 305–319.

Jensen, J.R. 2007. *Remote Sensing of the Environment: An Earth Resource Perspective*. Upper Saddle River, New Jersey: Prentice-Hall, 2nd edition.

Johannes, R.E. 1978. Reproductive strategies of coastal marine fishes in the tropics. *Environmental Biology of Fishes* **3**, 65–84.

Johnston, S.V., Rosario, A., Rivera, J.A., Timko, M.A., Nealson, P.A. & Kumagai, K.K. 2006. Hydroacoustic evaluation of spawning red hind (*Epinephelus guttatus*) aggregations along the coast of Puerto Rico in 2002 and 2003. In *Emerging Technologies for Reef Fisheries Research and Management*, J.C. Taylor (ed.). *NOAA Professional Paper NMFS Series* **5**, 10–17.

Jones, G.P., Almany, G.R., Russ, G.R., Sale, P.F., Steneck, R.S., van Oppen, M.J.H. & Willis, B.L. 2009. Larval retention and connectivity among populations of corals and reef fishes: history, advances and challenges. *Coral Reefs* **28**(2), 307–325.

Jones, G.P., Milicich, M.J., Emslie, M.J. & Lunow, C. 1999. Self-recruitment in a coral reef fish population. *Nature* **402**, 802–804.

Jones, G.P., Planes, S. & Thorrold, S.R. 2005. Coral reef fish larvae settle close to home. *Current Biology* **15**, 1314–1318.

Jonsson, N., Hansen, L.P. & Jonsson, B. 1991. Variation in age, size and repeat spawning of adult Atlantic salmon in relation to river discharge. *Journal of Animal Ecology* **60**, 937–947.

Kadison, E., Nemeth, R.S. & Blondeau, J. 2009a. Assessment of an unprotected red hind (*Epinephelus guttatus*) spawning aggregation on Saba Bank in the Netherlands Antilles. *Bulletin of Marine Science* **85**, 101–118.

Kadison, E., Nemeth, R.S., Blondeau, J., Smith, T. & Calnan, J. 2009b. Nassau grouper (*Epinephelus striatus*) in St. Thomas, U.S. Virgin Islands, with evidence for a spawning aggregation site recovery. *Proceedings of the Gulf and Caribbean Fisheries Institute* **62**, 273–279.

Kadison, E., Nemeth, R.S., Herzlieb, S. & Blondeau, J. 2006. Temporal and spatial dynamics of *Lutjanus cyanopterus* and *L. jocu* (Pisces: Lutjanidae) spawning aggregations on a multi-species spawning site in the USVI. *Revista Biologia Tropical* **54**, 69–78.

Karnauskas, M., Chérubin, L.M. & Paris, C.B. 2011. Adaptive significance of the formation of multi-species fish spawning aggregations near submerged capes. *PLoS one* **6**(7), e22067.

Kingsford, M.J., Wolanski, E. & Choat, J.H. 1991. Influence of tidally induced fronts and Langmuir circulations on distribution and movements of presettlement fishes around a coral reef. *Marine Biology* **109**, 167–180.

Kjerfve, B. 1981. Tides of the Caribbean Sea. *Journal of Geophysical Research* **86**, 4243–4247.

Kobara, S. & Heyman, W.D. 2006. Caribbean-wide geospatial analysis of the location of transient reef fish spawning aggregation sites using remote sensing. *Proceedings of the Gulf and Caribbean Fisheries Institute* **59**, 463–465.

Kobara, S. & Heyman, W.D. 2008. Geomorphometric patterns of Nassau grouper (*Epinephelus striatus*) spawning aggregation sites in the Cayman Islands. *Marine Geodesy* **31**, 231–245.

Kobara, S. & Heyman, W.D. 2010. Sea bottom geomorphology of multi-species spawning aggregation sites in Belize. *Marine Ecology Progress Series* **405**, 231–242.

Koenig, C.C., Coleman, F.C., Collins, L.A., Sadovy, Y. & Colin, P.L. 1996. Reproduction in gag (*Mycteroperca microlepis*) (Pisces: Serranidae) in the eastern Gulf of Mexico and the consequences of fishing spawning aggregations. In *Biology, Fisheries and Culture of Tropical Groupers and Snappers*, F. Arreguin-Sanchez, et al. (eds). *ICLARM Conference Proceedings* **48**, 307–323.

Koenig, C.C., Coleman, F.C., Grimes, C.B., Fitzhugh, G.R., Scanlon, K.M., Gledhill, C.T. & Grance, M. 2000. Protection of fish spawning habitat for the conservation of warm-temperate reef-fish fisheries of shelf-edge reefs of Florida. *Bulletin of Marine Science* **66**, 593–616.

Kracker, L.M., Tayler, J.C., Ebert, E.F., Battista, T.A. & Mendaza, C. 2011. Integration of fisheries acoustics surveys and bathymetric mapping to characterize midwater-seafloor habitats of U.S. Virgin Islands and Puerto Rico (2008–2010). NOAA Technical Memorandum NOS NCCOS 130. Charleston, South Carolina: NOAA National Center for Coastal Ocean Science.

Lam, T.J. 1983. Environmental influences on gonadal activity in fish. In *Fish Physiology*, Vol. 9, W.S. Hoar et al. (eds). New York: Academic Press, 65–116.

Lauck, T., Clark, C.W., Mangel, M. & Munro, G.R. 1998. Implementing the precautionary principle in fisheries management through marine reserves. *Ecological Applications* **8**, S72–S78.

Levin, S.A. & Lubchenco, J. 2008. Resilience, robustness and marine ecosystem-based management. *BioScience* **58**, 27–32.

Lindeman, K.C., Pugliese, R., Waugh, G.T. & Ault, J.S. 2000. Developmental patterns within a multispecies reef fishery: management applications for essential fish habitats and protected areas. *Bulletin of Marine Science* **66**, 929–956.

Lobel, P.S. 1978. Diel, lunar, and seasonal periodicity in the reproductive behavior of the pomacanthid fish, *Centropyge potteri*, and some other reef fishes in Hawaii. *Pacific Science* **32**, 193–207.

Lobel, P.S. 1992. Sounds produced by spawning fishes. *Environmental Biology of Fishes* **33**, 351–358.

Lobel, P.S. 2005. Scuba bubble noise and fish behavior: a rationale for silent diving technology. In *Diving for Science 2005. Proceedings of the American Academy of Underwater Sciences 24th Annual Symposium*, J.M. Godfrey & S.E. Shumway (eds). Dauphin Island, Alabama: American Academy of Underwater Sciences, 49–59.

Lobel, P.S. & Lobel, L.K. 2008. Aspects of the biology and geomorphology of Johnston and Wake Atolls, Pacific Ocean. In *Coral Reefs of the USA*, B.M. Riegl & R.E. Dodge (eds). New York: Springer, 655–689.

Lobel, P.S. & Robinson, A.R. 1988. Larval fishes and zooplankton in a cyclonic eddy in Hawaiian waters. *Journal of Plankton Research* **10**, 1209–1223.

Locascio, J.V. & Mann, D.A. 2008. Diel periodicity of fish sound production in Charlotte Harbor, Florida. *Transactions of the American Fisheries Society* **137**, 606–615.

Love, R.H. 1977. Target strength of an individual fish at any aspect. *Journal of the Acoustical Society of America* **62**, 1397–1403.

Luckhurst, B.E. 1998. Site fidelity and return migration of tagged red hinds (*Epinephelus guttatus*) to a spawning aggregation site in Bermuda. *Proceedings of the Gulf and Caribbean Fisheries Institute* **50**, 750–763.

Luckhurst, B.E. 2010. Observations of a black grouper (*Mycteroperca bonaci*) spawning aggregation in Bermuda. *Gulf and Caribbean Research* **22**, 43–49.

Mann, D.A. & Lobel, P.S. 1995. Passive acoustic detection of sounds produced by the damselfish, *Dascyllus albisella* (Pomacentridae). *Bioacoustics* **6**, 199–213.

Mann, D.A., Locascio, J.V., Coleman, F.C. & Koenig, C.C. 2009. Goliath grouper *Epinephelus itajara* sound production and movement patterns on aggregation sites. *Endangered Species Research* **7**, 229–236.

Mann, D., Locascio, J., Schärer, M., Nemeth, M. & Appeldoorn, R. 2010. Sound production by red hind *Epinephelus guttatus* in spatially segregated spawning aggregations. *Aquatic Biology* **10**, 149–154.

Matos-Caraballo, D., Posada, J.M. & Luckhurst, B.E. 2006. Fishery-dependent evaluation of a spawning aggregation of tiger grouper (*Mycteroperca tigris*) at Vieques Island, Puerto Rico. *Bulletin of Marine Science* **79**, 1–16.

Mazeroll, A.I. & Montgomery, W.L. 1995. Structure and organization of local migrations in brown surgeonfish (*Acanthurus nigrofuscus*). *Ethology* **99**, 89–106.

Mazeroll, A.I. & Montgomery, W.L. 1998. Daily migrations of a coral reef fish in the Red Sea (Gulf of Aqaba, Israel): initiation and orientation. *Copeia* **1998**, 893–905.

McLeod, K.L., Lubchenco, J., Palumbi, S.R. & Rosenberg, A.A. 2005. *Scientific Consensus Statement on Marine Ecosystem-Based Management*. Washington, DC: Communication Partnership for Science and the Sea.

Mendes, J.M. & Woodley, J.D. 2002. Timing of reproduction in *Montastraea annularis*: relationship to environmental variables. *Marine Ecology Progress Series* **227**, 241–251.

Munro, J. & Blok, L. 2003. The status of stocks of groupers and hinds in the Northeastern Caribbean. *Proceedings of the Gulf and Caribbean Fisheries Institute* **56**, 283–294.

Nelson, M.D., Koenig, C.C., Coleman, F.C. & Mann, D.A. 2011. Sound production of red grouper *Epinephelus morio* on the West Florida Shelf. *Aquatic Biology* **12**, 97–108.

Nemeth, M.I., Schärer, M.T. & Appeldoorn, R.S. 2007. Observations from a small grouper spawning aggregation at Mona Island, Puerto Rico. *Proceedings of the Gulf and Caribbean Fishery Institute* **59**, 489–492.

Nemeth, R.S. 2005. Population characteristics of a recovering U.S. Virgin Islands red hind spawning aggregation following protection. *Marine Ecology Progress Series* **286**, 81–97.

Nemeth, R.S. 2009. Dynamics of reef fish and decapod crustacean spawning aggregations: underlying mechanisms, habitat linkages and trophic interactions. In *Ecological Interactions Among Tropical Coastal Ecosystems*, I. Negelkerken (ed.). Dordrecht, The Netherlands: Springer, 73–134.

Nemeth, R.S. 2012. Ecosystem aspects of species that aggregate to spawn. In *Reef Fish Spawning Aggregations: Biology, Research and Management*, Y. Sadovy de Mitcheson & P.L. Colin (eds). Fish & Fisheries Series **35**. Dordrecht, The Netherlands: Springer, 21–56.

Nemeth, R.S., Blondeau, J., Herzlieb, S. & Kadison, E. 2007. Spatial and temporal patterns of movement and migration at spawning aggregations of red hind, *Epinephelus guttatus*, in the U.S. Virgin Islands. *Environmental Biology of Fishes* **78**, 365–381.

Nemeth, R.S., Kadison, E., Blondeau, J.E., Idrisi, N., Watlington, R., Brown, K., Smith, T. & Carr, L. 2008. Regional coupling of red hind spawning aggregations to oceanographic processes in the eastern Caribbean. In *Caribbean Connectivity: Implications for Marine Protected Area Management. Proceedings of a Special Symposium, 9–11 November 2006, 59th Annual Meeting of the Gulf Caribbean Fisheries Institute, Belize City, Belize*, R. Grober-Dunsmore & B.D. Keller (eds). Marine Sanctuaries Conservation Series NMSP-08-07. Silver Spring, Maryland: U.S. Department of Commerce, National Oceanic and Atmospheric Administration, Office of National Marine Sanctuaries, 170–183.

Nemeth, R.S., Kadison, E., Herzlieb, S., Blondeau, J. & Whiteman, E. 2006. Status of a yellowfin grouper (*Mycteroperca venenosa*) spawning aggregation in the U.S. Virgin Islands with notes on other species. *Proceedings of the Gulf and Caribbean Fisheries Institute* **57**, 543–558.

Nemeth, R.S. & Quandt, A. 2005. Differences in fish assemblage structure following the establishment of the Marine Conservation District, St. Thomas U.S. Virgin Islands. *Proceedings of the Gulf and Caribbean Fisheries Institute* **56**, 367–381.

Ojeda-Serrano, E., Appeldoorn, R. & Ruiz-Valentine, I. 2007. Reef fish spawning aggregations of the Puerto Rican Shelf. Final Report to Caribbean Coral Reef Institute, La Parguera, Puerto Rico. Mayagüez, Puerto Rico: University of Puerto Rico. Online. http://ccri.uprm.edu/researcher/Ojeda/Ojeda_Final_Report_CCRI_SPAG%27s.pdf (accessed 8 April 2013).

Pankhurst, N.W. & Fitzgibbon, Q.P. 2006. Characteristics of spawning behavior in cultured greenback flounder *Rhombosolea tapirina*. *Aquaculture* **253**, 279–289.

Paris, C.B. & Cowen, R.K. 2004. Direct evidence of a biophysical retention mechanism for coral reef fish larvae. *Limnology and Oceanography* **49**, 1964–1979.

Paris, C.B., Cowen, R.K., Claro, R. & Lindeman, K.C. 2005. Larval transport pathways from Cuban snapper (Lutjanidae) spawning aggregations based on biophysical modeling. *Marine Ecology Progress Series* **296**, 93–106.

Paz, G. & Grimshaw, T. 2001. Status report on Nassau grouper aggregations in Belize, Central America. In *Proceedings of the First National Workshop on the Status of Nassau Groupers in Belize: Working Towards Sustainable Management, 30 July 2001*. Belize City, Belize: Green Reef Environmental Institute, 27–36.

Pet, J.S., Mous, P.J., Muljadi, A.H., Sadovy, Y.J. & Squire, L. 2005. Aggregations of *Plectropomus areolatus* and *Epinephelus fuscoguttatus* (groupers, Serranidae) in the Komodo National Park, Indonesia: monitoring and implications for management. *Environmental Biology of Fishes* **74**, 209–218.

Pittman, S.J. & Brown, K.A. 2011. Multi-scale approach for predicting fish species distributions across coral reef seascapes. *PLoS one* **6**(5), e20583.

Pittman, S.J., Costa, B. & Battista, T. 2009. Using Lidar bathymetry and boosted regression trees to predict the diversity and abundance of fish and corals. *Journal of Coastal Research* **S53**, 27–38.

Pittman, S.J. & Legare, B. 2010. *Preliminary Results from the Caribbean Acoustic Tracking Network (CATn): A Data Sharing Partnership for Acoustic Tracking and Movement Ecology of Marine Animals in the Caribbean Sea*. Silver Spring, Maryland: NOAA Biogeography Branch & St. Thomas, U.S. Virgin Islands: University of the Virgin Islands.

Pittman, S.J. & McAlpine, C.A. 2003. Movement of marine fish and decapod crustaceans: process, theory and application. *Advances in Marine Biology* **44**, 205–294.

Planes, S., Thorrold, S.R. & Jones, G.P. 2009. Larval dispersal connects fish populations in a network of marine protected areas. *Proceedings of the National Academy of Sciences of the United States of America* **106**, 5693–5697. doi:10.1073/pnas.0808007106.

Prior, D.B. & Hooper, J.R. 1999. Sea floor engineering geomorphology: recent achievements and future directions. *Geomorphology* **31**, 411–439.

Randall, J.E. & Randall, H.A. 1963. The spawning and early development of the Atlantic parrot fish, *Sparisoma rubripinne*, with notes on other scarid and labrid fishes. *Zoologica* **48**, 49–60.

Rhodes, K.L., McIlwain, J., Joseph, E. & Nemeth, R.S. 2012. Reproductive movement, residency and fisheries vulnerability of brown-marbled grouper, *Epinephelus fuscoguttatus* (Forsskål, 1775). *Coral Reefs* **31**, 443–453.

Rhodes, K.L. & Sadovy, Y. 2002. Temporal and spatial trends in spawning aggregations of camouflage grouper, *Epinephelus polyphekadion* (Bleeker 1849) in Pohnpei, Micronesia. *Environmental Biology of Fishes* **63**, 27–39.

Rhodes, K.L., Taylor, B.M. & McIlwain, J.L. 2011. Detailed demographic analysis of an *Epinephelus polyphekadion* spawning aggregation and fishery. *Marine Ecology Progress Series* **421**, 183–198.

Rhodes, K.L. & Tupper, M.H. 2007. A preliminary market-based analysis of the Pohnpei, Micronesia, grouper (Serranidae: Epinepheline) fishery reveals unsustainable fishing practices. *Coral Reefs* **26**, 335–344.

Rhodes, K.L. & Tupper, M.H. 2008. The vulnerability of reproductively active squaretail coral grouper (*Plectropomis areolatus*) to fishing. *Fishery Bulletin* **106**, 194–203.

Rivera, J.A., Prada, M.C., Arsenault, J.L., Moody, G. & Benoit, N. 2006. Detecting fish aggregations from reef habitats mapped with high resolution side scan sonar imagery. In *Emerging Technologies for Reef Fisheries Research and Management*, J.C. Taylor (ed.). *NOAA Professional Paper NMFS Series* **5**, 88–104.

Roberts, C.M. 1995. Rapid build-up of fish biomass in a Caribbean marine reserve. *Conservation Biology* **9**, 815–826.

Roberts, C.M. 2000. Selecting marine reserve locations: optimality versus opportunism. *Bulletin of Marine Science* **66**, 581–592.

Roberts, C.M., Bohnsack, J.A., Gell, F., Hawkins, J.P. & Goodridge, R. 2001. Effects of marine reserves on adjacent fisheries. *Science* **294**(5548), 1920–1923.

Robertson, D.R. 1991. The role of adult biology in the timing of spawning of tropical reef fishes. In *The Ecology of Fishes on Coral Reefs*, P.F. Sale (ed.). Sydney: Academic Press, 356–386.

Rochet, M.-J., Cadiou, J.-F. & Trenkel, V. 2006. Precision and accuracy of fish length measurements obtained with two visual underwater methods. *Fishery Bulletin* **104**, 1–9.

Rowell, T.J., Schärer, M.T., Appeldoorn, R.S., Nemeth, M.I., Mann, D.A. & Rivera, J.A. 2012. Sound production as an indicator of red hind density at a spawning aggregation. *Marine Ecology Progress Series* **462**, 241–250.

Sadovy, Y. 1997. The case of the disappearing grouper: *Epinephelus striatus* (Pisces: Serranidae). *Journal of Fish Biology* **46**, 961–976.

Sadovy, Y., Colin, P.L. & Domeier, M.L. 1994. Aggregation and spawning in the tiger grouper, *Mycteroperca tigris* (Pisces, Serranidae). *Copeia* **2**, 511–516.

Sadovy, Y. & Domeier, M. 2005. Are aggregation-fisheries sustainable? Reef fish fisheries as a case study. *Coral Reefs* **24**, 254–262.

Sadovy, Y. & Eklund, A.M. 1999. Synopsis of biological data on the Nassau grouper, *Epinephelus striatus* (Bloch, 1792), and the jewfish, *E. itajara* (Lichtenstein, 1822). NOAA Technical Report NMFS 146. Seattle, Washington: U.S. Department of Commerce.

Sadovy de Mitcheson, Y., Cornish, A., Domeier, M., Colin, P.L., Russell, M. & Lindeman, K.C. 2008. A global baseline for spawning aggregations of reef fishes. *Conservation Biology* **22**, 1233–1244.

Sadovy de Mitcheson, Y., Craig, M.T., Bertoncini, A.A., Carpenter, K.E., Cheung, W.L., Choat, J.H., Cornish, A.S., Fennessy, S.T., Ferreira, B.P., Heemstra, P.C., Liu, M., Myers, R.F., Pollard, D.A., Rhodes, K.L., Rocha, L.A., Russell, B.C., Samoilys, M.A. & Sanciangco, J. 2012. Fishing groupers towards extinction: a global assessment of threats and extinction risks in a billion dollar fishery. *Fish and Fisheries* doi:10.1111/j.1467–2979.2011.00455.x.

Sala, E., Ballesteros, E. & Starr, R.M. 2001. Rapid decline of Nassau grouper spawning aggregations in Belize: fishery management and conservation needs. *Fisheries* **26**(10), 23–30.

Sale, P.F. 2004. Connectivity, recruitment variation, and the structure of reef fish communities. *Integrative and Comparative Biology* **44**(5), 390–399.

Sale, P.F., Cowen, R.K., Danilowicz, B.S., Jones, G.P., Kritzer, J.P., Lindeman, K.C., Planes, S., Polunin, N.V.C., Russ, G.R., Sadovy, Y.J. & Steneck, R.S. 2005. Critical science gaps impede use of no-take fishery reserves. *Trends in Ecology and Evolution* **20**, 74–80.

Samoilys, M.A. 1997. Periodicity of spawning aggregations of coral trout (*Plectropomus leopardus*) on the Great Barrier Reef. *Marine Ecology Progress Series* **160**, 149–159.

Sancho, G., Petersen, C.W. & Lobel, P.S. 2000a. Predator-prey relations at a spawning aggregation site of coral reef fishes. *Marine Ecology Progress Series* **203**, 275–288.

Sancho, G., Solow, A.R. & Lobel, P.S. 2000b. Environmental influences on the diel timing of spawning in coral reef fishes. *Marine Ecology Progress Series* **206**, 193–212.

Schärer, M.T., Nemeth, M.I. & Appeldoorn, R.S. 2010. Protecting a multi-species spawning-aggregation site at Mona Island, Puerto Rico. *Proceedings of the Gulf and Caribbean Fisheries Institute* **62**, 252–259.

Schärer, M.T., Nemeth, M.I., Mann, D., Locascio, J., Appeldoorn, R.S. & Rowell, T.J. 2012a. Sound production and reproductive behavior of yellowfin grouper, *Mycteroperca venenosa* (Serranidae) at a spawning aggregation. *Copeia* **2012**, 135–144.

Schärer, M.T., Rowell, T.J., Nemeth, M.I. & Appeldoorn, R.S. 2012b. Sound production and associated reproductive behavior of Nassau grouper *Epinephelus striatus* (Pisces: Epinephelidae) at spawning aggregations. *Endangered Species Research* doi:10.3354/esr00457.

Semmens, B.X., Luke, K.E., Bush, P.G., McCoy, C. & Johnson, B.C. 2006. Isopod infestation of postspawning Nassau grouper around Little Cayman Island. *Journal of Fish Biology* **69**, 933–937.

Shapiro, D., Hensley, D. & Appeldoorn, R. 1988. Pelagic spawning and egg transport in coral-reef fishes: a skeptical overview. *Environmental Biology of Fishes* **22**, 3–14.

Shapiro, D.Y., Sadovy, Y. & McGehee, M.A. 1993. Size, composition, and spatial structure of the annual spawning aggregation of the red hind, *Epinephelus guttatus* (Pisces: Serranidae). *Copeia* **2**, 399–406.

Shcherbina, A.Y., Gawarkiewicz, G.G., Linder, C.A. & Thorrold, S.R. 2008. Mapping bathymetric and hydrographic features of Glover's Reef, Belize, with a REMUS autonomous underwater vehicle. *Limnology and Oceanography* **53**, 2264–2272.

Sims, D.W., Wearmouth, V.J., Genner, M.J., Southwood, A.J. & Hawkins, S.J. 2005. Low-temperature-driven early spawning migration of temperate marine fish. *Journal of Animal Ecology* **73**, 333–341.

Smith, C.L. 1971. A revision of the American groupers: *Epinephelus* and allied genera. *Bulletin of the American Museum of Natural History* **146**, 67–242.

Smith, C.L. 1972. A spawning aggregation of Nassau grouper, *Epinephelus striatus* (Bloch). *Transactions of the American Fisheries Society* **2**, 257–261.

Smith, T.B., Blondeau, J., Nemeth, R.S., Calnan, J.M., Kadison, E. & Gass, J. 2010. Benthic structure and cryptic mortality in a Caribbean mesophotic coral reef bank system, the Hind Bank Marine Conservation District, U.S. Virgin Islands. *Coral Reefs* **29**, 289–308.

Society for the Conservation of Reef Fish Aggregations (SCRFA). 2011. Spawning aggregation database. Hong Kong: Society for the Conservation of Reef Fish Aggregations. Online. http://www.scrfa.org/database/ (accessed January 2011).

Sosa-Cordero, E. & Cárdenas-Vidal, J.L. 1996. Estudio preliminar de la pesqueria del mero *Epinephelus striatus* del sur de Quintana Roo, México. *Proceedings of the Gulf and Caribbean Fisheries Institute* **44**, 56–72.

Stacey, N., Chojnacki, A., Narayana, A., Cole, T. & Murphy, C. 2003. Hormonally derived sex pheromones in fish: exogenous cues and signals from gonad to brain. *Canadian Journal of Physiology and Pharmacology* **81**, 329–341.

Starr, R.M., Sala, E., Ballesteros, E. & Zabala, M. 2007. Spatial dynamics of the Nassau grouper *Epinephelus striatus* in a Caribbean atoll. *Marine Ecology Progress Series* **343**, 239–249.

Stevick, P.T., Incze, L.S., Kraus, S.D., Rosen, S., Wolff, N. & Baukus, A. 2008. Trophic relationships and oceanography on a small offshore bank. *Marine Ecology Progress Series* **363**, 15–28.

Stumpf, R.P., Holderied, K. & Sinclair, M. 2003. Determination of water depth with high-resolution satellite imagery over variable bottom types. *Limnology and Oceanography* **48**, 547–556.

Swearer, S.E., Caselle, J.E., Lea, D.W. & Warner, R.R. 1999. Larval retention and recruitment in an island population of a coral-reef fish. *Nature* **402**, 799–802.

Symonds, D.J. & Rogers, S.I. 1995. The influence of spawning and nursery grounds on the distribution of sole *Solea solea* (L.) in the Irish Sea, Bristol Channel and adjacent areas. *Journal of Experimental Marine Biology and Ecology* **190**, 243–261.

Taylor, J.C., Eggleston, D.B. & Rand, P.S. 2006. Nassau grouper (*Epinephelus striatus*) spawning aggregations: hydroacoustic surveys and geostatistical analysis. In *Emerging Technologies for Reef Fisheries Research and Management*, J.C. Taylor (ed.). *NOAA Professional Paper NMFS Series* **5**, 18–25.

Thresher, R.E. 1984. *Reproduction in Reef Fishes*. Neptune City, New Jersey: TFH.

Tucker, J.W., Bush, P.G. & Slaybaugh, S.T. 1993. Reproductive patterns of Cayman Islands Nassau grouper (*Epinephelus striatus*) populations. *Bulletin of Marine Science* **52**, 961–969.

Vincent, A.C.J. & Sadovy, Y. 1998. Reproductive ecology in the conservation and management of fishes. In *Behavioural Ecology and Conservation Biology*, T. Caro (ed.). Oxford, UK: Oxford University Press, 209–245.

Wall, C.C., Lembke, C. & Mann, D.A. 2012. Shelf-scale mapping of sound production by fishes in the eastern Gulf of Mexico, using autonomous glider technology. *Marine Ecology Progress Series* **449**, 55–64.

Warner, R.R. 1995. Large mating aggregations and daily long-distance spawning migrations in the bluehead wrasse, *Thalassoma bifasciatum*. *Environmental Biology of Fishes* **44**, 337–345.

Watanabe, W.O., Ellis, S.C., Ellis, E.P., Head, W.D., Kelley, C.D., Moriwake, A., Lee, C.S. & Bienfang, P.K. 1995a. Progress in controlled breeding of Nassau grouper (*Epinephelus striatus*) broodstock by hormone induction. *Aquaculture* **138**, 205–219.

Watanabe, W.O., Lee, C.S., Ellis, S.C. & Ellis, E.P. 1995b. Hatchery study of the effects of temperature on eggs and yolksac larvae of the Nassau grouper *Epinephelus striatus*. *Aquaculture* **136**, 141–147.

Werner, F.E., Cowen, R.K. & Paris, C.B. 2007. Coupled biological and physical models. *Oceanography* **20**, 54–69.

Whaylen, L., Bush, P., Johnson, B., Luke, K.E., McCroy, C., Heppell, S., Semmens, B. & Boardman, M. 2006. Aggregation dynamics and lessons learned from five years of monitoring at a Nassau grouper (*Epinephelus striatus*) spawning aggregation in Little Cayman, Cayman Islands, BWI. *Proceedings of the Gulf and Caribbean Fisheries Institute* **59**, 479–488.

Whaylen, L., Pattengill-Semmens, C.V., Semmens, B.X., Bush, P.G. & Boardman, M.R. 2004. Observations of a Nassau grouper, *Epinephelus striatus*, spawning aggregation site in Little Cayman, Cayman Islands, including multi-species spawning information. *Environmental Biology of Fishes* **70**, 305–313.

White, D.B., Wyanski, D.M., Eleby, B.M. & Lilyestrom, C.G. 2002. Tiger grouper (*Mycteroperca tigris*): profile of a spawning aggregation. *Bulletin of Marine Science* **70**, 233–240.

Wright, D.J. & Heyman, W.D. 2008. Introduction to the special issue: marine and coastal GIS for geomorphology, habitat mapping and marine reserves. *Marine Geodesy* **31**, 223–230.

Zaiser, M.J. & Moyer, J.T. 1981. Notes on the reproductive behavior of the lizardfish *Synodus ulae* at Miyake-jima, Japan. *Japanese Journal of Ichthyology* **28**, 95–98.

Zeller, D.C. 1998. Spawning aggregations: patterns of movement of the coral trout *Plectropomus leopardus* (Serranidae) as determined by ultrasonic telemetry. *Marine Ecology Progress Series* **162**, 253–263.

Oceanography and Marine Biology: An Annual Review, 2013, **51**, 327-374
© Roger N. Hughes, David Hughes, and I. Philip Smith, Editors
Taylor & Francis

OCEAN ECOGEOCHEMISTRY: A REVIEW

KELTON W. MCMAHON[1,2,3], LI LING HAMADY[1] & SIMON R. THORROLD[1]

¹Biology Department, Woods Hole Oceanographic Institution,
MS50, Woods Hole, MA 02543, USA
E-mail: kmcmahon@whoi.edu (corresponding author), lhamady@whoi.edu
²Red Sea Research Center, King Abdullah University of Science and Technology,
Thuwal, Kingdom of Saudi Arabia
E-mail: sthorrold@whoi.edu
³Ocean Sciences Department, University of California-Santa Cruz, SantaCruz, CA 95064, USA

Animal movements and the acquisition and allocation of resources provide mechanisms for individual behavioural traits to propagate through population, community and ecosystem levels of biological organization. Recent developments in analytical geochemistry have provided ecologists with new opportunities to examine movements and trophic dynamics and their subsequent influence on the structure and functioning of animal communities. We refer to this approach as *ecogeochemistry*—the application of geochemical techniques to fundamental questions in population and community ecology. We used meta-analyses of published data to construct δ^2H, $\delta^{13}C$, $\delta^{15}N$, $\delta^{18}O$ and $\Delta^{14}C$ isoscapes throughout the world's oceans. These maps reveal substantial spatial variability in stable isotope values on regional and ocean-basin scales. We summarize distributions of dissolved metals commonly assayed in the calcified tissues of marine animals. Finally, we review stable isotope analysis (SIA) of amino acids and fatty acids. These analyses overcome many of the problems that prevent bulk SIA from providing sufficient geographic or trophic resolution in marine applications. We expect that ecologists will increasingly use ecogeochemistry approaches to estimate animal movements and trace nutrient pathways in ocean food webs. These studies will, in turn, help provide the scientific underpinning for ecosystem-based management strategies in marine environments.

Introduction

The acquisition and allocation of resources are fundamental requirements for all animals and significantly influence behaviour, population dynamics and ecosystem functioning. Animal movement plays a critical role in resource acquisition and the transfer of these resources among locations. Trophic and movement ecology are therefore inextricably linked across a range of spatiotemporal scales within and among food webs. This connection extends to the techniques used to study connections among habitats and trophic groups. Stable isotope analysis (SIA) and other geochemical methods have been used extensively in food web studies and, more recently, to trace animal movements across habitats with distinctive isotopic signatures (Hobson 1999, Boecklen et al. 2011). This convergence represents a new direction for the field of ecogeochemistry, a term first used by Mizutani et al. (1991) to describe the use of SIA to infer diets of bats and subsequently expanded to include a range of geochemical approaches applied to ecological studies of food web dynamics and movement (McMahon et al. 2013).

Ecologists have embraced the use of SIA in studies of marine food webs. A recent review found that nearly 60% of trophic ecology studies using SIAs published between 2007 and 2009 were conducted in marine or estuarine environments (Boecklen et al. 2011). However, while stable isotopes have been used in animal migration studies in terrestrial environments for several decades (Hobson 1999, Rubenstein & Hobson 2004), the approach has received far less attention in marine systems (Fry 1981, Schell et al. 1989, Best & Schell 1996). This lack of effort may be due, at least in part, to a failure to recognize the degree of geographic variation in isotope and element abundances across marine environments (Hobson 1999, Rubenstein & Hobson 2004). Compilations of maps showing spatial variation in isotope values have identified marine isoscapes that are clearly sufficient for use in movement studies over ocean-basin scales (West et al. 2010, McMahon et al. 2013).

The use of ecogeochemical approaches to examine trophic dynamics and movement patterns of animals offers significant advantages over traditional methods in marine environments. For instance, the use of stable isotopes has overcome at least some of the problems associated with stomach content analysis to determine diets (Michener & Schell 1994). Ecogeochemistry has also been employed to overcome problems associated with conventional tagging methods of the early life-history stages of marine animals (Thorrold et al. 2002, Becker et al. 2007). Finally, in some instances isotope analyses of ancient calcified tissues have provided a means of investigating ecological processes over millennial timescales (Limburg et al. 2011). Taken together, ecogeochemistry may allow for significant progress in a number of important, but as yet unresolved, questions in ocean ecology.

In this review, we outline the processes controlling isotope and elemental fractionation and summarize geographic gradients in isotope and elemental distributions in ocean and estuarine environments. We assemble global ocean isoscapes for key elements in marine ecogeochemistry, including seawater hydrogen ($\delta^2 H_{SW}$), dissolved inorganic carbon (DIC) ($\delta^{13}C_{DIC}$), seawater radiocarbon ($\Delta^{14}C_{SW}$), plankton carbon ($\delta^{13}C_{PLK}$), plankton nitrogen ($\delta^{15}N_{PLK}$) and seawater oxygen ($\delta^{18}O_{SW}$). We summarize distributions of those minor and trace elements that are consistently and accurately analysed in the calcified tissues of marine fish and invertebrates and used as natural geochemical tags of natal origin (Thorrold et al. 1997). Finally, we highlight the potential for compound-specific stable isotope analyses, acknowledging that more research is needed in terms of understanding the processes controlling stable isotope fractionation of individual amino acids and fatty acids.

Data sources and isoscape methods

In this review, we have assembled isoscapes for a number of key elements in the marine environment. The data used to generate the isoscapes were collected from meta-analyses of published isotope data. For $\delta^2 H_{SW}$ and $\delta^{18}O_{SW}$, all data were available on the Global Oxygen-18 Database (Schmidt et al. 1999) on the National Aeronautics and Space Administration website (http://data. giss.nasa.gov/o18data/). Similarly, seawater radiocarbon ($\Delta^{14}C_{SW}$) data were mined from the Global Data Analysis Project (GLODAP) (Key et al. 2004). Seawater DIC $\delta^{13}C_{DIC}$ data were collected from GLODAP (Key et al. 2004), the Open Access library Pangaea (http://www.pangaea.de), and extensive searches of Google Scholar and Web of Science. Date for both $\Delta^{14}C_{SW}$ and $\delta^{13}C_{DIC}$ were predominantly more recent than the 1990s. The $\delta^{13}C_{DIC}$ data in the horizontal isoscape were from the top 100 m of the world's oceans. Horizontal isoscapes of plankton $\delta^{13}C_{PLK}$ and $\delta^{15}N_{PLK}$ values were mined from extensive searches of Google Scholar, Web of Science, and several online data repositories, including Pangaea. We limited the plankton isoscape search to samples described as net plankton (<1 mm) collected in the euphotic zone (<150 m depth) and not preserved in formalin. The plankton isoscapes comprise a range of species but consist predominantly of copepods and similar zooplankton. To achieve the best spatial coverage, no attempts were made to sort data temporally. However, most data presented are more recent than the 1990s. In addition to papers cited individually elsewhere in this review, data were obtained from the work of Sackett et al. (1965);

Degens et al. (1968); Wada & Hattori (1976); Fontugne & Duplessy (1978); Rau et al. (1982, 1983, 2003); Shadsky et al. (1982); Fry et al. (1983); Thayer et al. (1983); Macko et al. (1984); Mullin et al. (1984); Rodelli et al. (1984); Checkley & Entzeroth (1985); Peterson & Howarth (1987); Wada et al. (1987); Fry (1988); Libes & Deuser (1988); Checkley & Miller (1989); Dunton et al. (1989); Altabet & Small (1990); Hobson & Montevecchi (1991); Sholto-Douglas et al. (1991); Mackensen et al. (1993, 1996); Fry & Quinones (1994); Hobson et al. (1994, 1995, 2002); Keeling & Guenther (1994); Matsura & Wada (1994); Laws et al. (1995); Yamamuro et al. (1995); Boon et al. (1997); Sydeman et al. (1997); Bentaleb et al. (1998); France et al. (1998); Millero et al. (1998); Schell et al. (1998); Gruber et al. (1999); Popp et al. (1999); Sigman et al. (1999); van Woesik et al. (1999); Wu et al. (1999); Calvert (2000); Hofmann et al. (2000); Kaehler et al. (2000); Koppelmann & Weikert (2000); Pinnegar et al. (2000); Tittlemier et al. (2000); Villinski et al. (2000); Waser et al. (2000); Dunton (2001); Lesage et al. (2001); Mackensen (2001); Polunin et al. (2001); Stuck et al. (2001); Devenport & Bax (2002); Hoekstra et al. (2002, 2003); Nyssen et al. (2002); Sato et al. (2002); Schlitzer (2002); Smith et al. (2002); Bode et al. (2003, 2004, 2007); Das et al. (2003); Estrada et al. (2003); Jennings & Warr (2003); Kang et al. (2003); McClelland et al. (2003); Quay et al. (2003); Schmidt et al. (2003); Corbisier et al. (2004); Mahaffey et al. (2004); Abed-Navandi & Dworschak (2005); Iken et al. (2005); Kiriakoulakis et al. (2005); Le Loc'h & Hily (2005); Quillfeldt et al. (2005); Sommer et al. (2005); Galimov et al. (2006); Goni et al. (2006); Tamelander et al. (2006); Carlier et al. (2007); Holl et al. (2007); Cianco et al. (2008); Harmelin-Vivien et al. (2008); Lamb & Swart (2008); Le Loc'h et al. (2008); Petursdottir et al. (2008, 2010); Fanelli et al. (2009, 2011); Frederich et al. (2009); Hirch (2009); Lysiak (2009); Richoux & Froneman (2009); Laakmann & Auel (2010); Miller et al. (2010); Olson et al. (2010); Pajuelo et al. (2010); Forest et al. (2011); Hill & McQuaid (2011); Kohler et al. (2011); Kolasinski et al. (2011); Kurten et al. (2011); Pomerleau et al. (2011); Stowasser et al. (2012); and Wyatt (2011).

Isoscapes were generated using Ocean Data View (ODV) version 4.5.0 (Brown 1998, Schlitzer 2002, http://odv.awi.de/). Data are displayed as colour-shaded maps based on contouring of the original data using the Data Interpolating Variational Analysis (DIVA) gridding software. DIVA software was designed to interpolate data spatially (Barth et al. 2010). DIVA gridding takes into account coastlines, subbasins and advection. Calculations are highly optimized and rely on a finite-element resolution. In particular, the finite-element method takes into account the distance between analysis and data (observation constraint), the regularity of the analysis (smoothness constraint) and physical laws (behaviour constraint). Information about the DIVA gridding software used by ODV can be found at http://modb.oce.ulg.ac.be/mediawiki/index.php/DIVA. Details of the algorithms employed by DIVA gridding can be found at http://modb.oce.ulg.ac.be/mediawiki/index.php/DIVA_method.

Systematics

To be successfully applied in the field, an ecogeochemistry approach must do each of the following (Hobson et al. 2010): (1) establish a baseline isoscape that characterizes distinct geochemical signatures in different habitats or food web end members; (2) constrain tissue isotope turnover rates that determine the period of integration of geochemical signatures for a particular tissue; and (3) identify isotope or elemental fractionation factors between consumer and diet, or between animals and the ambient environment, that offset animal geochemical signatures from the baseline isotope values. The isotopic composition of elements in the marine environment is influenced by a variety of physical, chemical and biological processes that together produce unique geographic distributions, termed isoscapes (West et al. 2010). Next, we provide a brief overview of the dominant sources of fractionation for elements commonly used in ecogeochemistry studies and discuss the resulting geographic distributions in the marine and estuarine environments. Common reference standards for the elements discussed can be found in Table 1.

Table 1 Common isotope ratios, their percentage abundances, reference standards, substance types, and δ values

Isotope ratio	Abundance (%)	Standard	Type	Value
$^2H/^1H$	0.01/99.98	SMOW (Standard Mean Ocean Water)	Water	δD (SMOW) = 0.00‰
$^2H/^1H$	0.01/99.98	SLAP (Standard Light Antarctic Precipitation)	Water	δD (SMOW) = –428.00‰
$^{13}C/^{12}C$	1.10/98.90	PDB (PeeDee Belemnite)	Calcite	$δ^{13}C$ (PDB) = 0.00‰
$^{13}C/^{12}C$	1.10/98.90	NBS-19 (National Bureau of Standards-19)	Calcite	$δ^{13}C$ (PDB) =1.95‰
$^{14}C/^{12}C$	<0.01/98.90	NBS HOx1 (National Bureau of Standards Oxalic Acid I)	Oxalic acid	
$^{15}N/^{14}N$	0.36/99.64	Atmospheric AIR	Air	$δ^{15}N$ (AIR) = 0.00‰
$^{18}O/^{16}O$	0.20/99.76	SMOW (Standard Mean Ocean Water)	Water	$δ^{18}O$ (SMOW) = 0.00‰
$^{18}O/^{16}O$	0.20/99.76	SLAP (Standard Light Antarctic Precipitation)	Water	$δ^{18}O$ (SMOW) = –55.50‰
$^{18}O/^{16}O$	0.20/99.76	PDB (PeeDee Belemnite)	Calcite	$δ^{18}O$ (SMOW) = 0.27‰
$^{18}O/^{16}O$	0.20/99.76	NBS-19 (National Bureau of Standards-19)	Calcite	$δ^{18}O$ (SMOW) = 28.64‰
$^{34}S/^{32}S$	4.21/95.02	CDT (Canyon Diablo Troilite)	Iron sulphide	$δ^{18}O$ (SMOW) = 0.00‰

Note: For a complete set of standards, see Coplen et al. (2002).

Hydrogen isotopes

Hydrogen isotope fractionation

Hydrogen has the largest mass difference between isotopes relative to mass and hence the greatest variability of δ values of any element on Earth. Physical-chemical processes governing hydrogen isotope fractionation in the marine environment include evaporation, precipitation, mixing and exchange reactions (Friedman 1953). Evaporation from a number of sources, including clouds, water bodies, soil and plant and animal respiration, forms an important basis for fractionation of hydrogen isotopes in the hydrologic cycle. Fractionation of hydrogen isotopes during evaporation is largely a kinetic process that depends on a number of factors and can, in turn, be quite large. Vapour pressures of the isotopologues of water decrease with decreasing molecular weight; therefore, the heavy isotope (^2H) will favour that part of the system in which it is more strongly bound, that is, the liquid phase (Gat 1996). The same is true for condensation; therefore, during evaporation and condensation in an open system, the liquid phase of water becomes progressively enriched in ^2H, while the vapour phase becomes progressively depleted. This process is known as Rayleigh distillation (Gat 1996). Rayleigh distillation has important implications for latitudinal gradients in ocean $δ^2H$ values of seawater ($δ^2H_{SW}$) as described in the next section. Hydrogen isotope fractionation factors for phase transitions of water are so large that even in high-temperature systems like hydrothermal vents, significant variation in $δ^2H_{SW}$ can be seen (Horita & Wesolowski 1994). Further fractionation of hydrogen isotopes can take place at hydrothermal vents and deep-water cold seeps during oxidation of H_2 and CH_4 to H_2O and CO_2. The pressure (depth) at which these reactions take place can also have an impact on hydrogen isotope fractionation (Horita 1999).

Hydrogen isotopes can also be fractionated by biological processes, including bacterially mediated production of hydrogen gas and methane, both of which tend to be depleted in ^2H (Krichevsky et al. 1961). Hydrogen isotope fractionation during photosynthesis typically results in organic matter that is relatively depleted in ^2H; however, the individual steps involved in fractionation are still unclear (White 1989). It is important to note that at low temperatures water hydrogen will exchange quickly and reversibly with labile organic hydrogen bound in organic nitrogen, sulphur and oxygen compounds (Werstiuk & Ju 1989). Water at neutral pH and low temperature in the absence of a catalyst, however, does not readily exchange with most carbon-bound hydrogen (Sternberg 1988), particularly hydrocarbons and lipids. Therefore, it is important to choose appropriate tissues that contain non-exchangeable hydrogen or correct for exchange with suitable standards when attempting to examine animal migration with stable hydrogen isotopes (Kelly et al. 2009).

Hydrogen isotope geographic variability

Surface δ^2H_{SW} values range from approximately –60‰ in the Arctic Ocean to approximately 12‰ in parts of the Mediterranean Sea (Figure 1). Rayleigh distillation is a fundamental process controlling latitudinal gradients in δ^2H_{SW}. As water vapour travels polewards and cools, some of the vapour condenses out as enriched precipitation, leaving the remaining vapour further depleted relative to its source. The water vapour becomes more depleted as the fraction of vapour remaining becomes smaller, until it is finally deposited as highly ^2H-depleted snow at the poles. This depletion can be modelled as follows:

$$^2H/^1H_V = {}^2H/^1H_{V_0} f^{(\alpha-1)},$$

where f is the fraction of vapour remaining, α is the equilibrium fractionation factor for the water-vapour phase transition, and $^2H/^1H_V$ and $^2H/^1H_{V_0}$ are the hydrogen isotope ratios of the current and initial water vapour fractions, respectively. This process results in a gradient of δ^2H_{SW} values that are more negative with increasing latitude. δ^2H_{SW} values that are more negative also occur in regions of river run-off from large drainage basins. Riverine and groundwater sources typically have δ^2H values that reflect the average isotope composition of precipitation that fell relatively recently into watersheds or recharge sites for that area (Kendall & Coplen 2001). These freshwater sources introduce unique δ^2H signatures into coastal areas that are consistently lower than δ^2H_{SW} values. The mouths of large rivers like the Amazon can introduce anomalously low δ^2H values that penetrate hundreds of kilometres into the Atlantic Ocean. Similarly, deuterium can be a valuable tracer of subsurface groundwater sources (Sonntag et al. 1983), resulting in unique nearshore δ^2H_{SW} signatures that could be used to track animal migration within and among nearshore habitats. More positive δ^2H_{SW} values are typically observed in highly evaporative sites, such as the subtropical gyres, the Mediterranean Sea and the Arabian Sea. Vertical profiles of δ^2H_{SW} generally tend to show less variation than the horizontal surface variation and covary with salinity.

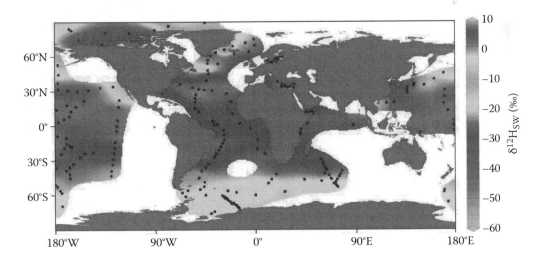

Figure 1 (See also colour figure in the insert) Horizontal isoscape of published seawater δ^2H_{SW} values in the surface waters (top 100 m) of the world's oceans ($n = 360$ data points). Data were collected from the Global Seawater Oxygen-18 database (Schmidt et al. 1999). Isoscapes were generated in Ocean Data View (Schlitzer 2011). Black dots indicate the sample locations.

It is interesting to note two exceptions to the patterns described. In the Pacific Ocean, the California Current exhibits δ^2H_{SW} values that are more negative than would be predicted based on latitude due to advection of ^{2}H-depleted subpolar water towards the equator along the eastern Pacific Ocean. Conversely, the north-eastern North Atlantic exhibits enriched δ^2H_{SW} values for its latitude due to polewards advection of ^{2}H-enriched low-latitude water via the Gulf Stream.

Carbon isotopes

Carbon-13 (^{13}C)

^{13}C fractionation   Carbon isotopes are fractionated during several reactions as they move through the environment and into food webs. The first of these fractionation processes is caused by equilibration of CO_2 between the atmosphere and the surface ocean DIC pool (Boutton 1991). The dissolution of CO_2 in seawater occurs through the following reactions:

$$CO_{2(atm)} \rightleftharpoons CO_{2(aq)} + H_2O \rightleftharpoons H_2CO_2 \rightleftharpoons H^+ + HCO_3^- \rightleftharpoons 2H^+ + CO_3^{2-}$$

Although the equilibration of CO_2 alone is relatively fast (approximately 1 year), exchange between atmospheric CO_2 and DIC that includes all the inorganic carbon species is a slower process, on the order of about 10 years (Broecker & Peng 1974). At a typical surface seawater pH of 8.2, seawater DIC has 91% bicarbonate ion, 8% carbonate ion, and 1% dissolved CO_2. There is also significant temperature-dependent fractionation of carbon isotopes among the various dissolved carbon species (Zeebe & Wolf-Gladrow 2001). The δ^{13}C value of $CO_{2(atm)}$ is approximately $-8‰$ (Keeling et al. 2005). At 25°C, carbon isotope fractionation between $CO_{2(atm)}$ and $CO_{2(aq)}$ is 1.3‰, and fractionation between $CO_{2(aq)}$ and bicarbonate ions is $-9‰$ (Mook 1986, Zhang et al. 1995). Finally, fractionation between bicarbonate and carbonate ions was reported by Mook (1986) to be 0.4‰, although the value is not particularly well constrained (Zeebe & Wolf-Gladrow 2001).

Photosynthetic marine organisms take up the lighter carbon isotope (^{12}C) at a faster rate, with a δ^{13}C fractionation value of approximately $-19‰$ between $\delta^{13}C_{DIC}$ and fixed organic carbon (Lynch-Stieglitz et al. 1995). Internal biological parameters of primary producers, such as biosynthesis rate, enzymatic activity and cell lipid content, can significantly influence their $\delta^{13}C_{PLK}$ values (Fry & Wainright 1991, Hinga et al. 1994). The most prominent example of this differential fractionation during photosynthesis can be seen in the δ^{13}C values of C_3 and C_4 plants. Plants using the C_4 photosynthetic pathway typically have δ^{13}C values that are more positive, ranging from $-8‰$ to $-12‰$, than C_3 plants (-22‰ to $-28‰$) (see review by Farquhar et al. 1989). This difference is due to C_4 plants using a different primary carboxylating enzyme, phosphenolpyruvate (PEP) carboxylase, and a different species of inorganic carbon than C_3 plants that use ribulose-bisphosphate carboxylase-oxygenase (Rubisco). During photosynthesis, phytoplankton take up DIC, and after moderate kinetic fractionation, phytoplankton obtains a δ^{13}C value between $-16‰$ and $-24‰$ (Peterson & Fry 1987).

^{13}C geographic variability   Temperature is a key physical parameter that influences δ^{13}C variability of CO_2 globally. The dissolved CO_2 concentration of the surface mixed layer $[CO_2]_{aq}$ is inversely related to sea-surface temperature (SST) (Weiss 1974), and cold waters with higher $[CO_2]_{aq}$ tend to have lower δ^{13}C values than warm waters. This process establishes a strong latitudinal gradient in surface ocean δ^{13}C values of CO_2 and DIC (Figure 2). Introduction of ^{13}C-depleted atmospheric CO_2 in regions of the ocean with prominent CO_2 invasion, such as the North Atlantic, results in relatively low surface water $\delta^{13}C_{DIC}$ values. Conversely, outgassing of CO_2 in equatorial upwelling zones gives surface waters more positive δ^{13}C values (Lynch-Stieglitz et al. 1995). Organic material is remineralized as it sinks; therefore, water masses at depth are low in $\delta^{13}C_{DIC}$ value relative to surface waters,

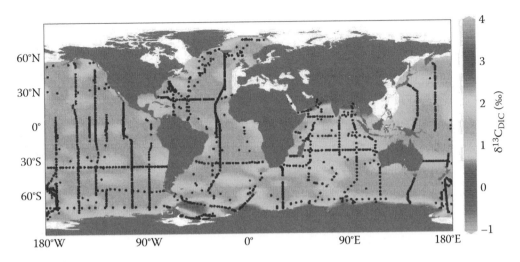

Figure 2 (See also colour figure in the insert) Horizontal isoscape of published seawater dissolved inorganic carbon (DIC) $\delta^{13}C_{DIC}$ values in the surface waters (top 100 m) of the world's oceans ($n = 5501$ data points). Data were collected from the GLODAP database (Key et al. 2004). Isoscapes were generated in Ocean Data View (Schlitzer 2011). Black dots indicate the sample locations.

with values typically approaching 0‰ (Figure 3). Upwelling events can also be a significant source of anomalously low surface $\delta^{13}C_{DIC}$ values (Kroopnick 1985).

The $\delta^{13}C$ value of primary producers ($\delta^{13}C_{PLK}$) is strongly influenced by the $\delta^{13}C$ value of the local DIC pool; thus, spatial variability in $\delta^{13}C_{PLK}$ is primarily driven by the same physical parameters (temperature and $[CO_2]_{aq}$) that influence $\delta^{13}C_{DIC}$ values (Figure 4). For example, Goericke & Fry (1994) demonstrated that phytoplankton $\delta^{13}C$ values generally decreased with increasing latitude as the $\delta^{13}C$ of particulate organic carbon (POC) weakly tracked with temperature on a global scale. However, as mentioned, biological processes can also influence phytoplankton $\delta^{13}C_{PLK}$ values and thus decouple these patterns in some situations (Fry & Wainright 1991, Hinga et al. 1994, Kelly 2000).

Nearshore and benthic systems are typically more ^{13}C-enriched than oceanic systems due to higher nutrient concentrations near shore causing greater overall productivity (France 1995). In addition, tighter terrestrial and benthic-pelagic coupling in nearshore systems can increase inputs from ^{13}C-enriched benthic macrophytes and C_4 marsh plants (France 1995). In contrast, pelagic waters are ^{13}C-depleted owing to lower nutrient availability, lower phytoplankton growth rates and overall productivity and reduced contributions from benthic macrophytes. As a result, there are often steep gradients in $\delta^{13}C_{base}$ values from nearshore to offshore and benthic to pelagic habitats. This gradient can be particularly pronounced in regions of strong upwelling or seasonal coastal phytoplankton blooms (Pancost et al. 1997).

There can also be significant seasonal variability in $\delta^{13}C$ values at the base of marine food webs ($\delta^{13}C_{base}$) in many ocean ecosystems (Gearing et al. 1984, Cifuentes et al. 1988, Goering et al. 1990, Ostrom et al. 1997). There are a number of factors that can contribute to seasonal variability in $\delta^{13}C_{base}$, including seasonal changes in water mass properties and fluctuations in terrestrial run-off, temperature and associated $[CO_2]_{aq}$, phytoplankton productivity and growth rate and primary producer species composition. In general, seasonal variability is larger at high latitudes, with large variations in temperature and productivity, compared to low-latitude, tropical systems. Cifuentes et al. (1988) showed that the pattern of $\delta^{13}C$ in suspended particulate matter varied by nearly 9‰ on seasonal timescales in the Delaware Estuary. Goering et al. (1990) found a similar pattern with more than a 4‰ difference in net phytoplankton $\delta^{13}C_{PLK}$ values between April and May alone

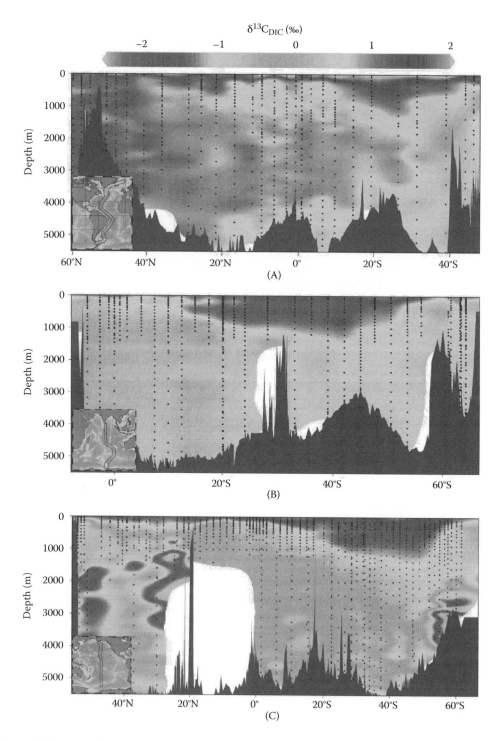

Figure 3 (See also colour figure in the insert) Vertical isoscapes of published seawater dissolved inorganic carbon (DIC) $\delta^{13}C_{DIC}$ values from (A) Atlantic Ocean ($n = 659$ data points); (B) Indian Ocean ($n = 809$ data points); and (C) Pacific Ocean ($n = 1353$ data points). Data were collected from an extensive search on Web of Knowledge and the GLODAP database (Key et al. 2004). Isoscapes were generated in Ocean Data View (Schlitzer 2011). Black dots indicate the sample locations.

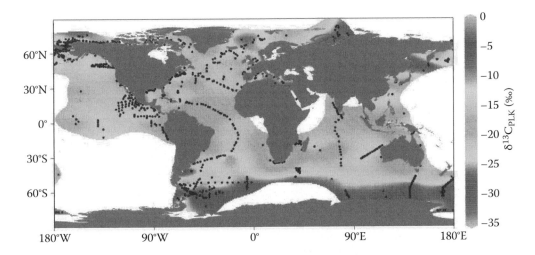

Figure 4 (See also colour figure in the insert) Horizontal isoscape of published surface plankton $\delta^{13}C_{PLK}$ values in the surface waters (top 200 m) of the world's oceans ($n = 1434$ data points). Data were collected from an extensive search on Web of Knowledge and several online data repositories. Isoscapes were generated in Ocean Data View (Schlitzer 2011). Black dots indicate the sample locations.

during the seasonal spring bloom in coastal Alaska. These seasonal fluctuations in $\delta^{13}C$ can be passed on to higher trophic levels as well, particularly for primary consumers with fast turnover rates (Gearing et al. 1984, Simenstad & Wissmar 1985, Goering et al. 1990, Riera & Richard 1997). The transfer of such variability to consumers at the upper trophic level tends to diminish with increasing trophic level as a result of time averaging due to slower tissue turnover rates and feeding on multiple food sources, potentially across multiple food webs for highly mobile species (Goering et al. 1990, O'Reilly et al. 2002). For instance, in the Goering et al. (1990) study, zooplankton had a similar, although smaller, seasonal variability ($\delta^{13}C_{PLK}$ about 3‰) to net phytoplankton.

Carbon-14 (^{14}C)

^{14}C fractionation Carbon has a naturally occurring, radiogenic isotope, ^{14}C (radiocarbon), that may be a useful tracer of habitat use in the marine environment. Radiocarbon is created by cosmic ray bombardment of nitrogen in the atmosphere. In living organisms, radiocarbon exists at levels in isotopic equilibrium with their surroundings; when an organism dies, the ^{14}C begins to decay at a predictable and measurable rate. This makes radiocarbon analysis useful for dating organic matter. In modern times, several anthropogenic perturbations have altered natural radiocarbon levels. Fossil fuel emissions introduced old, 'dead' radiocarbon into the atmosphere and led to a decrease of about 20‰ in atmospheric $\Delta^{14}C$ values from 1890 to 1950 (Suess 1955, Levin & Hesshaimer 2000). In addition, atmospheric testing of atomic bombs in the 1950s and early 1960s resulted in a rapid and well-documented increase in radiocarbon in the atmosphere, leading to disequilibrium with the world's oceans and biosphere (Druffel & Linick 1978). The initial rise of bomb radiocarbon in surface ocean waters from prebomb levels (approximately –50‰ in the Pacific Ocean and –65‰ to –45‰ in the Atlantic Ocean) occurred in 1959 ± 1 year, and ^{14}C levels rose relatively rapidly to peak $\Delta^{14}C$ values between 1967 and 1970 (approximately 210‰ in the Pacific Ocean and 270‰ in the Atlantic Ocean), with a subsequent slow but steady declining trend since then (Ostlund et al. 1974, Stuiver et al. 1981, Nydal 1998). This bomb radiocarbon chronology is almost synchronous around the world in biogenic carbonates, such as coral skeletons, bivalve shells and fish otoliths (Druffel & Linick 1978, Kalish 1993, Weidman & Jones 1993), thus serving as a dated marker in calcified structures exhibiting periodic growth bands (Figure 5).

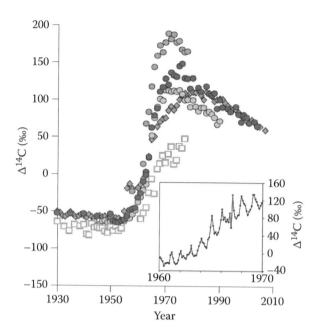

Figure 5 (See also colour figure in the insert)   Radiocarbon (Δ¹⁴C) chronologies from marine corals in the world's oceans from 1930 to 2010. Blue circles: *Porites* sp., French Frigate Shoals, Pacific Ocean (Druffel 1987). Blue squares: *Pavona clavus*, Galapagos Islands, Pacific Ocean (Druffel 1981). Blue diamonds: *Porites lutea*, Palmyra Atoll, Pacific Ocean (Druffel-Rodriguez et al. 2012). Red circles: *Porites lutea*, Puerto Rico, Atlantic Ocean (Kilbourne et al. 2007). Yellow circles: *Porites* sp., Mentawai Islands, Indian Ocean (Grumet et al. 2004). Inset: seasonal variability in *Porites lutea* Δ¹⁴C values from the South Makassar Strait, Indonesia, between 1960 and 1970 (Fallon & Guilderson 2008).

¹⁴C geographic variability   The distribution of ¹⁴C$_{DIC}$ in the ocean is largely determined by air-sea exchange of CO_2 and ocean circulation (Siegenthaler 1989). Due to initial asymmetrical atmospheric input, the Δ¹⁴C$_{SW}$ maximum values peaked 1–2 years earlier in the Northern Hemisphere compared to the Southern Hemisphere (Linick 1978). Deep-water masses that are isolated from the atmosphere and transported to depth via thermohaline circulation have Δ¹⁴C$_{SW}$ values that become more negative with increasing residence time in the ocean (Broecker et al. 1985, Jain et al. 1995) (Figure 6). For instance, the relatively young deep waters of the Atlantic Ocean have Δ¹⁴C$_{SW}$ values on the order of –140‰, while the Δ¹⁴C$_{SW}$ values of older deep North Pacific waters are approximately –250‰ (Siegenthaler 1989, but see Druffel & Williams 1990).

Seawater Δ¹⁴C values also can vary on regional scales, both horizontally and vertically. Subtropical gyres that entrain water at the surface and possess a strong thermocline with limited vertical mixing typically have high sea-surface Δ¹⁴C$_{SW}$ values. Conversely, subpolar gyres and areas of divergence and upwelling bring older, more negative Δ¹⁴C$_{SW}$ waters to mix at the surface. In the Pacific, maximal Δ¹⁴C values occur in midlatitudes around 30° north and south; at the peak, these regions registered 210‰ or 260‰ above prebomb levels, whereas equatorial waters reached 50‰ or 110‰ above prebomb levels (Linick 1978). These Δ¹⁴C$_{SW}$ surface patterns translate down through the mixed layer, although values drop off quickly below that and remain relatively constant below about 1000 m in all areas (Key et al. 2004). The vertical Δ¹⁴C$_{SW}$ gradients observed in the North Pacific and North Atlantic may potentially be used as a depth tracer in the context of animal movements if the analysed tissue is primarily derived from DIC (Pearcy & Stuiver 1983, Rau et al. 1986). Upwelling can cause significant variability in sea-surface Δ¹⁴C$_{SW}$ values on regional and seasonal scales (Figure 5). For instance, strong upwelling around the Galapagos Islands brings old

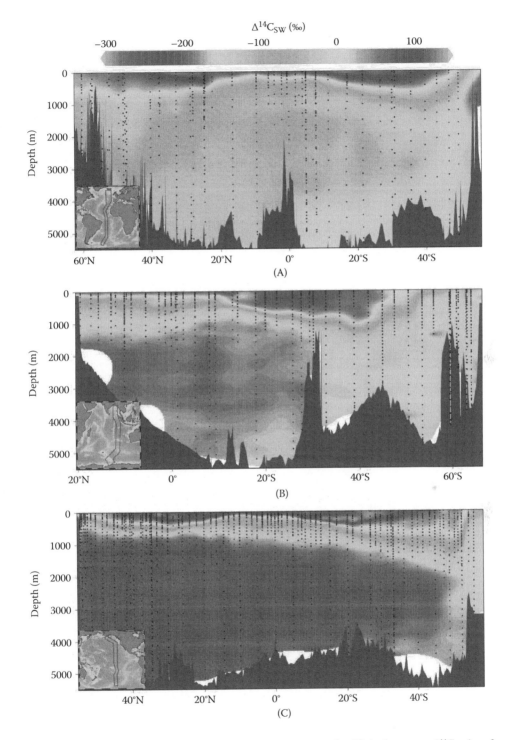

Figure 6 (See also colour figure in the insert) Vertical isoscapes of published seawater $\Delta^{14}C$ values from (A) Atlantic Ocean (n = 645 data points); (B) Indian Ocean (n = 1026 data points); and (C) Pacific Ocean (n = 1878 data points). Data were collected from the GLODAP database (Key et al. 2004). Isoscapes were generated in Ocean Data View (Schlitzer 2011). Black dots indicate the sample locations.

'dead' carbon up from depth that significantly reduces regional surface $\Delta^{14}C$ values (Druffel 1981). Similarly, the South Makassar Strait in Indonesia exhibits significant seasonal variability, in the range of 10‰ to 20‰ seasonally, associated with upwelling that overlays the interannual trend of increasing $\Delta^{14}C$ value with time after the initial bomb spike (Fallon & Guilderson 2008).

Nitrogen isotopes

Nitrogen isotope fractionation

Although only a small fraction of the global nitrogen reservoir is contained in living matter, organic nitrogen is of tremendous importance for nitrogen isotope distributions because almost all nitrogen isotope fractionation results from metabolically related processes (Hübner 1986). Fractionation associated with nitrogen fixation is typically small, with the average isotope effect between atmospheric N_2 and fixed nitrogen near 0‰ (Hoering & Ford 1960, Fogel & Cifuentes 1993). Mineralization of organic matter to ammonium also has a relatively small fractionation factor (0‰ ± 1‰), but concurrent nitrification of ammonium to nitrate, particularly in vent and seep environments with high concentrations of ammonium, can result in large fractionations (–18‰ to –42‰) (Hoch et al. 1992, Fogel & Cifuentes 1993). Assimilation of NH_4^+, NO_2^-, and NO_3^- by microorganisms can cause significant and often highly variable fractionation (–27‰ to 0‰), regulated by nitrogen availability and reaction rates (Fogel & Cifuentes 1993). Denitrification produces N_2 gas, which if lost to the atmosphere by diffusional processes can produce large isotope effects owing to classic Rayleigh fractionation (as discussed previously). Denitrification produces N_2 gas that can be upwards of 40‰ lower in $\delta^{15}N$ relative to dissolved nitrate, leaving the remaining nitrate relatively ^{15}N-enriched (Cline & Kaplan 1975).

Nitrogen isotope geographic variability

The $\delta^{15}N$ of nitrogenous species in seawater is determined by the balance of nitrogen sources and sinks, as well as fractionation resulting from biologically mediated processes. Sources of nitrogen to the marine environment include river run-off, atmospheric deposition, and N_2 fixation by cyanobacteria, while the major sinks are burial in sediments and denitrification. The Atlantic and Pacific Oceans both show large-scale, albeit opposite, geographic relationships between $\delta^{15}N$ of plankton ($\delta^{15}N_{PLK}$) and latitude (Figure 7). A meta-analysis of published zooplankton $\delta^{15}N_{PLK}$ values from the upper ocean of the North Atlantic shows a pattern of enrichment with increasing latitude. The lowest $\delta^{15}N_{PLK}$ values are found in the oligotrophic gyres, particularly the Sargasso Sea, where diazotrophic cyanobacteria fix N_2 (0‰) into organic nitrogen (Montoya et al. 2002). $\delta^{15}N_{PLK}$ values increase with increasing latitude as NO_3^- (5‰) becomes the major fixed nitrogen source for marine phytoplankton. In the Pacific Ocean, however, the $\delta^{15}N_{PLK}$-latitude correlation is reversed, with the highest $\delta^{15}N_{PLK}$ values recorded in the eastern tropical and central gyre (Saino & Hattori 1987). This is because (1) the Pacific is generally iron limited and thus lacks significant N_2 fixation in the surface ocean and (2) year-round stratification and large oxygen minimum zones result in significant amounts of denitrification and thus nitrate ^{15}N enrichment (Saino & Hattori 1987). For example, the $\delta^{15}N$ of dissolved nitrate in Antarctic Intermediate Water can be upwards of 12.5‰ lower than that of active denitrification zones in the North Pacific Ocean (Cline & Kaplan 1975). Particulate organic matter decomposition and respiration, resulting in faster losses in ^{14}N, can create a gradient of increased $\delta^{15}N$ with depth in the ocean (Saino & Hattori 1980). This is particularly evident over areas of high productivity, where large diatoms at the base of the euphotic zone may substantially affect vertical $\delta^{15}N$ gradients (Kalansky et al. 2011).

There can also be significant variability in $\delta^{15}N$ values on smaller spatial scales. Anthropogenic sources of nitrogen, including fertilizers, sewage and agricultural animal waste, and atmospheric deposition via fossil fuel burning, are all important point sources that can have a significant

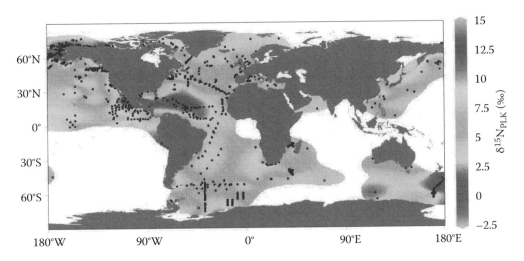

Figure 7 (See also colour figure in the insert)   Horizontal isoscape of published surface plankton $\delta^{15}N_{PLK}$ values in the surface waters (top 200 m) of the world's oceans ($n = 1088$ data points). Data were collected from an extensive search on Web of Knowledge and several online data repositories. Isoscapes were generated in Ocean Data View (Schlitzer 2011). Black dots indicate the sample locations.

impact on coastal $\delta^{15}N_{POM}$ (Heaton 1986). For instance, sewage discharge into coastal estuaries has provided enriched $\delta^{15}N$ isotopic point sources due to isotope fractionation during treatment, which is reflected in the $\delta^{15}N$ values of resident organisms (Hansson et al. 1997, Dierking et al. 2012). Similarly, excess nutrients in wastewater associated with diffuse source anthropogenic activities, such as urban run-off and lawn/field fertilization, have led to eutrophication in coastal bays (McClelland et al. 1997). This can result in an increase in primary production and subsequent denitrification, both of which also provide an enriched $\delta^{15}N$ isotopic signal that is reflected in the tissue $\delta^{15}N$ values of local fishes and invertebrates (Griffin & Valiela 2001).

Plankton $\delta^{15}N_{PLK}$ values can also vary temporally, particularly on seasonal timescales, due to changes in primary productivity associated with shifts in nutrient sources and concentrations, microbial nitrogen cycling and phytoplankton species growth rates and composition (Cifuentes et al. 1988, Goering et al. 1990, Ostrom et al. 1997, Vizzini & Mazzola 2003). Seasonal changes in $\delta^{15}N$ at the base of food webs ($\delta^{15}N_{base}$) can be quite large. Cifuentes et al. (1988) found that the $\delta^{15}N$ value of suspended particulate matter in the Delaware Estuary in winter alone ranged from +5.5‰ to +12.2‰. The authors observed $\delta^{15}N$ values as low as +2.3‰ in early spring, and just 3 weeks later, a $\delta^{15}N$ maximum of +18.7‰ was located in the central portion of the estuary. This large seasonal variability was associated with seasonal shifts in available nitrogen sources, as NH_4^+ utilization far exceeded NO_3^- in the winter, and with increases in productivity and decreases in nutrient availability during the spring bloom. Large shifts in baseline stable isotope values will have a cascading effect on upper trophic levels. As a result, seasonal variation must be considered when constructing and using isoscapes to address questions of connectivity and trophic dynamics in the marine environment.

Oxygen isotopes

Oxygen isotope fractionation

Fractionation of oxygen isotopes is temperature dependent (Urey 1947, Gat 1996), and $\delta^{18}O$ analyses of marine carbonates have routinely been used as a proxy for temperature in both palaeo- and modern applications (Aharon 1991, Fairbanks et al. 1997, Thorrold et al. 1997). However, this temperature-dependent fractionation effect is small (~0.2‰°C^{-1}) relative to the processes of evaporation and precipitation that control ocean basin-scale variation in seawater δ^{18}O values (δ^{18}O$_{sw}$). Rayleigh distillation plays an important role in determining the fractionation of oxygen isotopes in the hydrologic cycle (Gat 1996). These processes are largely the same as those regulating hydrogen isotopes, which can be seen in the meteoric water line (MWL):

$$\delta^2H = 8*\delta^{18}O + d,$$

where d is the 'deuterium excess' (d = 10‰ for the global MWL; Dansgaard 1964). As a result, δ^{18}O values of seawater exhibit geographic variation in the world's oceans that typically covary with deuterium and salinity.

Many calcified tissues, including otoliths (Kalish 1991a, Thorrold et al. 1997), bones (Barrick et al. 1992), teeth (Kolodny et al. 1983), and shells (Mook & Vogel 1968) are precipitated in oxygen isotope equilibrium with ambient water. Some biogenic carbonates, however, exhibit kinetic effects that result in δ^{18}O values out of equilibrium with ambient water (McConnaughey 1989a,b).

Oxygen isotope geographic variability

Oxygen isotope values of ocean water on regional, short-term spatiotemporal scales can reflect a mass balance between evaporation E, precipitation P, advection A, mixing M, and river run-off R (Figure 8), which can be modelled as (Benway & Mix 2004)

$$\delta^{18}O_{SW} = \frac{\left[\left(F_p*\delta_p\right) - \left(F_E*\delta_E\right) + \left(F_A*\delta_A\right) + \left(F_M*\delta_M\right) + \left(F_R*\delta_R\right)\right]}{\left[\left(F_p\right) - \left(F_E\right) + \left(F_A\right) + \left(F_M\right) + \left(F_R\right)\right]},$$

where F is the fraction, and δ is the isotopic value of each component contributing to the balance. δ^{18}O$_{SW}$ values that are more positive are observed in highly evaporative subtropical gyres and low-latitude shallow seas, including the Mediterranean Sea (maximum δ^{18}O$_{SW}$ 1.7‰; Rohling & Rijk 1999) and the Red Sea (maximum δ^{18}O$_{SW}$ about 1.6‰; Ganssen & Kroon 1991). The δ^{18}O$_{SW}$

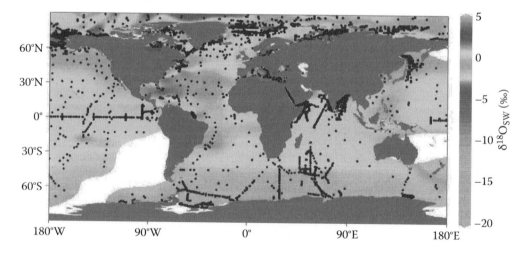

Figure 8 (See also colour figure in the insert)   Horizontal isoscape of published seawater δ^{18}O$_{SW}$ values in the surface waters (top 100 m) of the world's oceans (n = 12,344 data points). Data were collected from the Global Seawater Oxygen-18 database (Schmidt et al. 1999). Isoscapes were generated in Ocean Data View (Schlitzer 2011). Black dots indicate the sample locations.

values that are most negative are found at high latitudes (nearly –20‰ in the Arctic Ocean) and regions of extensive freshwater input. Freshwater discharge lowers $\delta^{18}O_{SW}$ values of coastal ocean waters and, in the case of large rivers like the Amazon and the Orinoco in the tropics and the MacKenzie and Ob in the Arctic, can produce anomalously low $\delta^{18}O_{SW}$ values that penetrate hundreds of kilometres into the ocean (e.g., Cooper et al. 2005).

There are several notable exceptions to the general pattern of decreasing $\delta^{18}O_{SW}$ values with latitude. Advection of ^{18}O-depleted subpolar water towards the equator via the California Current results in anomalously low $\delta^{18}O_{SW}$ values along the eastern boundary of the North Pacific Ocean. Conversely, advection of ^{18}O-enriched low-latitude water via the Gulf Stream causes the western boundary of the North Atlantic to have relatively high $\delta^{18}O_{SW}$ values for its latitude. Vertical profiles of $\delta^{18}O$ trend towards 0‰ with depth and typically show less variation than the horizontal isoscapes (Schmidt et al. 1999).

Sulphur isotopes

Sulphur isotope fractionation

Major sources of sulphur in the marine environment include hydrothermal processes and riverine input. Removal of sulphur in the marine environment is mostly through evaporite deposits, pyrite or organic compound burial and formation of carbonate compounds (Bottrell & Newton 2006). Sulphate is the most common and biologically available species of sulphur in the open ocean. The fractionation of sulphate between minerals and dissolved sulphate is approximately zero; therefore, neither the creation nor the dissolution of evaporites has a significant effect on sulphate $\delta^{34}S$ values. Bacterial sulphate reduction, which takes place in sediments and other anoxic environments, is the major source of biological fractionation of sulphur in the marine environment (Bottrell & Newton 2006), resulting in a 30‰ to 70‰ isotope effect (Peterson & Fry 1987). Bacterial sulphate reduction is a kinetic process that produces sulphides and organic matter that are depleted in ^{34}S relative to the sulphate being reduced. Rainwater sulphates are also potential sources of sulphur for intertidal plants (Fry et al. 1982). The equilibrium between oxidized and reduced sulphur species is typically only found at very high temperatures, characteristic of hydrothermal vent systems (Krouse et al. 1988).

The $\delta^{34}S$ values of phytoplankton, upland plants and marsh grasses are often quite distinct because of their use of different sources of inorganic sulphur (Peterson et al. 1985). Marine phytoplankton and seaweeds utilize marine sulphate (21‰; Rees et al. 1978) and fractionate it little during uptake and assimilation into organic sulphur compounds. Upland plants in aerobic soils also fractionate sulphate little during uptake and assimilation, but they obtain sulphate from precipitation with a lower $\delta^{34}S$ value (2‰ to 8‰). Marsh plants and other primary producers living in anoxic conditions often use sulphides with much lower $\delta^{34}S$ values for at least some of their sulphur requirements, resulting in organic matter that is equally depleted in ^{34}S. The $\delta^{34}S$ signal established by primary producers is passed on to higher trophic levels in the food web because the essential sulphur-bearing compounds are typically incorporated into consumer tissues with little to no trophic fractionation (Peterson et al. 1986, Florin et al. 2011, but see Tanz & Schmidt 2010).

Sulphur isotope geographic variability

Sulphur isotope distributions in the marine environment vary with distribution of sulphides and sulphates, quality of growing conditions (aerobic versus anaerobic), atmospheric deposition from natural sources and point sources from pollution. The residence time of sulphate in seawater is 2×10^7 years; thus, sulphate is well mixed in the marine environment, maintaining a relatively constant $\delta^{34}S_{SW}$ value of 21‰ throughout the open ocean (Rees et al. 1978). In general, benthic and nearshore habitats, including estuaries and marshes, are typically more anoxic than pelagic, offshore ecosystems and thus experience elevated levels of sulphate reduction with correspondingly higher $\delta^{34}S_{SW}$ sulphate values. There can be times when ocean sulphate values do change significantly. For example, during periods of intense weathering, particularly of shales, ocean sulphate $\delta^{34}S$ can decrease significantly. Conversely, anoxic zones, such as parts of the Black Sea and regions of high primary productivity, may experience increased ocean sulphate $\delta^{34}S_{SW}$ values due to elevated bacterial sulphate reduction (Neretin et al. 2003). On geological timescales, the isotope composition of seawater sulphate varies from 10‰ to 33‰ as a result of changes in the magnitude of sulphur fluxes into and out of the marine environment as well as changes in the isotope fractionation between sulphate and buried sulphide (Peterson & Fry 1987). The $\delta^{34}S$ value of terrestrial sulphur is highly variable and dependent on rock type and climate-dependent weathering patterns (Krouse & Grinenko 1991). The $\delta^{34}S$ value of river water sulphate (global mean 7‰) varies regionally according to bedrock lithology, anthropogenic inputs and atmospheric deposition from natural sources (Thode et al. 1961). As a result, terrestrial run-off, rivers and groundwater inputs can be major sources of coastal $\delta^{34}S$ variability. Sulphur isotopes are thus most useful in coastal and estuarine environments, where highly variable freshwater inputs mix with relatively constant marine values to create steep gradients in sulphate $\delta^{34}S$ (Peterson & Fry 1987, Fry 2002).

Human-induced perturbations to the natural sulphur cycle have markedly increased since industrialization, and the majority of sulphur emitted into the atmosphere is now likely of anthropogenic origin (Peterson & Fry 1987). Anthropogenic sources of sulphur, most notably from fossil fuel burning, can overwhelm natural variability on small scales. These sources can produce unique point sources to track movement and residence patterns in the marine environment. In addition, eutrophication as a result of sewage inputs to coastal habitats can promote anoxic conditions, which support sulphate reduction and the generation of low-$\delta^{34}S$ sulphides.

Minor and trace elements in calcified tissues

The chemical composition of oceanic minerals has been used to determine environmental conditions in palaeo-oceans for decades (Kastner 1999). More recently, ocean ecogeochemistry applications have focused on inferring movement patterns of fish and invertebrates from the elemental chemistry of aragonitic otoliths, shells and statoliths (e.g., Campana et al. 1999, Zacherl et al. 2003a, Arkhipkin et al. 2004, Becker et al. 2005, Elsdon et al. 2008, Walther & Limburg 2012). While a number of elements are found in biogenic aragonite, most researchers to date have focused on some combination of six elements that both substitute for calcium in the aragonite matrix (and therefore are more likely to record ambient dissolved concentrations) and are sufficiently abundant and free from isobaric interferences to allow for quantification using inductively coupled plasma mass spectrometry (Thorrold & Swearer 2009). These elements are characterized by conservative-type (lithium, magnesium, strontium), nutrient-type (barium) and scavenged-type (manganese, lead) distributions in the oceans and are most usefully reported as ratios to calcium in both ambient waters and in the calcified tissues.

Minor and trace element fractionation

Minor and trace element composition of calcified tissues are correlated with ambient dissolved concentrations in at least some instances. For instance, concentrations of strontium and barium in otoliths (Bath et al. 2000, Elsdon & Gillanders 2004, Dorval et al. 2007) and gastropod protoconchs (Zacherl et al. 2003b) appear to reflect environmental parameters and serve as valuable tracers of juvenile movements and larval dispersal, respectively. However, strontium, barium and indeed almost all metals, with the notable exception of manganese (Elsdon & Gillanders 2003), are found at significantly lower concentrations in calcified tissues than in the ambient environment (Campana & Thorrold 2001). This fractionation is primarily controlled by biological processes that occur during ion transport from seawater to the internal fluid from which the calcified tissue precipitates (e.g., Melancon et al. 2009). Ion exchange across intestinal or gill membranes, ionic exchange between blood plasma and endolymph and the partition coefficients of ions at the otolith growth surface all likely play some role in regulating the composition of the precipitating fluid (Kalish 1991b).

Temperature typically has a positive influence on Sr:Ca in low-strontium aragonite in fish otoliths and mollusc protoconchs (Bath et al. 2000, Elsdon & Gillanders 2002, Zacherl et al. 2003b) and a negative effect on Sr:Ca in high-strontium aragonite, including coral skeletons and mollusc statoliths (Beck et al. 1992, Zacherl et al. 2003b, Cohen & Thorrold 2007). The effects of temperature on barium, manganese and magnesium are more variable and less conclusive (Elsdon & Gillanders 2002, Zacherl et al. 2003b, Martin & Thorrold 2005); a recent study found that the Li:Ca ratio was positively correlated with temperature in the otolith of the flatfish *Solea solea* (Tanner et al. 2013).

Minor and trace element geographic variability

The long residence times of conservative elements leads to generally uniform distributions throughout the world's oceans. For instance, lithium has a residence time of 1.5 million years and a global Li:Ca ratio of about 2.5 mmol mol^{-1} (Huh et al. 1998). Dissolved lithium values in river waters are considerably more variable and generally lower than in seawater, ranging from 30 nM to 11.7 µM, with concomitant Li:Ca ratios ranging from 77.8 µmol mol^{-1} to 15.7 mmol mol^{-1} (Huh et al. 1998). A significant correlation between otolith Li:Ca and salinity suggests that lithium may be a useful tracer of movement between marine and freshwater habitats (Hicks et al. 2010), although the range of potential values in river waters means that freshwater end members would need to be characterized first. Magnesium is also conservative in seawater with a mean Mg:Ca value of 5.14 mol mol^{-1} (Bruland & Lohan 2004). Riverine water ratios are almost invariably lower than that of seawater. With a global average value for freshwater of 0.45 mol mol^{-1}, Mg:Ca ratios are potentially a useful tracer of salinity (Surge & Lohmann 2002). Finally, the global seawater Sr:Ca ratio is approximately 8.5 mmol mol^{-1} (de Villiers 1999) and is relatively invariant throughout the oceans. Dissolved strontium values in freshwater are largely controlled by surrounding bedrock geological composition, both rock type and weathering efficiency, and are often nearly an order of magnitude lower than seawater values (Bricker & Jones 1995, Limburg 1995, Capo et al. 1998). Freshwater values show significant geographic and temporal variability, with Sr:Ca ratios ranging from 0.27 to 19.18 mmol mol^{-1} (Brown & Severin 2009). Strontium isotope values in ocean waters are relatively invariant with ^{87}Sr:^{86}Sr = 0.70918 (Ando et al. 2010), while fluvial ^{87}Sr:^{86}Sr ratios typically vary from about 0.704 in basaltic drainages to at least 0.75 in older, highly radiogenic granites (Barnett-Johnson et al. 2010, Muhlfeld et al. 2012). Biological fractionation of strontium isotopes is minimal; therefore, ^{87}Sr:^{86}Sr ratios represent an excellent tracer of movements between freshwater and ocean environments (McMahon et al. 2013).

Barium follows a nutrient-type distribution in seawater, with typical surface ocean values of 0.01 to 0.02 µM, increasing to 0.03 and 0.09 µM at a depth of 3000 m in the North Atlantic and North Pacific, respectively (Bruland & Lohan 2004). Barium concentrations in riverine and coastal areas are relatively high compared to slope and oceanic waters (Shen & Stanford 1990), varying by nearly an order of magnitude globally around a worldwide riverine average of 0.10 µM (Gaillardet et al. 2003). Upwelling of cold, nutrient-rich deep water can be a significant secondary source of relatively high barium concentrations to the ocean surface waters (Lea et al. 1989).

Both manganese and lead are scavenged elements characterized by strong interactions with particles that lead to very short oceanic residence times of less than 1000 years (Donat & Bruland 1995). Atmospheric dust is a major source of manganese and lead in oceanic environments. For example, peak concentrations of dissolved manganese in North Atlantic Ocean surface waters occur at 20° north, which coincides with the zone of maximum dust deposition from the Sahara Desert

(Bergquist & Boyle 2006). However, in estuarine settings, manganese and lead fluxes from porewaters often overwhelm atmospheric and fluvial inputs (Rivera-Duarte & Flegal 1994, Warnken et al. 2001). The vertical distributions of dissolved manganese in both the Pacific and the Atlantic Ocean are characterized by a surface maximum driven by atmospheric deposition and photoreduction of manganese oxides (Sunda & Huntsman 1988), a subsurface minimum and a second maximum coincident with the oxygen minimum layer and presumably generated by redox dissolution (Landing & Bruland 1987, Boye et al. 2012). Most of the dissolved lead in the oceans comes from anthropogenic sources, in particular from the use of leaded gasoline in the United States and Europe (Boyle 2001). The introduction of leaded gasoline in the 1920s led to a marked increase in dissolved lead levels in surface waters of the North Atlantic and, to a lesser extent, the North Pacific that peaked in the early 1970s. This lead spike provided a dated marker that was recorded in the skeletons of corals in Bermuda (Shen & Boyle 1987), sclerosponges in the Bahamas (Swart et al. 2002), and bivalve shells in the North Atlantic (Krause-Nehring et al. 2012).

Previous studies have found little variation in manganese, strontium and barium concentrations from estuarine waters on monthly to seasonal timescales. However, water chemistry in dynamic environments with large tidal ranges may vary over shorter daily or tidal scales (Dorval & Jones 2005, Elsdon & Gillanders 2006). For example, Elsdon & Gillanders (2006) found significant differences in manganese, strontium and barium concentrations between water samples collected on different days within three small (<10 km) tidal estuaries that accounted for up to 64% of the total variation on scales of days, weeks, months and seasons.

Biological tissues

Ecogeochemistry rests fundamentally on the assumption that the composition of a tissue will reflect the isotopic or elemental composition of the source from which an element is obtained and some degree of fractionation (Tieszen et al. 1983, Gannes et al. 1997). In the previous section, we discussed the processes that create geographic variability in stable isotope and trace element values and the resulting isoscapes. In this section, we address the issues of constraining tissue-specific isotope turnover rates and discrimination factors that control the offset between the baseline isoscape and the consumer isotope or element values.

Element turnover rates

Isotope turnover rate plays an important role in determining the temporal scales over which a tissue records an isotopic signature of residence or diet (Dalerum & Angerbjörn 2005). Isotope turnover rates can vary from hours to years depending on a number of factors, including tissue type, the metabolic turnover or growth rate, and the taxa studied (Boecklen et al. 2011). Analysis of faeces or gut contents provides short-term information about an organism's diet, ranging from hours for zooplankton to a few days for large mammals. More metabolically active tissues, including liver and blood, typically have faster turnover rates (weeks) than less metabolically active tissues, such as muscle (months) or bone (years) (Tieszen et al. 1983, Buchheister & Latour 2009, Malpica-Cruz et al. 2011). Similarly, the isotope composition of whole tissues represents an integration of the isotopic values of the tissue's constituents (e.g., proteins, lipids, carbohydrates), each with characteristic turnover rates that may differ from the bulk tissue turnover rate.

Tissues that are metabolically inert after formation, including hair, baleen, claws and otoliths, preserve a permanent record of source isotope composition (Rubenstein & Hobson 2004). Calcified, accretionary tissues with density or optical bands corresponding to daily, seasonal or annual patterns, including otoliths, bivalve shells, teeth, claws, scales, vertebrae and baleen, may also provide a chronological record of lifetime animal diet and movement (Richardson 1988, Schell et al. 1989, Campana & Thorrold 2001, Campana et al. 2002). Otoliths are particularly valuable tissues for retrospective studies of diet and movement because they grow continuously through successive addition of daily and annual aragonitic growth bands on a proteinaceous matrix, and they are metabolically inert postdeposition (Degens et al. 1969, Campana & Neilson 1985, Campana 1999). Otoliths therefore preserve a chronological record of a fish's metabolic activity and the physical and chemical characteristics of the water in which the fish resided during the time of deposition (Thorrold et al. 1997). Researchers have similarly used the elemental composition of other calcified tissues, including bivalve (Becker et al. 2007) and gastropod shells (Zacherl et al. 2003a), elasmobranch vertebrae (Hale et al. 2006, Tillett et al. 2011), and squid statoliths (Arkhipkin et al. 2004), to determine movement patterns during specific life-history stages of a number of marine species.

Element turnover rate also varies by taxon. A literature review by Boecklen et al. (2011) found significant variation in tissue-specific carbon turnover rates among taxa. For example, the isotopic half-life (time for a tissue isotope value to change halfway from initial to its new equilibrium value) of muscle in mammals (1 to 3 months) was considerably longer than that for fish (2 to 8 weeks) and birds (1 to 3 weeks). One mechanism generating differences in isotope turnover rate among taxa can be seen in relative contributions of metabolic turnover rate and growth rate to isotope turnover rate in teleost fishes and elasmobranchs. Isotope turnover rates tend to be strongly correlated with growth rate and accretion of biomass in teleost fishes (Herzka 2005, Logan et al. 2006). Fish that are growing quickly (often during early life-history stages) tend to reach isotopic equilibrium with their diet much faster than older, slower-growing fish (Herzka 2005). Conversely, in elasmobranchs isotopic turnover rate appears to be closely linked to metabolic turnover rate (Logan & Lutcavage 2010, Malpica-Cruz et al. 2011).

Trophic discrimination factors

We define trophic discrimination between diet and consumer to include kinetic fractionation associated with enzymatic reactions during metabolism as well as differences in stable isotope values due to differences in tissue composition. As with isotope turnover rate, trophic discrimination can vary widely among tissue types (Gannes et al. 1997, Vander Zanden & Rasmussen 2001, McCutchan et al. 2003, Olive et al. 2003). In this section, we focus primarily on bulk trophic discrimination of carbon (Δ^{13}C) and nitrogen (Δ^{15}N) between diet and consumer as they are the primary elements used to assess trophic dynamics in ecogeochemistry studies. However, it is important to note that other elements undergo varying degrees of trophic discrimination. For example, sulphur isotopes exhibit little or no trophic discrimination; thus, δ^{34}S values of consumers reflect the baseline signatures (Peterson et al. 1986). There are mixed results regarding the degree of trophic discrimination (Δ^2H) in non-exchangable hydrogen. Some studies report significant differences between diet and consumer δ^2H values, suggesting some degree of trophic fractionation (Macko et al. 1983, Birchall et al. 2005). However, other studies found that trophic fractionation was negligible and suggested that some non-exchangeable hydrogen in consumer tissue may come from ingestion or diffusion from ambient water (Solomon et al. 2009).

Marine carbon is typically thought to be conservatively fractionated (Δ^{13}C = 0‰ to 1‰) as it continues to move through food webs (DeNiro & Epstein 1978). This small trophic discrimination is the basis for using δ^{13}C to track diet sources and carbon flow through food webs as well as migration among isotopically distinct habitats (see reviews by Hobson 1999, Kelly 2000, Rubenstein & Hobson 2004). However, there can be significant variability in Δ^{13}C, from –3‰ to 5‰, among tissues and taxa owing to differential digestion or fractionation during assimilation and metabolic processing (Vander Zanden & Rasmussen 2001, Post 2002, McCutchan et al. 2003). Trophic discrimination is typically larger for animals with higher rates of respiration relative to growth, such as birds and mammals, compared to fish and invertebrates. In addition, herbivores that must convert plant biomass into animal biomass often have higher bulk Δ^{13}C values than carnivores and omnivores (Elsdon et al. 2010). The form of excreted waste may also affect trophic discrimination, as urea and uric acid contain carbon, while ammonia does not (however, this concept has received greater attention for its effects on $\Delta^{15}N$, discussed separately below). Analysing whole tissues often results in larger diet-to-consumer $\Delta^{13}C$ values compared to muscle due to the inclusion of lipids during whole-body analysis. For example, Malpica-Cruz et al. (2011) found significant differences in tissue-specific $\Delta^{13}C$ values for laboratory-reared leopard sharks (*Triakis semifasciata*). Liver, which had the highest lipid content, showed the lowest $\Delta^{13}C$ values, fins and cartilage had the highest $\Delta^{13}C$ values, and muscle and blood were intermediate. This variability can pose a significant confounding variable when comparing $\delta^{13}C$ values from large consumers, typically analysed as muscle, and small consumers, which are often analysed whole.

The isotope value of consumer tissue may not always follow bulk diet isotope values, causing further complications for the interpretation of bulk stable isotope data in an ecogeochemistry context. Much of the variability among tissue-specific trophic discrimination factors is attributed to differences in tissue composition and lipid content (Malpica-Cruz et al. 2011). The carbon skeletons of different dietary components (proteins, lipids, and carbohydrates), which are often isotopically distinct from each other, can be routed to different tissue constituents in a process termed *isotopic routing* (Schwarcz 1991). Several studies have emphasized the problems that isotopic routing poses to the interpretation of bulk stable isotope data in diet reconstructions (Parkington 1991, Schwarcz & Schoeninger 1991, Ambrose & Norr 1993, Elsdon et al. 2010, McMahon et al. 2010). Studies have shown that changes in amino acid and lipid composition among tissues or ontogenetically within tissues can obscure changes in $\delta^{13}C$ associated with diet or location shifts (Watabe et al. 1982, Murayama 2000, Hüssy et al. 2004). To control for tissue composition differences, it is often desirable to analyse the same tissue type across all samples. This is not always possible, particularly for large food web reconstructions that may require analysing a wide range of tissues.

Another common, although often debated, practice used to control for tissue composition differences is to normalize tissue lipid content, through either chemical extraction or mathematical correction (Post et al. 2007, Logan et al. 2008, Boecklen et al. 2011). While the change in $\delta^{13}C$ values with lipid removal is expected, chemical extraction often affects nitrogen isotope values as well (Sotiropoulos et al. 2004, Logan et al. 2008). Given that most lipids do not contain nitrogen, these findings indicate that we do not fully understand the changes in tissue composition that occur during chemical lipid extraction. Several mathematical models have been developed to correct non-lipid extracted tissue isotope values for lipid contribution *a posteriori* (Post et al. 2007, Logan et al. 2008). Most models use elemental carbon-to-nitrogen ratios (C:N) of bulk tissue as a proxy for lipid content and a protein-lipid $\delta^{13}C$ discrimination factor (Sweeting et al. 2006, Post et al. 2007). However, these parameters are not well constrained and may vary among species or higher taxa, resulting in a large range in the predictive power of the approach ($0.25 < R^2 < 0.96$) (Post et al. 2007, Logan et al. 2008, Tarroux et al. 2010).

Nitrogen isotopes typically exhibit a 3‰ to 4‰ trophic discrimination ($\Delta^{15}N$) between diet and consumer (DeNiro & Epstein 1981, Minagawa & Wada 1984). This enrichment stems from a combination of fractionation during assimilation and protein synthesis as well as the preferential excretion of light isotopes as waste during metabolism (see review by Kelly 2000). As with carbon, there can be significant variability in trophic discrimination around the commonly accepted mean. Estimates of $\Delta^{15}N$ range from –1‰ to 9‰ as a function of dietary protein content, consumer species, tissue type, physiological stress and biochemical form of nitrogenous waste (see reviews by Minagawa & Wada 1984, Michener & Schell 1994, Vander Zanden & Rasmussen 2001, McCutchan et al. 2003, Vanderklift & Ponsard 2003). For instance, animals feeding on high-protein diets often exhibit significantly higher $\Delta^{15}N$ values compared to those feeding on low-protein diets (Vander Zanden & Rasmussen 2001). Thus, diet quality and composition can have a significant impact on $\Delta^{15}N$ values within and among taxa. Vanderklift & Ponsard (2003) reviewed nitrogen trophic discrimination in animals with a variety of forms of nitrogen excretion (e.g., ammonia, urea, uric acid). They found that animals excreting urea typically exhibited significantly larger mean trophic discrimination factors ($\Delta^{15}N = 3‰$) than ammonia-excreting animals (2‰). Many ureoletic elasmobranchs show relatively low $\Delta^{15}N$ values of 1‰ to 2‰ (Hüssy et al. 2010, Malpica-Cruz et al. 2011), although this pattern is not ubiquitous (Logan & Lutcavage 2010). These examples illustrate the complex contributions of the bulk trophic discrimination of nitrogen isotopes.

Nitrogen isotopes are commonly used to calculate the trophic position of consumers in the marine environment. The simplest model for calculating trophic position *TP* using bulk SIA is as follows:

$$TP_{bulk} = \lambda + (\delta^{15}N_{con} - \delta^{15}N_{base})/\Delta^{15}N,$$

where $\delta^{15}N_{con}$ is the nitrogen isotope value of the consumer, $\delta^{15}N_{base}$ is the nitrogen isotope value of the baseline consumer, λ is the trophic position of the baseline consumer, and $\Delta^{15}N$ is the trophic discrimination between diet and consumer. Typically, $\delta^{15}N_{con}$ is measured directly, and $\Delta^{15}N$ is assumed to be between 3‰ and 4‰, despite the large range discussed previously. Choosing a suitable $\delta^{15}N_{base}$ is one of the most challenging, and thus limiting, factors in trophic estimation using bulk SIA. There can be significant temporal variability in $\delta^{15}N_{base}$ associated with the typically much faster turnover rates and thus shorter integration times of basal food web components relative to longer-lived consumers in the upper trophic level (Hannides et al. 2009). In marine environments, the microalgae that support marine food webs typically have $\delta^{15}N$ values that change spatially and seasonally due to incomplete utilization of nitrogenous nutrients (Altabet & Francois 2001, Lourey et al. 2003) and differential utilization of nitrogen sources (nitrate, ammonium, N_2) in space and time (Dugdale & Goering 1967, Dore et al. 2002). Additional complications arise when organisms feed in multiple food webs with different $\delta^{15}N_{base}$ sources. As was the case for bulk $\delta^{13}C$ interpretations, differences in tissue composition and metabolic processing can make interpreting bulk $\delta^{15}N$ values challenging. For instance, Schmidt et al. (2004) found that variability in bulk $\delta^{15}N$ values between euphausiid sexes and tissues (digestive glands and abdominal muscle) were driven by differences in the relative proportions of amino acids (up to 5 mol%) and their $\delta^{15}N$ variability (up to 11‰), as well as differences in tissue metabolism, primarily protein synthesis and degradation for energy supply. The authors showed that, despite the offset in bulk $\delta^{15}N$ values between female and male euphausiids (1.3‰), both sexes were in fact feeding at the same trophic level, and the tissue composition and metabolism differences actually confound trophic-level interpretations of bulk $\delta^{15}N$ values.

One of the biggest challenges of interpreting bulk tissue stable isotope values is the confounding effect of changes in trophic position with variations in isotope values at the base of the food web ($\delta^{13}C_{base}$ and $\delta^{15}N_{base}$; Post 2002). It can be difficult to determine whether changes in a consumer's stable isotope value are due to changes in its diet or trophic position, changes in the baseline food web stable isotope value, or both. This can be particularly problematic when studying the diet and movement of highly migratory marine organisms that may change diet and trophic position as well as habitats throughout ontogeny (Graham et al. 2010, McMahon et al. 2013). The factors described can make interpretations of bulk tissue SIA challenging for studies of diet and migration. As a result, there have been calls for more studies to examine the biochemical and physiological basis of stable isotope ratios in ecology (Gannes et al. 1997, Gannes et al. 1998, Karasov & Martínez del Rio 2007).

Compound-specific stable isotope analysis

Thanks in large part to advances in mass spectrometry, including gas chromatograph/combustion/ isotope ratio monitoring-mass spectrometry (GC/C/irm-MS) (Merritt et al. 1994, Meier-Augenstein 1999, Sessions 2006) and more recently the Finnigan LC IsoLink (McCullagh et al. 2006) and moving wire interface (Krummen et al. 2004, Sessions et al. 2005), it is now possible to obtain precise and accurate stable isotope measurements from individual biological compounds, including amino acids and fatty acids. Compound-specific SIA has the potential to increase the specificity of ecogeochemistry studies significantly and avoid many of the confounding variables that make it challenging to interpret bulk stable isotope values. Specifically, the metabolic and physiological processes that affect the isotopic values of individual compounds are better constrained and often better understood than the numerous variables affecting bulk tissue stable isotope values. While fractionation between bulk compounds is typically in the range of 1‰ to 5‰, fractionation between individual amino acids can be greater than 20‰ (Macko et al. 1987, Keil & Fogel 2001, McMahon et al. 2010). While the use of compound-specific SIA in the marine environment is still relatively new, the technique has been applied to a variety of tissues, including blood, muscle, bone, and otoliths, to assess changes in diet and habitat use (Hare et al. 1991, Popp et al. 2007, Lorrain et al. 2009, McMahon et al. 2011a,b). In the following sections, we discuss the processes that result in fractionation of individual compounds (amino acids and fatty acids). We also highlight several key advantages of compound-specific SIA for ecogeochemistry studies, as well as current limitations and the direction of the field.

Amino acids

Carbon

Amino acids have conventionally been classified into two categories with regard to carbon metabolism, essential (indispensable) and non-essential (dispensable), relating to their synthesis by various organisms (Table 2). Borman et al. (1946) termed as indispensable those amino acids that cannot be synthesized by an organism from materials normally available to the cells at a speed adequate with the demands for normal growth. However, this definition emphasizes that there will be some variability in how amino acids are parsed into each category depending on the metabolic capabilities and demands of the organism. There are nine amino acids that are classified as truly essential, meaning that while plants and bacteria can synthesize them *de novo*, animals have lost the

Table 2 Classification of dietary amino acids according to their dietary requirements by animals

| | | Carbon | |
		Essential	Non-essential
Nitrogen	Source	Lysine (primary amine)	Glycine*
		Phenylalanine (aromatic ring)	Serine
		Threonine (secondary alcohol)	Tyrosine*
		Methionine (secondary thiol)	
	Trophic	Isoleucine (branched aliphatic side chain)	Alanine
		Leucine (branched aliphatic side chain)	Aspartic acid
		Valine (branched aliphatic side chain)	Glutamic acid
			Proline*
	Unknown	Histidine (imidazole ring)	Arginine*
		Tryptophan (indole ring)	Asparagine
			Taurine*
			Cystine*

Source: Modified from Karasov & Martinez del Rio (2007).

Note: Amino acids are divided into essential, non-essential and conditionally essential (designated with *) for carbon isotopes and source, trophic and unknown for nitrogen isotopes. The structures that are likely to render certain amino acids indispensable are in parentheses next to the essential amino acids (Reeds 2000).

enzymatic pathways to synthesize these amino acids and thus must acquire them directly from diet. Another seven amino acids are considered conditionally essential because their rate of synthesis is limited for certain species or conditions. Typically, the synthesis of these amino acids is limited by the availability of a precursor to donate carbon or accessory groups, such as sulphur. In some cases, synthesis is limited to certain tissues (e.g., proline and arginine in the intestines; Wakabayashi et al. 1994) or by physiological demands (e.g., arginine as a precursor for ornithine used to detoxify ammonia in some carnivores; Morris 1985). Non-essential amino acids can be synthesized by all species in sufficient quantities to maintain normal growth.

Modest bulk tissue $\Delta^{13}C$ values often reflect little-to-no trophic discrimination for all essential amino acids and relatively large trophic discrimination factors for many non-essential amino acids (Hare et al. 1991, Howland et al. 2003, Jim et al. 2006, McMahon et al. 2010) (Figure 9A). As a result, essential amino acid $\Delta^{13}C$ values between diet and animal consumers are typically near 0‰. Consumer essential amino acid $\delta^{13}C$ values therefore represent the isotopic signature of primary producers at the base of the food web ($\delta^{13}C_{base}$) without the confounding variable of trophic discrimination. The inherent metabolic diversity within and among prokaryotes and plants generates distinct patterns of essential amino acid $\delta^{13}C$ values and profiles that can be used to identify the origin of amino acids even when there is considerable variability in bulk $\delta^{13}C$ (Abraham & Hesse 2003, Scott et al. 2006). Unlike bulk SIA, which only relies on differences in $\delta^{13}C$ value among end members, compound-specific SIA also makes use of differences in amino acid profiles, which can arise from different biosynthetic pathways used by various groups or from different isotope effects during the biosynthetic process. For example, plants and fungi have unique pathways for lysine and leucine biosynthesis, leading to isotopically distinct carbon isotope signatures compared to bacteria (Hagelstein et al. 1997, Hudson et al. 2005). Larsen et al. (2009) showed that the isotopic difference between isoleucine and leucine was much larger for plants and fungi (6‰ to 12‰) than bacteria (–3‰ to 2‰), making them particularly valuable in distinguishing plant and fungal carbon from bacterially derived carbon. In contrast, the pathways for synthesis of alanine, valine, leucine and isoleucine in the pyruvate family are similar across fungi, bacteria and plants (Hagelstein et al. 1997), and it is the taxon-specific isotope effects associated with pyruvate dehydrogenase that cause differential enrichment of pyruvate available for biosynthesis (Blair et al. 1985). Amino acids in the pyruvate family can be useful for distinguishing the biosynthetic origin of amino acids in plants from fungi and bacteria (Larsen et al. 2009). Given the power of essential amino acid $\delta^{13}C$ profiles to discriminate among key primary producers and the fact that those isotopic signatures are transferred to consumers at the upper trophic level with little to no trophic discrimination, compound-specific SIA provides a promising tool for identifying $\delta^{13}C_{base}$ values to track animal movement through isotopically distinct food webs and to trace carbon flow pathways in marine food webs.

McMahon et al. (2011a,b) explored the potential for otolith amino acid geochemistry in snapper (family Lutjanidae) to identify diet and residency patterns in juvenile nursery habitats. The technique relies on natural geographic variations in $\delta^{13}C$ at the base of food webs among mangrove habitats, coral reefs and seagrass beds that are permanently recorded by otolith amino acids. McMahon et al. (2011b) found that while bulk inorganic otolith $\delta^{13}C$ and $\delta^{18}O$ values differed significantly between snapper from seagrass-dominated Red Sea coastal wetlands and the mangrove-dominated sites on the Pacific coast of Panama, it failed to distinguish nursery residence on local scales. Essential amino acid $\delta^{13}C$ values in otoliths, on the other hand, varied as a function of habitat type and provided a better tracer of residence in different juvenile nursery habitats than conventional bulk otolith SIA alone. By targeting individual amino acids, McMahon et al. (2011b) avoided many of the confounding variables inherent in bulk otolith SIA, such as DIC masking dietary signatures. This study presented robust tracers of juvenile nursery residence that are crucial for reconstructing ontogenetic migration patterns of fishes among coastal wetlands and coral reefs. McMahon et al. (2012) used a compound-specific ecogeochemistry approach to identify essential coral reef fish habitats and

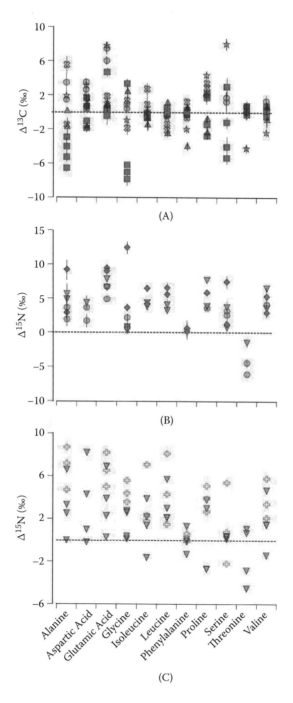

Figure 9 (See also colour figure in the insert) A compilation of individual amino acid fractionation factors between diet and consumer, including (A) $\Delta^{13}C$ from controlled feeding experiments, (B) $\Delta^{15}N$ from controlled feeding experiments, and (C) $\Delta^{15}N$ from natural field samples. Data sources as follows: squares, McMahon et al. (2010); triangles, Fantle et al. (1999); X, Jim et al. (2006); circles, Hare et al. (1991); stars, Howland et al. (2003); inverted triangles, McClelland & Montoya (2002); crosses, Chikaraishi et al. (2007); diamonds, Chikaraishi et al. (2009). Blue symbols: terrestrial vertebrates. Red symbols: aquatic vertebrates. Green symbols: aquatic invertebrates. Multiple symbols of the same shape and colour represent animals fed different diets.

connectivity within a tropical Red Sea seascape. The authors characterized unique $\delta^{13}C$ signatures from five potential juvenile nursery habitats by analysing five essential amino acid $\delta^{13}C$ values from a commercially and ecologically important snapper species, *Lutjanus ehrenbergii*. The authors then quantified the relative contribution of coastal wetland and reef habitats to *L. ehrenbergii* populations on coastal, shelf, and oceanic coral reefs by classifying the juvenile core of adult fish otoliths to one of the five potential nursery habitat signatures using the multivariate amino acid $\delta^{13}C$ data. The results provided the first direct measurements of the juvenile snappers' remarkable migrations of over 30 km between nurseries and reefs. This study found that seascape configuration played a critical but heretofore-unrecognized role in determining connectivity among habitats.

The correlation between consumer essential amino acid $\delta^{13}C$ value and the carbon isotope value at the base of the food web is not always predictable, particularly for consumers with an extensive microbial gut community. Newsome et al. (2011) conducted a controlled feeding experiment on Nile tilapia (*Oreochromis niloticus*) reared on diets in which the percentage protein and $\delta^{13}C$ value of macronutrients (protein, lipids and carbohydrates) varied significantly. The authors found that when tilapia were fed high-protein diets, the $\delta^{13}C$ values of their essential amino acids closely resembled those of their diet, as expected. However, in the low-protein diet treatment, tilapia essential amino acid $\delta^{13}C$ values were significantly higher than their corresponding dietary amino acids. This pattern indicated *in vivo* synthesis of essential amino acids from the bulk carbohydrate pool by microbes in the gut. The microbial contribution of essential macronutrients (vitamins and essential amino acids) to the host's nutrition has been well studied in ruminants (Kung & Rode 1996, Karasov & Carey 2008). It is becoming clear that non-ruminant consumers may also rely on microbial gut contributions under certain conditions. More research into the role of the gut microbial community in consumer amino acid metabolism is warranted, particularly for species such as sea turtles, dugongs and herbivorous fishes that are known to have extensive gut microbe communities (e.g., Mountfort et al. 2002, Andre et al. 2005).

Non-essential amino acid $\Delta^{13}C$ values often exhibit significant deviations from $\Delta^{13}C = 0‰$ and much greater variability among amino acids, diet types and species than essential amino acids (Hare et al. 1991, Howland et al. 2003, Jim et al. 2006, McMahon et al. 2010) (Figure 9A). This variability reflects the influence of the varied metabolic processes that shape the isotopic values of non-essential amino acids during metabolic processing. Patterns in non-essential amino acid $\Delta^{13}C$ values are less clear than those for essential amino acids but show evidence of both *de novo* biosynthesis from bulk dietary carbon pools as well as direct isotopic routing from dietary protein. The relative contributions of isotopic routing versus biosynthesis of non-essential amino acids has been attributed to variability in protein content and amino acid composition of the diet as well as differential utilization of dietary constituents contributing to the bulk carbon pool (O'Brien et al. 2003, Jim et al. 2006, McMahon et al. 2010, Newsome et al. 2011).

Isotopic routing of individual amino acids from diet to consumer is an important contributor to the divergence of consumer bulk stable isotope values from those of its whole diet. If significant isotopic routing of dietary amino acids into consumer protein occurs, then consumer tissue $\delta^{13}C$ values will significantly underrepresent the non-protein macronutrient content of the diet. Isotopic routing of non-essential amino acids is predicted to occur when consumers feed on high-protein diets, as this is far more energetically efficient than *de novo* biosynthesis. Jim et al. (2006) hypothesized that a threshold percentage of protein (5% to 12%) exists in the diet, with bone collagen $\delta^{13}C$ values representing those contributed by dietary protein. Several previous studies have estimated a routing of 50% to 65% of dietary amino acids to bone collagen when the diet supplied an excess of each amino acid (Ambrose & Norr 1993, Ambrose et al. 1997). Much of this work was conducted on terrestrial vertebrates, and several studies have found important deviations from these patterns in aquatic vertebrates. In a controlled feeding experiment rearing common mummichogs (*Fundulus heteroclitus*) on four isotopically distinct diets, McMahon et al. (2010) found a high degree of biosynthesis of non-essential amino acids, despite being fed high-protein diets. The authors suggested that since fish use a significant portion of dietary protein for energetic purposes (Dosdat et al. 1996), it is possible that they exhibit a lower degree of dietary routing than terrestrial vertebrates. Newsome et al. (2011) found a similar trend for Nile tilapia (*Oreochromis niloticus*) in the controlled feeding experiment discussed above. The authors showed that even when fed high-protein diets, non-protein dietary sources (carbohydrates and lipids) contributed a significant amount of carbon to the biosynthesis of non-essential amino acids in the proteinaceous tissues.

Variability in non-essential amino acid $\Delta^{13}C$ values also reflects differences in utilization of the bulk carbon pool from diet (O'Brien et al. 2003, Jim et al. 2006, McMahon et al. 2010, Newsome et al. 2011). For example, catabolizing lipids as a significant energy source provides a very ^{13}C-depleted carbon pool from which non-essential amino acids can be biosynthesized. McMahon et al. (2010) showed that the impact of a lipid-rich diet on the non-essential amino acid $\delta^{13}C$ values of fish appears to be greatest near the source of carbon entering glycolysis and becomes diluted or altered as carbon flows through the tricarboxylic acid (TCA) cycle. Conversely, amino acids such as alanine that are synthesized from pyruvate become enriched in fish that are synthesizing large quantities of lipids (Gaye-Siessegger et al. 2011). This is because the pyruvate dehydrogenase complex heavily fractionates pyruvate when it splits acetyl coenzyme A (CoA) and CO_2 as a precursor for lipid synthesis, thus leaving the remaining pyruvate enriched. Aspartate and glutamate are biosynthesized from oxaloacetate and α-ketoglutarate, respectively, which in turn are generated by a variety of precursors in the Krebs cycle. Multiple cycling of metabolic intermediates through the Krebs cycle likely causes large fractionation during metabolic processing. As a result, Newsome et al. (2011) suggested that the $\delta^{13}C$ analysis of aspartate and glutamate may be particularly valuable for reconstructing bulk diet. SIA of non-essential amino acids may provide better resolution of metabolic processing and carbon utilization than conventional bulk SIA. However, additional controlled feeding experiments to examine the underlying mechanisms behind non-essential amino acid fractionation are warranted. These studies are necessary to determine how much information on diet and metabolic processing we can glean from non-essential amino acid stable isotope values.

Nitrogen

As for carbon, amino acids have recently been classified into two categories, source and trophic (Popp et al. 2007), relating to the degree of $\delta^{15}N$ fractionation between diet and consumer during nitrogen metabolism. It is important to note that while source and essential amino acids all show little-to-no trophic fractionation between diet and consumer, they are not necessarily the same suite of amino acids, as is the case for trophic and non-essential amino acids displaying large trophic discrimination factors (Table 2). The dominant metabolic-processing routes of source amino acids do not significantly fractionate nitrogen because those reactions do not form or break bonds of nitrogen atoms. For example, there are no nitrogen-associated reactions in the conversion of methionine to S-adenosylmethionine or phenylalanine to tyrosine. As a result, $\delta^{15}N$ values of source amino acids in consumers reflect $\delta^{15}N_{base}$ without the confounding variable of trophic discrimination (McClelland & Montoya 2002, Chikaraishi et al. 2007, Popp et al. 2007) (Figures 9B, 9C). Sherwood et al. (2011) examined historical nutrient regime shifts in the western North Atlantic Ocean using source amino acid $\delta^{15}N$ variability in deep-sea gorgonian corals. The authors were able to interpret coral amino acid nitrogen isotope values as a proxy for nitrate source and suggested that nutrient variability in this region was correlated with recent climate change events.

Trophic amino acids, on the other hand, undergo significant fractionation during nitrogen metabolism (Figures 9B, 9C). The removal and translocation of the amine functional group during deamination and transamination are the dominant metabolic processes in the formation of new amino acids via corresponding keto acids. These metabolic processes likely cause nitrogen isotope discrimination between the metabolized and remaining amino acids for many trophic amino acids, including alanine, valine, leucine, isoleucine, and glutamic acid (Macko et al. 1986). Variations in the magnitude of trophic amino acid fractionation between diet and consumer should reflect the isotope effect and relative flux of the deamination/transamination process for each amino acid (Gaebler et al. 1966). Amino acids such as arginine, lysine and histidine that contain multiple nitrogen atoms typically have more variable isotopic compositions due to their dependence on multiple nitrogen reservoirs and enzymatic inhibition reactions (Macko et al. 1987). Glutamic acid plays a key role in both the synthesis of several other amino acids (Lehninger 1975) and the excretion of ammonia in many marine taxa (Claybrook 1983). Nitrogen is transferred from glutamate via transamination to valine, isoleucine, leucine, tyrosine, phenylalanine and aspartic acid, leaving glutamate isotopically heavier. Thus, it is not surprising that glutamic acid exhibits a large $\Delta^{15}N$ between diet and consumer.

Compound-specific SIA provides an opportunity for more refined estimates of trophic position that avoid many of the confounding variables of bulk SIA, particularly variable $\Delta^{15}N$ values and uncertainty in $\delta^{15}N_{base}$. Compound-specific SIA makes use of the differences in fractionation of trophic and source amino acid to provide an internally indexed indicator of trophic position that normalizes for differences in $\delta^{15}N_{base}$. The general equation for trophic-level estimation with compound-specific SIA is as follows:

$$TP_{TA-SA} = 1 + (\delta^{15}N_{TA} - \delta^{15}N_{SA} + \beta)/\Delta^{15}N_{TA},$$

where $\delta^{15}N_{TA}$ and $\delta^{15}N_{SA}$ represent the nitrogen isotope values of the consumer trophic and source amino acids, respectively; β represents the difference in $\delta^{15}N$ between the trophic and source amino acids of primary producers (e.g., $-3‰$ to $-4‰$ for aquatic cyanobacteria and algae, $+8.4‰$ for terrestrial C_3 plants, and $-0.4‰$ for terrestrial C_4 plants; McClelland & Montoya 2002, Chikaraishi et al. 2010); and $\Delta^{15}N_{TA}$ represents the trophic discrimination factor for the trophic amino acid. Phenylalanine consistently shows little-to-no fractionation across multiple marine and terrestrial taxa in feeding experiments and natural samples, making it an ideal source amino acid. Glutamic acid is typically the chosen trophic amino acid for trophic position calculations, although the magnitude of $\Delta^{15}N$ can vary among taxa.

McClelland & Montoya (2002) and Chikaraishi et al. (2007) suggested that large ^{15}N enrichment in trophic amino acids (e.g., glutamic acid $\Delta^{15}N$ = about $7‰$) between diet and consumer provides a greater capacity for defining trophic level than moderate changes in bulk material ($\Delta^{15}N$ = about $3.4‰$). In addition, minimal fractionation of source amino acids (e.g., phenylalanine $\Delta^{15}N$ = $0‰$) provides information on $\delta^{15}N_{base}$, as discussed previously. Hence, a single analysis of amino acid $\delta^{15}N$ values from a consumer tissue provides concurrent information about trophic fractionation and $\delta^{15}N_{base}$ that is not possible using bulk SIA. However, it is important to note that, as was the case with bulk trophic position estimates, there are several important assumptions involved with the compound-specific trophic position equation. In particular, variability in β and $\Delta^{15}N_{TA}$ are not well constrained as yet. Additional controlled experiments to determine the variability in these two parameters is necessary to fully realize the potential of the compound-specific trophic position equation.

One of the biggest challenges of interpreting bulk tissue stable isotope values is the confounding effect of changes in trophic position with variations in $\delta^{13}C_{base}$ and $\delta^{15}N_{base}$ (Post 2002). Compound-specific SIA provides an effective tool to tease apart these confounding variables. For instance, Dale et al. (2011) used a compound-specific ecogeochemistry approach, coupled with conventional stomach content analysis and bulk SIA, to examine the foraging ecology and habitat use of brown stingrays (*Dasyatis lata*) in Kane'ohe Bay, Hawaii. The authors found a counterintuitive trend of decreasing bulk $\delta^{15}N$ values as a function of size, with juvenile stingrays having significantly higher $\delta^{15}N$ values than adults. The authors posed two competing hypotheses to explain this trend: (1) Stingrays of all sizes were feeding in isotopically similar habitats but decreased in trophic position as they moved out of the bay as adults; or (2) the adult stingrays feeding outside the bay were feeding in a system with a distinct $\delta^{15}N_{base}$ value. Using the amino acid trophic position equation discussed previously, the authors showed that trophic position increased with size despite the decrease in bulk $\delta^{15}N$ value and confirmed a foraging habitat shift between the bay and deeper water coincident with the onset of sexual maturity.

Lorrain et al. (2009) used compound-specific SIA to examine trophic dynamics of penguins in the Indian and Southern Oceans. Conventional bulk stable isotope values suggested that king (*Aptenodytes patagonicus*) and Adélie (*Pygoscelis adeliae*) penguins occupied the highest trophic level, southern rockhopper penguins (*Eudyptes chrysocome chrysocome*) occupied the lowest trophic level, and northern rockhopper penguins (*E. chrysocome moseleyi*) were intermediate. The amino acid $\delta^{15}N$ data, however, indicated that king penguins had a higher trophic level compared to the other species than was predicted from bulk SIA. Furthermore, northern rockhoppers had a higher trophic level than the Adélie penguins. However, trophic position alone could not explain the patterns in bulk $\delta^{15}N$ values of penguins in this study. Significant differences were found in $\delta^{15}N$ values of a source amino acid (phenylalanine) among penguin species, suggesting that northern and southern rockhopper penguins were not foraging in the same oceanic regions, and that the differences in their bulk $\delta^{15}N$ values were due, in part, to $\delta^{15}N_{base}$ differences.

Other elements

The majority of compound-specific research in ecogeochemistry has been directed at carbon and nitrogen. However, recent work on δ^2H of amino acids suggests that compound-specific deuterium analysis may be a valuable new avenue for studies of movement and foraging. Fogel et al. (2010) used bacterial cultures grown on deuterium-labelled water and growth media to show that δ^2H of essential amino acids corresponded to the δ^2H of diet, but the δ^2H of non-essential amino acids reflected that of the supplied water (Fogel et al. 2010). The large geographic variation in hydrogen isoscapes suggests that δ^2H analysis of amino acids may provide a valuable new tracer for studies of movement and foraging ecology.

Fatty acids

Fatty acids represent the main constituent of the majority of lipids found in all organisms. Unlike proteins that are broken down during digestion, fatty acids of carbon chain length 14 or more are not degraded once they are released from dietary lipid molecules during digestion. Once fatty acids have been incorporated into consumer tissue, they are either used for energy or reesterified and stored in adipose tissue, often as triacylglycerols. Thus, fatty acids are generally deposited into adipose tissue in predictable patterns with little modification, providing an integrated record of diet (Iverson et al. 2004). Marine organisms have a diverse suite of long-chain, polyunsaturated fatty acids that originate from various microorganisms, phytoplankton, and higher plants (Ackman 1980). A number of studies have shown that specific fatty acid patterns are passed from diet to consumer for a variety of taxa, from zooplankton and benthic macrofauna to pinnipeds and cetaceans (see Iverson et al. 2004, Budge et al. 2006 and references therein). Given that the pattern of fatty acids found in some plants and in many fish and invertebrates can be used to identify individual species accurately (Iverson et al. 1997, Budge et al. 2002), fatty acid profiles have become a powerful tool for quantitative assessment of predator diets (Iverson et al. 2004).

Much of the previous research using fatty acid signatures to examine spatial or temporal variations in diet and trophic ecology has been qualitative examination of changes in consumer fatty acid signatures alone (e.g., Iverson et al. 1997, McMahon et al. 2006). However, Iverson et al. (2004) developed a quantitative statistical model to estimate the contributions of prey species to the diets of predators using fatty acid signatures. This method computes the most likely combination of diet fatty acid signatures that matches the consumer, after accounting for consumer fatty acid metabolism (Budge et al. 2012). To be successful, this method requires that the fatty acid compositions of all important diet sources must be known, and there must be sufficient within-species sampling to assess variability in fatty acid signatures with ecological and demographic factors (e.g., Budge et al. 2002).

Compound-specific SIA of fatty acids has the advantage of providing both fatty acid profiles and isotopic information for dietary studies (e.g., Uhle et al. 1997, McLeod & Wing 2007, Budge et al. 2008). Stable carbon isotope analysis of fatty acids shows similar patterns to those discussed for amino acids, with some fatty acids showing significant diet-to-consumer discrimination and others showing little-to-no isotopic change. While the $\delta^{13}C$ value of pooled fatty acids is similar to that of the bulk carbon pool, individual storage fatty acids in consumers differ from dietary fatty acid due to chain elongation and dehydrogenation as well as metabolic turnover processes (Stott et al. 1997, Hammer et al. 1998). The kinetic isotope effect resulting from metabolic processing of non-essential fatty acids may hold valuable information about carbon utilization, similar to non-essential amino acids (Uhle et al. 1997). Conversely, essential fatty acids, such as omega fatty acids (e.g., linoleic acid 18:2n-6), are directly incorporated from diet into consumer tissue (Stott et al. 1997). Isotopic routing of essential fatty acids provides a record of the isotopic signature of the dietary source preserved in consumer tissues, much like essential amino acids.

Budge et al. (2008) found that ice algae and phytoplankton, the two dominant forms of primary production fuelling Arctic food webs, had distinct differences in fatty acid profiles and unique $\delta^{13}C$ signatures of two individual fatty acids, 16:4n-1 (-24.0‰ ± 2.4‰ and –30.7‰ ± 0.8‰, respectively) and 20:5n-3 (-18.3‰ ± 2.0‰ and –26.9‰ ± 0.7‰, respectively). The authors used these differences in base of the food web end members to track carbon flow pathways to consumers at the upper trophic level, including fish, seabirds, pinnipeds and cetaceans. They found that although ice algae were only available to consumers for a short period of time (April–May), ice algae-derived carbon contributed up to 24% of the carbon passed on to upper trophic levels.

Cholesterol has also been shown to be an indicator of short-term, whole-diet $\delta^{13}C$ values in several controlled feeding experiments on terrestrial vertebrates (Stott et al. 1997, Howland et al. 2003, Jim et al. 2004). Howland et al. (2003) found that pig bone cholesterol $\delta^{13}C$ values were 3.4‰ depleted relative to whole diet, owing to a kinetic isotope effect resulting from oxidation of pyruvate to acetyl-CoA by the enzyme pyruvate dehydrogenase. This offset indicates that even though cholesterol was present in the diet, the bone cholesterol was biosynthesized from a bulk carbon pool rather than isotopically routed directly from the diet.

Conclusions and future directions

Ecogeochemistry relies, in large part, on isoscapes that integrate chemical, physical and biological processes that ultimately determine the isotope composition of marine animals. To enhance the use of isoscapes, we need continued efforts to collect and analyse isotope data throughout the world's oceans. Marine systems are inherently dynamic, and the generation of temporally explicit isoscapes will greatly enhance the accuracy and scope of ecogeochemistry studies. This is particularly important in light of the growing effects of climate change and ocean acidification on the biological, chemical and physical processes in our oceans (Bowen 2010). For example, the effects of temperature on productivity and the frequency and distribution of hypoxic events will potentially shift baseline isoscapes and change patterns of variability across spatial and temporal scales. In addition, we need increased modelling efforts that address the complex ecosystem processes driving geographic variability in isotope distributions. This requires new process-based research to help explain the underlying mechanisms driving spatiotemporal variability in isotopes (Schmittner et al. 2008, Somes et al. 2010).

Enhanced knowledge of isotopic routing, tissue turnover rates and fractionation factors is necessary to fully realize the potential of ecogeochemistry. We are confident that compound-specific SIA will improve the resolution of studies investigating trophic dynamics and movements. However, we clearly need additional controlled feeding experiments to understand the mechanisms that control non-essential and trophic amino acid stable isotope values. Improving instrument sensitivity is likely to reduce sample size requirements and increase temporal resolution from analyses of accretionary tissues. Gains in instrument sensitivity will be particularly helpful when applied to the analysis of individual compounds. Finally, we need improved networking to enhance the dissemination and exchange of data among ecogeochemists. This will require increased collaboration among geochemists, ecologists, geostatisticians and software developers, as well as the establishment of easily accessible public databases.

Acknowledgements

We were supported by funding from the National Science Foundation (Division of Ocean Sciences, 0825148 to S.R.T.), Award No. USA 00002 and KSA 00011 from the King Abdullah University of Science and Technology (to S.R.T.) and a National Science Foundation Graduate Research Fellowship (to L.H.). We thank all of the researchers who contributed published data to the meta-analyses used to generate our isoscapes, N. Lysiak and G. Lawson at Woods Hole Oceanographic Institution for providing unpublished zooplankton samples for the organic isoscapes, and B. Fry for constructive comments on the manuscript.

References

Abed-Navandi, D. & Dworschak, P.C. 2005. Food sources of tropical thalassinidean shrimps: a stable-isotope study. *Marine Ecology Progress Series* **291**, 159–168.

Abraham, W.R. & Hesse, C. 2003. Isotope fractionation in the biosynthesis of cell components by different fungi: a basis for environmental carbon flux studies. *Microbial Ecology* **46**,121–128.

Ackman, R.G. 1980. Fish lipids, part 1. In *Advances in Fish Science and Technology*, J.J. Connell (ed.). Oxford, UK: Fishing News Books, 86–103.

Aharon, P. 1991. Recorders of reef environment histories. *Coral Reefs* **10**, 71–90.

Altabet, M.A. & Francois, R. 2001. Nitrogen isotope biogeochemistry of the Antarctic Polar Frontal Zone at 170°W. *Deep-Sea Research II* **48**, 4247–4273.

Altabet, M.A. & Small, L.F. 1990. Nitrogen isotopic ratios in fecal pellets produced by marine zooplankton. *Geochimica et Cosmochimica Acta* **54**, 155–163.

Ambrose, S.H., Butler, B.M., Hanson, D.B., Hunter-Anderson, R.L. & Krueger, H.W. 1997. Stable isotopic analysis of human diet in the Marianas Archipelago, Western Pacific. *American Journal of Physical Anthropology* **104**, 343–361.

Ambrose, S.H. & Norr, L. 1993. Carbon isotopic evidence for routing of dietary protein to bone collagen, and whole diet to bone apatite carbonate: purified diet growth experiments. In *Molecular Archaeology of Prehistoric Human Bone*, J. Lambert & G. Grupe (eds). Berlin: Springer-Verlag, 1–37.

Ando, A., Nakano, T., Kawahata, H., Yokoyama, Y. & Khim, B.-K. 2010. Testing seawater Sr isotopic variability on a glacial–interglacial timescale: an application of latest high-precision thermal ionization mass spectrometry. *Geochemical Journal* **44**, 347–357.

Andre, J., Gyuris, E. & Lawler, I.R. 2005. Comparison of the diets of sympatric dugongs and green turtles on the Orman Reefs, Torres Strait, Australia. *Wildlife Research* **32**, 53–62.

Arkhipkin, A., Campana, S.E., FitzGerald, J. & Thorrold, S.R. 2004. Spatial and temporal variation in elemental signatures of statoliths from the Patagonian longfin squid (*Loligo gahi*). *Canadian Journal of Fisheries and Aquatic Sciences* **61**, 1212–1224.

Barnett-Johnson, R., Teel, D.J. & Castillas, E. 2010. Genetic and otolith isotopic markers identify salmon populations in the Columbia River at broad and fine geographic scales. *Environmental Biology of Fishes* **89**, 533–546.

Barrick, R.E., Fischer, A.G., Kolodny, Y., Luz, B. & Bohaska, D. 1992. Cetacean bone oxygen isotopes as proxies for Miocene ocean composition and glaciation. *Palaios* **7**, 521–531.

Barth, A., Alvera, A., Troupin, C., Ouberdous, M. & Beckers, J.M. 2010. A web interface for griding arbitrarily distributed *in situ* data based on Data-Interpolating Variational Analysis (DIVA). *Advances in Geosciences* **28**, 29–37.

Bath, G.E., Thorrold, S.R., Jones, C.M., Campana, S.E., McLaren, J.W. & Lam, J.W.H. 2000. Strontium and barium uptake in aragonitic otoliths of marine fish. *Geochimica et Cosmochimica Acta* **64**, 1705–1714.

Beck, J.W., Edwards, R.L., Ito, M., Taylor, F.W., Rougerie, F., Joannot, P. & Henin, C. 1992. Sea-surface temperature from coral skeletal strontium/calcium ratios. *Science* **257**, 644–647.

Becker, B.J., Fodrie, F.J., Macmillan, P.A. & Levin, L.A. 2005. Spatial and temporal variation in trace element fingerprints of mytilid mussel shells: a precursor to invertebrate larval tracking. *Limnology and Oceanography* **50**, 48–61.

Becker, B.J., Levin, L.A., Frodie, J. & McMillan, P.A. 2007. Complex larval connectivity patterns among marine invertebrate populations. *Proceedings of the National Academy of Sciences USA* **104**, 3267–3272.

Bentaleb, I., Fontugne, M., Descolas-Gros, C., Girardin, C., Mariotti, A., Pierre, C., Brunet, C. & Poisson, A. 1998. Carbon isotope fractionation by plankton in the Southern Indian Ocean: relationship between $\delta^{13}C$ of particulate organic carbon and dissolved carbon dioxide. *Journal of Marine Systems* **17**, 39–58.

Benway, H.M. & Mix, A. 2004. Oxygen isotopes, upper-ocean salinity, and precipitation in the eastern tropical Pacific. *Earth and Planetary Science Letters* **224**, 493–507.

Bergquist, B.A. & Boyle, E.A. 2006. Dissolved iron in the tropical and subtropical Atlantic Ocean. *Global Biogeochemical Cycles* **20**, GB1015, doi:10.1029/2005GB002505.

Best, P.B. & Schell, D.M. 1996. Stable isotopes in southern right whale (*Eubalaena australis*) baleen as indicators of seasonal movements, feeding and growth. *Marine Biology* **124**, 483–494.

Birchall, J., O'Connell, T.C., Heaton, T.H.E. & Hedges, R.E.M. 2005. Hydrogen isotope ratios in animal body protein reflect trophic level. *Journal of Animal Ecology* **74**, 877–881.

Blair, N., Leu, A., Munoz, E., Olsen, J., Kwong, E. & Desmarais, D. 1985. Carbon isotopic fractionation in heterotrophic microbial-metabolism. *Applied and Environmental Microbiology* **50**, 996–1001.

Bode, A.M., Alvarez-Ossorio, M.T., Carrera, P. & Lorenzo, J. 2004. Reconstruction of trophic pathways between plankton and the North Iberian sardine (*Sardina pilchardus*) using stable isotopes. *Scientia Marina* **68**, 165–178.

Bode, A., Alvarez-Ossorio, M.T., Cunha, M.E., Garrido, S., Peleteiro, P.B., Valdes, L. & Varela, M. 2007. Stable nitrogen isotope studies of the pelagic food web on the Atlantic shelf of the Iberian Peninsula. *Progress in Oceanography* **74**, 115–131.

Bode, A., Cerrera, P. & Lens, S. 2003. The pelagic foodweb in the upwelling ecosystem of Galicia (NW Spain) during spring: natural abundance of stable carbon and nitrogen isotopes. *ICES Journal of Marine Science* **60**, 11–22.

Boecklen, W.J., Yarnes, C.T., Cook, B.A. & James, A.C. 2011. On the use of stable isotopes in trophic ecology. *Annual Review of Marine Science* **42**, 411–440.

Boon, P.I., Bird, F.L. & Bunn, S.E. 1997. Diet of the intertidal callianassid shrimps *Biffarius arenosus* and *Trypea australiensis* (Decapoda: Thalassinidea) in Western Port (southern Australia), determined with multiple stable-isotope analyses. *Marine and Freshwater Research* **48**, 503–511.

Borman, A., Wood, T.R., Black, H.C., Anderson, E.G., Oesterling, M.J., Womack, M. & Rose, W.C. 1946. The role of arginine in growth with some observations on the effects of argininic acid. *Journal of Biological Chemistry* **166**, 585–594.

Bottrell, S.H. & Newton, R.J. 2006. Reconstruction of changes in global sulfur cycling from marine sulfate isotopes. *Earth Science Reviews* **75**, 59–83.

Boutton, T.W. 1991. Stable carbon isotope ratios of natural materials: II. Atmospheric, terrestrial, marine and freshwater environments. In *Carbon Isotope Techniques*, D.C. Coleman & B. Fry (eds). San Diego, CA: Academic Press, 173–186.

Bowen, G.J. 2010. Isoscapes: spatial pattern in isotopic biogeochemistry. *Annual Review of Earth and Planetary Science* **38**, 161–187.

Boye, M., Wake, B.D., Lopez Garcia, P., Bown, J., Baker, A.R., Achterberg, E.P. 2012. Distributions of dissolved trace metals (Cd, Cu, Mn, Pb Ag) in the southeast Atlantic and the Southern Ocean. *Biogeosciences Discussions* **9**, 3579–3613.

Boyle, E.A. 2001. Anthropogenic trace elements in the ocean. In *Encyclopedia of Ocean Sciences*, J.H. Steel & K.K. Turekian (eds). London: Academic Press, 162–169.

Bricker, O.P. & Jones, B.F. 1995. Main factors affecting the composition of natural waters. In *Trace Elements in Natural Waters*, B. Salbu & E. Steinnes (eds). Boca Raton, FL: CRC Press, 1–20.

Broecker, W.S. & Peng, T.H. 1974. Gas exchange rates between air and sea. *Tellus* **26**, 21–35.

Broecker, W.S., Peng, T.H., Ostlund, G. & Stuiver, M. 1985. The distribution of bomb radiocarbon in the ocean. *Journal of Geophysical Research* **90**, 6953–6970.

Brown, M. 1998. Ocean data view 4.0. *Oceanography* **11**, 19–21.

Brown, R.J. & Severin, K.P. 2009. Otolith chemistry analyses indicate that water Sr:Ca is the primary factor influencing otolith Sr:Ca for freshwater and diadromous fish but not for marine fish. *Canadian Journal of Fisheries and Aquatic Sciences* **66**,1790–1808.

Bruland, K.W. & Lohan, M.C. (2004). The control of trace metals in seawater. In *Treatise on Geochemistry*, H.D. Holland & K.K. Turekian (eds). Amsterdam: Elsevier, 23–47.

Buchheister, A. & Latour, R.J. 2009. Turnover and fractionation of carbon and nitrogen stable isotopes in tissues of a migratory coastal predator, summer flounder (*Paralichthys dentatus*). *Canadian Journal of Fisheries and Aquatic Sciences* **67**, 445–461.

Budge, S.M., Iverson, S.J., Bowen, W.D. & Ackman, R.G. 2002. Among- and within-species variation in fatty acid signatures of marine fish and invertebrates on the Scotian Shelf, Georges Bank and southern Gulf of St. Lawrence. *Canadian Journal of Fisheries and Aquatic Sciences* **59**, 886–898.

Budge, S.M., Iverson, S.J. & Koopman, H.N. 2006. Studying trophic ecology in marine ecosystems using fatty acids: a primer on analysis and interpretation. *Marine Mammal Science* **22**, 759–801.

Budge, S.M., Penney, S.N. & Lall, S.P. 2012. Estimating diets of Atlantic salmon (*Salmo trutta*) using fatty acid signature analyses; validation with controlled feeding studies. *Canadian Journal of Fisheries and Aquatic Sciences* **69**, 1033–1046.

Budge, S.M., Wooller, M.J., Springer, A.M., Iverson, S.J., McRoy, C.P. & Divoky, G.J. 2008. Tracing carbon flow in an Arctic marine food web using fatty acid-stable isotope analysis. *Oecologia* **157**, 117–229.

Calvert, S.E. 2000. Stable isotope data from sediment traps and net tows in the Atlantic and Northeast Pacific Oceans. JGOFS Canada Data Sets 1989–1998. Vancouver, British Columbia, Canada: Marine Environmental Data Service, Department of Fisheries and Oceans.

Campana, S.E. 1999. Chemistry and composition of fish otoliths: pathways, mechanisms and applications. *Marine Ecology Progress Series* **188**, 263–297.

Campana, S.E., Chouinard, G.A., Hanson, J.M. & Frechet, A. 1999. Mixing and migration of overwintering Atlantic cod (*Gadus morhua*) stocks near the mouth of the Gulf of St. Lawrence. *Canadian Journal of Fisheries and Aquatic Sciences* **56**, 1873–1881.

Campana, S.E., Natanson, L.J. & Myklevoll, S. 2002. Bomb dating and age determination of large pelagic sharks. *Canadian Journal of Fisheries and Aquatic Sciences* **59**, 450–455.

Campana, S.E. & Neilson, J.D. 1985. Microstructure of fish otoliths. *Canadian Journal of Fisheries and Aquatic Sciences* **42**, 1014–1032.

Campana, S.E. & Thorrold, S.R. 2001. Otoliths, increments, and elements: keys to a comprehensive understanding of fish populations? *Canadian Journal of Fisheries and Aquatic Sciences* **58**, 30–38.

Capo, R.C., Stewart, B.W. & Chadwick, O.A. 1998. Strontium isotopes as tracers of ecosystem processes: theory and methods. *Geoderma* **82**, 197–225.

Carlier, A., Riera, P., Amouroux, J.M., Bodiou, J.Y., Escoubeyrou, K., Desmalades, M., Caparros, J. & Gremare, A. 2007. A seasonal survey of the food web in the Lapalme Lagoon (northwestern Mediterranean) assessed by carbon and nitrogen stable isotope analysis. *Estuarine, Coastal and Shelf Science* **73**, 299–315.

Checkley, C.M. & Entzeroth, L.C. 1985. Elemental and isotopic fractionation of carbon and nitrogen by marine, planktonic copepods and implications to the marine nitrogen cycle. *Journal of Plankton Research* **7**, 553–568.

Checkley, D.M. & Miller, C.A. 1989. Nitrogen isotope fractionation by oceanic zooplankton. *Deep-Sea Research* **36**, 1449–1456.

Chikaraishi, Y., Kashiyama, Y., Ogawa, N.O., Kitazato, H. & Ohkouchi, N. 2007. Metabolic control of nitrogen isotope composition of amino acids in macroalgae and gastropods: implications for aquatic food web studies. *Marine Ecology Progress Series* **342**, 85–90.

Chikaraishi, Y., Ogawa, N.O., Kashiyama, Y., Takano, Y., Suga, H., Tomitani, A., Miyashita, H., Kitazato, H. & Ohkouchi, N. 2009. Determination of aquatic food-web structure based on compound-specific nitrogen isotopic composition of amino acids. *Limnology and Oceanography: Methods* **7**, 740–750.

Chikaraishi, Y., Ogawa, N.O. & Ohkouchi, N. 2010. Further evaluation of the trophic level estimation based on nitrogen isotopic composition of amino acids. In *Earth, Life, and Isotopes*, N. Ohokouchi, I. Tayasu & K. Koba (eds). Kyoto: Kyoto University Press, 37–51.

Cianco, J.E., Pascual, M.A., Botto, G., Frere, E. & Iribarne, O. 2008. Trophic relationships of exotic anadromous salmonids in the southern Patagonian Shelf as inferred from stable isotopes. *Limnology and Oceanography* **53**, 788–798.

Cifuentes, L.A., Sharp, J.H. & Fogel, M.L. 1988. Stable carbon and nitrogen isotope biogeochemistry in the Delaware Estuary. *Limnology and Oceanography* **33**, 1102–1115.

Claybrook, D.L. 1983. Nitrogen metabolism. In *The Biology of Crustacea, Vol. 5. Internal Anatomy and Physiological Regulation*, L.H. Mantel (ed.). New York: Academic Press, 163–213.

Cline, J.D. & Kaplan, I.R. 1975. Isotopic fractionation of dissolved nitrate during denitrification in the eastern tropical North Pacific Ocean. *Marine Chemistry* **3**, 271–299.

Cohen, A.L. & Thorrold, S.R. 2007. Recovery of temperature records from slow-growing corals by fine scale sampling of skeletons. *Geophysical Research Letters* **34**, L17706, doi:10.1029/2007GL030967.

Cooper, L.W., Benner, R., McClelland, J.W., Peterson, B.J., Holmes, R.M., Raymond, P.A., Hansell, D.A., Brebmeier, J.M. & Codispoti, L.A. 2005. Linkages among runoff, dissolved organic carbon, and the stable oxygen isotope composition of seawater and other water mass indicators in the Arctic Ocean. *Journal of Geophysical Research* **110**, G02013, doi:10.1029/2005JG000031.

Coplen, T.B., Hopple, J.A., Bohlke, J.K., Peiser, H.S., Rieder, S.E., Krouse, H.R., Rosman, K.J.R., Ding, T., Vocke, R.D.J., Revesz, K.M., Lamberty, A., Taylor, P. & DeBievre, P. 2002. Complication of minimum and maximum isotope ratios of selected elements in naturally occurring terrestrial materials and reagents. United States Geological Survey, Reston, Virginia.

Corbisier, T., Petti, M.V., Skowronski, R.P. & Brito, T.S. 2004. Trophic relationships in the nearshore zone of Martel Inlet (King George Island, Antarctic): $\delta^{13}C$ stable-isotope analysis. *Polar Biology* **27**, 75–82.

Dale, J.J., Wallsgrove, N.J., Popp, B.N. & Holland, K.N. 2011. Nursery habitat use and foraging ecology of the brown stingray *Dasyatis lata* from stomach contents, bulk and amino acid stable isotopes. *Marine Ecology Progress Series* **433**, 221–236.

Dalerum, F. & Angerbjörn, A. 2005. Resolving temporal variation in vertebrate diets using naturally occurring stable isotopes. *Oecologia* **144**, 647–658.

Dansgaard, W. 1964. Stable isotopes in precipitation. *Tellus* **16**, 436–468.

Das, K., Lepoint, G., Yann, L. & Bouquegneau, J.M. 2003. Marine mammals from the southern North Sea: feeding ecology from $\delta^{13}C$ and $\delta^{15}N$ measurements. *Marine Ecology Progress Series* **263**, 287–298.

Degens, E.T., Behrendt, M., Goddhardt, B. & Reppmann, E. 1968. Metabolic fractionation of carbon isotopes in marine plankton II. Data on samples collected off the coast of Peru and Ecuador. *Deep-Sea Research* **15**, 11–20.

Degens, E.T., Deuser, W.G. & Haedrich, R.L. 1969. Molecular structure and composition of fish otoliths. *Marine Biology* **2**, 105–113.

DeNiro, M.J. & Epstein, S. 1978. Influence of diet on distribution of carbon isotopes in animals. *Geochimica et Cosmochimica Acta* **42**, 495–506.

DeNiro, M.J. & Epstein, S. 1981. Influence of diet on the distribution of nitrogen isotopes in animals. *Geochimica et Cosmochimica Acta* **45**, 341–351.

Devenport, S.R. & Bax, N.J. 2002. A trophic study of a marine ecosystem off southeastern Australia using stable isotopes of carbon and nitrogen. *Canadian Journal of Fisheries and Aquatic Sciences* **59**, 514–530.

de Villiers, S. 1999. Seawater strontium and Sr/Ca variability in the Atlantic and Pacific oceans. *Earth and Planetary Science Letters* **171**, 623–634.

Dierking, J., Morat, F., Letourneur, Y. & Harmelin-Vivien, M. 2012. Fingerprints of lagoonal life: migration of the marine flatfish *Solea solea* assessed by stable isotopes and otolith microchemistry. *Estuarine, Coastal and Shelf Science* **104–105**, 23–32.

Donat, J.R. & Bruland, K.W. 1995. Trace elements in the oceans. In *Trace Elements in Natural Waters*, B. Salbu & E. Steinnes (eds). Boca Raton, Florida: CRC Press, 247–281.

Dore, J.E., Brum, J.R., Tupas, L.M. & Karl, D.M. 2002. Seasonal and interannual variability in sources of nitrogen supporting export in the oligotrophic subtropical North Pacific Ocean. *Limnology and Oceanography* **47**, 1595–1607.

Dorval, E. & Jones, C.M. 2005. Chemistry of surface waters: distinguishing fine-scale differences in sea grass habitats of Chesapeake Bay. *Limnology and Oceanography* **50**, 1073–1083.

Dorval, E., Jones, C.M., Hannigan, R. & van Montfrans, J. 2007. Relating otolith chemistry to surface water chemistry in a coastal plain estuary. *Canadian Journal of Fisheries and Aquatic Sciences* **64**, 1–14.

Dosdat, A., Servais, F., Metailler, R., Huelvan, C. & Desbruyeres, E. 1996. Comparison of nitrogenous losses in five teleost fish species. *Aquaculture* **141**, 107–127.

Druffel, E.M. & Linick T.W. 1978. Radiocarbon in annual coral rings of Florida. *Geophysical Research Letters* **5**, 913–916.

Druffel, E.M. & Williams, P.M. 1990. Identification of a deep marine source of particulate organic carbon using bomb ^{14}C. *Nature* **347**, 172–174.

Druffel, E.R.M. 1981. Radiocarbon in annual coral rings from the eastern tropical Pacific Ocean. *Geophysical Research Letters* **8**, 59–62.

Druffel, E.R.M. 1987. Bomb radiocarbon in the Pacific: annual and seasonal timescale variations. *Journal of Marine Research* **45**, 667–698.

Druffel-Rodriguez, K.C., Vetter, D., Griffin, S., Druffel, E.R.M., Dunbar, R.B., Mucciarone, D.A., Ziolkowski, L.A. & Sanchez-Cabeza, J.A. 2012. Radiocarbon and stable isotopes in Palmyra corals during the past century. *Geochimica et Cosmochimica Acta* **82**, 154–162.

Dugdale, R.C. & Goering, J.J. 1967. Uptake of new and regenerated forms of nitrogen in primary productivity. *Limnology and Oceanography* **12**, 196–206.

Dunton, K.H. 2001. δ^{15}N and δ^{13}C measurements of Antarctic peninsula fauna: trophic relationships and assimilation of benthic seaweeds. *American Zoologist* **41**, 99–112.

Dunton, K.H., Saupe, S.M., Golikov, A.N., Schell, D.M. & Schonberg, S.V. 1989. Trophic relationships and isotopic gradients among arctic and subarctic marine fauna. *Marine Ecology Progress Series* **56**, 89–97.

Elsdon, T.S., Ayvazian, S., McMahon, K.W. & Thorrold, S.R. 2010. Experimental evaluation of stable isotope fractionation in fish muscle and otoliths. *Marine Ecology Progress Series* **408**, 195–205.

Elsdon, T.S. & Gillanders, B.M. 2002. Interactive effects of temperature and salinity on otolith chemistry: challenges for determining environmental histories of fish. *Canadian Journal of Fisheries and Aquatic Sciences* **59**, 1796–1808.

Elsdon, T.S. & Gillanders, B.M. 2003. Relationship between water and otolith elemental concentrations in juvenile black bream *Acanthopagrus butcheri*. *Marine Ecology Progress Series* **260**, 263–272.

Elsdon, T.S. & Gillanders, B.M. 2004. Fish otolith chemistry influenced by exposure to multiple environmental variables. *Journal of Experimental Marine Biology and Ecology* **313**, 269–284.

Elsdon, T.S. & Gillanders, B.M. 2006. Temporal variability of elemental concentrations in coastal and estuarine waters. *Estuarine, Coastal and Shelf Science* **66**, 147–156.

Elsdon, T.S., Wells, B.K., Campana, S.E., Gillanders, B.M., Jones, C.M., Limburg, K.E., Secor, D.H., Thorrold, S.R. & Walther, B.D. 2008. Otolith chemistry to describe movements and life-history parameters of fishes: hypotheses, assumptions, limitations and inferences. *Oceanography and Marine Biology: an Annual Review* **46**, 297–330.

Estrada, J.A., Rice, A.N., Lutcavage, M.E. & Skomal, G.B. 2003. Predicting trophic position in sharks of the north-west Atlantic Ocean using stable isotope analysis. *Journal of the Marine Biological Association of the UK* **83**, 1347–1350.

Fairbanks, R.G., Evans, M.N., Rubenstone, J.L., Mortlock, R.A., Broad, K., Moore, M.D. & Charles, C.D. 1997. Evaluating climate indices and their geochemical proxies measured in corals. *Coral Reefs* **16**, S93–S100.

Fallon, S.J. & Guilderson, T.P. 2008. Surface water processes in the Indonesian throughflow as documented by a high-resolution coral Δ^{14}C record. *Journal of Geophysical Research* **113**, C09001.

Fanelli, E., Cartes, J.E. & Papiol, V. 2011. Food web structure of deep-sea macrozooplankton and micronekton off the Catalan slope: insight from stable isotopes. *Journal of Marine Systems* **87**, 79–89.

Fanelli, E., Cartes, J.E., Rumolo, P. & Sprovieri, M. 2009. Food-web structure and trophodynamics of meopelagic-suprabenthic bathyal macrofauna of the Algerian Basin based on stable isotopes of carbon and nitrogen. *Deep-Sea Research I* **56**, 1504–1520.

Fantle, M.S., Dittel, A.I., Schwalm, S.M., Epifanio, C.E. & Fogel, M.L. 1999. A food web analysis of the juvenile blue crab, *Callinectes sapidus*, using stable isotopes in whole animals and individual amino acids. *Oecologia* **120**, 416–426.

Farquhar, G.D., Ehleringer, J.R. & Hubick, K.T. 1989. Carbon isotope discrimination and photosynthesis. *Annual Review of Plant Physiology and Plant Molecular Biology* **40**, 503–537.

Florin, S.T., Felicetti, L.A. & Robbins, C.T. 2011. The biological basis for understanding and predicting dietary-induced variations in nitrogen and sulphur isotope ratio discrimination. *Functional Ecology* **25**, 519–526.

Fogel, M.L. & Cifuentes, L.A. 1993. Isotopic fractionation during primary production. In *Organic Geochemistry*, M.H. Engel & S.A. Macko (eds). New York: Plenum Press, 73–98.

Fogel, M.L., Griffin, P. & Newsome, S.D. 2010. Hydrogen isotopes in amino acids trace food and water. Presented at the 7th International Conference on Applications of Stable Isotope Techniques to Ecological Studies, August 2010, Fairbanks, Alaska.

Fontugne, M. & Duplessy, J.C. 1978. Carbon isotope ratio of marine plankton related to surface water masses. *Earth and Planetary Science Letters* **41**, 365–371.

Forest, A., Galindo, V., Darnis, G., Lalande, C., Pineault, S., Tremblay, J.E. & Fortier, L. 2011. Carbon biomass, elemental ratios (C:N) and stable isotopic composition (δ^{13}C, δ^{15}N) of dominant calanoid copepods during the winter-to-summer transition in the Amundsen Gulf (Arctic Ocean). *Journal of Plankton Research* **33**, 161–178.

France, R., Loret, J., Mathews, R. & Springer, J. 1998. Longitudinal variation in zooplankton δ^{13}C through the Northwest Passage: inference for incorporation of sea-ice POM into pelagic foodwebs. *Polar Biology* **20**, 335–341.

France, R.L. 1995. Carbon-13 enrichment in benthic compared to planktonic algae: foodweb implications. *Marine Ecology Progress Series* **124**, 307–312.

Frederich, B., Fabri, G., Lepoint, G., Vandewalle, P. & Parmentier, E. 2009. Trophic niches of thirteen damselfishes (Pomacentridae) at the Grand Recif of Toliara, Madagascar. *Ichthyological Research* **56**, 10–17.

Friedman, I. 1953. Deuterium content of natural waters and other substances. *Geochimica et Cosmochimica Acta* **4**, 89–103.

Fry, B. 1981. Natural stable carbon isotope tag traces Texas shrimp migrations. *Fisheries Bulletin* **79**, 337–345.

Fry, B. 1988. Food web structure on Georges Bank from stable C, N, and S isotopic compositions. *Limnology and Oceanography* **33**, 1182–1190.

Fry, B. 2002. Conservative mixing of stable isotopes across estuarine salinity gradients: a conceptual framework for monitoring watershed influences on downstream fisheries production. *Estuaries* **25**, 264–271.

Fry, B. & Quinones, R.B. 1994. Biomass spectra and stable isotope indicators of trophic level in zooplankton of the northwest Atlantic. *Marine Ecology Progress Series* **112**, 201–204.

Fry, B., Scalan, R.S. & Parker, P.L. 1983. ^{13}C/^{12}C ratios in marine food webs of the Torres Strait, Queensland. *Australian Journal of Marine and Freshwater Research* **34**, 707–715.

Fry, B., Scalan, R.S., Winters, J.K. & Parker, P.L. 1982. Sulphur uptake by salt grasses, mangroves and seagrasses in anaerobic sediments. *Geochimica et Cosmochimica Acta* **46**, 1121–1124.

Fry, B. & Wainright, S.C. 1991. Diatom sources of ^{13}C-rich carbon in marine food webs. *Marine Ecology Progress Series* **76**, 149–157.

Gaebler, O.H., Vitti, T.G. & Vukmirovich, R. 1966. Isotope effects on metabolism of ^{14}N and ^{15}N from unlabeled dietary proteins. *Canadian Journal of Biochemistry* **44**, 1249–1257.

Gaillardet, J., Viers, J. & Dupré, B. 2003. Trace elements in river waters. In *Treatise on Geochemistry*, D.H. Heinrich & K.T. Karl (eds). Oxford, UK: Pergamon, 225–272.

Galimov, E.M., Kodina, L.A., Stepanets, O.V. & Korobeinik, G.S. 2006. Composition of phytoplankton and carbon isotopic composition of plankton and dissolved bicarbonate ions from the subsurface water layer along the Yenisei River–Kara Sea meridional profile (70–77° N). *Geochemistry International* **44**, 1053–1104.

Gannes, L.Z., Martínez del Rio, C. & Koch, P. 1998. Natural abundance variations in stable isotopes and their potential uses in animal physiological ecology. *Comparative Biochemistry and Physiology* **119A**, 725–737.

Gannes, L.Z., O'Brien, D.M. & Martínez del Rio, C. 1997. Stable isotopes in animal ecology: assumptions, caveats, and a call for more laboratory experiments. *Ecology* **78**, 1271–1276.

Ganssen, G. & Kroon, D. 1991. Evidence for Red Sea surface circulation from oxygen isotopes of modern surface waters and planktonic foraminiferal tests. *Paleoceanography* **6**, 73–82.

Gat, J.R. 1996. Oxygen and hydrogen isotopes in the hydrologic cycle. *Annual Review of Earth and Planetary Science* **24**, 225–262.

Gaye-Siessegger, J., McCullagh, J.S.O. & Focken, U. 2011. The effect of dietary amino acid abundance and isotopic composition on the growth rate, metabolism and tissue $\delta^{13}C$ of rainbow trout. *British Journal of Nutrition* **105**, 1764–1771.

Gearing, J.N., Hearing, P.J., Rudnick, D.T., Requejo, A.G. & Hutchins, M.J. 1984. Isotopic variability of organic carbon in a phytoplankton-based, temperate estuary. *Geochimica et Cosmochimica Acta* **48**, 1089–1098.

Goericke, R. & Fry, B. 1994. Variations of marine plankton $d^{13}C$ with latitude, temperature, and dissolved CO_2 in the World Ocean. *Global Biogeochemical Cycles* **8**, 85–90.

Goering, J., Alexander, V. & Haubenstock, N. 1990. Seasonal variability of stable carbon and nitrogen isotope ratios of organisms in a North Pacific bay. *Estuarine, Coastal and Shelf Science* **30**, 239–260.

Goni, M.A., Monacci, N., Gisewhite, R., Ogston, A., Crockett, J. & Nittrouer, C. 2006. Distribution and sources of particular organic matter in the water column and sediments of the Fly River Delta, Gulf of Papua (Papua New Guinea). *Estuarine, Coastal and Shelf Science* **69**, 225–245.

Graham, B.S., Koch, P.L., Newsome, S.D., McMahon, K.W. & Aurioles, D. 2010. Using isoscapes to trace the movements and foraging behavior of top predators in oceanic ecosystems. In *Isoscapes: Understanding Movement, Pattern and Process on Earth Through Isotope Mapping*, J. West et al. (eds). New York: Springer-Verlag, 299–318.

Griffin, M.P.A. & Valiela, I. 2001. $\delta^{15}N$ isotope studies of life history and trophic position of *Fundulus heteroclitus* and *Menidia menidia*. *Marine Ecology Progress Series* **214**, 299–305.

Gruber, N., Keeling, C.D., Bacastow, R.B., Guenther, P.R., Lueker, T.J., Wahlen, M., Meijer, H.A.J., Mook, W.G. & Stocker, T.F. 1999. Spatiotemporal patterns of carbon-13 in the global surface oceans and the oceanic suess effect. *Global Biogeochemical Cycles* **13**, 307–335.

Grumet, N.S., Abram, N.J., Beck, J.W., Dunbar, R.B., Gagan, M.K., Guilderson, T.P., Hantoro, W.S. & Suwargadi, B.W. 2004. Coral radiocarbon records of Indian Ocean water mass mixing and wind-induced upwelling along the coast of Sumatra, Indonesia. *Journal of Geophysical Research* **109**, C05003.

Hagelstein, P., Sieve, B., Klein, M., Jans, H. & Schultz, G. 1997. Leucine synthesis in chloroplasts: leucine/isoleucine aminotransferase and valine aminotransferase are different enzymes in spinach chloroplasts. *Journal of Plant Physiology* **150**, 23–30.

Hale, L.F., Dudgeon, J.V., Mason, A.Z. & Lowe, C.G. 2006. Elemental signatures in the vertebral cartilage of the round stingray, *Urobatis halleri*, from Seal Beach, California. *Environmental Biology of Fishes* **77**, 317–325.

Hammer, B.T., Fogel, M.L. & Hoering, T.C. 1998. Stable carbon isotope ratios of fatty acids in seagrass and redhead ducks. *Chemical Geology* **152**, 29–41.

Hannides, C.C.S., Popp, B.N., Landry, M.R. & Graham, B.S. 2009. Quantification of zooplankton trophic position in the North Pacific Subtropical Gyre using stable nitrogen isotopes. *Limnology and Oceanography* **54**, 50–61.

Hansson, S., Hobbie, J.E., Elmgren, R., Larsson, U., Fry, B. & Johansson, S. 1997. The stable nitrogen isotope ratio as a marker of food-web interactions and fish migration. *Ecology* **78**, 2249–2257.

Hare, P.E., Fogel, M.L., Stafford, T.W., Mitchell, A.D. & Hoering, T.C. 1991. The isotopic composition of carbon and nitrogen in individual amino acids isolated from modern and fossil proteins. *Journal of Archaeological Science* **18**, 277–292.

Harmelin-Vivien, M., Loizeau, V., Mellon, C., Beker, B., Arlhac, D., Bodiguel, X., Ferraton, F., Hermand, R., Philippon, X. & Salen-Picard, C. 2008. Comparison of C and N stable isotope ratios between surface particulate organic matter and microphytoplankton in the Gulf of Lions (NW Mediterranean). *Continental Shelf Research* **28**, 1911–1919.

Heaton, T.H.E. 1986. Isotopic studies of nitrogen pollution in the hydrosphere and atmosphere: a review. *Chemical Geology* **59**, 87–102.

Herzka, S.Z. 2005. Assessing connectivity of estuarine fishes based on stable isotope ratio analysis. *Estuarine, Coastal and Shelf Science* **64**, 58–69.

Hicks, A.S., Closs, G.P. & Swearer, S.E. 2010. Otolith microchemistry of two amphidromous galaxiids across an experimental salinity gradient: a multi-element approach for tracking diadromous migrations. *Journal of Experimental Marine Biology and Ecology* **394**, 86–97.

Hill, J.M. & McQuaid, C.D. 2011. Stable isotope methods: the effect of gut contents on isotopic ratios of zooplankton. *Estuarine, Coastal and Shelf Science* **92**, 480–485.

Hinga, K.R., Arthur, M.A., Pilson, M.E. & Whitaker, D. 1994. Carbon isotope fractionation by marine phytoplankton in culture: the effects of CO_2 concentration, pH, temperature and species. *Global Biogeochemical Cycles* **8**, 91–102.

Hirch, S. 2009. *Trophic interactions at seamounts*. PhD thesis, Departments Biologie der Fakultät für Mathematik, Informatik & Naturwissenschaften der Universität Hamburg, University of Hamburg, Germany.

Hobson, K.A. 1999. Tracing origins and migration of wildlife using stable isotopes: a review. *Oecologia* **120**, 314–326.

Hobson, K.A., Ambrose, W.G. & Renaud, P.E. 1995. Sources of primary production, benthic-pelagic coupling, and trophic relationships within the Northeast Water Polynya: insights from $\delta^{13}C$ and $\delta^{15}N$ analysis. *Marine Ecology Progress Series* **128**, 1–10.

Hobson, K.A., Barnett-Johnson, R. & Cerling, T. 2010. Using isoscapes to track animal migration. In *Isoscapes: Understanding Movement, Pattern and Process on Earth Through Isotope Mapping*, J. West et al. (eds). New York: Springer-Verlag, 273–298.

Hobson, K.A., Fisk, A., Karnovsky, N., Holst, M., Gagnon, J.M. & Fortier, M. 2002. A stable isotope ($\delta^{13}C$, $\delta^{15}N$) model for the North Water food web: implications for evaluating trophodynamics and the flow of energy and contaminants. *Deep-Sea Research II* **49**, 5131–5150.

Hobson, K.A. & Montevecchi, W.A. 1991. Stable isotopic determination of the trophic relationships of Great Auks. *Oecologia* **87**, 528–531.

Hobson, K.A., Piatt, J.F. & Pitocchelli, J. 1994. Using stable isotopes to determine seabird trophic relationships. *Journal of Animal Ecology* **63**, 786–798.

Hoch, M.P., Fogel, M.L. & Kirchman, D.L. 1992. Isotope fractionation associated with ammonium uptake by a marine bacterium. *Limnology and Oceanography* **37**, 1447–1459.

Hoekstra, P.F., Dehn, L.A., George, J.C., Solomon, K.R., Muir, D.C.G. & O'Hara, T.M. 2002. Trophic ecology of bowhead whales (*Balaena mysticetus*) compared with that of other Arctic marine biota as interpreted from carbon-, nitrogen, and sulfur-isotope signatures. *Canadian Journal of Zoology* **80**, 223–231.

Hoekstra, P.F., O'Hara, T.M., Fisk, A.T., Borga, K., Solomon, K.R. & Muir, D.C.G. 2003. Trophic transfer of persistent organochlorine contaminants (OCs) within an Arctic marine food web from the southern Beaufort-Chukchi Seas. *Environmental Pollution* **124**, 509–522.

Hoering, T.C. & Ford, H.T. 1960. The isotope effect in the fixation of nitrogen by *Azotobacter*. *Journal of the American Chemical Society* **82**, 376–378.

Hofmann, M., Wolf-Gladrow, D.A., Takahashi, T., Sutherland, S.C., Six, K.D. & Maier-Reimer, E. 2000. Stable carbon isotope distribution of particulate organic matter in the ocean: a model study. *Marine Chemistry* **72**, 131–150.

Holl, C.M., Villareal, T.A., Payyne, C.D., Clayton, T.D., Hart, C. & Montoya, J.P. 2007. *Trichodesmium* in the western Gulf of Mexico: $^{15}N_2$-fixation and natural abundance stable isotope evidence. *Limnology and Oceanography* **52**, 2249–2259.

Horita, J. 1999. Abiogenic methane formation and isotopic fractionation under hydrothermal conditions. *Science* **285**, 1055–1057.

Horita, J. & Wesolowski, D.J. 1994. Liquid-vapor fraction of oxygen and hydrogen isotopes of water from the freezing to the critical temperature. *Geochimica et Cosmochimica Acta* **58**, 3425–3437.

Howland, M.R., Corr, L.T., Young, S.M.M., Jones, V., Jim, S., van der Merwe, N.J., Mitchell, A.D. & Evershed, R.P. 2003. Expression of the dietary isotope signal in the compound-specific $\delta^{13}C$ values of pig bone lipids and amino acids. *International Journal of Osteoarchaeology* **13**, 54–65.

Hübner, H. 1986. Isotope effects of nitrogen in the soil and biosphere. *Handbook of Environmental Isotope Geochemistry* **2**, 361–425.

Hudson, A.O., Bless, C., Macedo, P., Chatterjee, S.P., Singh, B.K., Gilvarg, C. & Leustek, T. 2005. Biosynthesis of lysine in plants: evidence for a variant of the known bacterial pathways. *Biochimica et Biophysica Acta* **1721**, 27–36.

Huh, Y., Chan, L.-H., Zhang, L. & Edmond, J.M. 1998. Lithium and its isotopes in major world rivers: implications for weathering and the oceanic budget. *Geochimica et Cosmochimica Acta* **62**, 2039–2051.

Hüssy, N.E., MacNeil, M.A. & Fisk, A.T. 2010. The requirement for accurate diet-tissue discrimination factors for interpreting stable isotopes in sharks. *Hydrobiologia* **654**, 1–5.

Hüssy, K., Mosegaard, H. & Jessen, F. 2004. Effect of age and temperature on amino acid composition and the content of different protein types of juvenile Atlantic cod (*Gadus morhua*) otoliths. *Canadian Journal of Fisheries and Aquatic Sciences* **61**, 1012–1020.

Iken, K., Bluhm, B.A. & Gradinger, R. 2005. Food web structure in the high Arctic Canada Basin: evidence from $\delta^{13}C$ and $\delta^{15}N$ analysis. *Polar Biology* **28**, 238–249.

Iverson, S.J., Field, C., Don Bowen, W. & Blanchard, W. 2004. Quantitative fatty acid signature analysis: a new method of estimating predator diets. *Ecological Monographs* **74**, 211–235.

Iverson, S.J., Frost, K.J. & Lowry, L.L. 1997. Fatty acid signatures reveal fine scale structure of foraging distribution of harbor seals and their prey in Prince William Sound, Alaska. *Marine Ecology Progress Series* **151**, 255–271.

Jain, A.K., Kheshgi, H.S., Hoffert, M.I. & Wuebbles, D.J. 1995. Distribution of radiocarbon as a test of global carbon cycle models. *Global Biogeochemical Cycles* **9**, 153–166.

Jennings, S. & Warr, K.J. 2003. Environmental correlates of large-scale spatial variation in the $\delta^{15}N$ of marine animals. *Marine Biology* **142**, 1131–1140.

Jim, S., Ambrose, S.H. & Evershed, R.P. 2004. Stable carbon isotopic evidence for differences in the dietary origin of bone cholesterol, collagen and apatite: implications for their use in palaeodietary reconstruction. *Geochimica et Cosmochimica Acta* **68**, 61–72.

Jim, S̃., Jones, V., Ambrose, S.H. & Evershed, R.P. 2006. Quantifying dietary macronutrient sources of carbon for bone collagen biosynthesis using natural abundance stable carbon isotope analysis. *British Journal of Nutrition* **95**, 1055–1062.

Kaehler, S., Pakhomov, E.A. & McQuaid, C.D. 2000. Trophic structure of the marine food web at the Prince Edward Islands (Southern Ocean) determined by $\delta^{13}C$ and $\delta^{15}N$ analysis. *Marine Ecology Progress Series* **208**, 13–20.

Kalansky, J.F., Robinson, R.S. & Popp, B.N. 2011. Insights into nitrogen cycling in the western Gulf of California from nitrogen isotopic composition of diatom-bound organic matter. *Geochemistry Geophysics Geosystems* **12**, Q06015, doi:10.1029/2010GC003437.

Kalish, J.M. 1991a. Oxygen and carbon stable isotopes in the otoliths of wild and laboratory-reared Australian salmon (*Arripis trutta*). *Marine Biology* **110**, 37–47.

Kalish, J.M. 1991b. Determinants of otolith chemistry: seasonal variation in the composition of blood plasma, endolymph and otoliths of bearded rock cod *Pseudophycis barbatus*. *Marine Ecology Progress Series* **74**, 137–159.

Kalish, J.M. 1993. Pre- and post-bomb radiocarbon in fish otoliths. *Earth and Planetary Science Letters* **114**, 549–554.

Kang, C.K., Kim, J.B., Lee, K.S., Kim, J.B., Lee, P.Y. & Hong, J.S. 2003. Trophic importance of benthic microalgae to macrozoobenthos in coastal bay systems in Korea: dual stable C and N isotope analyses. *Marine Ecology Progress Series* **259**, 79–92.

Karasov, W.H. & Carey, H.V. 2008. Metabolic teamwork between gut microbes and hosts. *Microbe* **4**, 323–328.

Karasov, W.H. & Martínez del Rio, C. 2007. *Physiological Ecology*. Princeton, NJ: Princeton University Press.

Kastner, M. 1999. Oceanic minerals: their origin, nature of their environment, and significance. *Proceedings of the National Academy of Sciences USA* **96**, 3380–3387.

Keeling, C.D. & Guenther, P. 1994. Shore based carbon analysis: duplicate carbon measurements made by the Carbon Dioxide Research Group during the 1983–1988 cruises, Scripps Institution of Oceanography, University of California, San Diego. Online. http://cdiac.esd.ornl.gov/ftp/oceans/keeling.data/. Carbon Dioxide Information Analysis Center, Oak Ridge National Laboratory, U.S. Department of Energy, Oak Ridge, Tennessee. doi:10.3334/CDIAC/otg.KEELING_DATA.

Keeling, C.D., Piper, S.C., Bacastow, R.B., Wahlen, M., Whorf, T.P., Heimann, M. & Meijer, H.A. 2005. Atmospheric CO_2 and $^{13}CO_2$ exchange with the terrestrial biosphere and oceans from 1978 to 2000: observations and carbon cycle implications. In *A History of Atmospheric CO$_2$ and Its Effects on Plants, Animals, and Ecosystems*, J.R. Ehleringer, T.E. Cerling & M.D. Dearing (eds). New York: Springer-Verlag, 83–113.

Keil, R.G. & Fogel, M.L. 2001. Reworking of amino acid in marine sediments: stable carbon isotopic composition of amino acids in sediments along the Washington Coast. *Limnology and Oceanography* **46**, 14–23.

Kelly, J.F. 2000. Stable isotopes of carbon and nitrogen in the study of avian and mammalian trophic ecology. *Canadian Journal of Zoology* **78**, 1–27.

Kelly, J.F., Bridge, B.S., Fudickar, A.M. & Wassenaar, L.I. 2009. A test of comparative equilibration for determining non-exchangeable stable hydrogen isotope values in complex organic materials. *Rapid Communications in Mass Spectrometry* **23**, 2316–2320.

Kendall, C. & Coplen, T.B. 2001. Distribution of oxygen-18 and deuterium in river waters across the United States. *Hydrological Processes* **15**, 1363–1393.

Key, R.M., Kozyr, A., Sabine, C.L., Lee, K., Wanninkhof, R., Bullister, J.L., Freely, R.A., Millero, F.J., Mordy, C. & Peng, T.-H. 2004. A global ocean carbon climatology: results from Global Data Analysis Project (GLODAP). *Global Biogeochemical Cycles*, **18**, GB4031.

Kilbourne, K.H., Quinn, T.M., Guilderson, T.P., Webb, R.S. & Taylor, F.W. 2007. Decadal to interannual-scale source water variations in the Caribbean Sea recorded by Puerto Rican coral radiocarbon. *Climate Dynamics* **29**, 51–62.

Kiriakoulakis, K., Fisher, E., Wolff, G.A., Freiwald, A., Grehan, A. & Roberts, J.M. 2005. Lipids and nitrogen isotopes of two deep-water corals from the North-East Atlantic: initial results and implications for their nutrition. In *Cold-Water Corals and Ecosystems,* A. Freiwald & J.M. Roberts (eds). Berlin: Springer-Verlag, 715–729.

Kohler, S.A., Connan, M., Hill, J.M., Mablouke, C., Bonnevie, B., Ludynia, K., Kemper, J., Huisamen, J., Underhill, L.G., Cherel, Y., McQuaid, C.D. & Jaquemet, S. 2011. Geographic variation in the trophic ecology of an avian rocky shore predator, the African black oystercatcher, along the southern African coastline. *Marine Ecology Progress Series* **435**, 235–249.

Kolasinski, J., Rogers, K., Cuet, P., Barry, B. & Prouin, P. 2011. Sources of particulate organic matter at the ecosystem scale: a stable isotope and trace element study in a tropical coral reef. *Marine Ecology Progress Series* **443**, 77–93.

Kolodny, Y., Luz, B. & Navon, O. 1983. Oxygen isotope variations in phosphate of biogenic apatites, I. Fish bone apatite-rechecking the rules of the game. *Earth and Planetary Science Letters* **64**, 398–404.

Koppelmann, R. & Weikert, H. 2000. Transfer of organic matter in the deep Arabian Sea zooplankton community: insights from $\delta^{15}N$ analysis. *Deep-Sea Research II* **47**, 2653–2672.

Krause-Nehring, J., Brey, T. & Thorrold, S.R. 2012. Centennial records of lead contamination in northern Atlantic bivalves (*Arctica islandica*). *Marine Pollution Bulletin* **64**, 233–240.

Krichevsky, M.I., Friedman, I., Newell, M.F. & Sisler, F.D. 1961. Deuterium fractionation during molecular hydrogen formation in a marine pseudomonad. *Journal of Biological Chemistry* **236**, 2520–2525.

Kroopnick, P.M. 1985. The distribution of ^{13}C of ΣCO_2 in the world oceans. *Deep-Sea Research* **32**, 57–84.

Krouse, H.R. & Grinenko, V.A. 1991. *Stable Isotopes: Natural and Anthropogenic Sulphur in the Environment.* New York: Wiley.

Krouse, H.R., Viau, C.A., Eliuk, L.S., Ueda, A. & Halas, S. 1988. Chemical and isotopic evidence of thermochemical sulfate reduction by light hydrocarbon gases in deep carbonate reservoirs. *Nature* **333**, 415–419.

Krummen, M., Hilkert, A.W., Juchelka, D., Duhr, A., Schluter, H.J. & Pesch, R. 2004. A new concept for isotope ratio monitoring liquid chromatography/mass spectrometry. *Rapid Communications in Mass Spectrometry* **18**, 2260–2266.

Kung, L. & Rode, L.M. 1996. Amino acid metabolism in ruminants. *Animal Feed Science and Technology* **59**, 167–172.

Kurten, B., Painting, S.J., Struck, U., Polunin, N.V.C. & Middelburg, J.J. 2011. Tracking seasonal changes in North Sea zooplankton dynamics using stable isotopes. *Biogeochemistry* doi:10.1007/s10533-011-9630-y.

Laakmann, S. & Auel, H. 2010. Longitudinal and vertical trends in stable isotope signatures ($\delta^{13}C$ and $\delta^{15}N$) of omnivorous and carnivorous copepods across the South Atlantic Ocean. *Marine Biology* **157**, 463–471.

Lamb, K. & Swart, P.K. 2008. The carbon and nitrogen isotopic values of particulate organic material from the Florida Keys: a temporal and spatial study. *Coral Reefs* **27**, 351–362.

Landing, W.M. & Bruland, K.W. 1987. The contrasting biogeochemistry of iron and manganese in the Pacific Ocean. *Geochimica et Cosmochimica Acta* **51**, 29–43.

Larsen, T., Taylor, D.L., Leigh, M.B. & O'Brien, D.M. 2009. Stable isotope fingerprinting: a novel method for identifying plant, fungal, or bacterial origins of amino acids. *Ecology* **90**, 3526–3535.

Laws, E.A., Popp, B.N., Bidigare, R.R., Kennicutt, M.C. & Macko, S.A. 1995. Dependence of phytoplankton carbon isotopic composition on growth rate and $[CO_2]_{aq}$: theoretical and experimental results. *Geochemica et Cosmochimica Acta* **59**, 1131–1138.

Lea, D.W., Shen, G.T. & Boyle, E.A. 1989. Coralline barium records temporal variability in equatorial Pacific upwelling. *Nature* **340**, 373–376.

Lehninger, A.H. 1975. *Biochemistry*. New York: Worth, 2nd edition.

Le Loc'h, F. & Hily, C. 2005. Stable carbon and nitrogen isotope analysis of *Nephrops norvegicus/Merluccius merluccius* fishing grounds in the Bay of Biscay (Northeast Atlantic). *Canadian Journal of Fisheries and Aquatic Sciences* **62**, 123–132.

Le Loc'h, F., Hily, C. & Grall, J. 2008. Benthic community and food web structure on the continental shelf of the Bay of Biscay (north eastern Atlantic) revealed by stable isotope analysis. *Journal of Marine Systems* **72**, 17–34.

Lesage, V., Hammill, M.O. & Kovacs, K.M. 2001. Diet-tissue fractionation of stable carbon and nitrogen isotopes in phocid seals. *Marine Mammal Science* **18**, 182–193.

Levin, I. & Hesshaimer, V. 2000. Radiocarbon—a unique tracer of global carbon cycle dynamics. *Radiocarbon* **42**, 69–80.

Libes, S.M. & Deuser, W.G. 1988. The isotope geochemistry of particulate nitrogen in the Peru Upwelling Area and Gulf of Maine. *Deep-Sea Research* **35**, 517–533.

Limburg, K.E. 1995. Otolith strontium traces environmental history of subyearling American shad *Alosa sapidissima*. *Marine Ecology Progress Series* **119**, 25–35.

Limburg, K.E., Olson, C., Walther, Y., Dale, D., Slomp, C.P. & Hoie, H. 2011. Tracking Baltic hypoxia and cod migration over millennia with natural tags. *Proceedings of the National Academy of Sciences USA* **108**, E177–E182.

Linick, T.W. 1978. La Jolla measurements of radiocarbon in the oceans. *Radiocarbon* **20**, 333–359.

Logan, J.M., Haas, H., Deegan, L. & Gaines, E. 2006. Turnover rates of nitrogen stable isotopes in the salt marsh mummichog, *Fundulus heteroclitus*, following a laboratory diet switch. *Oecologia* **147**, 391–395.

Logan, J.M., Jardine, T.D., Miller, T.J., Bunn, S.E., Cunjak, R.A. & Lutcavage, M.E. 2008. Lipid corrections in carbon and nitrogen stable isotope analysis: comparison of chemical extraction and modeling methods. *Journal of Animal Ecology* **77**, 838–846.

Logan, J.M. & Lutcavage, M.E. 2010. Stable isotope dynamics in elasmobranch fishes. *Hydrobiologia* **644**, 231–244.

Lorrain, A., Graham, B., Menard, F., Popp, B., Bouillon, S., van Breugel, P. & Cherel, Y. 2009. Nitrogen and carbon isotope values of individual amino acids: a tool to study foraging ecology of penguins in the Southern Ocean. *Marine Ecology Progress Series* **391**, 293–306.

Lourey, M.J., Trull, T.W. & Sigman, D.M. 2003. Sensitivity of $\delta^{15}N$ of nitrate, surface suspended and deep sinking particulate nitrogen to seasonal nitrate depletion in the Southern Ocean. *Global Biogeochemical Cycles* **17**, G1081, doi:10.1029/2002GB001973.

Lynch-Stieglitz, J., Stocker, T.F., Broecker, W.S. & Fairbanks, R.G. 1995. The influence of air-sea exchange on the isotopic composition of oceanic carbon: observations and modeling. *Global Biogeochemical Cycles* **9**, 653–665.

Lysiak, N.S. 2009. *Investigating the migration and foraging ecology of North Atlantic right whales with stable isotope geochemistry of baleen and zooplankton*. PhD thesis, Biology Department, Boston University.

Mackensen, A. 2001. Oxygen and carbon stable isotope tracers of Weddell Sea water masses: new data and some paleoceanographic implications. *Deep-Sea Research I* **48**, 1401–1422.

Mackensen, A., Hubberten, H.-W., Bickert, T., Fischer, G. & Fütterer D.K. 1993. The $\delta^{13}C$ in benthic foraminiferal tests of *Fontbotia wuellerstorfi* (Schwager) relative to the $\delta^{13}C$ of dissolved inorganic carbon in Southern Ocean Deep Water: implications for glacial ocean circulation models. *Paleoceanography* **8**, 587–610.

Mackensen, A., Hubberten, H.-W., Scheele, N. & Schlitzer, R. 1996. Decoupling of $\delta^{13}C_{\Sigma CO2}$ and phosphate in recent Weddell Sea deep and bottom water: implications for glacial Southern Ocean paleoceanography. *Paleoceanography* **11**, 203–215.

Macko, S.A., Entzeroth, L. & Parker, P.L. 1984. Regional differences in nitrogen and carbon isotopes on the continental shelf of the Gulf of Mexico. *Naturwissenschaften* **71**, 374–375.

Macko, S.A., Estep, M.L.F., Engel, M.H. & Hare, P.E. 1986. Kinetic fractionation of stable nitrogen isotopes during amino acid transamination. *Geochimica et Cosmochimica Acta* **50**, 2143–2146.

Macko, S.A., Estep, M.L.F., Hare, P.E. & Hoering, T.C. 1987. Isotopic fractionation of nitrogen and carbon in the synthesis of amino acids by microorganisms. *Chemical Geology* **65**, 79–92.

Macko, S.A., Estep, M.L.F. & Lee, W.Y. 1983. Stable hydrogen isotope analysis of food webs on laboratory and field populations of marine amphipods. *Journal of Experimental Marine Biology and Ecology* **72**, 243–249.

Mahaffey, C., Williams, R.G. & Wolff, G.A. 2004. Physical supply of nitrogen to phytoplankton in the Atlantic Ocean. *Global Biogeochemical Cycles* **18**, GB1034 doi:10.1029/2003GB002129.

Malpica-Cruz, L., Herzka, S.Z., Sosa-Nishizaki, O. & Pablo Laso, J. 2011. Tissue-specific isotope trophic discrimination factors and turnover rates in a marine elasmobranch: empirical and modeling results. *Canadian Journal of Fisheries and Aquatic Sciences* **69**, 551–564.

Martin, G.B. & Thorrold, S.R. 2005. Temperature and salinity effect on magnesium, manganese, and barium incorporation in otoliths of larval and early juvenile spot *Leiostomus xanthurus*. *Marine Ecology Progress Series* **293**, 223–232.

Matsura, Y. & Wada, E. 1994. Carbon and nitrogen isotope ratios in marine organic matters of the coastal ecosystem in Ubatuba, southern Brazil. *Ciencia e Cultura* **46**, 141–146.

McClelland, J.W., Holl, C.M. & Montoya J.P. 2003. Relating low $\delta^{15}N$ values of zooplankton to N_2-fixation in the tropical North Atlantic: insights provided by stable isotope ratios of amino acids. *Deep-Sea Research I* **50**, 849–861.

McClelland, J.W. & Montoya, J.P. 2002. Trophic relationships and the nitrogen isotopic composition of amino acids in plankton. *Ecology* **83**, 2173–2180.

McClelland, J.W., Valiela, I. & Michener, R.H. 1997. Nitrogen stable isotope signatures in estuarine food webs: a record of increasing urbanization in coastal watersheds. *Limnology and Oceanography* **42**, 930–937.

McConnaughey, T. 1989a. ^{13}C and ^{18}O isotopic disequilibrium in biological carbonates: I. patterns. *Geochimica et Cosmochimica Acta* **53**, 151–162.

McConnaughey, T. 1989b. ^{13}C and ^{18}O isotopic disequilibrium in biological carbonates: II. *in vitro* simulation of kinetic isotope effects. *Geochimica et Cosmochimica Acta* **53**, 163–171.

McCullagh, J.S.O., Juchelka, D. & Hedges, R.E.M. 2006. Analysis of amino acid ^{13}C abundance from human and faunal bone collagen using liquid chromatography/isotope ratio mass spectrometry. *Rapid Communications in Mass Spectrometry* **20**, 2761–2768.

McCutchan, J.H., Jr., Lewis, W.M., Jr., Kendall, C. & McGrath, C.C. 2003. Variation in trophic shift for stable isotope ratios of carbon, nitrogen, and sulfur. *Oikos* **102**, 378–390.

McLeod, R.J. & Wing, S.R. 2007. Hagfish in the New Zealand fjords are supported by chemoautotrophy of forest carbon. *Ecology* **88**, 809–816.

McMahon, K.W., Ambrose, W.G., Jr., Johnson, B.J., Sun, M.Y., Lopez, G.R., Clough, L.M. & Carroll, M.L. 2006. Benthic community response to ice algae and phytoplankton in Ny Alesund, Svalbard. *Marine Ecology Progress Series* **310**, 1–14.

McMahon, K.W., Berumen, M.L., Mateo, I., Elsdon, T.S. & Thorrold, S.R. 2011a. Carbon isotopes in otolith amino acids identify residency of juvenile snapper (Family: Lutjanidae) in coastal nurseries. *Coral Reefs* **30**, 1135–1145.

McMahon, K.W., Berumen, M.L. & Thorrold, S.R. 2012. Linking habitat mosaics and connectivity in a coral reef seascape. *Proceedings of the National Academy of Sciences USA* **109**, 15372–15376.

McMahon, K.W., Fogel, M.L., Elsdon, T. & Thorrold, S.R. 2010. Carbon isotope fractionation of amino acids in fish muscle reflects biosynthesis and isotopic routing from dietary protein. *Journal of Animal Ecology* **79**, 1132–1141.

McMahon, K.W., Fogel, M.L., Johnson, B.J., Houghton, L.A. & Thorrold, S.R. 2011b. A new method to reconstruct fish diet and movement patterns from $\delta^{13}C$ values in otolith amino acids. *Canadian Journal of Fisheries and Aquatic Sciences* **68**, 1330–1340.

McMahon, K.W., Hamady, L.L., Thorrold, S.R. 2013. A review of ecogeochemistry approaches to estimating movements of marine animals. *Limnology and Oceanography* **58**, 697–714.

Meier-Augenstein, W. 1999. Applied gas chromatography coupled to isotope ratio mass spectrometry. *Journal of Chromatography* **842**, 351–371.

Melancon, S., Fryer, B.J. & Markham, J.L. 2009. Chemical analysis of endolymph and the growing otolith: fractionation of metals in freshwater fish species. *Environmental Toxicology and Chemistry* **28**, 1279–1287.

Merritt, D.A., Brand, W.A. & Hayes, J.M. 1994. Isotope-ratio-monitoring gas chromatography-mass spectrometry: methods for isotopic calibration. *Organic Geochemistry* **21**, 573–583.

Michener, R.H. & Schell, D.M. 1994. Stable isotope ratios as tracers in marine aquatic food webs. In *Stable Isotopes in Ecology and Environmental Science*, K. Lajtha & R.H. Michener (eds). Boston: Blackwell Scientific, 138–157.

Miller, T.W., Omori, K., Hamaoka, H., Shibata, J.Y. & Hidejiro, O. 2010. Tracing anthropogenic inputs to production in the Seto Island Sea, Japan—a stable isotope approach. *Marine Pollution Bulletin* **60**, 1803–1809.

Millero, F., Lee, K. & Wanninkhof, R. 1998. Carbon dioxide, hydrographic, and chemical data obtained during the R/V Ronald Brown in the Atlantic Ocean during WOCE Section AR01(A05) (23 January–24 February, 1998). Online. http://cdiac.ornl.gov/ftp/oceans/ar01woce/. Carbon Dioxide Information Analysis Center, Oak Ridge National Laboratory, U.S. Department of Energy, Oak Ridge, Tennessee. doi:10.3334/CDIAC/otg.WOCE_A05_1998.

Minagawa, M. & Wada, E. 1984. Stepwise enrichment of ^{15}N along food chains: further evidence and the relations between δ^{15}N and animal age. *Geochimica et Cosmochimica Acta* **48**, 1135–1140.

Mizutani, H., Kabaya, Y. & Wada, E. 1991. Nitrogen and carbon isotope compositions relate linearly in cormorant tissues and its diet. *Isotopenpraxis* **27**, 166–168.

Montoya, J.P., Carpenter, E.J. & Capone, D.G. 2002. Nitrogen fixation and nitrogen isotope abundances in zooplankton of the oligotrophic North Atlantic. *Limnology and Oceanography* **47**, 1617–1628.

Mook, W.G. 1986. ^{13}C in atmospheric CO_2. *Netherlands Journal of Sea Research* **20**, 211–223.

Mook, W.G. & Vogel, J.C. 1968. Isotopic equilibrium between shells and their environment. *Science* **159**, 874–875.

Morris, J.G. 1985. Nutritional and metabolic responses to arginine deficiency in carnivores. *Journal of Nutrition* **115**, 524–531.

Mountfort, D.O., Campbell, J. & Clements, K.D. 2002. Hindgut fermentation in three species of marine herbivorous fish. *Applied and Environmental Microbiology* **68**, 1374–1380.

Muhlfeld, C.C., Thorrold, S.R., McMahon, T.E. & Marotz, B. 2012. Estimating trout movements in a river network using strontium isoscapes. *Canadian Journal of Fisheries and Aquatic Sciences* **69**, 906–915.

Mullin, M.M., Rau, G.H. & Eppley, R.W. 1984. Stable nitrogen isotopes in zooplankton: some geographic and temporal variations in the North Pacific. *Limnology and Oceanography* **29**, 1267–1273.

Murayama, E. 2000. Review of the growth regulation processes of otolith daily increment formation. *Fisheries Research* **46**, 53–67.

Neretin, L.N., Bottcher, M.E. & Grinenko, V.A. 2003. Sulfur isotope geochemistry of the Black Sea water column. *Chemical Geology* **200**, 59–69.

Newsome, S.D., Fogel, M.L., Kelly, L. & Martinez del Rio, C. 2011. Contribution of direct incorporation from diet and microbial amino acids to protein synthesis in Nile tilapia. *Functional Ecology* **25**, 1051–1062.

Nydal, R. 1998. Carbon-14 measurements in surface water CO_2 from the Atlantic, Indian and Pacific Oceans, 1965–1994. In *Carbon Dioxide Information Analysis Center*, A.L. Brenkert & T.A. Boden (eds). Oak Ridge, Tennessee: Oak Ridge National Laboratory, ORNL/CDIAC-104, NDP-057A.

Nyssen, F., Brey, T., Lepoint, G., Bouquegneau, J.M., De Broyer, C. & Dauby, P. 2002. A stable isotope approach to the eastern Weddell Sea trophic web: focus on benthic amphipods. *Polar Biology* **25**, 280–287.

O'Brien, D.M., Boggs, C.L. & Fogel, M.L. 2003. Pollen feeding in the butterfly *Heliconius charitonia*: isotopic evidence for essential amino acid transfer from pollen to eggs. *Proceedings of the Royal Society of London B* **270**, 2631–2636.

Olive, P.J.W., Pinnegar, K.J., Polunin, N.V.C., Richards, G. & Welch, R. 2003. Isotope trophic-step fractionation: a dynamic equilibrium model. *Journal of Animal Ecology* **72**, 608–617.

Olson, R.J., Popp, B.N., Graham, B.S., López-Ibarra, G.A., Galván-Magaña, F., Lennert-Cody, C.E., Bocanegra-Castillo, N., Wallsgrove, N.J., Gier, E., Alatorre-Ramírez, V., Ballance, L.T. & Fry, B. 2010. Food web inferences of stable isotope spatial patterns in copepods and yellowfin tuna in the pelagic eastern Pacific Ocean. *Progress in Oceanography* **86**, 124–138.

O'Reilly, C.M., Hecky, R.E., Cohen, A.S. & Plisnier, P.D. 2002. Interpreting stable isotopes in food webs: recognizing the role of time averaging at different trophic levels. *Limnology and Oceanography* **47**, 306–309.

Ostlund, H.G., Doresy, H.G. & Rooth, C.G. 1974. GEOSECS North Atlantic radiocarbon and tritium results. *Earth and Planetary Science Letters* **23**, 69–86.

Ostrom, N.E., Macko, S.A., Deibel, D. & Thrompson, R.J. 1997. Seasonal variation in the stable carbon and nitrogen isotope biogeochemistry of a coastal cold ocean environment. *Geochimica et Cosmochimica Acta* **61**, 2929–2942.

Pajuelo, M., Bjorndal, K.A., Alfaro-Shigueto, J., Seminoff, J.A., Mangel, J.C. & Bolten, A.B. 2010. Stable isotope variation in loggerhead turtles reveals Pacific-Atlantic oceanographic differences. *Marine Ecology Progress Series* **417**, 277–285.

Pancost, R.D., Freeman, K.H., Wakeham, S.G. & Roberston, C.Y. 1997. Controls on carbon isotope fractionation by diatoms in the Peru upwelling region. *Geochimica et Cosmochimica Acta* **61**, 4983–4991.

Parkington, J. 1991. Approaches to dietary reconstruction in the Western Cape: are you what you have eaten? *Journal of Archaeological Science* **18**, 331–342.

Pearcy, W.G. & Stuiver, M. 1983. Vertical transport of carbon-14 into deep-sea food webs. *Deep-Sea Research* **30**, 427–440.

Peterson, B.J. & Fry, B. 1987. Stable isotopes in ecosystem studies. *Annual Review of Ecology and Systematics* **18**, 293–320.

Peterson, B.J. & Howarth, R.W. 1987. Sulfur, carbon, and nitrogen isotopes used to trace organic matter flow in the salt-marsh estuaries of Sapelo Island, Georgia. *Limnology and Oceanography* **32**, 1195–1213.

Peterson, B.J., Howarth, R.W. & Garritt, R.H. 1985. Multiple stable isotopes used to trace the flow of organic matter in estuarine food webs. *Science* **227**, 1361–1363.

Peterson, B.J., Howarth, R.W. & Garritt, R.H. 1986. Sulfur and carbon isotopes as tracers of salt-marsh organic matter flow. *Ecology* **67**, 865–874.

Petursdottir, H., Falk-Petersen, S., Hop, H. & Gislason, A. 2010. *Calanus finmarchicus* along the northern Mid-Atlantic Ridge: variation in fatty acid and alcohol profiles and stable isotope values, $\delta^{15}N$ and $\delta^{13}C$. *Journal of Plankton Research* **32**, 1067–1077.

Petursdottir, H., Gislason, A., Falk-Petersen, S., Hop, H. & Svavarsson, J. 2008. Trophic interactions of the pelagic ecosystem over the Reykjanes Ridge as evaluated by fatty acids and stable isotope analyses. *Deep-Sea Research II* **55**, 83–93.

Pinnegar, J.K., Polunin, N.V.C., Francour, P., Badalamenti, F., Chemello, R., Harmelin-Vivien, M.-L., Hereu, B., Milazzo, M., Zabala, M., D'Anna, G. & Pipitone, C. 2000. Trophic cascades in benthic marine ecosystems: lessons for fisheries and protected-area management. *Environmental Conservation* **27**, 179–200.

Polunin, N.V.C., Morales-Nin, B., Pawsey, W.E., Cartes, J.E., Pinnegar, J.K. & Moranta, J. 2001. Feeding relationships in Mediterranean bathyal assemblages elucidated by stable nitrogen and carbon isotope data. *Marine Ecology Progress Series* **220**, 13–23.

Pomerleau, C., Winkler, G., Sastri, A.R., Nelson, R.J., Vagle, S., Lesage, V. & Ferguson, S.H. 2011. Spatial patterns in zooplankton communities across the eastern Canadian sub-Arctic and Arctic waters: insights from stable carbon ($\delta^{13}C$) and nitrogen ($\delta^{15}N$) isotope ratios. *Journal of Plankton Research* **33**, 1779–1792.

Popp, B.N., Graham, B.S., Olson, R.J., Hannides, C.C.S., Lott, M.J., Lopez-Ibarra, A., Galvan-Magana, F. & Fry, B. 2007. Insight into the trophic ecology of yellowfin tuna, *Thunnus albacares*, from compound-specific nitrogen isotope analysis of proteinaceous amino acids. In *Stable Isotopes as Indicators of Ecological Change*, T.D. Dawson & R.T.W. Siegwolf (eds). Amsterdam: Elsevier/Academic Press, 173–190.

Popp, B.N., Trull, T., Kenig, F., Wakeham, S.G., Rust, T.M., Tilbrook, B., Griffiths, B., Wright, S.W., Marchant, H.J., Bidigare, R.R. & Laws, E.A. 1999. Controls on the carbon isotopic composition of Southern Ocean phytoplankton. *Global Biogeochemical Cycles* **13**, 827–843.

Post, D.M. 2002. Using stable isotopes to estimate trophic position: models, methods, and assumptions. *Ecology* **83**, 703–718.

Post, D.M., Layman, C.A., Albrey Arrington, D., Takimoto, G., Quattrochi, J. & Montaña, C.G. 2007. Getting to the fat of the matter: models, methods and assumptions for dealing with lipids in stable isotope analysis. *Oecologica* **152**, 179–189.

Quay, P., Sonnerup, R., Westby, T., Stutsman, J. & McNichol, A. 2003. Changes in the $^{13}C/^{12}C$ of dissolved inorganic carbon in the ocean as a tracer of anthropogenic CO_2 uptake. *Global Biogeochemical Cycles* **17**, G1004, doi:10.1029/2001GB001817.

Quillfeldt, P., McGill, R.A.R. & Furness, R.W. 2005. Diet and foraging areas of Southern Ocean seabirds and their prey inferred from stable isotopes: review and case study of Wilson's storm-petrel. *Marine Ecology Progress Series* **295**, 295–304.

Rau, G.H., Karl, D.M. & Carney, R.S. 1986. Does inorganic carbon assimilation cause ^{14}C depletion in deep-sea organisms? *Deep-Sea Research* **33**, 349–357.

Rau, G.H., Mearns, A.J., Young, D.R., Olson, R.J., Schager, H.A. & Kaplan, I.R. 1983. Animal $^{13}C/^{12}C$ correlates with trophic level in pelagic food webs. *Ecology* **64**, 1314–1318.

Rau, G.H., Ohman, M.D. & Pierrot-Bults, A. 2003. Linking nitrogen dynamics to climate variability off central California: a 51 year record based on $^{15}N/^{14}N$ in CalCOFI zooplankton. *Deep-Sea Research II* **50**, 2431–2477.

Rau, G.H., Sweeney, R.E. & Kaplan, I.R. 1982. Plankton $^{13}C:^{12}C$ ratio changes with latitude: differences between northern and southern oceans. *Deep-Sea Research* **29**, 1035–1039.

Reeds, P. 2000. Dispensable and indispensable amino acids for humans. *Journal of Nutrition* **130**, 1835S–1840S.

Rees, C.E., Jenkins, W.J. & Monster, J. 1978. The sulphur isotopic composition of ocean water sulphate. *Geochimica et Cosmochimica Acta* **42**, 377–381.

Richardson, C.A. 1988. Exogenous and endogenous rhythms of band formation in the shell of the clam *Tapes philippinarum* (Adams et Reeve, 1850). *Journal of Experimental Marine Biology and Ecology* **122**, 105–126.

Richoux, N.B. & Froneman, P.W. 2009. Plankton trophodynamics at the subtropical convergence, Southern Ocean. *Journal of Plankton Research* **31**, 1059–1073.

Riera, P. & Richard, P. 1997. Temporal variation of $\delta^{13}C$ in particulate organic matter and oyster *Crassostrea gigas* in Marennes-Oleron Bay (France): effect of freshwater inflow. *Marine Ecology Progress Series* **147**, 105–115.

Rivera-Duarte, I. & Flegal, A.R. 1994. Benthic lead fluxes in San Francisco Bay, California, USA. *Geochemica et Cosmochimica Acta* **58**, 3307–3313.

Rodelli, M.R., Gearing, J.N., Gearing, P.J., Marshall, N. & Sasekumar, A. 1984. Stable isotope ratio as a tracer of mangrove carbon in Malaysian ecosystems. *Oecologia* **61**, 326–333.

Rohling, E.J. & Rijk, S.D. 1999. Holocene climate optimum and last glacial maximum in the Mediterranean: the marine oxygen isotope record. *Marine Geology* **153**, 57–75.

Rubenstein, D.R. & Hobson, H.A. 2004. From birds to butterflies: animal movement patterns and stable isotopes. *Trends in Ecology and Evolution* **19**, 256–263.

Sackett, W.M., Eckelmann, W.R., Bender, M.L. & Be, A.W.H. 1965. Temperature dependence of carbon isotope composition in marine plankton and sediments. *Science* **148**, 235–237.

Saino, T. & Hattori, A. 1980. ^{15}N natural abundance in oceanic suspended particulate matter. *Nature* **283**, 752–754.

Saino, T. & Hattori, A. 1987. Geographical variation of the water column distribution of suspended particulate organic nitrogen and its ^{15}N natural abundance in the Pacific and its marginal seas. *Deep-Sea Research* **34**, 807–827.

Sato, T., Sasaki, H. & Fukuchi, M. 2002. Stable isotopic compositions of overwintering copepods in the Arctic and Subarctic waters and implications to the feeding history. *Journal of Marine Systems* **38**, 165–174.

Schell, D.M., Barnett, B.A. & Vinette, K.A. 1998. Carbon and nitrogen isotope ratios in zooplankton of the Bering, Chukchi and Beaufort seas. *Marine Ecology Progress Series* **162**, 11–23.

Schell, D.M., Saupe, S.M. & Haubenstock, N. 1989. Bowhead whale (*Balaena mysticetus*) growth and feeding as estimated by $\delta^{13}C$ techniques. *Marine Biology* **103**, 433–443.

Schlitzer, R. 2002. Interactive analysis and visualization of geoscience data with Ocean Data View. *Computers and Geosciences* **28**, 1211–1218.

Schlitzer, R. 2011. *Ocean Data View*. Paris: Alfred-Wegener-Institute for Polar and Marine Research. Online. http://odv.awi.de (accessed 1 August 2012).

Schmidt, K., Atkinson, A., Stubing, D., McClelland, J.W., Montoya, J.P. & Voss, M. 2003. Trophic relationships among Southern Ocean copepods and krill: some uses and limitation of a stable isotope approach. *Limnology and Oceanography* **48**, 277–289.

Schmidt, G.A., Bigg, G.R. & Rohling, E.J. 1999. *Global Seawater Oxygen-18 Database—v1.21*. New York: National Aeronautics and Space Administration. Online. http://data.giss.nasa.gov/o18data (accessed 1 August 2012).

Schmidt, K., McClelland, J.W., Mente, E., Montoya, J.P., Atkinson, A. & Voss, M. 2004. Trophic-level interpretation based on $\delta^{15}N$ values: implications of tissue-specific fractionation and amino acid composition. *Marine Ecology Progress Series* **266**, 43–58.

Schmittner, A., Oschlies, A., Matthews, H.D. & Galbraith, E.D. 2008 Future changes in climate, ocean circulation, ecosystems, and biogeochemical cycling simulated for a business-as-usual CO_2 emission scenario until year 4000 AD. *Global Biogeochemical Cycles* **22**, GB1013.

Schwarcz, H.P. 1991. Some theoretical aspects of isotope paleodiet studies. *Journal of Archaeological Science* **18**, 261–275.

Schwarcz, H.P. & Schoeninger, M.J. 1991. Stable isotope analyses in human nutritional ecology. *American Journal of Physical Anthropology* **34**, 283–321.

Scott, J.H., O'Brien, D.M., Emerson, D., Sun, H., McDonald, G.D., Salgado, A. & Fogel, M.L. 2006. An examination of the carbon isotope effects associated with amino acid biosynthesis. *Astrobiology* **6**, 867–880.

Sessions, A.L. 2006. Isotope-ratio detection for gas chromatography. *Journal of Separation Science* **29**, 1946–1961.

Sessions, A.L., Sylva, S.P. & Hayes, J.M. 2005. Moving-wire device for carbon isotopic analyses of nanogram quantities of nonvolatile organic carbon. *Analytical Chemistry* **77**, 6519–6527.

Shadsky, I.P., Romankevich, E.A. & Grinchenko, Y.I. 1982. Isotopic composition of carbon in lipids from suspended matter and bottom sediments east of the Kuril Islands. *Oceanology* **22**, 301–304.

Shen, G.T. & Boyle, E.A. 1987. Lead in corals: reconstruction of historical industrial fluxes to the surface ocean. *Earth and Planetary Science Letters* **82**, 289–304.

Shen, G.T. & Stanford, C.L. 1990. Trace element indicators of climate change in annually-banded corals. In *Global Consequences of the 1982–83 El Nino*, P.W. Glynn (ed.). New York: Elsevier, 255–283.

Sherwood, O.A., Lehmann, M.F., Schubert, C.J., Scott, D.B. & McCarthy, M.D. 2011. Nutrient regime shift in the western North Atlantic indicated by compound-specific $\delta^{15}N$ of deep-sea gorgonian corals. *Proceedings of the National Academy of Sciences USA* **108**, 1011–1015.

Sholto-Douglas, A.D., Field, J.G. & James, A.G. 1991. $^{13}C/^{12}C$ and $^{15}N/^{14}N$ isotope ratios in the Southern Benguela ecosystem: indicators of food web relationships among different size-classes of plankton and pelagic fish; differences between fish muscle and bone collagen tissues. *Marine Ecology Progress Series* **78**, 23–31.

Siegenthaler, U. 1989. Carbon-14 in the oceans. In *Handbook of Environmental Isotope Geochemistry*, P. Fritz & J.C. Fontes (eds). New York: Elsevier, **3**, 75–137.

Sigman, D.M., Altabet, M.A., McCorkle, D.C., Francois, R. & Fisher, G. 1999. The $\delta^{15}N$ of nitrate in the Southern Ocean: consumption of nitrate in surface waters. *Global Biogeochemical Cycles* **113**, 1149–1166.

Simenstad, C.A. & Wissmar, R.C. 1985. $\delta^{13}C$ evidence of the origins and fates of organic carbon in estuarine and nearshore food webs. *Marine Ecology Progress Series* **22**, 141–152.

Smith, S.L., Hendrichs, S.M. & Rho, T. 2002. Stable C and N isotopic composition of sinking particles and zooplankton over the southeastern Bering Sea shelf. *Deep-Sea Research* **49**, 6031–6050.

Solomon, C.T., Cole, J.J., Doucett, R.R., Pace, M.L., Preston, N.D., Smith, L.E. & Weidel, B.C. 2009. The influence of environmental water on the hydrogen stable isotope ration in aquatic consumers. *Oecologia* **161**, 313–324.

Somes, C.J., Schmittner, A., Galbraith, E.D., Lehmann, M.F., Altabet, M.A., Montoya, J.P., Letelier, R.M., Mix, A.C., Bourbonnais, A. & Eby, M. 2010. Simulating the global distribution of nitrogen isotopes in the ocean. *Global Biogeochemical Cycles* **24**, GB4019.

Sommer, F., Saage, A., Santer, B., Hansen, T. & Sommer, U. 2005. Linking foraging strategies of marine calanoid copepods to patterns of nitrogen stable isotope signatures in a mesocosm study. *Marine Ecology Progress Series* **286**, 99–106.

Sonntag, C., Münnich, K.O., Jacob, H. & Rozanski, K. 1983. Variations of deuterium and oxygen-18 in continental precipitation and groundwater, and their causes. In *Variations in the Global Water Budget*, A. Street-Perrott & A. Beran (eds). Amsterdam: Reidel, 107–124.

Sotiropoulos, M.A., Tonn, W.M., Wassenaar, L.I. 2004. Effects of lipid extraction on stable carbon and nitrogen isotope analyses of fish tissues: potential consequences for food web studies. *Ecology of Freshwater Fish* **13**, 155–160.

Sternberg, L.S.L. 1988. D/H ratios of environmental water recorded by D/H ratios of plant lipids. *Nature* **333**, 59–61.

Stott, A.W., Evershed, R.P. & Tuross, N. 1997. Compound-specific approach to the $\delta^{13}C$ analysis of cholesterol in fossil bones. *Organic Geochemistry* **26**, 99–103.

Stowasser, G., Atkinson, A., McGill, R.A.R., Phillips, R.A., Collins, M.A. & Pond, D.W. 2012. Food web dynamics in the Scotia Sea in summer: a stable isotope study. *Deep-Sea Research II* **59–60**, 208–221.

Stuck, U., Emeis, K-C., Vob, M., Krom, M.D. & Rau, G.H. 2001. Biological productivity during sapropel S5 formation in the eastern Mediterranean Sea: evidence from stable isotopes of nitrogen and carbon. *Geochimica et Cosmochimica Acta* **65**, 3249–3266.

Stuiver, M., Ostlund, H.G. & McConnaughey, T.A. 1981. GEOSECS Atlantic and Pacific ^{14}C distribution. In *Carbon Cycle Modeling*, B. Bolin (ed.). New York: Wiley, 201–222.

Suess, H.E. 1955. Radiocarbon concentration in modern wood. *Science* **122**, 415–417.

Sunda, W.G. & Huntsman, S.A. 1988. Effects of sunlight on redox cycles of manganese in the southwestern Sargasso Sea. *Deep-Sea Research* **35**, 1297–1317.

Surge, D.M. & Lohmann, K.C. 2002. Temporal and spatial differences in salinity and water chemistry in SW Florida estuaries: effects of human-impacted watersheds. *Estuaries* **25**, 393–408.

Swart, P.K., Thorrold, S., Rosenheim, B., Eisenhauer, A., Harrison, C.G.A., Grammer, M. & Latkoczy, C. 2002. Intra-annual variation in stable oxygen and carbon and trace element composition in sclerosponges. *Paleoceanography* **17**,1045, doi:10.1029/2000PA000622.

Sweeting, C.J., Polunin, N.V.C. & Jennings, S. 2006. Effects of chemical lipid extraction and arithmetic lipid correction on stable isotope ratios of fish tissues. *Rapid Communications in Mass Spectrometry* **20**, 595–601.

Sydeman, W.J., Hobson, K.A., Pyle, P. & McLaren, E.B. 1997. Trophic relationships among seabirds in central California: combined stable isotope and conventional dietary approach. *The Condor* **99**, 327–336.

Tamelander, T., Renaud, P.E., Hop, H., Carroll, M.L., Ambrose, W.G. & Hobson, K.A. 2006. Trophic relationships and pelagic-benthic coupling during summer in the Barents Sea Marginal Ice Zone, revealed by stable carbon and nitrogen isotope measurements. *Marine Ecology Progress Series* **310**, 33–46.

Tanner, S.E., Reis-Santos, P., Vasconcelos, R.P., Fonseca, V., Franca, S., Cabral, H.N. & Thorrold, S.R. 2013. Does otolith geochemistry record ambient environmental conditions in a temperate tidal estuary? *Journal of Experimental Marine Biology and Ecology* **441**, 7–15.

Tanz, N. & Schmidt, H.L. 2010. $\delta^{34}S$ value measurements in food origin assignments and sulphur isotope fractionations in plants and animals. *Journal of Agricultural Food Chemistry* **58**, 3139–3146.

Tarroux, A., Ehrich, D., Lecomte, N., Jardine, T.D., Bety, J. & Berteaux, D. 2010. Sensitivity of stable isotope mixing models to variation in isotopic ratios: evaluating consequences of lipid extraction. *Methods in Ecology and Evolution* **1**, 231–241.

Thayer, G.W., Govoni, J.J. & Connally, D.W. 1983. Stable carbon isotope ratios of the planktonic food web in the northern Gulf of Mexico. *Bulletin of Marine Science* **33**, 247–256.

Thode, H.G., Monster, J. & Dunford, H.B. 1961. Sulphur isotope geochemistry. *Geochimica et Cosmochimica Acta* **25**, 150–174.

Thorrold, S.R., Campana, S.E., Jones, C.M. & Swart, P.K. 1997. Factors determining $\delta^{13}C$ and $\delta^{18}O$ fractionation in aragonitic otoliths of marine fish. *Geochimica et Cosmochimica Acta* **61**, 2909–2919.

Thorrold, S.R., Jones, G.P., Hellberg, M.E., Burton, R.S., Swearer, S.E., Niegel, J.E., Morgan, S.G. & Warner, R.R. 2002. Quantifying larval retention and connectivity in marine populations with artificial and natural markers. *Bulletin of Marine Science* **70**, 291–308.

Thorrold, S.R. & Swearer, S.E. 2009. Otolith chemistry. In *Tropical Fish Otoliths: Information for Assessment, Management and Ecology,* B.S. Green et al. (eds). Berlin: Springer, 249–295.

Tieszen, L.L., Boutton, T.W., Tesdahl, K.G. & Slade, N.A. 1983. Fractionation and turnover of stable carbon isotopes in animal tissues: implications for $\delta^{13}C$ analysis of diet. *Oecologia* **57**, 32–37.

Tillett, B.J., Meekan, M.G., Parry, D., Munksgaard, M., Field, I.C., Thorburn, D. & Bradshaw, C.J.A. 2011. Decoding fingerprints: elemental composition of vertebrae correlates to age-related habitat use in two morphologically similar sharks. *Marine Ecology Progress Series* **434**, 133–142.

Tittlemier, S.A., Risk, A.T., Hobson, K.A. & Norstrom, R.J. 2000. Examination of the bioaccumulation of halogenated dimethyl bipyrroles in an Arctic marine food web using stable nitrogen isotope analysis. *Environmental Pollution* **116**, 85–93.

Uhle, M.E., Macko, S.A., Spero, S.A., Engel, M.H. & Lea, D.W. 1997. Sources of carbon and nitrogen in modern plankton Foraminifera: the role of algal symbionts as determined by bulk and compound specific stable isotope analysis. *Organic Geochemistry.* **27**, 103–113.

Urey, H.C. 1947. The thermodynamic properties of isotopic substances. *Journal of the Chemical Society of London* **1947**, 562–581.

Vanderklift, M.A. & Ponsard, S. 2003. Sources of variation in consumer-diet $\delta^{15}N$ enrichment: a meta-analysis. *Oecologia* **136**, 169–182.

Vander Zanden, M.J. & Rasmussen, J.B. 2001. Trophic fractionation: implications for aquatic food web studies. *Limnology and Oceanography* **46**, 2061–2066.

van Woesik, R., Tomascik, T. & Blake, S. 1999. Coral assemblages and physico-chemical characteristics of the Whisunday Islands: evidence of recent community change. *Marine and Freshwater Research* **50**, 427–440.

Villinski, J.C., Dunbar, R.B. & Mucciarone, D.A. 2000. $Carbon^{12}/Carbon^{13}$ ratios of sedimentary organic matter from the Ross Sea, Antarctica: a record of phytoplankton bloom dynamics. *Journal of Geophysical Research* **105**, 14163–14172.

Vizzini, S. & Mazzola, A. 2003. Seasonal variations in the stable carbon and nitrogen isotope ratios. $^{13}C/^{12}C$ and $^{15}N/^{14}N$ of primary producers and consumers in a western Mediterranean coastal lagoon. *Marine Biology* **142**, 1009–1018.

Wada, E. & Hattori, A. 1976. Natural abundance of ^{15}N in particulate organic matter in the North Pacific Ocean. *Geochimica et Cosmochimica Acta* **40**, 249–251.

Wada, E., Terazaki, M., Kabaya, Y. & Nemoto, T. 1987. ^{15}N and ^{13}C abundances in the Antarctic Ocean with emphasis on the biogeochemical structure of the food web. *Deep-Sea Research A* **34**, 829–841.

Wakabayashi, Y., Yamada, E., Yoshida, T. & Takahashi, H. 1994. Arginine becomes an essential amino acid after massive resection of rat small intestine. *Journal of Biological Chemistry* **269**, 32667–32671.

Walther, B.D. & Limburg, K. 2012. The use of otolith chemistry to characterize diadromous migrations. *Journal of Fish Biology* **81**, 796–825.

Warnken, K.W., Gill, G.A., Griffin, L.L. & Santschi, P.H. 2001. Sediment-water exchange of Mn, Fe, Ni and Zn in Galveston Bay, Texas. *Marine Chemistry* **73**, 215–231.

Waser, N.A.D., Harrison, W.G., Head, E.J.H., Nielsen, B., Lutz, V.A. & Calvert, S.E. 2000. Geographic variations in the nitrogen isotope composition of surface particulate nitrogen and new nitrogen production across the North Atlantic Ocean. *Deep-Sea Research I* **27**, 1207–1226.

Watabe, N., Tanaka, K., Yamada, J. & Dean, J.M. 1982. Scanning electron microscope observations of the organic matrix in the otolith of the teleost fish *Fundulus heteroclitus* (Linnaeus) and *Tilapia nilotica* (Linnaeus). *Journal of Experimental Marine Biology and Ecology* **58**, 127–134.

Weidman, C.R. & Jones, G.A. 1993. A shell-derived time history of bomb ^{14}C on Georges Bank and its Labrador Sea implications. *Journal of Geophysical Research* **98**, 14577–14588.

Weiss, R.F. 1974. Carbon dioxide in water and seawater: the solubility of a non-ideal gas. *Marine Chemistry* **2**, 203–215.

Werstiuk, N.H. & Ju, C. 1989. Protium-deuterium exchange of benzo-substituted heterocycles in neutral D_2O at elevated temperatures. *Canadian Journal of Chemistry* **67**, 812–815.

West, J., Bowen, G.J., Dawson, T.E. & Tu, K.P. 2010. *Isoscapes: Understanding Movement, Pattern and Process on Earth Through Isotope Mapping*. New York: Springer-Verlag.

White, J.W.C. 1989. Stable hydrogen isotope ratios in planes: a review of current theory and some potential applications. In *Stable Isotopes in Ecological Research*, P.W. Rundel (ed.). New York: Springer-Verlag, 142–162.

Wu, J., Calvert, S.E., Wong, C.S. & Whitney, F.A. 1999. Carbon and nitrogen isotopic composition of sedimenting particulate material at Station Papa in the subarctic northeast Pacific. *Deep-Sea Research II* **46**, 2793–2832.

Wyatt, A.S.J. 2011. *Oceanographic ecology of coral reefs: the role of oceanographic processes in reef-level biogeochmistry and trophic ecology*. PhD thesis, Oceans Institute and School of Environmental Systems, University of Western Australia, Perth.

Yamamuro, M., Kayanne, H. & Minagawa, M. 1995. Carbon and nitrogen stable isotopes of primary producers in coral reef ecosystems. *Limnology and Oceanography* **40**, 617–621.

Zacherl, D.C., Manríquez, P.H., Paradis, G., Day, R.W., Castilla, J.C., Warner, R.R., Lea, D.W. & Gaines, S.G. 2003a. Trace elemental fingerprinting of gastropod statoliths to study larval dispersal trajectories. *Marine Ecology Progress Series* **248**, 297–303.

Zacherl, D.C., Paradis, G. & Lea, D.W. 2003b. Barium and strontium uptake in larval protoconch and statolith of the marine neogastropod *Kelletia kelletii*. *Geochimica et Cosmochimica Acta* **67**, 4091–4099.

Zeebe, R.E. & Wolf-Gladrow, D.A. 2001. *CO_2 in Seawater: Equilibrium, Kinetics, Isotopes*. Amsterdam: Elsevier.

Zhang, J., Quay, P.D. & Wilbur, D.O. 1995. Carbon isotope fractionation during gas-water exchange and dissolution of CO_2. *Geochimica et Cosmochimica Acta* **59**, 107–114.

Author Index

Page numbers referring to complete articles are given in **bold** type.

Andersson, A.J. *See* Lerman, A., 62
Anderwald, P. *See* Evans, P.G.H., 264
Andrada, M. *See* Diaz-Delgado, J., 263
André, C. *See* Panova, M., 64
Andre, M. *See* Tasker, M.L., 278
Andrews, J. *See* Walker, M., 68
Anestis, A., 157
Anganuzzi, A., 259
Angelidis, P. *See* Anestis, A., 157
Anger, K. *See* Walther, K., 190
Angers, B. *See* Gagnon, M.C., 59
Angliss, R.P. *See* Wade, P.R., 279
Angulo, F. *See* Scallan, E., 185
Angutikjuaq, I. *See* Gearheard, S., 167
Ansell, T.J. *See* Rayner, N.A., 183
Ansmann, I. *See* Evans, P.G.H., 264
Anthony, K. *See* Feely, R.A., 166
Anthony, K.R. *See* Diaz-Pulido, G., 164
Anthony, K.R.N., 157
    See Diaz-Pulido, G., 164
Antoine, L., 259
    See Goujon, M., 266
    See Hassani, S., 267
Anton, A. *See* O'Connor, M.I., 180
Antonioli, F. *See* Lambeck, K., 61
Antonov, J.I. *See* Levitus, S., 175
Anzidei, M. *See* Lambeck, K., 61
Aoki, I. *See* Takasuka, A., 188
Aonghusa, C.N. *See* McBreen, F., 63
Aporta, C. *See* Krupnik, I., 175
Apostolaki, E.T. *See* Fourqurean, J.W., 167
Appeldoorn, R., 316
    See Mann, D., 321
    See Ojeda-Serrano, E., 322
    See Shapiro, D., 324
Appeldoorn, R.S. *See* Nemeth, M.I., 321
    See Rowell, T.J., 323
    See Schärer, M.T., 323
Appenzeller, T., 55
Araújo, M.B. *See* Pereira, H.M., 181
Arbelo, M. *See* Diaz-Delgado, J., 263
    See Fernández, A., 264
    See Jaber, J.R., 268
    See Jepson, P.D., 269
Arbic, B.K., 55
Archer, C.L., 157
Archer, D. *See* Clark, P.U., 57
Archer, S.K., 316
    See Heppell, S., 318
Arico, S. *See* Hobbs, R.J., 171
Arkema, K.K. *See* Reed, D.C., 183
Armstrong, R.A., 316
Armstrup, S.C. *See* Regehr, E.V., 183
Arndt, D.S. *See* Blunden, J., 159
Arnold, L. *See* Petraglia, M., 65
Arnold, M. *See* Bard, E., 56
Arnold, P. *See* Amaral, A.R., 259
    See Valentine, P.S., 279
Arnone, R. *See* Glenn, S., 168
Aronson, J. *See* Hobbs, R.J., 171
Aronson, R.B. *See* Carpenter, K.E., 161
    See Dudgeon, S.R., 164

Arsenault, J.L. *See* Rivera, J.A., 323
Artale, V. *See* Bindoff, N., 56
    See Bindoff, N.L., 159
Arzel, O., 157
Asaro, F. *See* Alvarez, L.W., 55
Asche, F. *See* Torrissen, O., 188
Ashford, D. *See* Dobos, K., 164
Ashjian, C.J. *See* Cooper, L.W., 163
Assinder, D.J. *See* Scourse, J.D., 66
Atkinson, A. *See* Miller, J., 178
Atkinson, C. *See* Kanzow, T., 174
Atkinson, D.E. *See* Loring, P.A., 176
Atkinson, J. *See* Berrow, S., 260
Attrill, M.J. *See* Kaiser, M.J., 61
Au, W.W.L., 259
Auad, G., **71–192**
Aubail, A. *See* Caurant, F., 261
Auer, C., **71–192**
Auffret, J.-P. *See* Lericolais, G., 62
Augeraud-Véron, E. *See* Mannocci, L., 271
Ault, J.S. *See* Humston, R., 172
    See Lindeman, K.C., 320
Aumont, O. *See* Orr, J.C., 181
Aurioles, D. *See* Graham, B.S., 266
Auster, P.J., 157
Austin, D., 157
Austin, R. *See* Stephens, G.L., 187
Austin, R.A. *See* Scourse, J.D., 66
Austin, R.M., 55
Austin, W.E.N., 55
    Scourse, J.D., 66
    See Evans, J.R., 58
    See Marret, F., 62
    See Scott, G.A., 65
    See Scourse, J.D., 66
Aydin, K.Y. *See* King, J.R., 174
Azueta, J. *See* Heyman, W.D., 319

B

Babij, E., **71–192**
Backer, L.C. *See* Moore, S.K., 179
Bäcklin, B.-M., 259
Bacon, S., 259
Baden, D.G. *See* Bossart, G.D., 159
    See Flewelling, L.J., 166
Badjeck, M., 157
Badts, V. *See* Morizur, Y., 272
Baer, H., 157, 158
Bage, H. *See* Westlund, L., 190
Bahri, T. *See* Cochrane, K., 162
Bailey, G.N. *See* Lambeck, K., 61
Bailey, H. *See* Bograd, S.J., 159
    See Shillinger, G., 186
    See Thompson, P.M., 278
Bailey, M.A. *See* Lesser, M.P., 175
Bailey, R.M. *See* O'Cofaigh, C., 64
Bailleul, F. *See* Cherel, Y., 262
Baillie, J.E.M. *See* Butchart, S.H.M., 160
Baillie, M.G.L. *See* Reimer, P.J., 65
Bain, D.E. *See* Williams, R., 280
Baines, M. *See* Evans, P.G.H., 264

SYSTEMATIC INDEX

SUBJECT INDEX

SUBJECT INDEX

Y

Z